Principles of Evolutionary Genetics

Principles of Evolutionary Genetics

Edited by Roger Greer

Syrawood Publishing House
New York

Published by Syrawood Publishing House,
750 Third Avenue, 9th Floor,
New York, NY 10017, USA
www.syrawoodpublishinghouse.com

Principles of Evolutionary Genetics
Edited by Roger Greer

International Standard Book Number: 978-1-68286-595-8 (Hardback)

Cataloging-in-Publication Data

Principles of evolutionary genetics / edited by Roger Greer.
p. cm.
Includes bibliographical references and index.
ISBN 978-1-68286-595-8
1. Evolutionary genetics. 2. Evolution (Biology). 3. Genetics. I. Greer, Roger.
QH390 .P75 2018
572.838--dc23

Printed in China.

TABLE OF CONTENTS

PREFACE

This book has been a concerted effort by a group of academicians, researchers and scientists, who have contributed their research works for the realization of the book. This book has materialized in the wake of emerging advancements and innovations in this field. Therefore, the need of the hour was to compile all the required researches and disseminate the knowledge to a broad spectrum of people comprising of students, researchers and specialists of the field.

Evolutionary genetics studies new mutations that still occur within a species. It derives its primary principles from evolutionary milestones that have been highlighted by the theory of natural selection. There are multiple micro-evolutionary changes that occur within the various groups of a species and they manifest as macro-evolutionary patterns across a period of time, resulting in varying individual and social behavioral actions. Evolutionary genetics has been applied in sub-fields such as ecological genetics, population genetics and quantitative genetics. This book strives to provide a fair idea about this discipline and to help develop a better understanding of the latest advances within this field. It will serve as a reference to a broad spectrum of readers.

At the end of the preface, I would like to thank the authors for their brilliant chapters and the publisher for guiding us all-through the making of the book till its final stage. Also, I would like to thank my family for providing the support and encouragement throughout my academic career and research projects.

Editor

1

AST: An Automated Sequence-Sampling Method for Improving the Taxonomic Diversity of Gene Phylogenetic Trees

Chan Zhou[1¤a], Fenglou Mao[1], Yanbin Yin[1¤b], Jinling Huang[2], Johann Peter Gogarten[3], Ying Xu[1,4]*

1 Computational Systems Biology Laboratory, Department of Biochemistry and Molecular Biology and Institute of Bioinformatics, University of Georgia, Athens, Georgia, United States of America, **2** Department of Biology, East Carolina University, Greenville, North Carolina, United States of America, **3** Department of Molecular and Cell Biology, University of Connecticut, Storrs, Connecticut, United States of America, **4** College of Computer Science and Technology, Jilin University, Changchun, China

Abstract

A challenge in phylogenetic inference of gene trees is how to properly sample a large pool of homologous sequences to derive a good representative subset of sequences. Such a need arises in various applications, e.g. when (1) accuracy-oriented phylogenetic reconstruction methods may not be able to deal with a large pool of sequences due to their high demand in computing resources; (2) applications analyzing a collection of gene trees may prefer to use trees with fewer operational taxonomic units (OTUs), for instance for the detection of horizontal gene transfer events by identifying phylogenetic conflicts; and (3) the pool of available sequences is biased towards extensively studied species. In the past, the creation of subsamples often relied on manual selection. Here we present an Automated sequence-Sampling method for improving the Taxonomic diversity of gene phylogenetic trees, AST, to obtain representative sequences that maximize the taxonomic diversity of the sampled sequences. To demonstrate the effectiveness of AST, we have tested it to solve four problems, namely, inference of the evolutionary histories of the small ribosomal subunit protein S5 of *E. coli*, 16 S ribosomal RNAs and glycosyl-transferase gene family 8, and a study of ancient horizontal gene transfers from bacteria to plants. Our results show that the resolution of our computational results is almost as good as that of manual inference by domain experts, hence making the tool generally useful to phylogenetic studies by non-phylogeny specialists. The program is available at http://csbl.bmb.uga.edu/~zhouchan/AST.php.

Editor: Paul Jaak Janssen, Belgian Nuclear Research Centre SCK•CEN, Belgium

Funding: The work was supported from the National Science Foundation Assembling the Tree of Life program (DEB 0830024 to CZ, FLM, JLH and JPG) and was in part supported by the DOE grant for the BioEnergy Science Center (to YB and YX). The funders had no role in study design, data collection and analysis, decision to publish, or preparation of the manuscript.

Competing Interests: The authors have declared that no competing interests exist.

* E-mail: xyn@bmb.uga.edu

¤a Current address: Gastrointestinal Unit, Massachusetts General Hospital, Harvard Medical School, Boston, Massachusetts, United States of America
¤b Current address: Department of Biological Sciences, Northern Illinois University, DeKalb, Illinois, United States of America

Introduction

Reconstruction of gene trees represents a commonly encountered problem in evolutionary studies, such as inferring the evolutionary history of a gene (or a gene family) [1,2], finding the origin of a gene, discovering the function of a gene [3,4], and estimating species trees from gene trees [5,6,7,8,9]. Reconstructing the phylogenetic history of a gene (or gene family) generally involves three steps: 1) selection of homologous sequences (DNA, RNA, or protein sequences); 2) multiple sequence alignment (MSA); and 3) phylogenetic tree reconstruction. Selection of homologous sequences is one of the key steps [10,11,12,13]. Different strategies of sequence sampling may lead to different gene trees. The problem of sequence sampling for gene tree construction has been generally treated in a subjective manner, although the related problem of taxonomic sampling for determining species phylogenies has been extensively discussed and investigated in the past two decades [14,15,16,17,18,19,20,21]. Taxonomic sampling of species trees refers to sampling of taxa based on some genetic markers of taxa or whole genomes, rather than sequences of genes or proteins [22,23]. In this work, we study the sequence sampling problem for gene trees.

One reason for the necessity of sequence sampling is the rapidly increasing amount of genomic data due to the advancement of next-generation sequencing techniques. A huge dataset of homologous genes (e.g. 537,686 sequences of glycosyl transferase gene family 2 [24]) may prevent biologists from using accuracy-oriented MSA software tools, such as Muscle [25], Mafft [26], ClusterW [27], T-Coffee [28], SAT-é [29,30], and PRANK [31,32,33], and phylogenetic tree estimation methods, such as PhyML [34], MrBayes [35] and PhyloBayes [36], due to their high demands for computational resources, including both memory and time. Even software tools specifically designed for large datasets have to limit the size of an input dataset. For example, DACTAL [37] is shown to be useful for datasets up to 28, 000 sequences, but the accuracy of alignments and associated trees decreases as the number of sequences increases.

An additional reason for sequence sampling is to facilitate the detection of horizontal gene transfer (HGT) based on phylogenetic tree comparisons. This is because many phylogenetic tree-based

HGT detection approaches will be more applicable and accurate when smaller gene trees are used [38,39,40].

The problem of homologous sequence sampling for a subset of sequences with diverse coverage is not trivial owing to the current biased pool of sequenced genomes [41] i.e. model organisms and medically (or economically) important species tend to have more sequences in the current databanks than others. Simple-minded sampling strategies may lead to sequence datasets that are biased towards certain families of organisms.

Two general strategies have been adopted in sequence sampling when calculating gene phylogenies. One is to sample the most similar sequences (SS) of a query sequence, using either BLAST [42] or HMM-based methods [43]; another is manual selection (MS), which samples sequences based on phylogenetic expertise and knowledge about the evolutionary relationships among relevant organisms. That knowledge may come from a preliminary tree with as many homologous sequences as possible, or from domain experts, who manually sample the tree-associated sequences based on their expertise and experiences. The use of SS may generate a tree that lacks important lineages as a result of the low coverage of some taxonomic branches, while MS, although perhaps suitable for limited case studies, is not generally scalable. Here we focus on the question of sequence sampling to achieve a high taxonomic diversity.

To obtain gene trees with high taxonomic diversity, we developed an algorithm named as ***AST*** to **a**utomatically **s**elect representative homologous sequences over **t**axa. In this study we show that, for the same number of sampled sequences, AST gives rise to more diverse taxa as compared to the currently used methods.

To illustrate its effectiveness, we applied the AST method to resolve the following evolutionary questions: (i) can we infer the evolutionary history of the small ribosomal subunit (SSU) protein S5 (rpS5), 16 S ribosomal RNA (16 S rRNA) and glycosyl-transferase gene family 8 (GT8), and can we identify ancient HGTs from bacteria to eukaryotes.

Materials and Methods

2.1 Sampling algorithm

The AST algorithm samples m sequences from n non-redundant homologous sequences of a query sequence, based on the NCBI taxonomic distribution covered by the n sequences. AST ensures that the sampled sequences will have a high taxonomic diversity covered by the given pool of n homologous sequences and the most even distribution across taxa, $m<n$. Consider a taxon T having G sub-taxa $\{T_1, T_2\ldots, T_G\}$with T_i having n_i homologous sequences (to the query) so $\sum_{i=1}^{G} n_i = n$, $n_i \geq 0$. Here, the sub- taxa $\{T_1, T_2\ldots, T_G\}$are the children taxa of a taxon T, rather than all its descendants. The goal is to select m_i representatives from the n_i homologs from T_i such that m_i will be chosen to be the integer value that is closest to the average number of sequences (calculated as m/G') in each taxon among the G' taxa, where G' is the number of taxa having non-zero n homologs ($n>0$) of the query. Then all the $\{m_1, m_2, m_3,\ldots m_{G'}\}$ values will be the same value or differ by 1 and all the remaining m_i: $\{m_{G'+1},\ldots, m_G\}$ where taxa T_i does not have homologs (i.e. $n_i = 0$) are set to be zero (see Figure 1b as an example). Explicitly, if m/G' does not equal an integer, we sort the taxa descendingly by number of sequences in each taxon, then assign m_i as $\text{int}(m/G')+1$ for the first $m\text{-}G' \times \text{int}(m/G')$ taxa, as $\text{int}(m/G')$ for the remaining taxa, where $\text{int}(x)$ represents the integral part of x. It is easy to check that $\sum_{i=1}^{G} m_i = m$.

To maximize the taxonomic diversity, our algorithm searches exclusively among the hierarchy of NCBI taxa. It recursively solves this problem as follows. If $m_i>1$ and T_i has children taxa in the next level, then recursively call the above algorithm by setting n to n_i, m to m_i and T_i to the current taxon. If $m_i>1$ and T_i does not have sub-taxa (i.e., a leaf-taxon), sample m_i sequences with the highest sequence similarity to the query from the n_i sequences. The algorithm iterates until $m_i = 1$ or 0, or T_i becomes a leaf-taxon. Figure 1a illustrates a workflow of the algorithm (we refer readers to Figure 1b for an example).

2.2 Simulated data

We used the EvolveAGene [44] program to generate three types of simulated gene trees: symmetric trees, random trees and asymmetric trees, as each of these three types may exist in reality. EvolveAGene [44] can simulate the evolution of DNA sequences through mimicking mutation and natural selection, and generate the off-spring sequences, whose ultimate structures will be symmetric or random trees based on specified parameters. Here we take each simulated DNA sequence in each generation as the whole genome of an organism, with each node of the simulated trees as a taxon. The relationships among the taxa (nodes) are available from the output of EvolveAGene.

To generate the simulated trees for this study, we randomly chose the *xisC* gene of bacterium *Nostoc sp. PCC 7120* (GenBank accession: U08014) as the initial root sequence and generated 1,024 simulated sequences using the program, where 1,024 is used because it is a power of 2 as required for generating a symmetric tree, and this size is comparable to the order of magnitude of the number of currently sequenced genomes [45].

EvolveAGene provides an option for generating random trees without any specified tree topology. When using this program, we set the average branch length at 0.3 and the number of leaf taxa as 1,024 with all the other parameters set at their default values.

To generate asymmetric trees, we first generated a symmetric tree with 2,048 leafs, and then select 1,024 leafs to construct an asymmetric tree using the following procedure. We randomly chose x percentage of the selected 1,024 leafs from the left branch of the symmetric tree and $(1.0-x)$ percentage of 1,024 leafs from the right branch. Here we used x values equal to 0.1, 0.2, 0.3, and 0.4 to generate the trees.

2.3 Biological data

Amino acid sequences of the rpS5 proteins from 816 bacteria and 68 archaea were downloaded from the NCBI curated Protein Cluster DB (Oct, 2010) (http://www.ncbi.nlm.nih.gov/proteinclusters). They were identified as non-redundant homologs of the *Escherichia coli* rpS5 protein using pBLAST with an *E*-value <0.01. The rank of the similarity scores between these homologous proteins were based on the BLAST bitscore.

The 918 GT8 protein sequences and their pair-wise similarity scores were provided by the authors of [1].

2.4 Phylogenetic analyses

To construct the phylogeny for rpS5 of *E. coli* and other related organisms, we performed multiple sequence alignments using MAFFT (version 6.603) [46], employing the L-INS-I model, which adopts local pair-wise alignments by the Smith-Waterman algorithm and is considered to be one of the most accurate multiple sequence alignment methods currently available [47,48]. Then a phylogenetic tree was constructed using the FastTree program (version 2.1.3) [49], which implements a superfast but fairly accurate approximate maximum likelihood method [49].

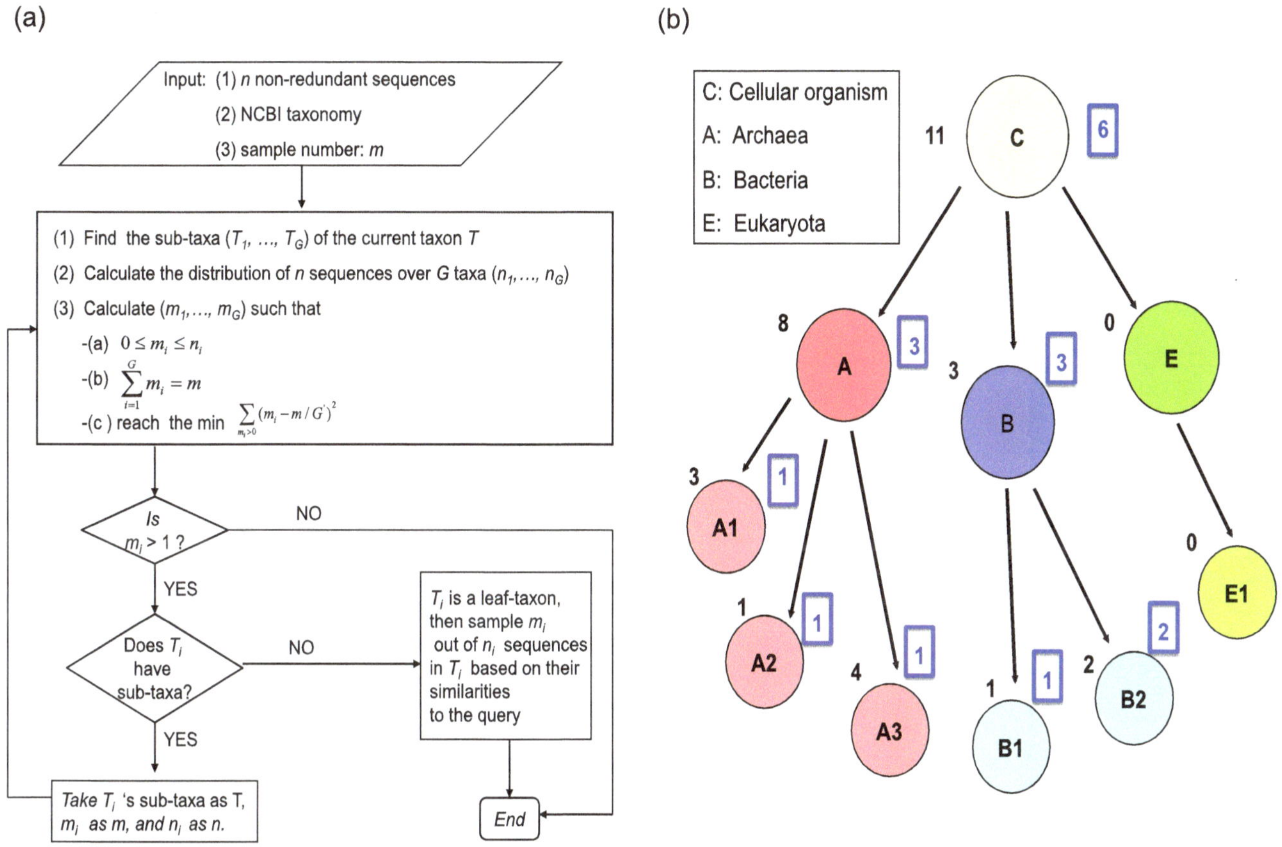

Figure 1. AST algorithm. (a) Workflow of the AST algorithm. (b) An example of the sampling procedure of AST. Each circle represents one taxon: C-all *Cellular Organism*; A-*Archaea*; B-*Bacteria*; A1 is an archaeal taxon labeled as A1, similar for A2, A3, B1, and B2. The number listed on the left shoulder of the circle (outside the rectangle) is the number of sequences from the taxon labeled in the circle, and the number listed on the right shoulder of each circle is the number of sampled sequences by AST from the taxon in the circle. In this example there are a total of 11 homologous sequences in all cellular organisms, among which 8 belong to archaea, 3 from bacteria and none from eukaryotes.

To study the phylogeny and horizontal gene transfer in the class-I of glycosyl-transferase gene family 8 (GT8), we adopted a rigorous PhyML [50] analysis as used in previous analyses of GT8 [1]. For the PhyML analyses, trees were built with the JTT substitution model [51] along with the following parameters: estimated proportion of invariable sites, four rate categories, estimated gamma distribution, and optimized starting BIONJ tree [52]. Bootstrapping was performed using 100 replications. MrBayes [53] analyses were used with a mixed amino acid model estimated in the run, an estimated proportion of invariable sites, an estimated gamma distribution parameter, and one million of generations.

2.5 AST software package

Currently, two versions of the AST program are provided at *http://csbl.bmb.uga.edu/~zhouchan/AST.php* a basic version, and a more advanced version. The basic AST suite consists of the core method of AST and deals with a user pre-prepared input file with the following information: a list of IDs of non-redundant homologous sequences, their taxon IDs and similarity scores. The advanced suite does not require a prepared list of IDs of homologs. Instead, it only needs the BLAST report (in xml format, -m7 output) and will generate the input file based on the BLAST report automatically. The program has a set of default parameters, such as the BLAST bitscore and *E*-value cutoffs, but users can adjust these values if needed (see the README file of the program package for details).

Results

We assessed the performance of AST on both simulated and real biological data, and compared the results of AST with those by SS, random sampling (RS) methods and MS (that is, if results were available in the literatures [1,24] using the MS method). Here, the SS method samples *m non-redundant* homologous sequences that are most similar to the query, while the RS method randomly samples *m non-redundant* homologous sequences. In this study we show that the trees generated by AST indeed have more taxonomic diversities than those by SS and RS, and are comparable with the taxonomic diversity of the whole gene trees that are generated with all available homologs.

3.1 Comparative analyses of tree construction on simulated data

We compared the taxonomic coverage of sequences sampled by AST, SS, and RS on three types of simulated trees: symmetric, random, and asymmetric trees. One hundred trees were generated for random trees and for each of the asymmetric trees with a bias index $x = 0.1$, 0.2, 0.3, 0.4, respectively (see Materials and Methods for details). Only one symmetric tree is generated since such trees always have the same topology.

The following summarizes the performance of the three methods on asymmetric trees in terms of the taxonomic coverage at each taxonomic level when sampling with $m = 50$, 100, 200, 300, 400, 500 from $n = 1{,}024$ sequences which are the simulated non-redundant homologs of the root-sequence (see Materials and Methods for details). Here *taxonomic level* refers to the level (the relative position) in a taxonomic hierarchy with the root taxon being at level 1, the direct children taxa being at level 2 and so on.

Here we use the asymmetric trees with bias index $x = 0.1$ as an example. Table 1 summarizes the taxonomic coverage for sub-trees as well as the whole tree at the 8th taxonomic level by the three methods. The sub-trees sampled by AST cover significantly more taxa than those sampled by RS (*P*-value = 0.025) and SS (*P*-value = 0.0017), respectively, across all the *m* values defined above (Table 1), as determined using Mann-Whitney tests. In addition, when the number of sampled sequences is larger than 200, the sub-trees sampled by AST cover all the taxa (~116) at the 8th taxonomic level of the whole tree whereas the RS and SS miss large numbers of taxonomic lineages (Table 1). Similar comparative results were obtained at all the other taxonomic levels, except for the 1st and 2nd levels where all the three methods sampled all the taxa. Highly similar comparative performances were observed on asymmetric trees generated using bias index $x = 0.2$, 0.3 and 0.4.

On the symmetric and random trees, we also obtained highly comparative performance results (see Table S1 and S2): in all cases the sub-trees sampled by AST cover more taxa than those sampled by SS and RS.

3.2 Inference of the evolutionary history of a gene or gene family

Inference of the evolutionary history for a gene (or gene family) can help to derive its detailed functions (e.g. orthologs *vs* paralogs), as well as its possible origin. In the following, we show that the global phylogenies of the *E. coli* rpS5 and 16 S bacterial ribosomal RNAs inferred based on the sub-tree sampled with AST are very similar to that inferred by the tree built on all homologs, highlighting the ability of our method to preserve the key evolutionary information of the whole phylogeny in a smaller tree. We also did a similar analysis on cell wall synthesis-related glycosyl-transferase family 8 (GT8), and the same level of high-quality phylogeny was obtained.

3.2.1 Comparative analyses of the rpS5 trees by three methods. A total of 884 rpS5 proteins (i.e. 816 and 68 of bacterial and archaeal origin, respectively) were identified as non-redundant homologs of the *E. coli* rpS5 protein (see Materials and Methods). To demonstrate that AST generates a pool of more taxonomically diverse representatives than the other two methods, we compared the trees with sequences sampled by AST, SS and RS, with the whole tree based on all 884 sequences.

We note that the sub-trees sampled by AST reflect the whole tree much better than the sub-trees sampled by RS and SS. Specifically, the sub-trees built on sequences sampled using AST contain all the 19 phyla represented in the whole tree (Figure 2 and Table S3) when *m* ranges from 50 to 400. In contrast, the sub-trees sampled by RS contain 10 out of 19 phyla and the sub-tree sampled by SS only covers one of 19 phyla at $m = 50$.

It is worth noting that the advantage of AST becomes more obvious when measuring the coverage at the higher taxonomic levels (e.g. phylum (Figure 2a), class (Figure 2b) and order levels). Regardless of the number of sequences sampled, AST always gives rise to higher taxa coverage than the other two methods (Table S3).

Additionally, we evaluated the performances of AST on a published large empirical benchmark datasets for phylogeny estimation [54], which includes 38,905 16 S rRNA sequences of 34,917 bacteria. The results of the 16 S rRNA sampling (see Figure S1 and Table S4) are similar to that of the rpS5 protein.

3.2.2 Inferring the evolution of the GT8 gene family. The GT8 family is a large gene family with extensive gene duplications [55]. It has been shown to fall into three well-delineated functional classes, which have cyanobacterial sequences mixed with eukaryotic sequences [1]. Here we applied AST to do the phylogenetic analysis of the same set consisting of 918 GT8 protein sequences as in [1] to examine if the same result can be achieved using this simple procedure.

Figure 3 shows three phylogenetic trees based on AST, SS and RS sampling for $m = 300$ out of the 918 sequences. We can see that the AST tree (Figure 3a) is very similar to the Figure 2 published in [1]. Specifically, the AST tree classifies GT8 proteins into three well-delineated functional classes and class-I has the cyanobacterial sequences mixed with eukaryote ones, while the SS and RS trees exhibit large discrepancies with the trees in [1] (not shown here). Specifically, the SS tree (Figure 3b) has four rather than three classes. It consists of only homologs in plant (in red) and

Table 1. The coverage over taxa at taxonomic level 8 for asymmetric trees with sequences sampled by AST, RS and SS using bias index = 0.1.

m	AST [a]	RS [b]	SS [c]	*All-seq* [d]
50	49.98±0.14	37.84±2.26	4±0	*116.4±2.29*
100	96.55±1.51	59.23±3.42	7.32±0.47	
200	116.4±2.29	78.7±3.20	14.30±0.46	
300	116.4±2.29	88.22±3.30	21.26±0.46	
400	116.4±2.29	94.24±3.45	28.23±0.44	
500	116.4±2.29	99.55±3.13	35.14±0.40	

[a]AST: automated sampling homologs over taxa.
[b]SS: sampling homologs most similar to the query by a custom script.
[c]RS: random sampling by a custom script.
[d]All-seq: all the 1024 homologous sequences without sampling.
The first column *m* represents the number of sampled sequences and the remaining value represents the mean±std of the number of covered taxa by each method when sampling *m* sequences. The last column indicates the average along with standard deviations of the number of taxa covered by the number of all homolgous sequences for 100 asymmetric trees.

(a)

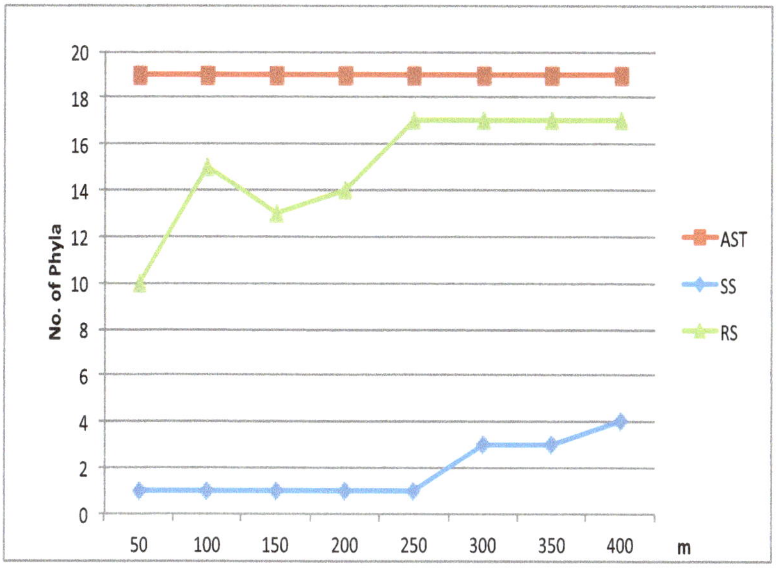

(b)

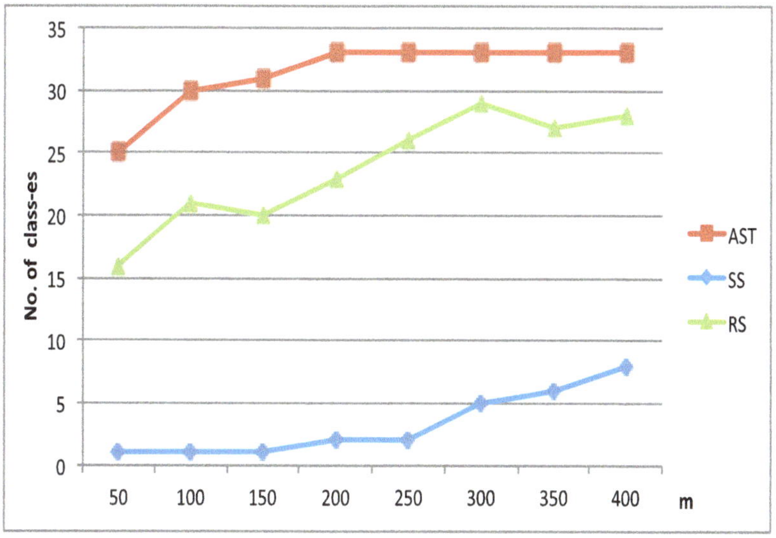

Figure 2. Taxonomic distributions at the phylum (a) and class level (b) for sub-trees of the rpS5 sequences sampled by AST, SS, and RS, respectively. The y-axis gives the number of phyla/classes covered by the sampled sequences, and the x-axis represents the number of sampled sequences *m*. The original non-redundant set covers 19 phyla and 33 classes (see Section 3.2.1 for details).

bacteria (excluding cyanobacteria in sienna), but misses considerable amount of information from other taxonomic lineages such as fungi (in green), virus (in yellow), metazoa (in purple) and cyanobacteria (in sienna). The RS tree covers more lineages (Figure 3c) than the SS tree, but it does not always include cyanobacteria in class-I. This is because each random sampling procedure gave rise to different RS trees and some RS trees may group cyanobateria in the class I while others may not include any cyanobateria.

We note that the tree structure with sequences sampled by AST is highly stable for $m >= 50$. The AST trees with $m = 50$, 100, 200 and 400 are given in Figures S2–S5.

3.3 Detection and rigorous testing of ancient HGT events

Ancient HGTs cannot be easily detected using a phylogenetic method due to lack of high-quality datasets. Here we applied the AST method to the detection of an ancient HGT from the ancestor of cyanobacteria to the ancestor of plants [1] to show that the AST method can reliably detect HGT events and provide a rigorous test through comparing the obtained results to those that were manually derived by domain experts.

When inferring the evolutionary history of GT8, our AST tree (Figure 3a) already shows that three cyanobacterial GT8 sequences appear among eukaryotic GT8 sequences. The cyanobacterial sequences in class-I are basal to sequences from plants and some other eukaryotic GT8 proteins. This observation suggests that either there is an ancient HGT from cyanobacteria to

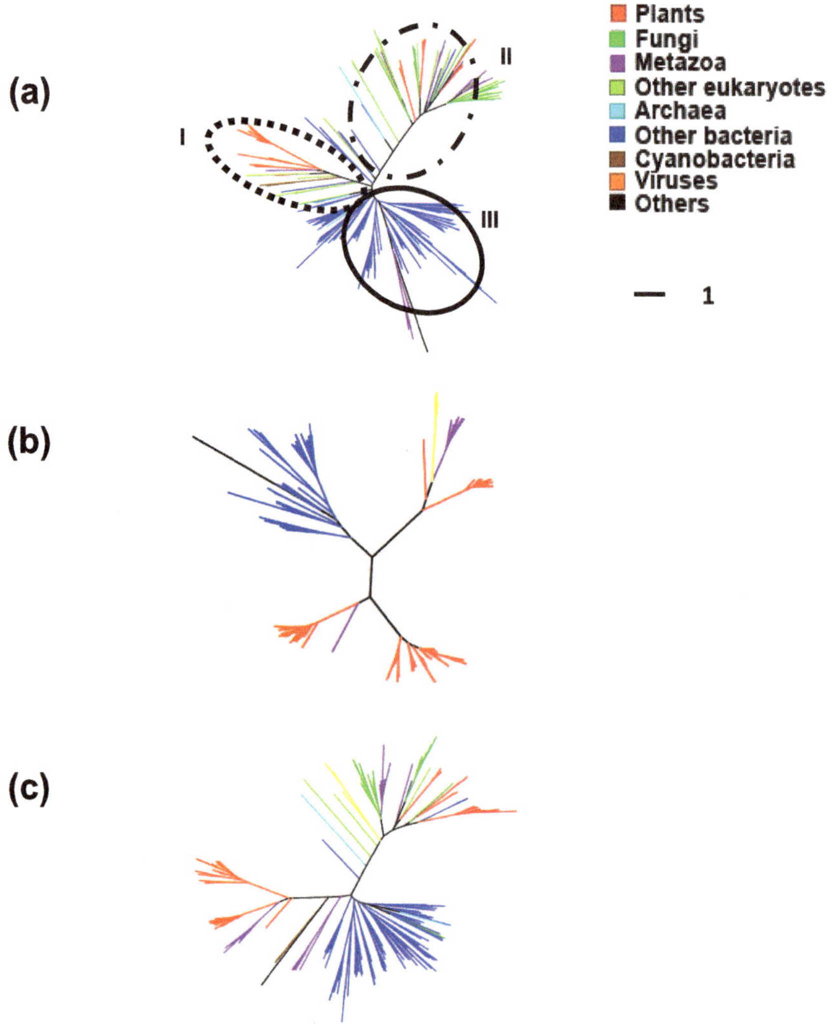

Figure 3. Phylogeny of 300 GT8 sequences sampled by (a) AST, (b) Similarity Sampling (SS) and (c) Random Sampling (RS) approaches, respectively. Three major functional classes are identified in (a), which is consistent with a previous publication [1]. In (b), the phylogenetic tree is composed of four classes, which are incorrect. In (c), cyanobacterial sequences (in sienna, GenBank gi: 254421706, 254423034, and 81299339) are incorrectly grouped with the out-group (in black). Color definitions: red for plants, green for fungi, purple for metazoa, olive-green for other eukaryotes, cyan for archaea, blue for other bacteria, sienna for cyanobacteria, orange for viruses and black for others.

eukaryote or cyanobacteria acquired their homolog from a eukaryote. To test these two hypotheses, a previous publication [1] manually selected 15 representative sequences based on the authors' prior knowledge, then constructed the phylogenetic trees of these 15 sequences through a computational procedure consisting of bootstrap and Bayesian analyses using PhyML [50] and MrBayes [53]. They found that in these well resolved phylogenetic trees the cyanobacterial sequence is indeed mixed with eukaryotes in class-I and groups the base of the plant homologs [1].

We directly applied AST to sample 15 sequences from all 268 sequences in class-I without any prior knowledge and complex computational procedure, and then used PhyML and MrBayes to perform bootstrap and Bayesian analyses with the same criteria as in [1]. The two phylogenetic trees, each generated by PhyML or MrBayes, are quite similar, although branch lengths and the statistical supports of some nodes are different (Figure 4a). The AST-based tree clusters all three cyanobacterial GT8 proteins (in sienna) and clusters them with the other class-I proteins from plants (in red) and other eukaryotes (in mignonette), except for some metazoan proteins (in purple), with strong statistical support values. This result is consistent with the results given in [1]. With regard to the RS- and SS-based trees (Figure 4b and 4c), no information about a possible HGT from cyanobacteria can be derived from either of them. AST is able to help rigorous HGT detection since the small trees sampled by AST will cover more taxa branches and hence will also cover the divergent recipients of HGT events, e.g. cyanobacteria in Figures 3 and 4.

Discussions

In addition to the aforementioned examples, AST can also be applied to infer local phylogenies of gene trees and detect recent HGTs. For the former analyses AST requires a well-prepared file as input: a list of IDs of homologs, which should be limited within the local taxonomic lineage under consideration. For example, if one would like to study the evolution of a gene across eukaryotes, it would be reasonable to prepare a list of IDs of homologous sequences mainly from eukaryotic lineages, then AST will sample

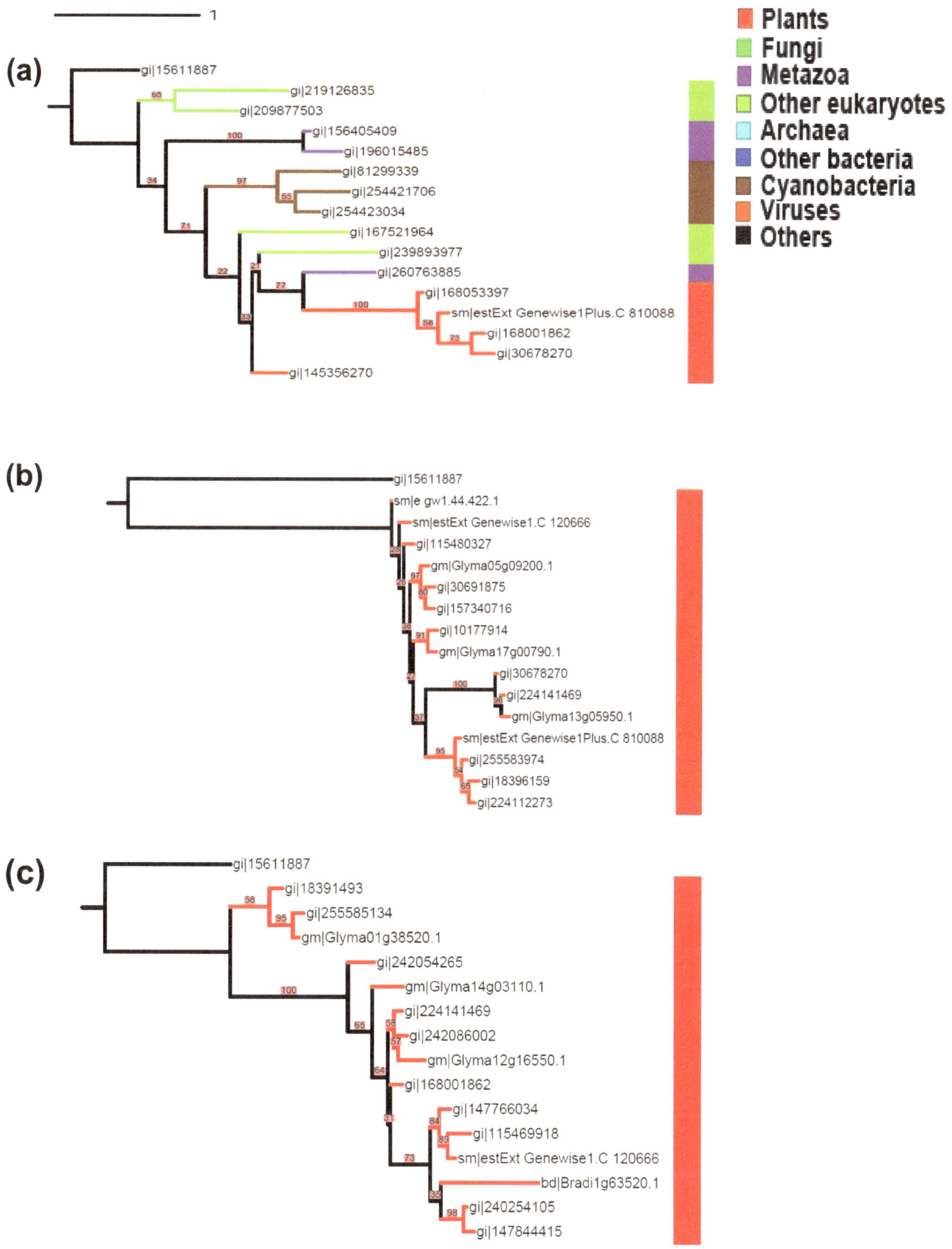

Figure 4. Phylogeny of 15 representative amino acid sequences from the GT8 class-I. To determine the roots of trees, we randomly selected a non-GT8 sequence (GI: 15611887) as an outgroup (in black). The sequences were sampled by (a) AST; (b) Random Sampling; and (c) Similarity Sampling approaches. The Bayesian posterior probability in grouping the 3 cyanobacterial sequences with metazoa and plants in (a) is 0.99

by using MrBayes. Tree (a) reflects the diversity of class-I with 15 sequences and also indicates a potential transfer between cyanobacteria and eukaryotes, while trees in (b) and (c) are only composed of plant sequences.

most sequences from eukaryotic and a few sequences from bacteria and/or archaea as out-groups if few (e.g. 1 or 2) of bacterial or archaeal sequences are included in the list of homologs. The detection of recent HGTs does not require sampling of homologs over all taxa. In a similar fashion as for the local gene phylogenies, AST requires a list of IDs of homologs within the concerned taxonomic group. For example, to infer recent HGTs within the proteobacteria group, only the homologs from the proteobacteria group are included in the input file. To detect ancient HGTs between distant organisms, which indeed requires sampling sequences over extensive taxa, the input file (a list of IDs of the homologs) should include homologs from all domains; otherwise the gene trees could not include branches from distant organisms which may have conflicts with their corresponding species tree, hence indicating a putative HGT.

If studying a gene family with many domain rearrangements, we suggest using the domain as a query to determine the homologous sequences instead of the entire gene sequence, and then apply the AST software to the homologous pool defined by that domain.

AST is designed to maximize taxonomic diversity, treating each branch equally and considering the topology of gene trees. In contrast, phylogenic diversity evaluates the quality of both branch length and topology [56]. If branch length is incorporated into the sampling procedure, then maximizing taxonomic diversity will also maximizing phylogenetic diversity [56,57] of gene trees. To expand applications of the AST software, we plan to introduce weights and branch lengths (as parameters) into the sampling procedure in a future upgrade, and will also take the uneven distribution of available sequences across different taxa into consideration. Last but not the least, while AST ensures a high taxonomic diversity, it still requires strong methods for multiple sequence alignment and tree estimation for reliable gene phylogenetic inference.

Supporting Information

Figure S1 Taxonomic distributions at the phylum (a) and class level (b) for sub-trees of 16 s ribosomal RNA sequences sampled by AST, SS, and RS, respectively. The *y*-axis gives the number of phyla/classes covered by the sampled sequences, and the *x*-axis represents the number of sampled sequences *m*. There are 26 phyla and 37 classes covered by the original non-redundant set and AST sampled sequences from each of all 26 phyla and each of all 37 classes.

Figure S2 Phylogenetic trees of 50 GT8 sequences sampled by AST, SS, and RS respectively. See the legend of Figure 3 for further details.

Figure S3 Phylogenetic trees of 100 GT8 sequences sampled by AST, SS, and RS respectively. See the legend of Figure 3 for further details.

Figure S4 Phylogenetic trees of 200 GT8 sequences sampled by AST, SS and RS respectively. See the legend of Figure 3 for further details.

Figure S5 Phylogenetic trees of 400 GT8 sequences sampled by AST, SS and RS respectively. See the legend of Figure 3 for further details.

Table S1 Taxonomic distributions for symmetric trees when sequences were sampled by AST, SS and RS, respectively.

Table S2 Taxonomic distributions for random trees when sequences were sampled by AST, SS and RS, respectively.

Table S3 Taxonomic distributions for rpS5 proteins with sequences sampled by AST, SS and RS at the super-phylum, order, family, genus, and species levels.

Table S4 Taxonomic distributions for 16 S rRNA sequences with sequences sampled by AST, SS and RS at the super-phylum, order, family, genus, and species levels.

Acknowledgments

We thank all the members of AToL projects at UGA, UConn and ECU for their stimulating discussion and suggestions, esp. Guiling Sun and David Williams. CZ thanks Amy Styer at UGA for improving the presentation. We also thank two anonymous referees for useful comments on the manuscript.

Author Contributions

Conceived and designed the experiments: CZ FM. Performed the experiments: CZ. Analyzed the data: CZ FM YY JPG. Contributed reagents/materials/analysis tools: CZ YY JH YX. Wrote the paper: CZ JPG YX. Suggested the need of this work: JH.

References

1. Yin Y, Chen H, Hahn MG, Mohnen D, Xu Y (2010) Evolution and function of the plant cell wall synthesis-related glycosyltransferase family 8. Plant Physiol 153: 1729–1746.
2. Chauve C, Doyon J-P, El-Mabrouk N (2007) Inferring a duplication, speciation and loss history from a gene tree. Comparative Genomics-Lecture Notes in Computer Science 4751: 45–57.
3. Eisen JA (1998) Phylogenomics: improving functional predictions for uncharacterized genes by evolutionary analysis. Genome Res 8: 163–167.
4. Zhou C, Yin Y, Dam P, Xu Y (2010) Identification of novel proteins involved in plant cell-wall synthesis based on protein-protein interaction data. J Proteome Res 9: 5025–5037.
5. Liu L, Yu L (2011) Estimating species trees from unrooted gene trees. Syst Biol 60: 661–667.
6. Wehe A, Bansal MS, Burleigh JG, Eulenstein O (2008) DupTree: a program for large-scale phylogenetic analyses using gene tree parsimony. Bioinformatics 24: 1540–1541.
7. Cranston KA, Hurwitz B, Ware D, Stein L, Wing RA (2009) Species trees from highly incongruent gene trees in rice. Syst Biol 58: 489–500.
8. Liu L, Pearl DK (2007) Species trees from gene trees: reconstructing Bayesian posterior distributions of a species phylogeny using estimated gene tree distributions. Syst Biol 56: 504–514.
9. Page RD (1998) GeneTree: comparing gene and species phylogenies using reconciled trees. Bioinformatics 14: 819–820.
10. Rosenberg MS, Kumar S (2001) Incomplete taxon sampling is not a problem for phylogenetic inference. Proc Natl Acad Sci U S A 98: 10751–10756.
11. Rannala B, Huelsenbeck JP, Yang Z, Nielsen R (1998) Taxon sampling and the accuracy of large phylogenies. Syst Biol 47: 702–710.

12. Dereeper A, Audic S, Claverie JM, Blanc G (2010) BLAST-EXPLORER helps you building datasets for phylogenetic analysis. BMC Evol Biol 10: 8.
13. Nabhan AR, Sarkar IN (2012) The impact of taxon sampling on phylogenetic inference: a review of two decades of controversy. Briefings in Bioinformatics 13: 122–134.
14. Townsend JP, Leuenberger C (2011) Taxon sampling and the optimal rates of evolution for phylogenetic inference. Syst Biol 60: 358–365.
15. Pick KS, Philippe H, Schreiber F, Erpenbeck D, Jackson DJ, et al. (2010) Improved phylogenomic taxon sampling noticeably affects nonbilaterian relationships. Mol Biol Evol 27: 1983–1987.
16. Townsend JP, Lopez-Giraldez F (2010) Optimal selection of gene and ingroup taxon sampling for resolving phylogenetic relationships. Syst Biol 59: 446–457.
17. Hillis DM (1998) Taxonomic sampling, phylogenetic accuracy, and investigator bias. Syst Biol 47: 3–8.
18. Kim J (1996) General inconsistency conditions for maximum parsimony: effects of branch lengths and increasing numbers of taxa. Syst Biol 45: 363–374.
19. Rosenberg MS, Kumar S (2003) Taxon sampling, bioinformatics, and phylogenomics. Syst Biol 52: 119–124.
20. Plazzi F, Ferrucci RR, Passamonti M (2010) Phylogenetic representativeness: a new method for evaluating taxon sampling in evolutionary studies. BMC Bioinformatics 11: 209.
21. Zwickl DJ, Hillis DM (2002) Increased taxon sampling greatly reduces phylogenetic error. Syst Biol 51: 588–598.
22. Pecon-Slattery J, Pearks Wilkerson AJ, Murphy WJ, O'Brien SJ (2004) Phylogenetic assessment of introns and SINEs within the Y chromosome using the cat family felidae as a species tree. Mol Biol Evol 21: 2299–2309.
23. Jacobsen F, Omland KE (2011) Species tree inference in a recent radiation of orioles (Genus Icterus): multiple markers and methods reveal cytonuclear discordance in the northern oriole group. Mol Phylogenet Evol 61: 460–469.
24. Yin Y, Mao X, Yang J, Chen X, Mao F, et al. (2012) dbCAN: a web resource for automated carbohydrate-active enzyme annotation. Nucleic Acids Res 40: W445–451.
25. Edgar RC (2004) MUSCLE: multiple sequence alignment with high accuracy and high throughput. Nucleic Acids Research 32: 1792–1797.
26. Katoh K, Kuma K, Toh H, Miyata T (2005) MAFFT version 5: improvement in accuracy of multiple sequence alignment. Nucleic Acids Res 33: 511–518.
27. Li KB (2003) ClustalW-MPI: ClustalW analysis using distributed and parallel computing. Bioinformatics 19: 1585–1586.
28. Notredame C, Higgins DG, Heringa J (2000) T-Coffee: A novel method for fast and accurate multiple sequence alignment. J Mol Biol 302: 205–217.
29. Liu K, Warnow TJ, Holder MT, Nelesen SM, Yu J, et al. (2012) SATe-II: very fast and accurate simultaneous estimation of multiple sequence alignments and phylogenetic trees. Syst Biol 61: 90–106.
30. Liu K, Raghavan S, Nelesen S, Linder CR, Warnow T (2009) Rapid and accurate large-scale coestimation of sequence alignments and phylogenetic trees. Science 324: 1561–1564.
31. Tamura K, Nei M (1993) Estimation of the number of nucleotide substitutions in the control region of mitochondrial DNA in humans and chimpanzees. Mol Biol Evol 10: 512–526.
32. Whelan S, Goldman N (2001) A general empirical model of protein evolution derived from multiple protein families using a maximum-likelihood approach. Mol Biol Evol 18: 691–699.
33. Loytynoja A, Goldman N (2005) An algorithm for progressive multiple alignment of sequences with insertions. Proc Natl Acad Sci U S A 102: 10557–10562.
34. Guindon S, Dufayard JF, Lefort V, Anisimova M, Hordijk W, et al. (2010) New Algorithms and Methods to Estimate Maximum-Likelihood Phylogenies: Assessing the Performance of PhyML 3.0. Systematic Biology 59: 307–321.
35. Ronquist F, Huelsenbeck JP (2003) MrBayes 3: Bayesian phylogenetic inference under mixed models. Bioinformatics 19: 1572–1574.
36. Lartillot N, Lepage T, Blanquart S (2009) PhyloBayes 3: a Bayesian software package for phylogenetic reconstruction and molecular dating. Bioinformatics 25: 2286–2288.
37. Nelesen S, Liu K, Wang LS, Linder CR, Warnow T (2012) DACTAL: divide-and-conquer trees (almost) without alignments. Bioinformatics 28: i274–282.
38. Poptsova M (2009) Testing phylogenetic methods to identify horizontal gene transfer. Methods Mol Biol 532: 227–240.
39. Poptsova MS, Gogarten JP (2007) The power of phylogenetic approaches to detect horizontally transferred genes. BMC Evol Biol 7: 45.
40. Syvanen M (2006) Some computational problems associated with horizontal gene transfer. In: Rigoutsos IaS, G., editor. Systems Biology, Volume I: Genomics. Oxford: Oxford University Press. pp. 248–268.
41. Benson DA, Karsch-Mizrachi I, Lipman DJ, Ostell J, Wheeler DL (2005) GenBank. Nucleic Acids Res 33: D34–38.
42. Altschul SF, Gish W, Miller W, Myers EW, Lipman DJ (1990) Basic Local Alignment Search Tool. Journal of Molecular Biology 215: 403–410.
43. Frickey T, Lupas AN (2004) PhyloGenie: automated phylome generation and analysis. Nucleic Acids Res 32: 5231–5238.
44. Hall BG (2008) Simulating DNA coding sequence evolution with EvolveAGene 3. Mol Biol Evol 25: 688–695.
45. Jenuth JP (2000) The NCBI. Publicly available tools and resources on the Web. Methods Mol Biol 132: 301–312.
46. Katoh K, Kuma K, Toh H, Miyata T (2005) MAFFT version 5: improvement in accuracy of multiple sequence alignment. Nucleic Acids Research 33: 511–518.
47. Ahola V, Aittokallio T, Vihinen M, Uusipaikka E (2006) A statistical score for assessing the quality of multiple sequence alignments. BMC Bioinformatics 7.
48. Elisabeth RM, Nuin PAS, Wang ZZ (2006) The accuracy of several multiple sequence alignment programs for proteins. BMC Bioinformatics 7.
49. Price MN, Dehal PS, Arkin AP (2009) FastTree: computing large minimum evolution trees with profiles instead of a distance matrix. Mol Biol Evol 26: 1641–1650.
50. Guindon S, Gascuel O (2003) A simple, fast, and accurate algorithm to estimate large phylogenies by maximum likelihood. Systematic Biology 52: 696–704.
51. Jones DT, Taylor WR, Thornton JM (1992) The rapid generation of mutation data matrices from protein sequences. Comput Appl Biosci 8: 275–282.
52. Gascuel O (1997) BIONJ: an improved version of the NJ algorithm based on a simple model of sequence data. Mol Biol Evol 14: 685–695.
53. Ronquist F, Huelsenbeck JP (2003) MrBayes 3: Bayesian phylogenetic inference under mixed models. Bioinformatics 19: 1572–1574.
54. Linder CR, Suri R, Liu K, Warnow T (2010) Benchmark datasets and software for developing and testing methods for large-scale multiple sequence alignment and phylogenetic inference. PLoS Curr 2: RRN1195.
55. Yin Y, Mohnen D, Gelineo-Albersheim I, Xu Y, Hahn GM (2011) Glycosyltransferases of the GT8 family. Plant Polysaccharides: Annual Plant Reviews. pp 167–212.
56. Faith PD (1992) Conservation evaluatio and phylogenetic diversity. Biological Conservation 61: 1–10.
57. Vane-Wright RI, Humphries CJ, Williams PH (1991) What to protect - systematics and the agony of choice. Biological Conservation 55: 235–254.

2

Analysis of Micro-Rearrangements in 25 Eukaryotic Species Pairs by SyntenyMapper

Stefanie Kaufmann[1], Dmitrij Frishman[1,2,3]*

1 Department of Genome Oriented Bioinformatics, Technische Universität München, Freising, Bavaria, Germany, **2** Institute of Bioinformatics and Systems Biology, German Research Center for Environmental Health, Neuherberg, Bavaria, Germany, **3** Department of Bioinformatics, St Petersburg State Polytechnical University, St Petersburg, Russia

Abstract

High-quality mapping of genomic regions and genes between two organisms is an indispensable prerequisite for evolutionary analyses and comparative genomics. Existing approaches to this problem focus on either delineating orthologs or finding extended sequence regions of common evolutionary origin (syntenic blocks). We propose SyntenyMapper, a novel tool for refining predefined syntenic regions. SyntenyMapper creates a set of blocks with conserved gene order between two genomes and finds all minor rearrangements that occurred since the evolutionary split of the two species considered. We also present TrackMapper, a SyntenyMapper-based tool that allows users to directly compare genome features, such as histone modifications, between two organisms, and identify genes with highly conserved features. We demonstrate SyntenyMapper's advantages by conducting a large-scale analysis of micro-rearrangements within syntenic regions of 25 eukaryotic species. Unsurprisingly, the number and length of syntenic regions is correlated with evolutionary distance, while the number of micro-rearrangements depends only on the size of the harboring region. On the other hand, the size of rearranged regions remains relatively constant regardless of the evolutionary distance between the organisms, implying a length constraint in the rearrangement process. SyntenyMapper is a useful software tool for both large-scale and gene-centric genome comparisons.

Editor: Shoba Ranganathan, Macquarie University, Australia

Funding: This work was supported by the German Research Foundation (DFG, www.dfg.de) International Research Training Group 'Regulation and Evolution of Cellular Systems' (GRK 1563). Publication was supported by the Technische Universität München within the funding programme Open Access Publishing. The funders had no role in study design, data collection and analysis, decision to publish, or preparation of the manuscript.

Competing Interests: The authors have declared that no competing interests exist.

* Email: d.frishman@wzw.tum.de

Introduction

The most basic step in comparative genomics is to find functional genetic elements (genes, pseudogenes, repeats, regulatory sequences) as well as entire genome regions that are conserved between two species. The finding of matching genomic regions is central to tracing the evolutionary history that led from a common ancestor to the contemporary genome sequences via a succession of evolutionary events, such as gene duplications and translocations. Establishing equivalent locations between genomes is also an important prerequisite for comparing position-specific functional, structural, and evolutionary features measured by modern high-throughput techniques, such as transcriptionally active regions, chromatin accessibility, replication domains and single nucleotide polymorphisms (SNPs).

Finding corresponding locations in two different genomes usually involves the identification of syntenic regions, which represent the longest sequence stretches of common evolutionary origin and consist of a number of conserved genome regions, often with interspersed short segments of lower or no similarity [1]. The order of these equivalent syntenic regions is different in both genomes due to an unknown number of large rearrangement events that occurred after the species diverged from the last common ancestor. Based on their extent, rearrangements are usually (somewhat arbitrarily) subdivided into two classes: i) macro-rearrangements, which involve multi-megabase sized intra- and interchromosomal relocation of large syntenic blocks, and ii) micro-rearrangements [1], i.e. re-ordering of smaller segments (below 1Mb) within a syntenic region.

The distinction between finding large regions of common origin and rearrangements of gene order at a much finer scale is intrinsically ill defined. In addition, each of these tasks has its own complications. While looking for long genomic blocks in two genomes that have evolved from the same sequence, one must allow for gaps and ignore micro-rearrangements. The search for orthologs is hampered by the presence of paralogous families and by local similarity hits covering individual domains of multi-domain proteins. Below we give a general overview of the most common types of methods for the identification of equivalent genomic locations using synteny, as illustrated in Figure 1.

Sequence-based methods

The most widely used approach to syntenic block detection is based on whole-genome sequence alignments. For example, ENSEMBL Compara [2,3] starts with short local alignment seeds with a perfect match in both organisms and elongates them until

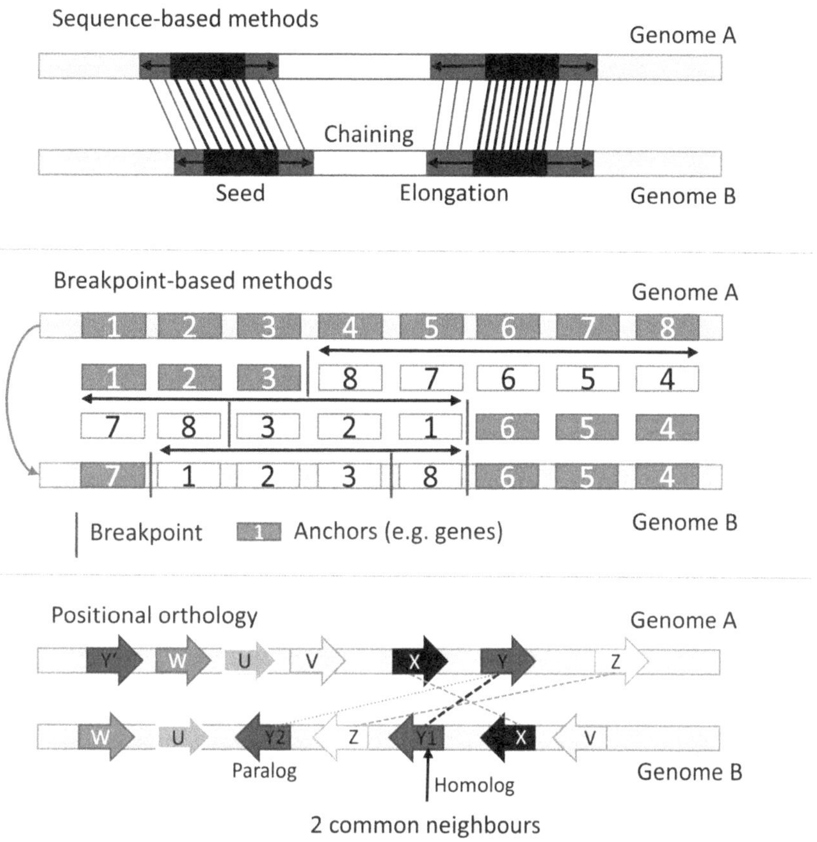

Figure 1. Overview of different approaches for identifying orthologous regions in two genomes. Sequence-based methods (e.g. ENSEMBL Compara) start with short local alignments that are extended to longest possible alignments over gaps. Breakpoint-based methods use orthologous elements (called 'anchors') to find the minimum number of rearrangements that transforms one genome into the other. Positional orthology tries to distinguish orthologs from paralogs by analyzing gene neighborhoods.

the similarity score falls below a predefined threshold. This procedure produces a set of medium length alignments with a low number of gaps and mismatches. Pairs of such alignments that are located sufficiently close to each other are then chained to create extended aligned stretches of common evolutionary origin. A similar approach was published by Liao et al. [4] who introduced pairs of unique 16-mers as a substitute for local alignments. This allows for a much faster searching for orthologous regions than whole-genome alignments. The method is designed to find the longest regions with a common ancestor in two genomes and is thus appropriate for analyzing large-scale genomic rearrangement events, but not for comparisons at the gene level.

Breakpoint-based methods

These methods [1] divide both genomes to be compared into homologous elements, e.g. genes that can be matched between them. The goal is to find the smallest combination of translocations, inversions and duplications explaining today's genomes. One example of a breakpoint-based method for the detection of syntenic blocks is the GRIMM genome rearrangements Web server [5], which is based on the popular algorithm by Hannenhalli and Pevzner [6,7]. GRIMM is able to distinguish between macro- and micro-rearrangements and relies on an initial alignment of orthologous elements. However, since GRIMM's main goal is the reconstruction of evolutionary events, it focuses on finding macro-rearrangements and the most important micro-rearrangements within them while discarding small rearrangements. While it is a well-suited method for the detection of macro-rearrangements, it fails to reconstruct a complete evolutionary history on the level of genes. Additionally, GRIMM is not able to deal with gene duplication in a genome (the so called 'word problem'). Another method, also based on the Hannenhalli-Pevzner algorithm, is Cinteny [8], which focuses on the detection of syntenic regions at different sensitivities.

Positional orthology

Positional orthology is an umbrella term for all ortholog prediction methods that rely on gene neighborhood, such as localSynteny [9] or MSOAR [10,11]. They aim to produce a one-to-one mapping of genes by taking the direct neighbors of all orthologs into account and attempt to delineate orthologs based on the evolutionary time point of their creation. This leads to a mapping with higher probabilities for the resulting orthologs to fulfill similar functions in their species. The main drawback of these methods is their definition of gene neighborhood, which is composed of the adjacent genes on both sides. Only a subset of direct neighbors is thus used to define the orthology of the genes in question, although consideration of a larger segment of genes, which lies in a conserved order in both genomes could greatly improve the prediction.

All types of approaches focus on one partial aspect of the evolutionary relationship between two species, even though they are highly interdependent, and no single approach, applied separately, is adequate for comparing quantitative or qualitative properties along eukaryotic chromosomes. Comparative genomics studies focusing on genes and their immediate environment (i.e. intergenic regions, regulatory elements, adjacent genes) usually only compare orthologs. If the immediate linear environment of the gene is important for the comparison of species, considering pairs of orthologs independently from each other is not sufficient. For example, the expression of Homeobox (Hox) genes is determined by their order on the chromosome [12,13]. Another example are lamina- associated domains (LAD) which, when located in the intergenic region between two genes, determine the subnuclear localization of the entire genomic region. Features such as LADs are known to correlate with hetero- and euchromatin [14] and hence also influence gene expression, which, again, demonstrates the importance of the linear environment in comparative genomics.

Syntenic regions, originally defined as segments of conserved gene order, would seem to be the perfect basis for comparing both genes and their neighborhoods between two species, but there are very few regions of continuous similarity between mammalian genomes. Instead, a more general term syntenic blocks is often used [1] to describe conserved regions that are interrupted by local micro-rearrangements and therefore do not lend themselves to straightforward comparison of genes and their environments. Refinement of such regions formed by macro-rearrangements and delineation of maximal length blocks of conserved gene order are required to allow for comparison of genes and regulatory elements in conserved neighborhoods.

In this work we propose a new method, SyntenyMapper, which aims to reconcile all three types of approaches discussed above by combining the search for completely conserved blocks of orthologous genes with the identification of microrearrangements within syntenic blocks. Using pre-calculated syntenic regions and orthologous genes as input, SyntenyMapper finds rearranged regions of conserved gene order within syntenic blocks. It is thus well suited for a gene-based comparison of genomes, since it not only allows for consideration of the genomic properties of orthologous genes, but also of their conserved gene neighborhood.

SyntenyMapper can be best compared with a class of orthology-based tools suitable for the identification of small blocks of conserved gene order, or collinear blocks, which usually work on a genome-wide scale. Similar to our tool, CYNTENATOR [15], MCScanX [16] and i-ADHoRe [17] identify regions of conserved gene order by applying alignment techniques to a set of ortholog anchors such as genes. These methods are in principle suitable for identifying all micro-rearrangements of gene order in a genome and are comparable to SyntenyMapper when applied to syntenic regions (see the Results section for a detailed comparison). However, in contrast to these methods, SyntenyMapper is designed to take into account the hierarchical structure of the genome and aims to find *all* exact micro-rearrangements within predefined syntenic blocks, independent of the number of elements they contain. CYNTENATOR, i-ADHoRe and MCScanX are less precise, allowing for gaps and mismatches of ortholog pairs. Our tool, however, delineates small blocks of perfectly conserved order that are ideal for comparison of closely related genomes and analyzing evolutionary rearrangement events in a region of interest. Through the use of predefined syntenic regions, the complexity of the task is greatly reduced and SyntenyMapper is thus able to find all small rearrangements within seconds.

Additionally, SyntenyMapper implements a preprocessing step to create a set of syntenic one-to-one orthologs. It thus also has a lot in common with positional orthology methods. However, SyntenyMapper relies on known orthology relationships and filters many-to-many groups according to gene order on the chromosome. The conservation of gene order in one large and conserved segment is thus the main guide for creating a one-to-one orthology mapping. SyntenyMapper expands and combines existing approaches for a more detailed assessment of evolutionary history.

SyntenyMapper is available as a stand-alone software repository via the Galaxy [18,19,20] platform's tool shed (https://toolshed.g2.bx.psu.edu, repository name "synteny_mapper") and as source code on our website (http://webclu.bio.wzw.tum.de/syntenymapper). Pre-computed results for the syntenic regions and orthologs from the ENSEMBL Compara database [2,3] can be accessed and downloaded at our website as well.

Material and Methods

SyntenyMapper

SyntenyMapper takes as input a set of syntenic regions and a set of one-to-one orthologs between two genomes of interest. Duplications of genes leads to presence of one-to-many and many-to-many co-orthologs, which are converted to one-to-one orthologs using the following pre-processing procedure (see Figure 2):

- Genes are grouped into one-to-many and many-to-many ortholog groups
- Each one-to-many group is reduced to a single one-to-one ortholog pair with the highest sequence identity
- Asymmetric many-to-many groups (those containing n genes in the genome A and m genes in the genome B, with $n \neq m$) are converted to symmetric many-to-many groups. For example, if $n = m + \delta$, exactly δ genes will be removed from the genome A based on an ascending ranking according to the average percent sequence identity of each gene to all other genes in the group.
- Many-to-many groups, all of which are now symmetric, are split into individual one-to-one orthologous pairs. For any many-to-many group consisting of n genes $\{a_1, a_2, ..., a_n\}$ in genome A and m genes $\{b_1, b_2, ..., b_m\}$ in genome B this is achieved by considering only orthology relationships between the genes with the same sequential number (i.e. a_1 with b_1, a_2 with b_2, etc.).

This pre-processing procedure aims to identify syntenic orthologs among groups of orthologous genes, i.e. those genes among a set of duplicated genes that are located at equivalent genomic positions with respect to their neighbors. For this reason, gene order is considered as the main factor for resolving many-to-many groups. In one-to-many groups sequence identity is used as a rough measure to identify the most recent ancestor gene.

As an example, Figure 2A shows two syntenic regions, the genes they harbor, and the pairwise orthologous relationships obtained from an external source, such as ENSEMBL [2,3]. Genes without orthologs (e.g. gene a_2) are excluded from consideration. Gene a_{10} from genome A has two orthologs in the genome B (one-to-many group). Genes a_4, a_8 and a_9 in the genome A as well as genes b_3 and b_5 in the genome B form asymmetric many-to-many orthologous groups. All remaining genes have only one ortholog. After conducting the pre-processing steps described above, the same syntenic regions look as depicted in Figure 2B. Gene a_2 has been removed and gene a_{10} now only has one ortholog with the highest sequence identity (gene b_9). The asymmetric many-to-many group has first been converted to a symmetric one by removing the gene a_9, and then split into ortholog pairs according to the gene order, resulting in one-to-one orthologous gene pairs b_3-$a_4$and b_5-a_8.

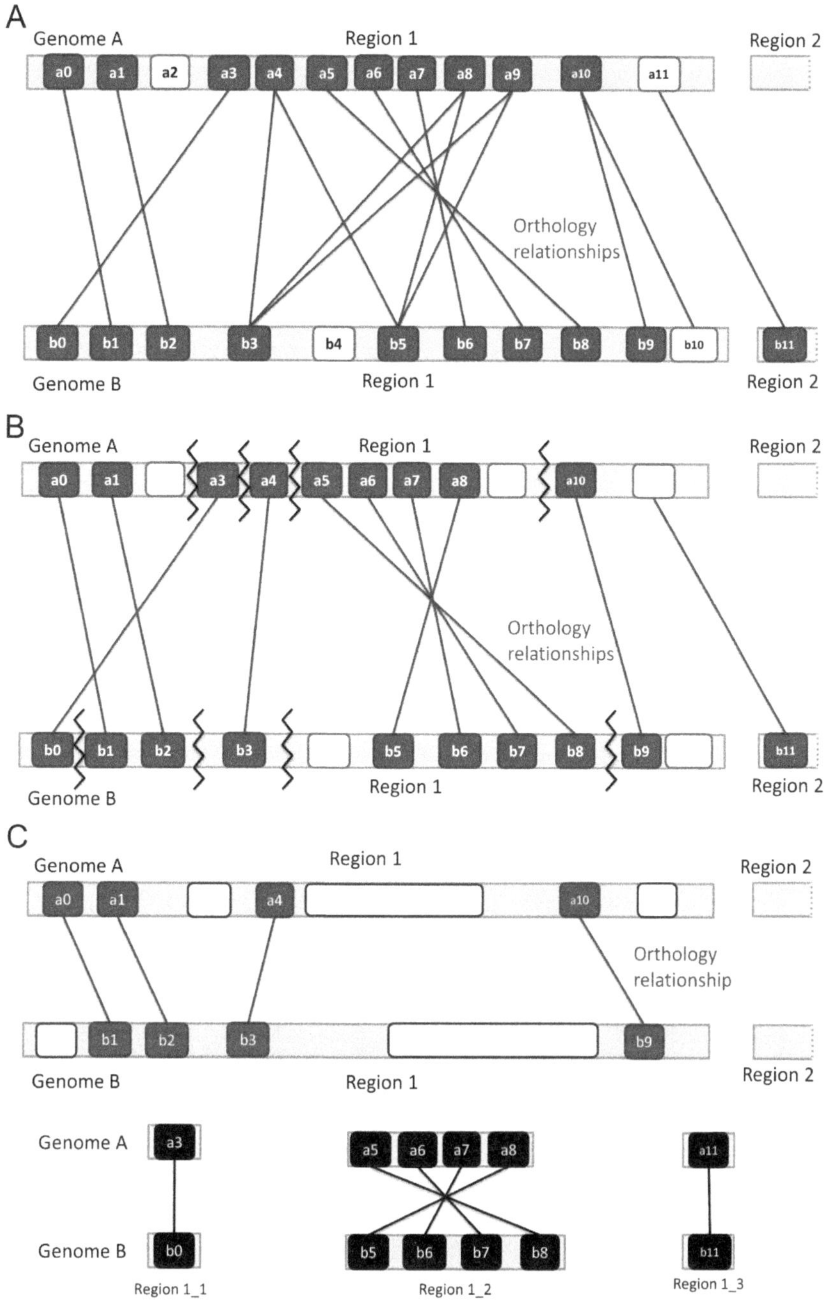

Figure 2. Detection of collinear blocks by SyntenyMapper. A: Illustration of a syntenic region between two species, with numbered boxes representing genes and connecting lines representing orthology relationships. Gene a_{11} and gene b_{11} have no orthologs in their syntenic regions, but are orthologous to each other. Genes a_2 and b_4 have no orthologs. B: During pre-processing one-to-many (genes a_{10} and b_9, b_{10}) and asymmetric many-to-many (genes a_5,a_6,a_7,a_8,a_9 and b_5,b_6,b_7,b_8), groups are first converted into symmetric groups by excluding genes with the lowest sequence identity to the rest of the group (genes a_9 and b_{10}), and subsequently paired as one-to-one orthologs based on gene order. Breakpoints (zig-zag lines) are identified as described in the Methods section. C: Using breakpoints, SyntenyMapper defines rearranged segments, shown in black, as new syntenic regions 1_1 to 1_3 within the long original region.

Given the resulting dataset of one-to-one orthologs, SyntenyMapper identifies two types of evolutionary events - translocation (genes a_3 and b_0 in Figure 2) and inversion of gene order (genes a_5, a_6, a_7, a_8 and b_5, b_6, b_7, b_8). A **translocated** segment is a chromosomal region that breaks off from its original position and reinserts into the genome at another position in one species; during an **inversion** the region inverts its direction before reinsertion. SyntenyMapper finds these events by searching for breakpoints, or locations of unconserved neighborhoods of two genes, in the reference genome *A* by using orthology relationships to genome *B* as a guide.

A **breakpoint** $(a_i/b_j,a_{i+1}/b_l)$ is defined by two orthologous gene pairs a_i, b_j and a_{i+1}, b_l if $j \pm 1 \neq l$, where genes a_i, a_{i+1} are from the reference genome *A* and genes b_j, b_l are from the genome *B*. In Figure 2B, breakpoints are marked in both genomes. When a rearranging genomic region reinserts into genome B at a new position, **two breakpoints** emerge in A, one at each end, as illustrated in Figure 3. Consequently, the genes between two subsequent breakpoints in A and their orthologs in B either form a so-called **block**, i.e. a set of genes within a translocated or inversed segment, or belong to a non-rearranged part of the original syntenic region that is enclosed between two blocks. To distinguish blocks from syntenic regions with conserved gene order that were not subject to rearrangement, SyntenyMapper looks up the two breakpoints preceding and following $(a_i/b_j,a_{i+1}/b_l)$ in the genome B. If there is no breakpoint preceding and/or following $(a_i/b_j,a_{i+1}/b_l)$, the start and/or end positions of the syntenic region are taken as reference points. The set of these three breakpoints or reference points defines two adjacent segments of genes. The length of both segments is compared and the shorter segment is defined as a new block and a new syntenic region, which lies within the longer original syntenic region.

The choice of a reference genome is arbitrary as each translocated segment in B with respect to A is also a translocated segment in A with respect to B. As seen in Figure S1, breakpoint definition with B as a reference genome leads to the definition of the same block as was found using A as a reference genome (Figure 3).

Within translocated blocks, the direction of genes is identical in both species, while inversed blocks show opposite orientation of genes. The pseudocode of the SyntenyMapper algorithm to find breakpoints and blocks is shown in Figure S2.

In the only step that involves all synteny regions, distant translocations, the so called 'external orthologs', are identified by searching for orthologs of those genes without orthologs in their own syntenic blocks. In Figures 2B, C gene a_{11} in the syntenic region 1 of the genome A and gene b_{11} in the syntenic region 2 of the genome B do not have orthologs in their own syntenic regions, but are orthologous of each other. These two genes are paired and excluded from their respective syntenic regions to define a new one.

Figure 2C shows syntenic regions refined by SyntenyMapper. Each gene now has a single syntenic orthologous partner while inversed and translocated blocks are redefined as new syntenic regions. Genes or groups of genes that were removed from their syntenic region during one of the prior steps will be termed 'excluded regions' in the following sections.

SyntenyMapper is implemented in Java and integrated in the Galaxy platform [18,19,20] for easy use and accessibility. Since each syntenic region is treated separately, SyntenyMapper computes micro-rearrangements fast and is able to analyze the human and mouse genomes in under one minute on a standard Linux workstation. Pre-computed results for ENSEMBL Compara syntenic regions data can be downloaded from our website (http://webclu.bio.wzw.tum.de/syntenymapper). The user can also analyze own data using the SyntenyMapper tool within the Galaxy environment if they match the required format described in detail in the help section. Syntenic regions from sources other than ENSEMBL should be defined in a similar manner, i.e. they should represent the longest possible regions evolved from a common sequence in the ancestor. Refined syntenic regions can be visualized using the linear or circular representations implemented in Circos [21] using the tool provided within the SyntenyMapper's Galaxy repository. Additionally, SyntenyMapper generates two output files containing a) coordinates of all syntenic regions, both original and newly defined during the mapping, marked as caused by either an internal or an external micro-rearrangement, and b) syntenic orthologs for each syntenic region.

TrackMapper

TrackMapper, also implemented in Java, allows users to directly compare so called feature tracks from two species, given a mapping from SyntenyMapper. The tool accepts as input any positional genomic features, such as binding sites or signal intensities, in the generally accepted BED format. In a BED file each line corresponds to one genomic element and contains genomic position (chromosome number, start and end coordinates) as well as an optional signal weight in columns 1–4, respectively. For each gene left after SyntenyMapper preprocessing, TrackMapper calculates the average coverage, *i.e.* the percentage of the gene's base pairs overlapping with the feature (e.g. a LINE repeat) if no signal weight is given, or the average value of the feature if a signal weight is given. These values often represent the strength of a signal, such as the height of a histone modification peak or the expression value. The vector of average coverage values for each gene is then converted into Z-scores by subtracting the mean and dividing by standard deviation. Let z_A and z_B be the vectors of feature Z-scores for syntenic one-to-one orthologs found by SyntenyMapper in genomes A and B, respectively. The measure of similarity between any two feature tracks, calculated as $z=|z_A-z_B|$, can be downloaded or further analyzed with other Galaxy Tools, which provide a wide range of plotting tools, such as distribution histograms, and statistical

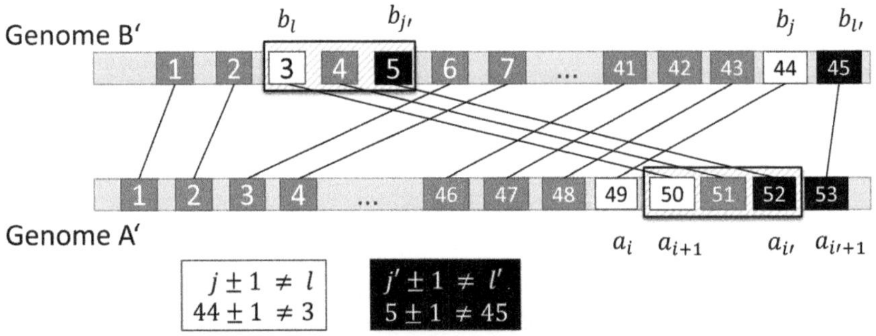

Figure 3. Breakpoint definition in SyntenyMapper. Illustration of two breakpoints emerging at both ends of a translocated segment a_3,a_4,a_5 in genome A and b_{50},b_{51},b_{52} in genome B (hatched box). By definition a breakpoint is constituted by two orthologous gene pairs a_i/b_j and a_{i+1},b_l if $j\pm1\neq l$, as shown in the boxes underneath the schema. The second breakpoint is described by $a_{i'}/b_{j'}$ and $a_{i'+1}/b_{l'}$. White and black boxes mark the four genes forming the first $(a_2/b_2,a_3/b_{50})$ and the second $(a_5/b_{52},a_6/b_3)$breakpoint, respectively. A is used as reference genome to define the block formed by a micro-rearrangement as the genes that lie between the adjacent breakpoints in A, in this case a_2, a_3 and a_5,a_6. The genes between these two breakpoints and their orthologs in B form a block.

analyses, such as correlation tests. Additionally, TrackMapper provides the vector mean as a compact similarity measure between the two tracks over all genes.

Currently, a widely used tool for mapping of tracks between species is LiftOver [22], which was not designed for the purpose of comparing two genomes and is also asymmetrical, in that it converts a feature track from one reference species to the other. By contrast, TrackMapper is able to directly compare feature tracks from two species on the gene level without defining one genome as a source and the other as a target.

Test data: syntenic regions and orthologs from human and mouse

Data on syntenic regions and orthologs in *Homo sapiens* and *Mus musculus* (assemblies hg19 and mm10, respectively) were obtained from ENSEMBL Compara (version 73) [3]. We downloaded 356 syntenic regions with a mean length of 7.63 mB and 6.89 mB in human and mouse, respectively, and the complete set of ENSEMBL protein-coding genes containing 23,618 and 22,769 unique genes in human and mouse with mean lengths of 59.8 kB and 44.3 kB, respectively. In *Homo sapiens*, an average of 55.10 genes fall into each synteny region, while in *Mus musculus* each syntenic region contains 58.40 genes on average. 27,453 pairwise orthology relationships between protein-coding genes were also downloaded from ENSEMBL.

To test the performance of TrackMapper in analyzing the conservation of genomic feature tracks we obtained ENCODE data for three histone modifications (H3k4me1, H3k4me3, H3k9ac) for human (Broad Institute, embryonic stem cells) and mouse (Ren Lab, ES-E14) [23]. Coordinates were converted to the most recent human (hg19) and mouse (mm10) genome assemblies using LiftOver [22]. For randomization we created n elements of the same length as the original peaks at random positions on each chromosome, with n being the number of histone modification peaks on this chromosome.

Calculation of sequence similarity between syntenic regions

We determined sequence similarity between ENSEMBL syntenic regions using our own implementation of the linear runtime trie-based sequence comparison algorithm proposed by Rieck and Laskov [24]. A trie is a tree structure similar to a suffix tree, with the distinction that for a finite set of strings, every string is represented in the path from the root to one of the leaves. Long sequences such as syntenic regions are converted into finite sets of strings by splitting them into overlapping k-mers (in our case k was set to 6). They are inserted into a trie by iterating over each 6-mers' characters and, starting from the root of the trie, following the edge with the corresponding character to the next node until no such edge exists and one or more new edges need to be inserted.

The algorithm then makes use of the trie's maximum depth of k by performing a parallel depth-first search on the tries of the two sequences. A distance score is calculated for each node using the inner function m, which compares the number of occurrences of the string represented by that node in both sequences. The score is accumulated over the whole trie using the outer function $\oplus$. In our implementation, we used the sum $\sum$ as outer function $\oplus$ and the Manhattan distance as inner function m (Equation 1).

$$m = |\phi(x) - \phi(y)| \quad (1)$$

where x,y are trie nodes and $\phi(x)$ returns the number of occurrences of the string readable from the root of the trie to x in the sequence.

Comparison with CYNTENATOR, i-ADHoRe and MCScanX

We compared SyntenyMapper with the latest versions of CYNTENATOR [15] (November 2012), i-ADHoRe 3.0 [17] and MCScanX [16] (May 2012). For each ENSEMBL syntenic region we used as input the actual gene order order and sequence identity values between the corresponding ENSEMBL orthologs. CYNTENATOR was run on each syntenic region with default parameters and homology type BLAST to consider identities, and a simple guide tree containing only human and mouse. We ran i-ADHoRe with the suggested gap size and cluster gap size of 15, probability cutoff 0.001, q-value 0.9, and anchor points 3. For MCScanX, the program MCScanX_h was used with default parameters.

Results

SyntenyMapper: a novel tool for refining syntenic orthologs

In this work we propose a new comparative genomics tool for the identification of micro-rearrangements within syntenic regions that were shaped by macro-rearrangements between species. SyntenyMapper complements the set of one-to-one orthologs from ENSEMBL by finding syntenic one-to-one orthologs among one-to-many/many-to-many orthologous groups and uses them to identify deletions, inversions, local and distant translocations within syntenic regions, further refining the definition of these regions. As a result, SyntenyMapper provides the user with a set of evolutionary building blocks with completely conserved gene order between two species.

The tool is easily accessible through the Galaxy Platform and can be used to analyze both user-supplied syntenic regions as well as those obtained from ENSEMBL Compara. The resulting refined syntenic regions and their associated annotation tracks can be interactively visualized using the circular Circos representation [21] or a linear plot created with an R script (for an example see Figure S3), and analyzed with our own track-comparison tool called TrackMapper.

Detection of micro-rearrangements between human and mouse genomes

To illustrate the performance of the SyntenyMapper approach, we applied it to syntenic regions and orthologs between the human and mouse genome obtained from ENSEMBL Compara. Syntenic region length is highly correlated between these species (Figure S4).

As seen in Table 1, the majority of protein-coding ENSEMBL genes (70.07%) have exactly one ortholog within their syntenic region. The second largest subset (12.30%) is constituted by those genes that do not have any orthologs and are hence not treated by SyntenyMapper. Genes in many-to-many groups within one syntenic region comprise only 2.30% of the total gene number; most of these groups are asymmetric. Finally, genes only having orthologs in another syntenic region (external orthologs) make up a mere 1.27% of the data. All other types of orthologous relationships, such as genes with orthologs in regions not covered by syntenic regions, represent less than 2% of all cases and are ignored by the current version of SyntenyMapper.

During the rearrangement detection process, SyntenyMapper implements a pre-processing step to find syntenic ortholog pairs among the set of orthologous genes. While converting one-to-many and many-to-many ortholog groups into one-to-one

Table 1. Frequency of different cases of orthologous relationships for a given gene in a syntenic region between human and mouse.

Type of orthologous relationship	#Genes
No ortholog	5,705 (12.30%)
One internal ortholog	32,505 (70.07%)
Many internal orthologs	1,062 (2.29%)
One external ortholog	487 (1.05%)
Many external orthologs	105 (0.23%)
Many in- and external orthologs	204 (0.44%)
One ortholog in a syntenic-block-free region	99 (0.21%)
Many orthologs in a syntenic-block-free region	15 (0.03%)
Many orthologs: internal, external, and in syntenic-block-free regions	166 (0.36%)

Italic font indicates classes of genes that are covered by SyntenyMapper. Internal orthologs: Orthologous genes that lie in the same syntenic region. External orthologs: Orthologous genes that lie in different syntenic regions. Syntenic-block-free region: Genomic region that is not covered by ENSEMBL syntenic regions.

orthologs, a total of 941 genes (2.1% of all human and mouse genes) are removed as non-syntenic. As all relationships of these genes are non-syntenic as well, a total of 10,840 relationships from the original set of 27,453 ENSEMBL Compara orthologs are eliminated with the non-syntenic genes that formed them.

SyntenyMapper finds 16,613 syntenic ortholog pairs between human and mouse and detects 2,898 blocks that were subject to micro-rearrangements within or between the original 356 syntenic regions. There are three different types of syntenic regions obtained after this analysis: i) original regions, which are identical to the input regions except for excluded non-syntenic genes (356, 10.94%), ii) internally translocated regions that represent blocks created by micro-rearrangements within the syntenic region (2,817, 86.57%), and iii) externally translocated regions that represent one or more neighboring genes that were translocated between syntenic regions (81, 2.49%).

Most frequent are internal micro-rearrangements, but blocks caused by these contain only 1.52 genes on average, while the majority of genes (73.68%) still lie within original syntenic regions. We can thus conclude that often blocks consisting of just one or two genes disturb the conservation of gene order within a syntenic region. Regions translocated over large distances are rare and, in the specific case of human/mouse comparison presented here, almost always contain only a single gene. Genes within one syntenic region lie within a linear distance of around 7 to 8mB from each other. Hence, the proximity of genes within a syntenic region leads to more frequent rearrangements compared to large linear distances. Linear proximity correlates with spatial proximity, which has been shown to be one of the triggers for rearrangements [25].

One major advantage of our approach is its ability to refine regions formed by macro-rearrangements, which are large-scale intra- or inter-chromosomal translocations, through the consideration of all micro-rearrangements, whereby short segments change their positions within syntenic regions [26]. SyntenyMapper uses pre-calculated syntenic regions, for example from ENSEMBL, which are created in such a way that their length is maximal, disregarding small rearrangements within them (see introduction). Identification of micro-rearrangements can greatly improve our understanding of evolutionary events shaping extant genomes and is instrumental for comparing the properties of regions of conserved gene order between two species.

Figure 4 shows a syntenic region with the same orientation in human and mouse, with most of the genes having syntenic one-to-one orthology relationships. However, the order of genes is disrupted by a translocation of a large block containing seven genes located in the beginning of the human region and in the end of the mouse region. The order and direction of genes within this block is preserved. Interestingly, there is a sizeable gap between syntenic orthologs and translocated genes on the human chromosome, which may imply that the translocation happened in *Homo sapiens* after human-mouse divergence.

SyntenyMapper is also able to identify a small-scale rearrangement with the majority of genes within the syntenic region in mouse lying on the reverse strand, even though the region is annotated as situated on the forward strand (Figure S5). In human, all genes lie on the forward strand, implying that an inversion occurred in mouse. Our approach recognizes this inversion of multiple genes and redefines it as a new syntenic region.

These examples illustrate the difficulties in comparing genome regions that evolved from the same sequence due to recent genome rearrangements that make any linear comparison impossible. Our method enables the user to refine syntenic regions and to discover micro-rearrangements at the level of individual genes. The analysis is not restricted to translocations within one region and also considers those genes that lie within other syntenic regions.

As an example, Figure S6 shows the relationship between two pairs of syntenic regions that share one orthologous gene pair. These rare external translocations (592, 1.27% of all genes) of genes out of their syntenic region almost always involve single genes in the human/mouse comparison. We hypothesize that translocations of single genes over large linear distances are favored by short spatial distances between the chromosomal regions involved, as has been shown for cancer cells [25]. All rearrangements illustrated in this section were not found by ENSEMBL Compara and cannot be directly inferred from orthology relationships, stressing the value of SyntenyMapper for comparative genomics.

To demonstrate the usefulness of TrackMapper for comparing feature tracks between genomes analyzed histone modifications, which are known to be largely conserved between mouse and human [27]. Using histone modification data from ENCODE we applied TrackMapper to calculate the measure that quantifies the difference in the extent to which genes overlap with histone

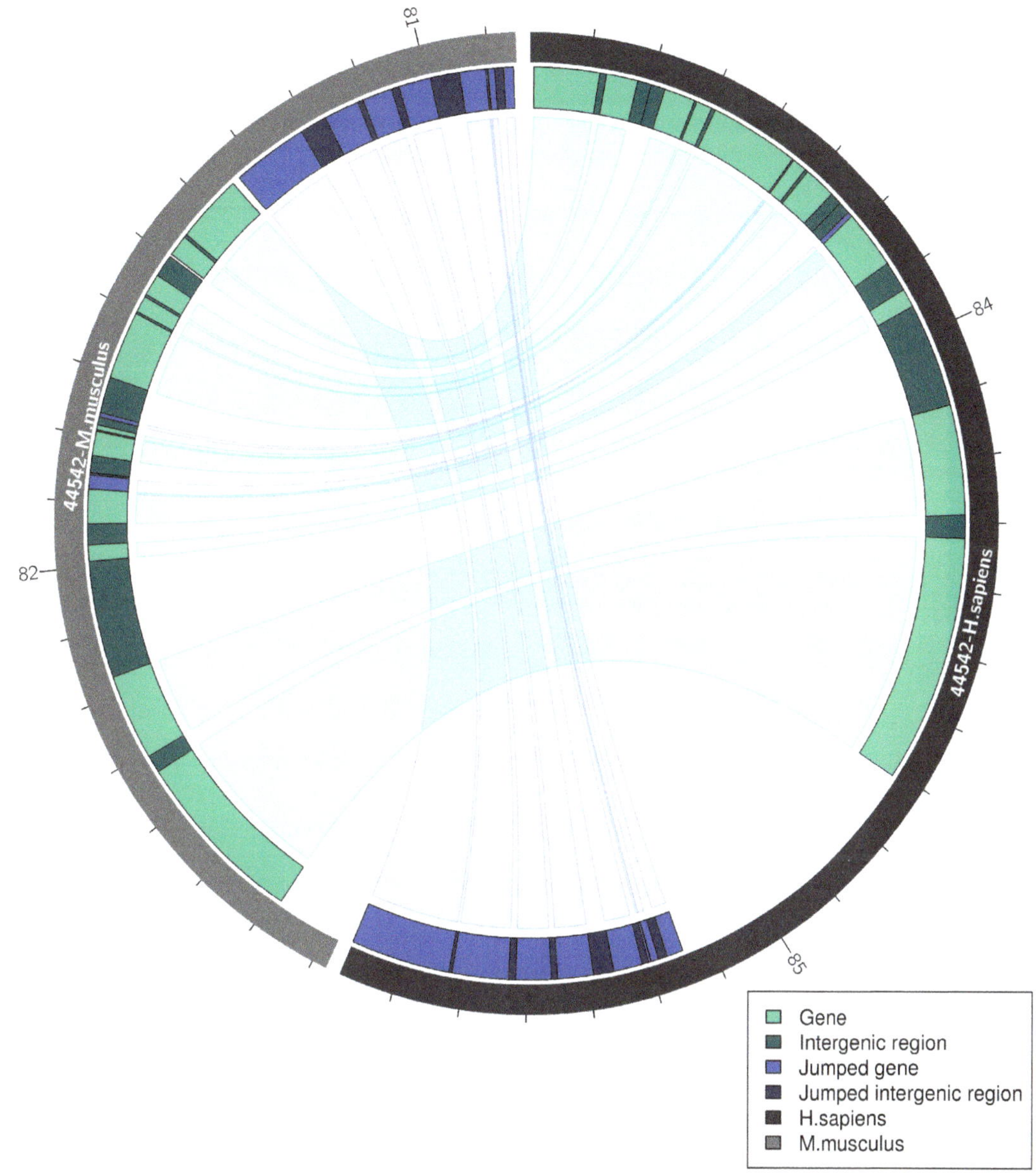

Figure 4. Visualization of SyntenyMapper results for a syntenic region in human and mouse. The region (ENSEMBL identifier 44542) is illustrated in human as a dark grey ideogram (right) and in mouse as a light grey ideogram (left). Ticks are placed at 100 kB distance and the numbers represent positions in mB on the human and mouse chromosomes 15 and 7, respectively. The Circos circular plot illustrates the positions of genes/ intergenic regions for one syntenic region in both species and the correspondence between them. Micro-rearrangements are illustrated by color-coding, with syntenic orthologs and out-of-order genes shown in grey and black, respectively, while the intergenic regions between syntenic orthologs and between out-of-order genes are shown in white. A large block of seven genes (black) was translocated in either human or mouse. In the Galaxy version of the plots, gene annotations are given as labels and as direct links to ENSEMBL through clicks onto the gene track.

modification peaks in human and mouse. For all three histone modifications considered here (see *Methods*) the difference measures are low, with a respective mean of 0.77, 0.63 and 0.70 for H3k4me1, H3k4me3 and H3k9ac. We ran TrackMapper on mouse data and randomized human histone modification peaks, to receive significantly higher difference value distributions (Wilcoxon rank sum test p-value $<2.2e-16$ in all three cases) with means of 0.98, 1.00 and 1.02, respectively. Of special interest are genes with very low difference scores, which can be easily identified using TrackMapper's tabular output (File S1). There are 171 ortholog pairs with a very low H3k4me3 difference score (below 0.01). TrackMapper is thus a convenient tool for in-depth analysis of feature conservation at the gene-level.

Pre-computed ENSEMBL Compara species comparison

ENSEMBL Compara provides syntenic regions and orthologs for 25 pairs of eukaryotic species, with 16 of them involving human. We conducted a large-scale analysis of micro-rearrangements within these syntenic regions using SyntenyMapper and calculated the number of micro-rearrangements within and between syntenic regions as well as the average number of genes involved in these events (Table S1).

Among the most closely related species pairs are human *vs* chimp (*P. troglodytes*) and mouse *vs* rat (*R. norvegicus*). Both of them share a high number of orthologs, however, the number of ENSEMBL syntenic regions differs vastly: while human and chimp share only 139 syntenic regions, there are 554 ones between mouse and rat. Because the average syntenic region length between human and chimp is about four times the size of syntenic regions in mouse and rat (22mB and 4.8mB, respectively), the percentage of the genomes covered is similar (human *vs* chimp: 87.76%, mouse *vs* rat: 94.64%).

Using branch lengths of the UCSC species tree [28] (available for all species pairs considered in this study except for those involving gorilla, pig, orangutan, common marmoset and turkey) we analyzed the correlation between the evolutionary distance and micro- and macro-rearrangement related genome features. In general, one would expect more closely related species to share a low number of very long syntenic regions, and an increase in the number and a decrease in the average length of syntenic regions with the growing evolutionary distance. Indeed, for species pairs separated by short or medium evolutionary distance, this expectation is true and the average syntenic region length and their number exhibit a negative exponential correlation (Figure 5A). However, more distantly related species, such as human and platypus, contain fewer ENSEMBL syntenic regions than would be expected based on this exponential correlation pattern. Strong deviation from this correlation pattern observed for the species pairs involving platypus as well as in the lizard vs chicken comparison is caused by the very low sequence similarity between these species, which leads to a low coverage of the genomes by syntenic regions (e.g. human *vs* platypus: only 21% of the genome is covered, see Table S2). Two other distant species pairs - human *vs* chicken and mouse *vs* chicken – are more in line with the general trend, but due to significant differences in genome size (human genome 3.1 gB, mouse genome 2.7 gB, chicken genome 1.1 gB), they also show a low genome synteny coverage of the respective longer genomes.

Because of SyntenyMappers's gene-based approach, the number of detected internal and external micro-rearrangements is dependent of the number of orthologs, which is higher for closely related species. As seen in Figure 5B, the average number of internal micro-rearrangements (per syntenic region), ranging between 0 and 30, increases with the increasing average syntenic region length and decreasing evolutionary distance. This implies that the number of micro-rearrangements depends on the size of the syntenic regions that harbor them, which is in turn correlated with evolutionary distance, as illustrated in Figure 5A.

While the total number of micro-rearrangements indirectly depends on the degree of relationship between the genomes, there is no correlation between the number of micro-rearrangements (internal or external) per mB and evolutionary distance (Figure S7). We also found only a slight positive correlation between sequence distance and the number of internal micro-rearrangements (Pearson correlation coefficient 0.22). In particular, low sequence similarity does not necessarily lead to a high number of micro-rearrangements (Figure S8).

The higher numbers of micro-rearranged regions per syntenic region in more closely related species are thus mainly due to the greater length of syntenic regions. By contrast, the size of micro-rearranged regions (i.e. the number of genes they contain) does not show any dependence on the syntenic region length and the evolutionary distance (Figure 5C), because the mechanisms of transposition, which is the main cause of small-scale translocations, are the same in all species.

External translocations between syntenic regions are less common, with the average number per syntenic region ranging between 0 and 4 for all species pairs. Overall there is a slight tendency for longer syntenic regions in more related species to contain a higher number of such external micro-rearrangements (Figure 5D). In almost all cases, only a single gene is subject to an external micro-rearrangement, regardless of the size of the synteny regions or the evolutionary distance (Figure 5E). However, with the exception of mouse and platypus, which harbor on average only 1.0 external micro-rearranged genes per syntenic region, all distantly related species pairs (mouse vs chicken, human vs platypus, human vs chicken, chicken vs lizard) have somewhat longer external translocations (average>1.05). Larger external micro-rearrangements (1.3 genes in external micro-rearrangements per syntenic region) are especially common between chicken (*G. gallus*) and wild turkey (*M. gallopavo*). In particular, SyntenyMapper was able to detect a group of six consecutive genes that was translocated between two syntenic regions in these organisms (see Figure S9). As the only closely related species pair with sharing relatively large external translocations, chicken and turkey genomes behave very differently in this aspect. Cases like this can be easily identified with SyntenyMapper, allowing for detailed genome comparison and further analysis based on the produced mapping.

Comparison with existing methods

SyntenyMapper is the first tool that was developed to detect micro-rearrangements in the form of blocks of conserved gene order within macro-rearranged regions. However, available software for detection of collinear blocks (i.e. sets of genes with conserved order in two or more species) can be applied to ENSEMBL syntenic regions as well, although no rearrangements between syntenic regions can be detected using this approach.

We detected collinear blocks between human and mouse with SyntenyMapper (discarding single-gene regions), CYNTENATOR, i-ADHoRe and MCScanX.

CYNTENATOR applies the Smith-Waterman algorithm to genomes represented by strings of genes, with homologies provided in the form of BLASTP output, and identifies local alignments, which represent blocks of conserved gene order in both species. The main advantage of this method is its handling of multiple genomes, when provided with a guide phylogenetic tree that can be obtained from other sources.

Since there is no gold standard annotation of syntenic regions, we have to rely on a qualitative comparison of the results. We have identified several reasons for discrepancies between CYNTENATOR and SyntenyMapper:

CYNTENATOR is not designed to detect external micro-rearrangements when applied to syntenic regions. However, this makes up for only a small percentage of the discrepancies: in only 8 of 356 syntenic regions (2.25%) CYNTENATOR misses just the external rearrangements. A larger problem is that CYNTENATOR, in contrast to our method, is unable to detect inversed collinear blocks of any size. Such inversions are common and occur in 170 out of 356 (47.75%) of all syntenic regions, including cases where the complete region is inversed in the second species (see Figure S10 for an example). In contrast, we have identified only one syntenic region where SyntenyMapper misses a collinear block detected by CYNTENATOR because two genes in this block overlap.

Other minor differences affect only the length of collinear blocks and are mainly caused by different definitions. Due to the usage of the Smith-Waterman alignment algorithm for aligning gene strings, CYNTENATOR sometimes mismatches non-orthologous

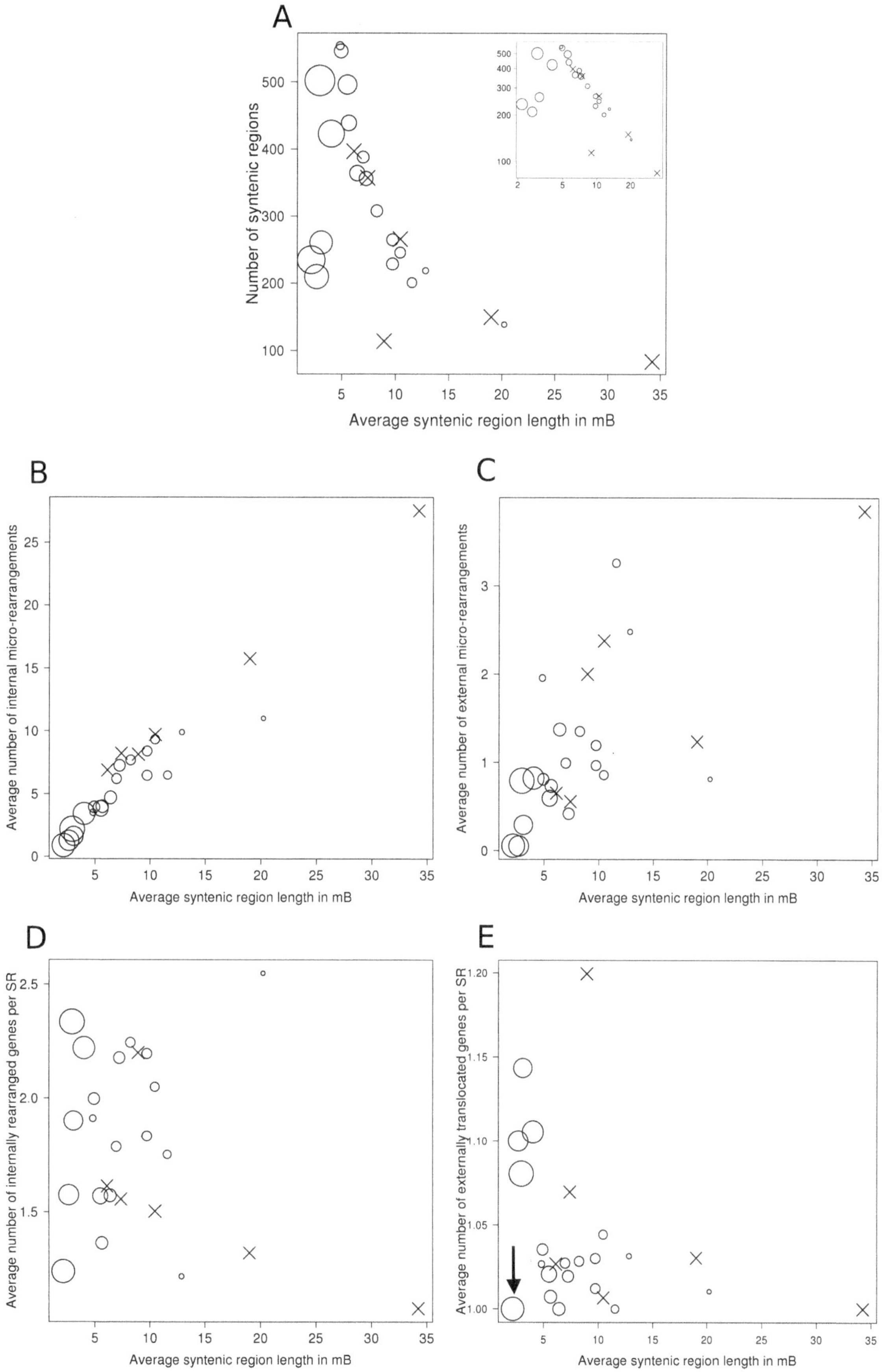
A
Number of syntenic regions
Average syntenic region length in mB
B
Average number of internal micro-rearrangements
Average syntenic region length in mB
C
Average number of external micro-rearrangements
Average syntenic region length in mB
D
Average number of internally rearranged genes per SR
Average syntenic region length in mB
E
Average number of externally translocated genes per SR
Average syntenic region length in mB

Figure 5. Relationship of syntenic region length, evolutionary distance and other features. Dependence of synteny features on the average syntenic region length (x axis) and evolutionary distance (circle size, inferred from branch lengths in Miller et al (2007), calculated as the average number of substitutions per site). Crosses correspond to the species pairs with no distance information available. A) Negative exponential correlation between the number of syntenic regions and their average length (inset: logarithmic axis scales). Closely related species (small circles) tend to have fewer, longer syntenic regions. Distant species (large circles) tend to have high numbers of very short syntenic regions. B) The average number of internal micro-rearrangements per syntenic region strongly correlates with syntenic region length and evolutionary distance. Evolutionary distant pairs of species share short syntenic regions with few internal micro-rearrangements. C) No clear correlation between syntenic region length and rearrangement number can be observed for external micro-rearrangements. D) The size of internal micro-rearrangements (average number of genes involved) does not correlate with the syntenic region length and evolutionary distance. E) Similarly, the number of genes involved in external translocations is generally independent of the syntenic region length. However, with the exception of mouse and platypus (average number of genes in externally translocated regions: 1.0, marked by arrow), distant species pairs tend to have somewhat longer externally translocated regions (average>1.05).

genes, and includes gaps when genes without orthologs occur. SyntenyMapper, while also allowing for the presence of genes without orthologs, will fragment a collinear block that is disrupted by a pair of non-orthologs into two blocks. Additionally, overlapping genes are excluded by our method. When we compare results from CYNTENATOR and SyntenyMapper, we thus often find apparently shorter but equivalent collinear blocks in the results from our method.

Another minor difference is caused by different definitions of collinear blocks. Our definition of a refined syntenic region provides for rearrangements that are embedded within a backbone of conserved gene order that ideally spans the whole syntenic region. In such a case, SyntenyMapper defines two collinear blocks, a small one embedded within a larger one, while CYNTENATOR will define three consecutive blocks.

Furthermore, CYNTENATOR fragments a large collinear block into two blocks if it is disturbed by external single-gene rearrangements, sometimes for no apparent reason (see Figure S11 for an example). SyntenyMapper is thus better suited for identifying all collinear blocks in a syntenic region, including inversions.

Another widely used tool for genome comparison is i-ADHoRe, which detects collinearity on the basis of a gene homology matrix. Blocks of conserved gene order are identified as dense diagonals and the corresponding genes are then aligned using the Needleman-Wunsch algorithm, a greedy graph strategy or a new algorithm named GG2. In this process, gaps are introduced if necessary to produce blocks with as many homologous pairs as possible. The software is well suitable for multi-genome comparisons.

Like CYNTENATOR, i-ADHoRe can be applied to syntenic regions and produces results that are comparable with SyntenyMapper. The main difference between the methods is the insertion of gaps and mismatches in i-ADHoRe alignments to create longer blocks, while our method creates a separate block if it is disrupted by a non-orthologous gene pair. The design goal of i-ADHoRe to detect collinear blocks that contain as many orthologs as possible comes at the expense of frequently missing small rearrangements that are partially embedded in larger ones.

For example, in a syntenic region comprising 11 genes, three of which are translocated within the region, SyntenyMapper correctly identifies the translocation (see Figure S12) while i-ADHoRe ignores it and defines it as a gap. Overall, both methods detect the same amount of collinear blocks in most syntenic regions (222, 62.36%), although mismatches in i-ADHoRe can result in longer detected regions. In the majority of the remaining cases (68, 19.10%) i-ADHoRe misses one or more regions due to mismatches (as described above) or inability to detect external micro-rearrangements.

We have investigated cases where SyntenyMapper detects fewer collinear blocks than i-ADHoRe (43, 12.08%). Besides the implementation differences, where our tool defines a backbone of genes with embedded micro-rearrangements while other tools define multiple adjacent blocks, we have also found that i-ADHoRe tends to disrupt collinear blocks if they are interrupted by a longer stretch of genes without orthologs. By contrast, SyntenyMapper ignores genes without orthologs when defining blocks of conserved gene order. From these results we conclude that, while both methods provide a similar functionality and largely consistent results, SyntenyMapper is better suited for a very detailed analysis of genome rearrangements that requires exact definitions of conserved gene order blocks, without mismatches and gaps. i-ADHoRe and CYNTENATOR, on the other hand, are applicable to more than two genomes and can provide a good overview of less stringent collinear blocks.

MCScanX is another software package that includes not only methods for the identification of collinear blocks. Like CYNTENATOR it requires BLASTP output or a list of orthologous genes as input, and uses a dynamic programming approach to find chains of collinear gene pairs in the two genomes. Its main advantage is the large set of downstream analysis tools, which includes four different visualization modes for the results, classification of duplicated genes into specific classes, and detection of whole genome duplication events.

When investigating reasons for discrepancies between SyntenyMapper and MCScanX, it became apparent that the latter method fails to handle duplicates correctly. For a one-to-many ortholog group it appears as if MCScanX detects multiple overlaying collinear blocks, where a single gene is contained in more than one such block and paired with different orthologs. SyntenyMapper uses a pre-processing procedure based on gene order and sequence similarity to identify one-to-one orthologs among sets of duplicated genes. As a consequence, each gene is associated with a single gene in the other genome, allowing for overlap-free definition of collinear blocks. MCScanX appears to keep one-to-many or many-to-many groups and defines blocks with genes in conserved order regardless of whether or not these genes are already present in other blocks. This behavior leads to significant overlapping between collinear blocks.

Per definition, a single gene cannot be part of two or even more collinear blocks. From these results we conclude that MCSCanX cannot provide the same functionality as SyntenyMapper, CYNTENATOR or i-ADHoRe.

Discussion

SyntenyMapper is a new rapid tool for comparative genomics that enables analysis of micro-rearrangements, detection of positional orthology, and direct comparison of genomic features between two species. Previously developed methods focus on finding regions caused by either macro- or micro-rearrangements on a global scale and use either whole-genome alignments or homologous elements alone to define these. By contrast, we identify micro-rearrangements within large synteny regions by focusing on the conservation of gene order. This combination of alignments and homologous elements allows us to find large blocks with conserved gene order that can be seen as the smallest evolutionary building blocks within largely conserved regions. The analysis of these numerous internal rearrangements can provide additional insights into evolutionary history and serves as a better basis for comparative genomics analyses than the original syntenic regions alone.

We have applied SyntenyMapper to 25 species pairs using syntenic regions and ortholog sets from ENSEMBL Compara. As expected, closely related species contain few but long syntenic regions, compared to a high number of short regions in distant pairs. The number of internal micro-rearrangements is proportional to the size of the synteny region they reside in, which in turn depends on the degree of relationship between the genomes. However, the density of micro-rearrangements, i.e. the number of micro-rearrangements per mB, does not correlate with the evolutionary distance. We found that similar sequences tend to have fewer micro-rearrangements, but the correlation is weak.

More distant species pairs tend to have short to medium syntenic regions harboring between 0 and 10 internal micro-rearrangements while more closely related organisms with long syntenic regions (over 10 mB length) contain between 10 and 25 micro-rearrangements. External syntenic regions are rare, with only 0 to 4 such translocations per syntenic region in all genome pairs, and show only a slight trend for a higher number in longer regions. We speculate that external translocations occur only under very specific circumstances, when distant genome regions are spatially close.

There is no correlation between the number of internally translocated genes in a micro-rearranged block and syntenic region length or evolutionary distance, implying that the transposition mechanism acts regardless of evolutionary history. While macro-rearrangements arise by complex processes involving chromosome double strand breaks [25], small-scale rearrangements are most likely caused by cut-and-paste DNA transposition [29]. Our results show that there are constraints that limit the total length of the translocated genome region. DNA transposons are often used as vector elements in biotechnology, and it has been reported that the efficiency of transposition decreases with increasing cargo size [30,31,32]. We show that the average cargo of transposons in higher eukaryotic genomes comprises between 1 and 2.5 genes for translocations over short linear distances, and between 1 and 1.2 genes for more distant translocations. This implies that the length constraints of transposable elements are also dependent on the linear distance between the source and the target position in the genome. This insight could be valuable for biotechnology, where ways to overcome the length limitations in transposons are searched for [32,33]. External translocations are also not correlated with the evolutionary distance or syntenic region length, but longer externally translocated regions appear to occur only in distant species pairs.

In general, we conclude that syntenic regions between the genomes of closely related species are longer and contain more micro-rearrangements of genes due to their length. More distant species pairs contain a higher number of short and less similar syntenic regions with fewer micro-rearrangements. The density and size of these is largely independent of the evolutionary distance and only slightly influenced by sequence similarity.

SyntenyMapper creates a one-to-one mapping of genes between species similar to positional orthology methods. Based on this mapping, TrackMapper is able to directly compare quantitative and qualitative features of genes, such as values representing the relative time point of replication or the overlap with long terminal repeats, and calculate a similarity measure for each pair of orthologs and the complete set of genes.

Beyond comparative genomics, identification of micro-rearrangements may be instrumental in understanding disease mechanisms. A number of diseases caused by micro-rearrangements are known, including the DiGeorge syndrome or Miller-Dieker lissencephaly [34].

Supporting Information

Figure S1 Effect of reversing the reference genome used in the example shown in Figure 5. The reference genome used here for the definition of breakpoints is A′ (= B in Figure 5). The detected breakpoints are (a_{49}/b_{44},a_{50}/b_3) (white boxes) and a_{52}/b_5,a_{53}/b_{45} (black boxes). Based on the adjacent breakpoints in the new reference genome A′, the same translocated segment (hatched box) is detected as in Figure 2.

Figure S2 Pseudocode of the SyntenyMapper algorithm to find rearranged segments. SyntenyMapper uses externally defined syntenic regions and one-to-one orthologs as input to find maximum length blocks of conserved gene order (termed blocks). It iterates over all genes in genome A and identifies breakpoints according to the definition given in the *Methods* section, using the helper function $j(i)$, which returns the index of gene i's ortholog in genome B. SyntenyMapper identifies the two breakpoints preceding and following the detected breakpoint ($a_{i-1}/b_{j(i-1)}$,$a_i/b_{j(i)}$), because each pair of adjacent breakpoints encloses either a rearranged genomic segment or the region between two rearranged segments, as described in *Methods*. To distinguish these two cases, lengths of both segments defined by the three adjacent breakpoints are compared and the shorter of the two is defined as a block resulting from a micro-rearrangement with respect to the longer original syntenic region. The type of the rearrangement (translocation or inversion) is detected based on gene order within this block.

Figure S3 Linear representation of a syntenic region (ENSEMBL identifier 44542) produced by SyntenyMapper. Upper half shows genes in mouse, lower half genes in human. Green genes are those with conserved order, translocated or overlapping genes are colored blue. To better see these overlapping genes, conserved order and translocated genes are depicted in independent lines and a combined line, marked in the legend on the right and left edge of the plot.

Figure S4 Comparison of syntenic region length in human and mouse. The regression line is shown as dotted line (Pearson correlation coefficient 0.9932).

Figure S5 A syntenic region (ENSEMBL identifier 44514) in human (dark grey) and mouse (light grey) containing a large inversed segment, illustrated in blue and only one ortholog with the same orientation as the syntenic region (green). Ticks are placed at 100 kB distance and the numbers show the position on chromosomes 13 (human) and 14 (mouse) in mB.

Figure S6 Translocation of a single gene from the human region 44801 to the mouse region 44598, shown with red line. Ticks are placed at 100 kB distance and the numbers show the positions in mB on chromosomes X in human and mouse (region 44801) as well as on chromosomes 19 in human and 7 in mouse (region 44598).

Figure S7 Average number of micro-rearrangements (internal and external) per megabase covered by syntenic regions versus evolutionary distance. No correlation can be observed.

Figure S8 Syntenic regions (SR) with low sequence similarity (high distance score) show a trend to contain more internal micro-rearrangements per megabase (Pearson correlation coefficient 0.22). Outliers with extremely high sequence distance are not shown.

Figure S9 A translocation involving six consecutive genes between two syntenic regions in chicken (*G. gallus*, dark grey) and wild turkey (*M. gallopavo*, light grey), marked by red lines. Ticks are placed at 100 kB distance and the numbers show the positions in mB on chromosomes 6 in chicken and 8 in turkey (region 47198) as well as on chromosomes 17 in chicken and 19 in turkey (region 47135).

Figure S10 An example of an inversed syntenic region between human and mouse where CYNTENATOR detects no collinear block despite eleven genes lying in the same order in both genomes.

Figure S11 An example of a syntenic region between human and mouse where CYNTENATOR fragments the existing collinear block into two blocks (location of split marked red in both species) despite lack of genes disrupting the gene order.

Figure S12 Example of a syntenic region (ENSEMBL identifier 44459) including a micro-rearrangement of three genes that is correctly identified by SyntenyMapper but ignored as a gap by i-ADHoRe.

Table S1 Statistics of pre-computed synteny mapping for ENSEMBL Compara (version 73).

Table S2 Evolutionary distance and genome coverage by syntenic regions for all species pairs. SR: Syntenic region.

File S1 List of human and mouse orthologs with significantly high similarity in H3k4me3 peak density (TrackMapper difference score below 0.01).

Acknowledgments

We thank Boris Joffe for useful discussions.

Author Contributions

Conceived and designed the experiments: SK DF. Performed the experiments: SK. Analyzed the data: SK. Wrote the paper: SK DF.

References

1. Pevzner P, Tesler G (2003) Genome rearrangements in mammalian evolution: lessons from human and mouse genomes. Genome Res 13: 37–45.
2. Vilella AJ, Severin J, Ureta-Vidal A, Heng L, Durbin R, et al. (2009) EnsemblCompara GeneTrees: Complete, duplication-aware phylogenetic trees in vertebrates. Genome Res 19: 327–335.
3. Flicek P, Ahmed IM, Amode MR, Barrell D, Beal K, et al. (2013) Ensembl 2013. Nucleic Acids Res 41: D48–55.
4. Liao BY, Chang YJ, Ho JM, Hwang MJ (2004) The UniMarker (UM) method for synteny mapping of large genomes. Bioinformatics 20: 3156–3165.
5. Tesler G (2002) GRIMM: genome rearrangements web server. Bioinformatics 18: 492–493.
6. Hannenhalli S, Pevzner PA (1995) Transforming men into mice (polynomical algorithm for genomic distance problem). In: 36th Annual Symposium on Foundations of Computer Science (Milwaukee, WI, 1995). Los Alamitos: IEEE Computer Society Press. pp.581–92.
7. Hannenhalli S, Pevzner PA (1999) Transforming cabbage into turnip: polynomial algorithm for sorting signed permutations by reversals. J ACM 46: 1–27.
8. Sinha AU, Meller J (2007) Cinteny: flexible analysis and visualization of synteny and genome rearrangements in multiple organisms. BMC Bioinformatics 8: 82.
9. Jun J, Mandoiu II, Nelson CE (2009) Identification of mammalian orthologs using local synteny. BMC Genomics 10: 630.
10. Fu Z, Chen X, Vacic V, Nan P, Zhong Y, et al. (2007) MSOAR: a high-throughput ortholog assignment system based on genome rearrangement. J Comput Biol 14: 1160–1175.
11. Shi G, Zhang L, Jiang T (2010) MSOAR 2.0: Incorporating tandem duplications into ortholog assignment based on genome rearrangement. BMC Bioinformatics 11: doi:10.1186/1471-2105-11-10.
12. Pearson JC, Lemons D, McGinnis W (2005) Modulating Hox gene functions during animal body patterning. Nat Rev Genet 6: 893–904.
13. Wellik DM (2007) Hox patterning of the vertebrate axial skeleton. Dev Dyn 236: 2454–2463.
14. Guelen L, Pagie L, Brasset E, Meuleman W, Faza MB, et al. (2008) Domain organization of human chromosomes revealed by mapping of nuclear lamina interactions. Nature 453: 948–951.
15. Rödelsperger C, Dieterich C (2010) CYNTENATOR: Progressive Gene Order Alignment of 17 Vertebrate Genomes. PLoS ONE 5: doi:10.1371/journal.pone.0008861.
16. Wang Y, Tang H, DeBarry JD, Tan X, Li J, et al. (2012) MCScanX: a toolkit for detection and evolutionary analysis of gene synteny and collinearity. Nucleic Acids Res 40: doi:10.1093/nar/gkr1293.
17. Proost S, Fostier J, De Witte D, Dhoedt B, Demeester P, et al. (2011) i-ADHoRe 3.0 – fast and sensitive detection of genomic homology in extremely large data sets. Nucleic Acids Research 40: doi:10.1093/nar/gkr955.
18. Giardine B, Riemer C, Hardison RC, Burhans R, Elnitski L, et al. (2005) Galaxy: a platform for interactive large-scale genome analysis. Genome Res 15: 1451–1455.
19. Blankenberg D, Von Kuster G, Coraor N, Ananda G, Lazarus R, et al. (2010) Galaxy: a web-based genome analysis tool for experimentalists. Curr Protoc Mol Biol 19: 1–21.
20. Goecks J, Nekrutenko A, Taylor J (2010) Galaxy: a comprehensive approach for supporting accessible, reproducible, and transparent computational research in the life sciences. Genome biology 11: doi: 10.1186/gb-2010-11-8-r86.
21. Krzywinski MI, Schein JE, Birol I, Connors J, Gascoyne R, et al. (2009) Circos: An information aesthetic for comparative genomics. Genome Research 19: doi: 10.1101/gr.092759.109.
22. Hinrichs AS, Karolchik D, Baertsch R, Barber GP, Bejerano G, et al. (2006) The UCSC Genome Browser Database: update 2006. Nucleic Acids Res 34(Database issue): 590–598.
23. Consortium TEP (2004) The ENCODE (ENCyclopedia Of DNA Elements) Project. Science 306: 636–640.
24. Rieck K, Laskov P (2008) Linear-Time Computation of Similarity Measures for Sequential Data. J Mach Learn Res 9: doi = 10.1.1.130.8262.
25. Mani RS, Chinnaiyan AM (2010) Triggers for genomic rearrangements: insights into genomic, cellular and environmental influences. Nat Rev Genet 11: 819–829.
26. Peng Q, Pevzner PA, Tesler G (2006) The fragile breakage versus random breakage models of chromosome evolution. PLoS Comput Biol 2: doi:10.1371/journal.pcbi.0020014.
27. Woo Y, Li W (2012) DNA replication timing and selection shape the landscape of nucleotide variation in cancer genomes. Nature communications 3: doi: 10.1038/ncomms1982.
28. Miller W, Rosenbloom K, Hardison RC, Hou M, Taylor J, et al. (2007) 28-Way vertebrate alignment and conservation track in the UCSC Genome Browser. Genome Res 17: 1797–1808.

29. Muñoz-López M, García-Pérez JL (2010) DNA Transposons: Nature and Applications in Genomics. Curr Genomics 11: 115–128.
30. Izsvak Z, Ivics Z, Plasterk RH (2000) Sleeping Beauty, a wide host-range transposon vector for genetic transformation in vertebrates. J Mol Biol 302: 93–102.
31. Ding S, Wu X, Li G, Han M, Zhuang Y, et al. (2005) Efficient transposition of the piggyBac(PB) transposon in mammalian cells and mice. Cell 122: 473–83.
32. Balciunas D, Wangensteen KJ, Wilber A, Bell J, Geurts A, et al. (2006) Harnessing a high cargo-capacity transposon for genetic applications in vertebrates. PLoS Genet 2: doi: 10.1371/journal.pgen.0020169.
33. Zayed H, Izsvak Z, Walisko O, Ivics Z (2004) Development of hyperactive sleeping beauty transposon vectors by mutational analysis. Mol Ther 9: 292–304.
34. Rosenberg C, Knijnenburg J, Bakker E, Vianna-Morgante AM, Sloos W, et al. (2006) Array-CGH detection of micro rearrangements in mentally retarded individuals: clinical significance of imbalances present both in affected children and normal parents. J Med Genet 43: 180–186.

3

Host Adaptation of *Chlamydia pecorum* towards Low Virulence Evident in Co-Evolution of the *ompA*, *incA*, and ORF663 Loci

Khalil Yousef Mohamad[1]*¤, Bernhard Kaltenboeck[2], Kh. Shamsur Rahman[2], Simone Magnino[3], Konrad Sachse[4], Annie Rodolakis[1]

1 INRA, UR1282 Infectiologie Animale et Santé Publique, Nouzilly, France, **2** Department of Pathobiology, College of Veterinary Medicine, Auburn University, Auburn, Alabama, United States of America, **3** Istituto Zooprofilattico Sperimentale della Lombardia e dell'Emilia Romagna "Bruno Ubertini", National Reference Laboratory for Animal Chlamydioses, Sezione Diagnostica di Pavia, Pavia, Italy, **4** Friedrich-Loeffler-Institut Jena, OIE and National Reference Laboratory for Chlamydiosis, Jena, Germany

Abstract

Chlamydia (*C.*) *pecorum*, an obligate intracellular bacterium, may cause severe diseases in ruminants, swine and koalas, although asymptomatic infections are the norm. Recently, we identified genetic polymorphisms in the *ompA*, *incA* and ORF663 genes that potentially differentiate between high-virulence *C. pecorum* isolates from diseased animals and low-virulence isolates from asymptomatic animals. Here, we expand these findings by including additional ruminant, swine, and koala strains. Coding tandem repeats (CTRs) at the *incA* locus encoded a variable number of repeats of APA or AGA amino acid motifs. Addition of any non-APA/AGA repeat motif, such as APEVPA, APAVPA, APE, or APAPE, associated with low virulence ($P<10^{-4}$), as did a high number of amino acids in all *incA* CTRs ($P=0.0028$). In ORF663, high numbers of 15-mer CTRs correlated with low virulence ($P=0.0001$). Correction for *ompA* phylogram position in ORF663 and *incA* abolished the correlation between genetic changes and virulence, demonstrating co-evolution of *ompA*, *incA*, and ORF663 towards low virulence. Pairwise divergence of *ompA*, *incA*, and ORF663 among isolates from healthy animals was significantly higher than among strains isolated from diseased animals ($P\leq10^{-5}$), confirming the longer evolutionary path traversed by low-virulence strains. All three markers combined identified 43 unique strains and 4 pairs of identical strains among all 57 isolates tested, demonstrating the suitability of these markers for epidemiological investigations.

Editor: Deborah Dean, University of California, San Francisco, University of California, Berkeley, and the Children's Hospital Oakland Research Institute, United States of America

Funding: The study was funded by the French Institut National de la Recherche Agronomique, and current funding by Auburn University. The funders had no role in study design, data collection and analysis, decision to publish, or preparation of the manuscript.

* Email: khalilyoussef@hotmail.com

¤ Current address: Laboratory of Animal Production - Pathology, Faculty of Agriculture, University of Al-Furat, Deir Ezzor, Syria

Background

Chlamydia (*C.*) *pecorum*, a Gram-negative obligate intracellular bacterium, is a species of the genus *Chlamydia* belonging to the family *Chlamydiaceae* [1]. *C. pecorum* strains have been isolated worldwide from ruminants and swine with conjunctivitis, encephalomyelitis, enteritis, pneumonia, polyarthritis, abortion, and reproductive or urinary tract diseases [2–4]. More recent studies have shown that wild animals may also be infected with *C. pecorum*, most prominently Australian marsupials, such as koalas, in which fertility is severely compromised by urogenital infections [5], and western barred bandicoots with conjunctivitis [6]. *C. pecorum* is also found in the conjunctiva, intestine, and vaginal mucus of clinically healthy ruminants and swine [7–9]. In fact, such asymptomatic *C. pecorum* infections are found very frequently, particularly in ruminants at high population density where prevalence rates can approach 100% [10,11], but also in pigs [12]. While high *C. pecorum* infectious loads associate significantly with disease symptoms [13], the majority of *C. pecorum* infections are asymptomatic and very low infectious loads are detected [8–10]. Nevertheless, even such asymptomatic infections of calves cause detectable lung dysfunction [14] and incur substantial reductions in weight gains [15]. Collectively, these observations raise the question if the parallel occurrence of asymptomatic and clinically manifest *C. pecorum* infections associates with virulence differences of strains that can be detected and characterized.

For several decades, serotyping using polyclonal or monoclonal antibodies in micro-immunofluorescence assays was used to characterize and classify individual chlamydial strains. Meanwhile, genotyping based on PCR and sequencing of *ompA* has gradually replaced serotyping. Indeed, several new methods were proposed, such as DNA microarray testing [16], multi-locus sequence typing (MLST) [17] and typing based on variable number tandem repeats (VNTR) [18]. However, none of these methods is congruent with the virulence of chlamydial isolates, although some parameters are correlated with clinical manifestations and serotyping. For *C. trachomatis*, in a study including 175 men and 135 women attending a sexually transmitted disease (STD) clinic, a correlation was reported between urethral discharge in men and serotypes H and J, and between lower abdominal pain in women and serotypes F and G [19]. Furthermore, 47.5% of asymptomatic patients were infected with *C. trachomatis* serovar E among 1,770 STD-infected women in China [20]. As to *C. psittaci*, serovar D strains induce the most severe disease in turkeys [21].

C. pecorum strains present many genetic and antigenic variations [22,23]. In earlier investigations, we found virulence-associated genetic differences among 19 *C. pecorum* strains by identifying different motifs of the variant coding tandem repeats (CTR) in *incA* of isolates from sick versus healthy ruminants [24]. By determining lower numbers of repetitions of the CTR in the hypothetical ORF663 in highly virulent *C. pecorum* strains than in low-virulence isolates, we further identified virulence-associated genetic polymorphisms of *C. pecorum* [25]. In addition, 6 out of 8 strains from diseased ruminants clustered to a single *ompA* sequence group [25].

In this study we further investigated the *C. pecorum* segregation by virulence in *ompA*, *incA*, and ORF663. These loci were sequenced for an expanded panel of *C. pecorum* isolates from most known hosts of *C. pecorum*, including 11 strains isolated from swine, 24 additional strains isolated from ruminants, and 3 strains isolated from koalas. Virulence associations of *incA* and ORF663 CTRs were confirmed and expanded to porcine and koala *C. pecorum* isolates, and low virulence significantly associated with evolutionary distance of *ompA*, *incA*, and ORF663 from the respective putative *C. pecorum* ancestor.

Results

Sequence analysis of *ompA*, *incA* and ORF663

All 32 strains yielded the expected amplification products of *ompA*, *incA*, and ORF663, except for two bovine strains (DC49 and DC55) that failed to give an *incA* and one porcine strain (R106) that failed to give an ORF663 amplicon (Table 1). Sequence analysis of *incA* showed that all 11 porcine strains had one encoded motif (APA) with 7 to 14 repetitions representing 7 variants (Table 1). Similarly, the 3 koala strains with recently deposited genomes showed 4–11 repetitions of the APA motif. In addition, a new motif of 9 nucleotides (GCTGGAGCC) encoding amino acids alanine and glycine AGA (Motif 5) was identified. This motif was detected only in 3 bovine strains isolated from different geographical areas and associated with different conditions (DC13, 2047, 66P130; Table 1).

Similar to the ruminant strains isolated from diseased animals, the 10 porcine strains and 3 koala strains, all from diseased hosts, possessed lower numbers of ORF663 CTR repetitions (<43 repeats) than most intestinal strains isolated from asymptomatic (healthy) ruminants (Table 1). Strains isolated from asymptomatic ruminants (n = 19) had more than 43 repeats, except for 6 strains isolated from cattle and one strain isolated from sheep (Table 1).

Correlation of *ompA* with virulence

The chlamydial *ompA* is one of the most polymorphic of all genes conserved throughout the genus *Chlamydia*, and therefore frequently used as surrogate for approximating overall evolution of chlamydial genomes. This diversity is based not on variation in repeat elements, but on frequent recombination within the 4 variable domains of the gene [25]. We therefore used *ompA* to estimate the correlation between *C. pecorum* genomic diversity with virulence. We first aligned the complete *ompA* sequences of all 57 *C. pecorum* strains used in this study (Figure 1; Figure S1) and performed neighbor-joining phylogenetic reconstruction of their evolution. The resultant phylogram in Figure 2 arranged the *ompA* genes into several clusters of closely related sequences (clades) that were separated from other clades by deep and strongly bootstrap-supported branches [25].

To estimate the association of *ompA* genotype with disease, the deeply separated branches of the *ompA* phylogram in Figure 2 were stratified into 5 high- and 4 low-virulence clades. "High-virulence" clades contained 50% or more strains isolated from diseased animals (strains E58-VB2, P787-DC13, iB2-HsLuRz, L1-1708, 2047-748/4), while "low-virulence" clades contained less than 50% strains isolated from diseased animals (strains DC47, 29531/1-IPTaLE, R69-iB5, 4283/3-iC2). In sum, the high-virulence clades contained 30 high-virulence strains among a total of 35, while the low-virulence groups differed highly significantly and contained 7 high-virulence strains among a total of 21 ($P = 0.0001$; two-tailed Fisher Exact Test).

We also used non-stratified *ompA* phylogenetic rank data to evaluate the relation between *ompA* evolution and virulence by logistic regression. All strains with a unique *ompA* genotype received a unique phylogenetic rank number between 1 and 35, based on their position in the phylogram in Figure 2. *C. pecorum* strains isolated from diseased animals were scored as "high virulent" versus the strains isolated from healthy animals scored as "low virulent". As evident in the highly significant regression plot ($P = 0.0066$; Figure 3), the probability of high virulence was high for low *ompA* rank, but dropped with increasing *ompA* rank. Thus, two analyses indicated that strain position on the *ompA* phylogenetic tree highly significantly correlates with virulence of *C. pecorum* isolates.

Correlation of *incA*-coding tandem repeat sequence motifs with virulence

Next, we sought to quantitatively assess the relationship between the numbers of repetitions of sequence motifs in *incA* and the virulence of the *C. pecorum* strains. While amino acid APA/AGA motifs are dominant, addition of different motifs (APEVPA, APAVPA, APE, or APAPE) highly significantly associated with low virulence, i.e. 10 of 17 low-virulence strains possessed such sequence motifs, while 30 of 31 high-virulence strains did not ($P<10^{-4}$; two-tailed Fisher Exact Test).

Similar to the *ompA* phylogenetic rank, we examined the correlation of the number of CTRs in *incA* with virulence by logistic regression. To account for the total amount of the different repeat motif insertions, we used the total number of amino acids encoded by these CTR codons. In logistic regression, high total CTR codon numbers highly significantly correlated with low virulence ($P = 0.0028$), with 50 codons representing a midpoint 50% probability of high virulence (Figure 4A).

A fundamental question in this analysis is whether molecular evolution of the *C. pecorum* strains and *incA* CTR codon numbers is co-linear, and whether, therefore, the correlation between *incA* CTR codons and virulence was confounded by the phylogenetic position of the isolates. To account for *C. pecorum* evolution, we created a corrected *incA* CTR codon dataset that was controlled for the position of the isolates in the *ompA* phylogram. This was achieved by creating standardized phylogram rank data (mean = 0, SD = 1), adjusting these data to positive by adding 1+ the absolute minimum standardized rank number (results in 1 as the minimum adjusted standardized rank number), and dividing the number of *incA* CTR codon of each strain by the respective adjusted standardized *ompA* rank number. Using the *ompA* rank-corrected *incA* CTR codon data, we repeated the logistic regression analysis, but failed to obtain a significant correlation to virulence ($P = 0.2785$; Figure 4B). Thus, both *incA* and *ompA* evolution progress in a co-linear fashion, linking the number of

Table 1. Sequence coding tandem repeat characteristics and accession numbers for all *C. pecorum* strains analyzed in this study.

Strain*	*incA* CTR numbers & amino acid motifs	ORF663 15-mer CTRs	*ompA* GenBank #	*incA* GenBank #	ORF663 GenBank #
E58[a]	12 APA	22	EU837071	EU837066	EU837072
LW613*[b]	9 APA	45	GQ228176	GQ228147	GQ228117
LW623*[b]	9 APA	45	GQ228177	GQ228148	GQ228118
LW679[c]	11 APA	45	EU684921	EU340821	EU684939
L14*[b]	11 APA	35	GQ228175	GQ228146	GQ228116
IPA[d]	7 APA	13	AZBD01000001.1	AZBD01000005.1	AZBD01000005.1
FC-Stra*[b]	7 APA	50	GQ228172	GQ228143	GQ228113
JP1751*[b]	10 APA	45	GQ228173	GQ228144	GQ228114
SBE[e]	12 APA	22	EU684916	EU340823	EU684934
AKT[f]	8 APA	24	EU684918	EU340816	EU684936
AB10[g]	9 APA	22	EU684917	EU340815	EU684935
VB2[f]	15 APA	24	EU684919	EU340824	EU684937
P787[h]	2 APA+10 APAPE	16	NC_022441.1	NC_022441.1	NC_022441.1
DBDeUG[d]	11 APA	30	AZBB01000001.1	AZBB01000008.1	AZBB01000004.1
L17*[i]	14 APA	28	GQ228181	GQ228152	GQ228123
PV3056/3[h]	9 APA+1 APE	27	NC_022439.1	NC_022439.1	NC_022439.1
DC13*[j]	12 AGA	25	GQ228171	GQ228142	GQ228111
iB2[g]	5 APA+3 APEVPA+4 APE	68	EU684925	EU340810	EU684943
iB1[g]	5 APA+3 APEVPA+3 APE	65	EU684924	EU340811	EU684942
C14*[j]	3 APA+8 APEVPA	46	GQ228169	GQ228140	GQ228109
824[g]	4 APA+8 APEVPA	41	EU684922	EU340809	EU684940
DC49*[j]	-	23	GQ228195	-	GQ228119
BE53[e]	8 APA	20	EU684923	EU340808	EU684941
L39*[i]	13 APA	10	GQ228182	GQ228153	GQ228124
L40*[i]	10 APA	10	GQ228184	GQ228155	GQ228126
L71*[b]	13 APA	10	GQ228185	GQ228156	GQ228127
HsLuRZ*[i]	12 APA	10	GQ228183	GQ228154	GQ228125
DC47*[j]	10 APA	7	GQ228193	GQ228164	GQ228135
L1*[i]	8 APA	42	GQ228174	GQ228145	GQ228115
R106*[i]	14 APA	-	GQ228197	GU014536	-
1886*[i]	10 APA	21	GQ228179	GQ228150	GQ228121
1920BRZ*[i]	7 APA	42	GQ228168	GQ228139	GQ228108
1710S*[b]	11 APA	42	GQ228167	GQ228138	GQ228107
1708*[b]	10 APA	15	GQ228194	GQ228165	GQ228136
29531/1*[k]	9 APA	5	GQ228189	GQ228160	GQ228131
M14[f]	22 APA	14	EU684920	EU340814	EU684938
5184/4*[k]	5 APA+3 APEVPA+6 APE	56	GQ228192	GQ228163	GQ228134
DC55*[j]	-	16	GQ228196	-	GQ228112
DC52*[j]	12 APA	15	GQ228178	GQ228149	GQ228120
iB3[g]	14 APA	53	EU684926	EU340827	EU684944
MC/MarsBar[d]	4 APA	11	AZBC01000001.1	AZBC01000014.1	AZBC01000008.1
iB4[g]	10 APA+4 APAPE	53	EU684927	EU340826	EU684945
IPTaLE[d]	10 APA	18	AZBE01000002.1	AZBE01000014.1	AZBE01000007.1
R69[l]	2 APA+8 APAPE	58	EU684930	EU340822	EU684948
W73[l]	2 APA+8 APAPE	58	EU684929	EU340825	EU684947
C4*[j]	3 APA+2 APAVPA	54	GQ228170	GQ228141	GQ228110
3257*[k]	2 APA	41	GQ228190	GQ228161	GQ228132
66P130*[b]	10 AGA	21	GQ228180	GQ228151	GQ228122
iB5[g]	12 APA	62	EU684928	EU340817	EU684946

Table 1. Cont.

Strain*	*incA* CTR numbers & amino acid motifs	ORF663 15-mer CTRs	*ompA* GenBank #	*incA* GenBank #	ORF663 GenBank #
2047*[k]	12 AGA	16	GQ228191	GQ228162	GQ228133
PV5*[k]	11 APA	31	GQ228166	GQ228137	GQ228106
748/4*[k]	3 APA	34	GQ228188	GQ228159	GQ228130
4283/3*[k]	3 APA	21	GQ228187	GQ228158	GQ228129
3638/3*[k]	3 APA	21	GQ228186	GQ228157	GQ228128
iC4[g]	6 APA+12 APAPE	60	EU684933	EU340819	EU684951
iC2[g]	6 APA+12 APAPE	52	EU684931	EU340818	EU684949
iC3[g]	6 APA+11 APAPE	59	EU684932	EU340820	EU684950

*Strains sequenced in this study. Strains not marked with an asterisk were sequenced in a preceding study [23], or posted as complete genomes [30–32].
[a]Referenced in [31].
[b]Referenced in [2].
[c]Referenced in [42].
[d]Referenced in [30].
[e]Isolated by M. Dawson, Virology Department, Central Veterinary Laboratory, Weybridge UK.
[f]Isolated at INRA, UR1282, Infectiologie Animale et Santé Publique, Centre de Recherche de Tours, France.
[g]Referenced in [43].
[h]Referenced in [32].
[i]Referenced in [3].
[j]Supplied by Konrad Sachse, Friedrich-Loeffler-Institut Jena, OIE and National Reference Laboratory for Chlamydiosis, 07743 Jena, Germany, 07743 Jena, Germany.
[k]Supplied by Simone Magnino, Istituto Zooprofilattico Sperimentale della Lombardia e dell'Emilia Romagna "Bruno Ubertini", National Reference Laboratory for Animal Chlamydioses, Sezione Diagnostica di Pavia, 27100 Pavia, Italy.
[l]Isolated by M.S. McNulty, Veterinary Research Laboratory, Stormont, Belfast, Ulster.
-not amplified by PCR.

incA CTR codons to the phylogenetic position of the *C. pecorum ompA*.

Correlation of coding tandem repeats in ORF663 with virulence

Similar to *incA*, we examined the correlation of the number of CTRs in ORF663 with virulence. The correlation of the number of CTRs in ORF663 with virulence was highly significant ($P = 0.0001$), again with low numbers of CTRs associating with high probability of high virulence, with 50% probability of high virulence at 43 repetitions (Figure 4C). Interestingly, there is a bimodal distribution of the CTR numbers in low-virulence strains, with 6 bovine strains isolated from healthy animals having less than 24 CTR repetitions, while all other isolates had 43 or more repetitions. When we tested for confounding by phylogenetic position using an *ompA* rank-corrected ORF663 CTR dataset, the correlation was lost ($P = 0.7565$; Figure 4D). Thus, analogous to *incA*, the number of CTRs in *C. pecorum* ORF663 and *ompA* evolution are closely linked.

Co-evolution of *ompA*, ORF663, and *incA* towards reduced virulence

As a final test for evolutionary linkage, we also tested for co-evolution of the CTR numbers in ORF663 and *incA* by creating an *incA* CTR codon number dataset that was corrected for ORF663 CTR numbers. Again, the ORF663 correction eliminated the correlation between *incA* CTR codons and virulence ($P = 0.4234$; data not shown), thus confirming *incA* and ORF663 co-evolution.

Collectively, these results suggested that the molecular evolution of *C. pecorum* progressed from ancestral strains with high virulence towards strains with low virulence, and that increased numbers of CTRs in *inc A and* ORF663, as well as recombination in *ompA* resulting in new *C. pecorum* serovars [26], were markers, if not mediators, of this progression towards low virulence. This hypothesis is testable with the present dataset since it implies that phylogenetic sequence divergence of *ompA* from the hypothetical ancestor inversely correlates with virulence of the extant *C. pecorum* strains. We assumed the root of the *ompA* phylogenetic tree at the connection to an outgroup composed of one *ompA* sequence of each of the eight remaining chlamydial species. This putative *C. pecorum* ancestor located to a set of short and weakly bootstrap-supported branches at the base of the phylogram (Figure 5). Association of virulence with evolutionary divergence of each strain from this ancestor was analyzed by logistic regression (Figure 5). While reduced probability of high virulence at long evolutionary distances was obvious from the placement of many strains isolated from healthy animals at the tips of long branches, this trend failed to reach significance ($P = 0.0770$; Figure 5).

Next, we examined *incA* for evidence of linkage between virulence and distance from the evolutionary ancestor. Alignment of genes that contain different numbers of CTRs, such as *incA*, is notoriously difficult and very sensitive to the choice of alignment parameters. We optimized the alignment by minimizing average pairwise sequence distance, mainly by setting a high penalty for gap opening, and, less so, for gap extension. The resultant alignment (Figure 6; Figure S2) was used for phylogenetic reconstruction (Figure 7), and the putative ancestor was placed where an outgroup of *incA* homologs from the other eight chlamydial species connected to the phylogram of the highly conserved N-terminal fragment of *C. pecorum incA* (the hyper-variable CTR region has no homolog). For *incA*, a highly significant inverse correlation between evolutionary divergence and probability of high virulence was found ($P = 0.0029$; Figure 7).

For ORF663, we used an approach similar to *incA* for alignment (Figure 8; Figure S3) and phylogenetic reconstruction (Figure 9). The relationship between long evolutionary distance

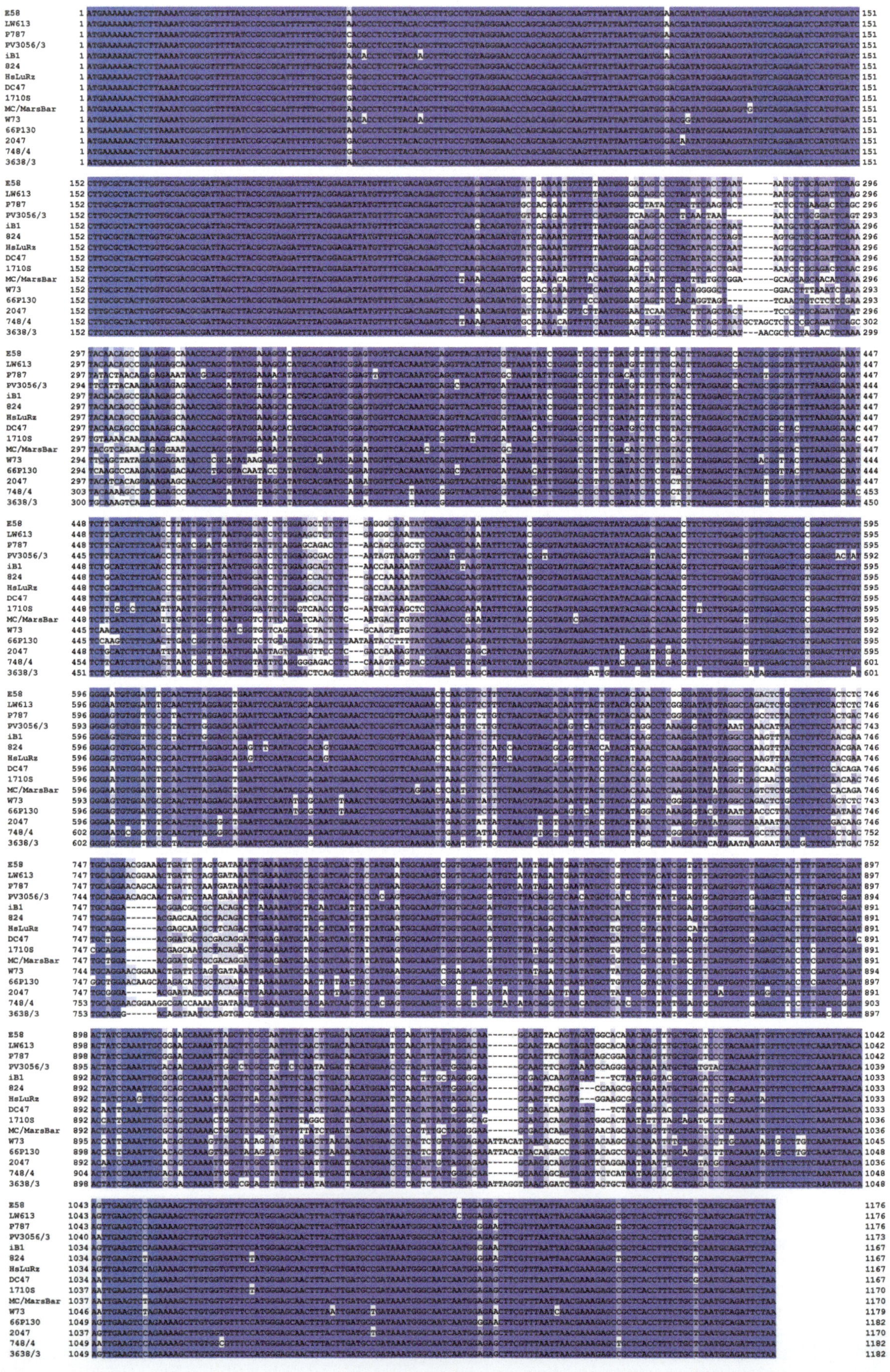
E58
LW613
P787
PV3056/3
iB1
824
HsLuRz
DC47
1710S
MC/MarsBar
W73
66P130
2047
748/4
3638/3

Figure 1. *C. pecorum ompA* alignment. A subset of the 57 analyzed *C. pecorum* strains was selected that represents all major clades of the phylogram in Figure 2. The corresponding sequence alignment of the complete *ompA* of all 57 strains was used to infer *C. pecorum ompA* evolution by construction of a phylogenetic tree. The alignment of the resultant amino acid sequences of all 57 *C. pecorum* OmpA proteins deduced from the nucleotide sequences is shown in Figure S1.

from the putative ancestor and low virulence was even more pronounced for ORF663 ($P = 0.0003$; Figure 9). Thus, based on phylogenetic modeling, *incA* and ORF663 highly significantly, and *ompA* marginally so, co-evolve towards low virulence, irrespective of the branch of the phylogram, on which a specific strain is located.

Confirmation of *C. pecorum* gene co-evolution towards low virulence by mean pairwise sequence divergence

If the notion of *C. pecorum* evolution towards low virulence were correct, then a consequence would be that low-virulence strains have travelled a longer evolutionary path than high-virulence strains. This implies that the mean pairwise sequence divergence between low-virulence strains must be higher than that of high-virulence strains, irrespective of the distance from the ancestor, thus providing an easily testable hypothesis. The mean pairwise distances between *ompA*, *incA*, and ORF663 of *C. pecorum* strains isolated from healthy or diseased animals, as well as those of all 57 strains are listed in Table 2. In fact, for all three genes the mean sequence distance between low-virulence isolates from healthy animals is highly significantly by 2–3% higher than that of high-virulence isolates from diseased animals (Table 2). These data provide unambiguous confirmation of *C. pecorum* evolution towards low virulence.

Discussion

We undertook the present study with the primary aim to identify genetic markers that would allow us to unambiguously discriminate between highly virulent ("pathogenic") and low-virulence or avirulent ("non-pathogenic") *C. pecorum* strains that presumably would occupy different branches (clades) of the *C. pecorum* phylogeny. What our results tell us, though, is a different story, in essence that the main driver of reduction in virulence of *C. pecorum* is the distance a strain has traversed in its evolution from the primordial *C. pecorum* strain, and not the phylogenetic position in a specific clade. While it is clear that certain branches of the *C. pecorum ompA* phylogram harbor more highly virulent strains than others, it is uncertain if this is a genetically fixed property of this clade or has more to do with the short evolutionary distance from the ancestor.

The finding of the association of evolutionary distance with virulence is not surprising, given the endemic nature of *C. pecorum* infections in ruminants, swine, and koalas [3,10,27,28], particularly in large herds [29]. Long-term coexistence of host and pathogen results in reduced virulence that is beneficial for the pathogen by maintaining a large host population. Effective adaptation to the host and the self-limiting nature of chlamydial intracellular multiplication may also explain the low number of isolates worldwide despite the ubiquity of *C. pecorum* infections. In addition, we assume that there is a bias towards isolation of *C. pecorum* from diseased animals rather than from healthy ones, because this is what diagnostic laboratories aim for, in particular given the high effort required for isolation of chlamydiae.

In consideration of the potential economic importance of these ubiquitous endemic bacteria [10,14,15], we collected a comprehensive set of DNAs of *C. pecorum* strains isolated worldwide from healthy as well as diseased mammalian livestock. Importantly, this study extended previous more limited analyses of ruminant *C. pecorum* strains to include unique sets of *C. pecorum* strains, isolated in Austria from diseased swine [2,3] and in Australia from diseased koalas [30]. Following a previous investigation, we chose the *ompA*, *incA* and ORF663 loci as targets of our genetic analysis, which have now been identified by genome comparison as being among the most polymorphic genes of the *C. pecorum* genome, which is otherwise more than 99% conserved among the *C. pecorum* strains from which the whole genome is known [30–32]. Among these 8 strains, i.e. ruminant *C. pecorum* type strain E58, and strains P787, W73, PV3056/3, and IPA, and koala strains DBDeUG, MC/MarsBar, and IPTaLE, *ompA* is up to 16% divergent, and *incA* and ORF663 up to 8%. This remarkable polymorphism is presumably driven by immunoselection acting on the encoded proteins, all of which have been found immunodominant and eliciting high antibody responses ([3,33], unpublished data).

The previously identified association of increasing numbers of CTRs in *incA* and ORF663 with reduced virulence [25] was highly significantly confirmed in this study. This finding is in agreement with a study that showed differences between environmental and clinical *Legionella pneumophila* strains in the repeat copy numbers of four genes [34]. Interestingly, six isolates from healthy animals in Germany, England, and the USA, had low numbers of ORF663 CTRs. This indicates that ORF663, as well as *incA* or *ompA*, cannot be used as the sole virulence marker. In *ompA*, specific sequence polymorphisms are not indicators of virulence, however in the context of the overall *C. pecorum ompA* phylogeny they are useful in quantifying distance from the root. As evident in Figure 4, the correlations of all three genes, *ompA*, *incA*, and ORF663 with virulence of the *C. pecorum* isolates are co-linear. Therefore, in combination these 3 genes may serve as probabilistic, but not absolute, markers of virulence.

For the practical use of such molecular markers, their genetic stability under non-selective culture conditions is important, and in fact they remain unchanged in laboratory maintenance of the isolates (data not shown), thus making these genes suitable for highly discriminatory epidemiological studies. At least one of these genes differed for two otherwise identical strains, except for 4 cases, namely LW613 and LW623, 3638/3 and 4283/3, L71 and L39, and E58 and SBE (Figure 2, Table 1), thus uniquely identifying 53 out of 57 *C. pecorum* strains.

The ability of *C. pecorum* to continuously evolve towards low virulence and generate successive allelic variants of *incA*, ORF663, and *ompA* may allow rapid adaptation to a host population and/or evasion of the host immune system. Changes in the repetitive coding regions (loss or gain), mediated by DNA replication error mechanisms, have been shown to cause phase variation in bacteria, which confer major defensive capabilities to the pathogen in order to escape from an aggressive host environment [35,36]. Similarly, *C. pecorum*, in the process of inserting increasing numbers of CTRs in *incA* and ORF663 and recombining *ompA*, generates new serovars [26] and evolves towards lower virulence. One can speculate that this *ompA* evolution and many CTR insertions in *incA* and ORF663 change the immunological signature of a *C. pecorum* strain. Equally possible, however, is a scenario in which the immunosignature evolution of these genes is accompanied by point mutations in other genes that alter their

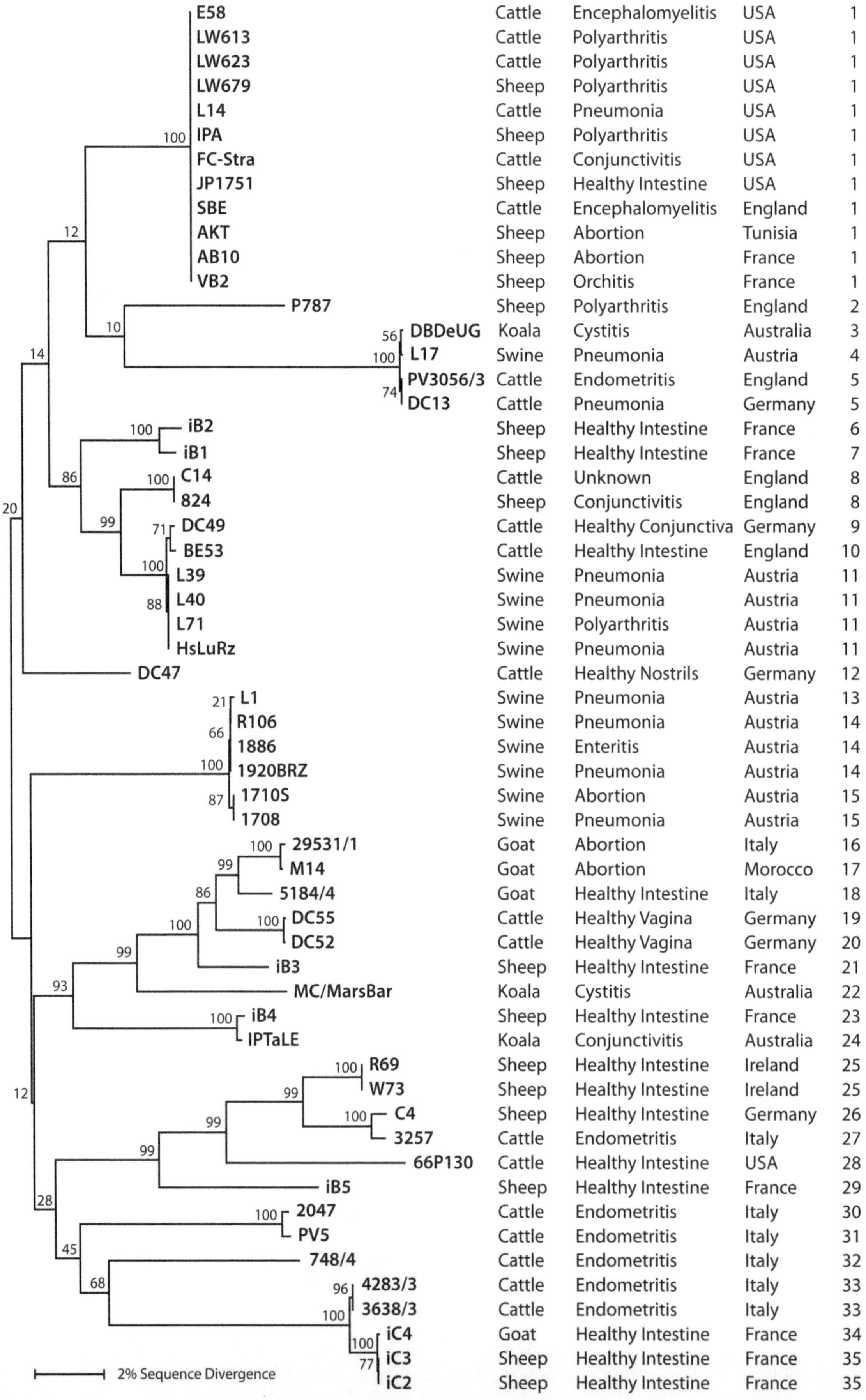
E58 Cattle Encephalomyelitis USA 1
LW613 Cattle Polyarthritis USA 1
LW623 Cattle Polyarthritis USA 1
LW679 Sheep Polyarthritis USA 1
L14 Cattle Pneumonia USA 1
IPA Sheep Polyarthritis USA 1
FC-Stra Cattle Conjunctivitis USA 1
JP1751 Sheep Healthy Intestine USA 1
SBE Cattle Encephalomyelitis England 1
AKT Sheep Abortion Tunisia 1
AB10 Sheep Abortion France 1
VB2 Sheep Orchitis France 1
P787 Sheep Polyarthritis England 2
DBDeUG Koala Cystitis Australia 3
L17 Swine Pneumonia Austria 4
PV3056/3 Cattle Endometritis England 5
DC13 Cattle Pneumonia Germany 5
iB2 Sheep Healthy Intestine France 6
iB1 Sheep Healthy Intestine France 7
C14 Cattle Unknown England 8
824 Sheep Conjunctivitis England 8
DC49 Cattle Healthy Conjunctiva Germany 9
BE53 Cattle Healthy Intestine England 10
L39 Swine Pneumonia Austria 11
L40 Swine Pneumonia Austria 11
L71 Swine Polyarthritis Austria 11
HsLuRz Swine Pneumonia Austria 11
DC47 Cattle Healthy Nostrils Germany 12
L1 Swine Pneumonia Austria 13
R106 Swine Pneumonia Austria 14
1886 Swine Enteritis Austria 14
1920BRZ Swine Pneumonia Austria 14
1710S Swine Abortion Austria 15
1708 Swine Pneumonia Austria 15
29531/1 Goat Abortion Italy 16
M14 Goat Abortion Morocco 17
5184/4 Goat Healthy Intestine Italy 18
DC55 Cattle Healthy Vagina Germany 19
DC52 Cattle Healthy Vagina Germany 20
iB3 Sheep Healthy Intestine France 21
MC/MarsBar Koala Cystitis Australia 22
iB4 Sheep Healthy Intestine France 23
IPTaLE Koala Conjunctivitis Australia 24
R69 Sheep Healthy Intestine Ireland 25
W73 Sheep Healthy Intestine Ireland 25
C4 Sheep Healthy Intestine Germany 26
3257 Cattle Endometritis Italy 27
66P130 Cattle Healthy Intestine USA 28
iB5 Sheep Healthy Intestine France 29
2047 Cattle Endometritis Italy 30
PV5 Cattle Endometritis Italy 31
748/4 Cattle Endometritis Italy 32
4283/3 Cattle Endometritis Italy 33
3638/3 Cattle Endometritis Italy 33
iC4 Goat Healthy Intestine France 34
iC3 Sheep Healthy Intestine France 35
iC2 Sheep Healthy Intestine France 35
2% Sequence Divergence

Figure 2. Unrooted neighbor-joining phylogram of *ompA* of 57 *C. pecorum* strains based on the nucleotide sequence alignment. Percentages of branching patterns in bootstrap analyses of the dataset (10,000 replications) are indicated left to the branches. Host animal species, disease association, country of origin, and *ompA* phylogenetic rank are indicated in the columns to the right of the strain names.

function, and in that way mediate reduced virulence. Or, alternatively, CTR insertion, as occurs aside from *incA* and ORF663 in multiple other *C. pecorum* proteins such as polymorphic membrane proteins, cytotoxins, and phospholipase D-like proteins [32], may alter both function and immunosignature and mediate virulence reduction by both mechanisms. Therefore, the simultaneous evolutionary changes in *ompA*, *incA*, and ORF663 may or may not be functional correlates of virulence.

A point of criticism of the present analysis of *C. pecorum* virulence may be the fact that the differentiation is based on a single clinical examination of the animal from which the isolate was recovered (Figure 2). The diagnosis in that case may be tenuous in an epidemiological setting with ubiquitous endemic infections of *C. pecorum* [10]. However, the high number of 57 isolates included in this study obtained by numerous investigators over a period of 50 years should alleviate concerns about diagnostic accuracy. If the clinical diagnoses had been widely aberrant, it is very unlikely that we would have been able to demonstrate co-linear correlation with virulence of three independent genetic markers in this study. In addition, Storz *et al.* [37] have experimentally confirmed this genetic differentiation in virulence long ago by experimental oral inoculation of calves, the original host, with *C. pecorum* isolates LW613 or 66P130. Highly virulent strain LW613 caused severe hemorrhagic diarrhea and polyarthritis with predominantly lethal outcome. In contrast, strain 66P130, isolated from feces of a healthy calf, caused only transient mild diarrhea. Thus, experimental inoculation of the original host may produce severe disease only with highly virulent isolates, while such isolates may also be detected in asymptomatic natural infections [15]. These asymptomatic infections, by high- as well as low-virulence *C. pecorum* strains, reduce growth rates in calves by eliciting a status of systemic inflammation [15]. Unraveling the contribution of low- and high-virulence *C.*

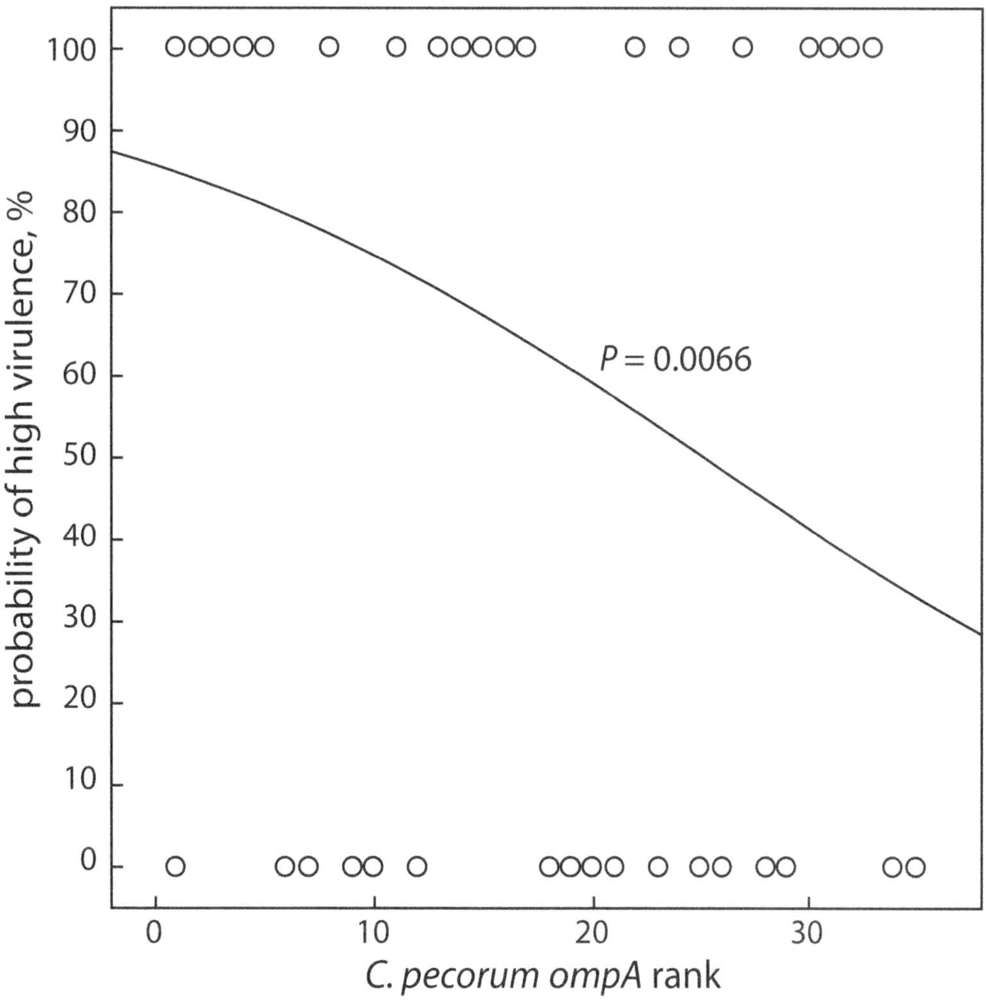

Figure 3. Relationship between virulence of *C. pecorum* strains and rank number in the *ompA* phylogram. The probability of high virulence was determined by logistic regression analysis of the virulence of *C. pecorum* isolates scored by host disease association (0 = healthy; 100 = diseased), and *ompA* rank numbers of the isolates were regressed against virulence. The probability of high virulence decreases highly significantly with increasing *ompA* rank number.

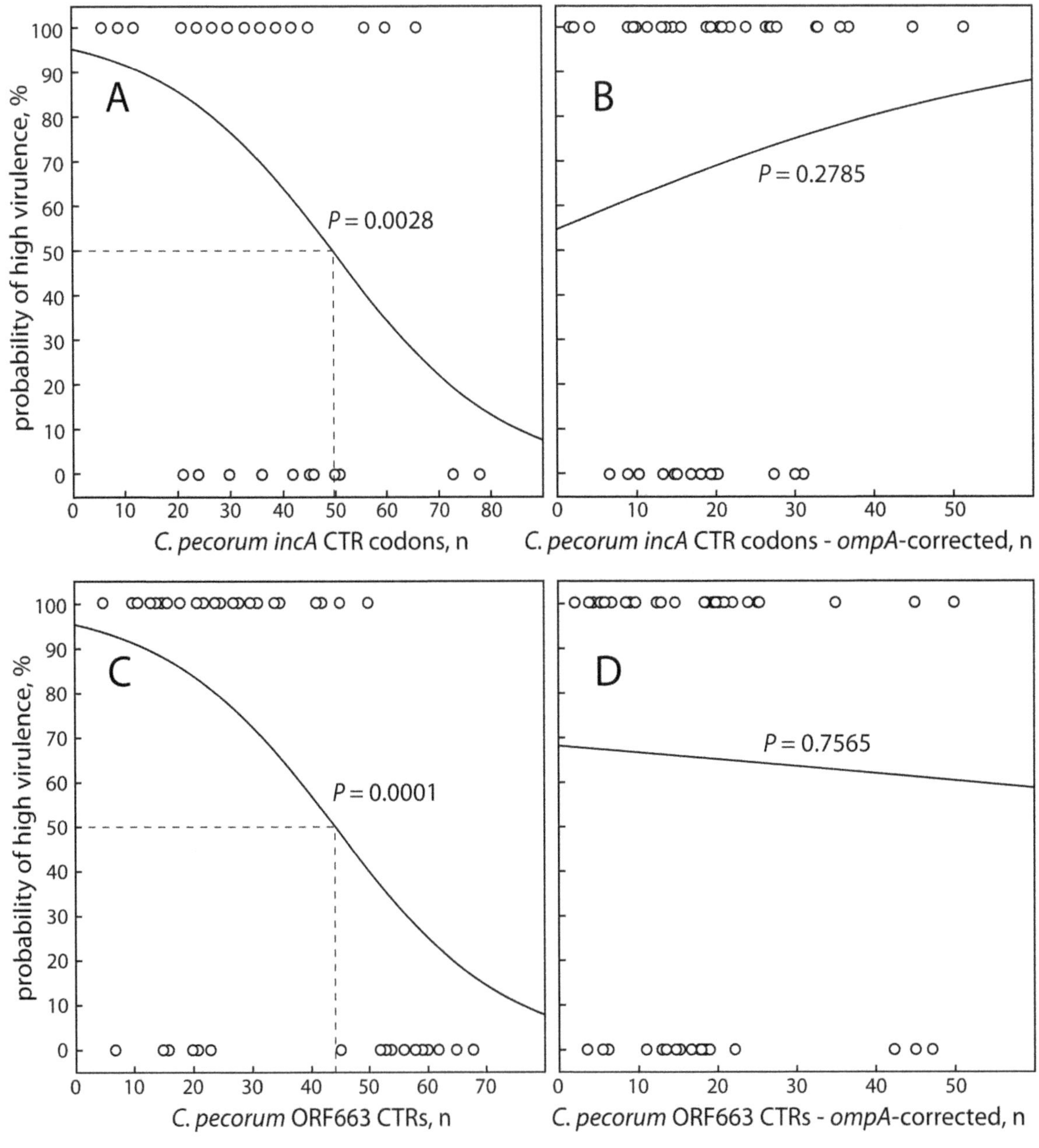

Figure 4. Relationship between virulence of *C. pecorum* strains and coding tandem repeats in *incA* and ORF663. The probability of high virulence was determined by logistic regression analysis of the virulence of *C. pecorum* isolates, scored as in Figure 3, and continuous parameters for CTR numbers of the isolates were regressed against virulence. (A) the number of codons encoded by the CTRs in *incA*; (B) the number of codons encoded by the CTRs in *incA* corrected for the rank number of each isolate in the *ompA* phylogram. (C) The number of CTRs in ORF663; and (D) the number of CTRs in ORF663 corrected for the *ompA* rank number of each isolate. The highly significant correlation between CTRs and probability of high virulence in both *incA* and ORF663 is abolished by correction for *ompA* rank, indicating co-evolution of *ompA*, *incA*, and ORF663.

pecorum strains to performance reduction in livestock will be of great scientific as well as economic interest.

Methods

Chlamydial isolates

Thirty two *C. pecorum* strains were newly analyzed in this study, while the remaining 25 isolates had been examined before [24,25] or published recently [30,32]. The strains were propagated in the yolk sac of chicken embryos and stored at −70°C as previously described [38]. All isolates were obtained from routine diagnostic specimens in veterinary diagnostics laboratories in Austria (11 isolates), England (1 isolate), Germany (6 isolates), Italy (8 isolates), the USA (6 isolates). The 17 strains from Austria and the USA were isolated between 1965 and 1970, when ethical regulations regarding animal specimens did not exist. The remaining isolates from England, Germany, and Italy were obtained between 1993 and 2006 in governmental veterinary diagnostic laboratories that strictly operated under ethics rules established in the respective countries. None of the specimens obtained caused suffering to the animals in addition to the suffering caused by the natural chlamydial infection.

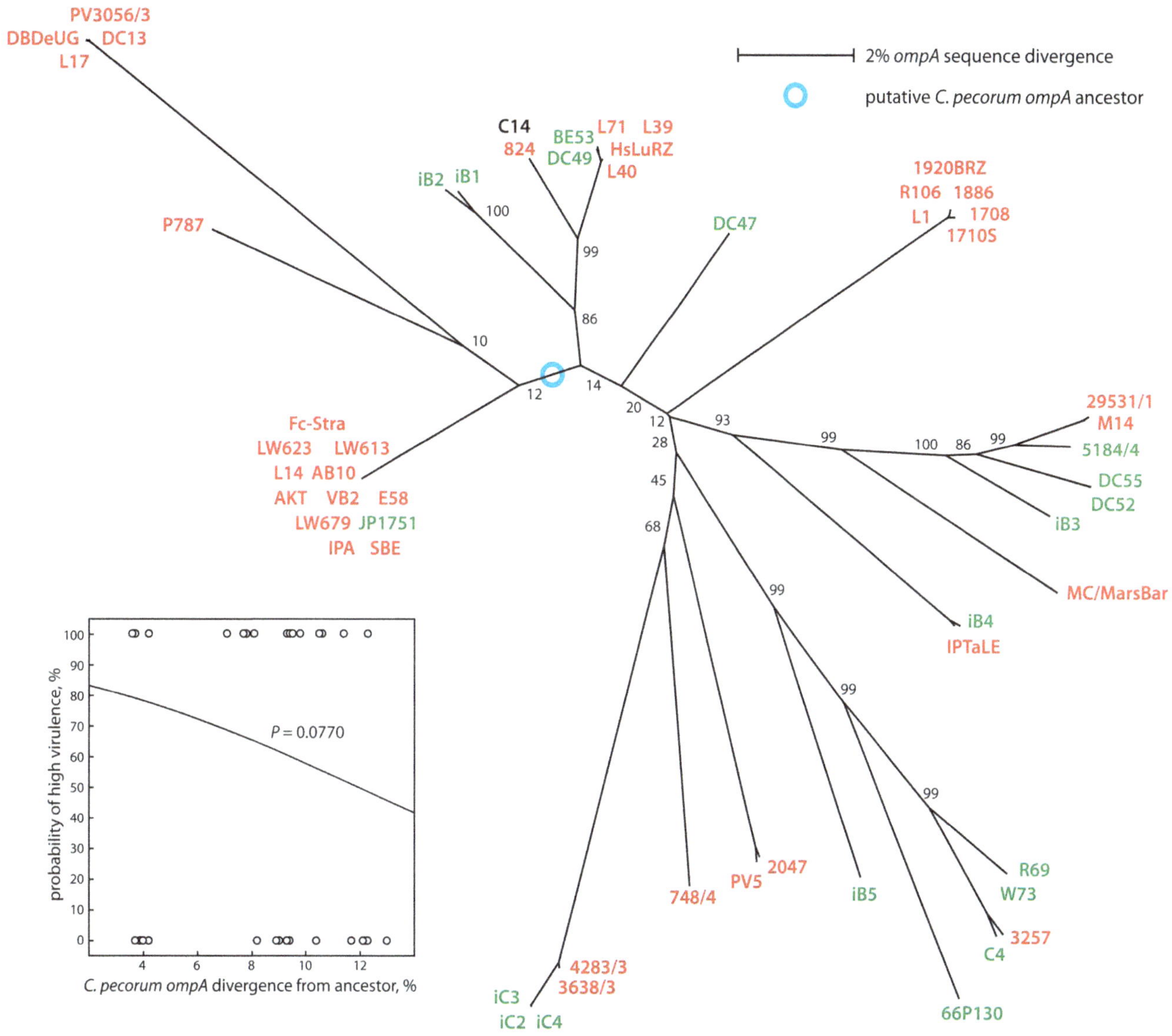

Figure 5. Evolutionary distance from the putative *ompA* ancestor in correlation to virulence of *C. pecorum* strains. A putative ancestral *ompA* was assumed at the connection of an outgroup, composed of one *ompA* sequence each of the 8 remaining chlamydial species, to the 57 *C. pecorum ompA* seqeunces (blue circle). This root is also consistent with an ancestor in the unrooted *ompA* neighbor-joining phylogram (Figure 2) at several poorly resolved and weakly bootstrap-supported branches that link the deep branches of the phylogenetic tree. Bootstrap support is indicated by numbers at branches, but not shown at terminal nodes of deep branches. Branch lengths are proportional to evolutionary *ompA* distance, with the bar indicating 2% sequence divergence (percent nucleotide substitutions). Low-virulence strains are indicated by green font, high-virulence strains by red font. **Inset**: The relationship between *ompA* evolutionary distance from the putative ancestor and the probability of high virulence of *C. pecorum* strains was determined by logistic regression analysis. Long evolutionary distance is correlated to low probability of high virulence, but fails to reach the $P<0.05$ significance threshold.

PCR conditions and sequencing

PCR was performed according to the GoTaq Flexi DNA Polymerase (Promega, Charbonnieres, France) protocol in a final volume of 50 μL, and consisted of DNA denaturation at 94°C for 5 min, followed by 30 cycles of amplification in a UNO II thermoblock (Biometra, Göttingen, Germany). Each cycle consisted of a denaturation step at 94°C for 30 sec, an annealing step at 55°C (for *ompA* and ORF663) or at 63°C (for *incA*) for 45 sec, an extension step at 72°C for 1 min, followed by a final chain elongation at 72°C for 7 min. The primer pairs used in this study except for forward *incA* primer b15-F (5′-CAAGAACAGTTGCGTCCTG-3′) have been described before [25]. The PCR products were sequenced by automated sequencing (Genome Express, Meylan, France). The complete DNA sequences of *ompA* genes and partial sequences of *incA* and ORF663 genes were deposited in GenBank under accession numbers listed in Table 1.

Sequence alignment and analysis

The number of repetitions of 15-mer CTR in ORF663 was identified using Tandem repeat finder software [39]. Deduced amino acid sequences were first aligned in the freeware MEGA6 [40] by use of the MUSCLE algorithm that considered for the nucleotide alignment all codon positions according to the Blosum 62 AA substitution matrix. Two obvious sequencing errors in the IPTaLE *incA* between positions 354–367 were manually corrected. Evolutionary distances were computed in MEGA6 in a

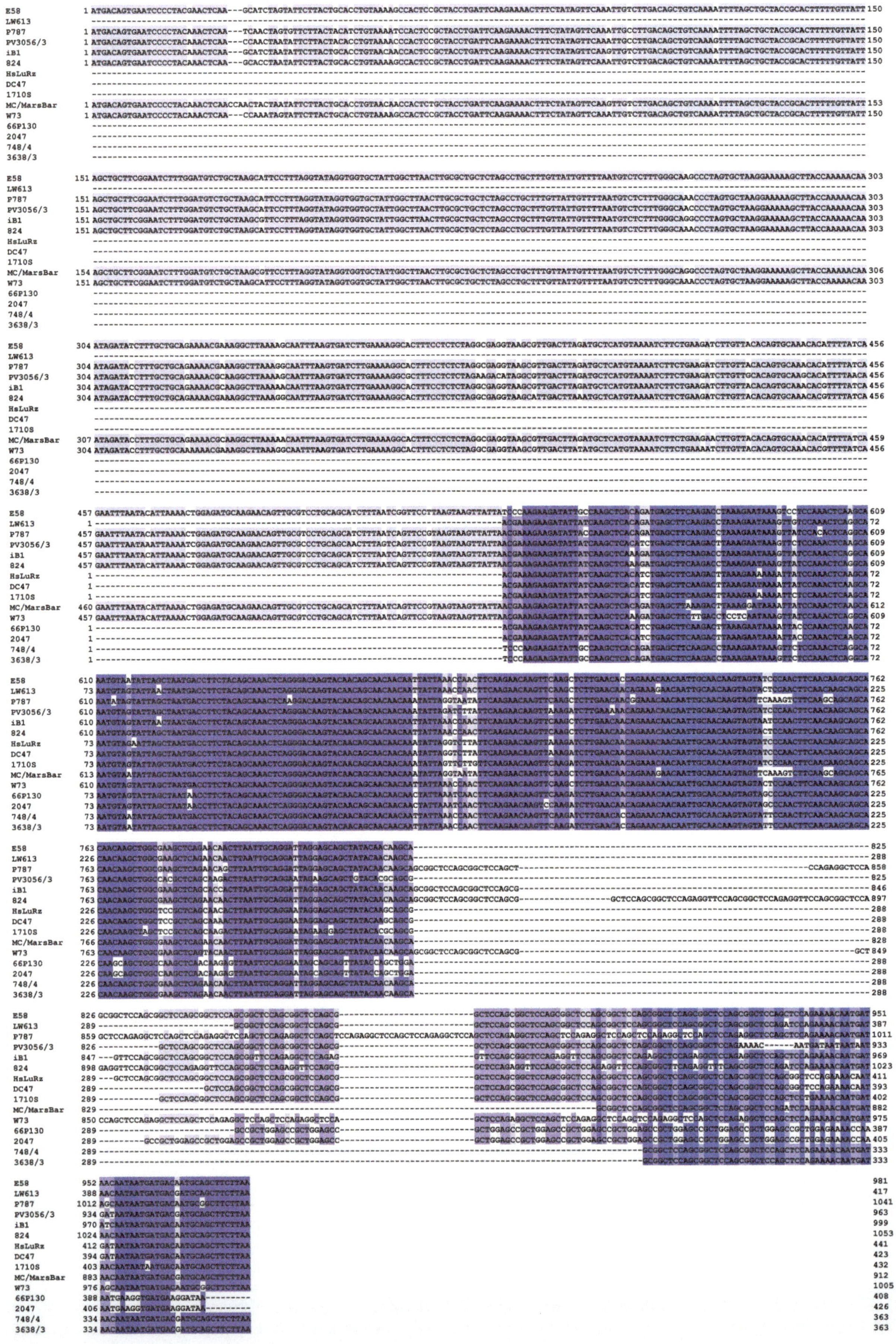

Figure 6. ***C. pecorum incA*** **alignment.** The strain subset used in Figure 1 is shown. The complete *incA* is shown for all strains for which the sequence is available, demonstrating highly conserved 5′ (position 1-825 of strain E58) and 3′ ends of the gene (position 906-end of strain E58), interrupted by a highly variable region of coding tandem repeats. The alignment of the PCR fragment sequences available for all 57 strains, corresponding to positions 537 through the 3′ end of strain E58, was optimized for minimal sequence divergence and used for construction of the *C. pecorum incA* phylogenetic tree in Figure 7. The alignment of the resultant amino acid sequences of all 57 partial *C. pecorum* IncA proteins deduced from the nucleotide sequences is shown in Figure S2.

maximum composite likelihood model as the number of base substitutions per site. Alignments were optimized by varying alignment parameters, in particular gap opening and extension penalties. A gap opening penalty of −5 and extension penalty of −1 resulted in minimum average pairwise sequence distances for all 3 genes, and was used to construct sequence alignments. Publication quality alignments were produced in freeware toolkit Jalview [41]. The evolutionary history was inferred by phylogenetic reconstruction by the neighbor-joining method in the freeware MEGA6 [40], with gaps removed from calculation by pairwise deletion.

Statistical analysis

Virulence association of *ompA* phylogram clades of and novel *incA* repeat sequences was analyzed by two-tailed Fisher Exact test. Correlation of *C. pecorum* strain virulence with *ompA* rank,

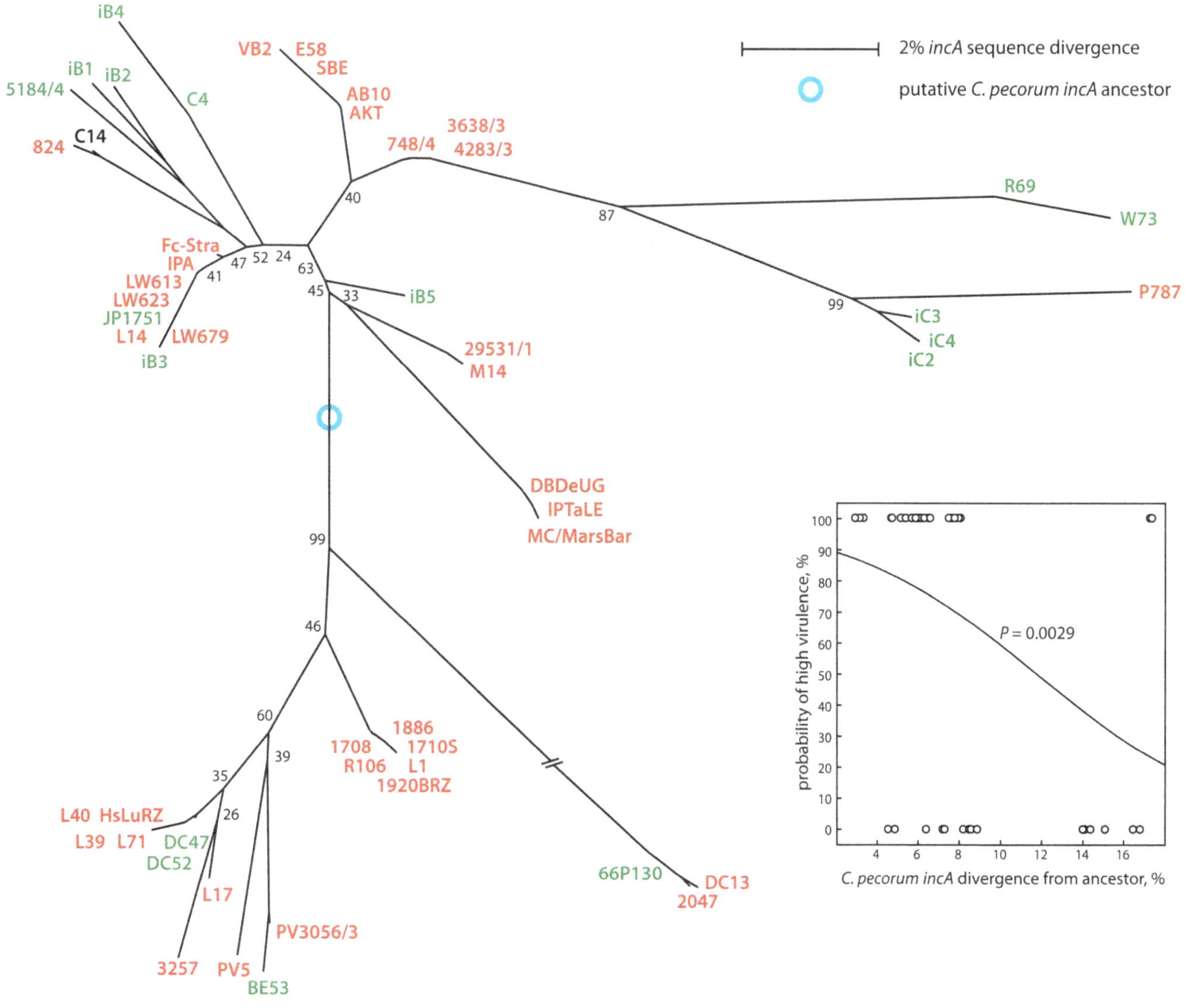

Figure 7. Evolutionary distance from the putative ***incA*** **ancestor correlates to virulence of** ***C. pecorum*** **strains.** A neighbor-joining phylogram (not shown) was constructed of the conserved 5′ portion of *incA* of all available *C. pecorum* sequences, and an outgroup composed of one *incA* sequence of each of the 8 remaining chlamydial species. In this unrooted phylogram based on the sequence alignment of the 3′ *incA* fragment available for all 57 *C. pecorum* strains in this study, the putative ancestral *incA* was assumed at the connection of this outgroup (blue circle). Indicators of bootstrap support, branch lengths, and strain virulence correspond to Figure 5. **Inset**: The relationship between *incA* evolutionary distance from the putative ancestor and the probability of high virulence of *C. pecorum* strains was determined by logistic regression analysis. Long evolutionary distance is highly significantly correlated to low probability of high virulence.

Figure 8. *C. pecorum* ORF663 alignment. The strain subset used in Figure 1 is shown. The complete ORF663 is shown for all strains for which the sequence is available, demonstrating a highly conserved 5′ portion (positions 1–493 of strain E58), followed by a highly variable region of coding tandem repeats containing a short conserved CTR fragment at position 748-786 of strain E58. The alignment of the PCR fragment sequences available for all 57 strains, corresponding to positions 424–786 of strain E58, was optimized for minimal sequence divergence and used for construction of the *C. pecorum* ORF663 phylogenetic tree in Figure 9. The alignment of the resultant amino acid sequences of all 57 full and partial *C. pecorum* IncA proteins deduced from the nucleotide sequences is shown in Figure S3.

incA CTR codon numbers, CTRs in ORF663, and the *ompA*, *incA*, and ORF663 nucleotide sequence divergence from the respective ancestral *C. pecorum* gene was determined by logistic regression analysis. Differences in mean pairwise sequence divergence were analysed by Student's t-test. All statistical analyses were performed by use of the Statistica 7.1 software package (Statsoft, Tulsa, Oklahoma, USA).

Supporting Information

Figure S1 *C. pecorum* OmpA protein alignment. Full-length peptide sequences of all 57 analyzed *C. pecorum* strains are shown. The background colors follow the Zappo color scheme for visualization of multi-peptide alignments [41], and correspond to alignment quality determined by amino acid identities and physicochemical similarities according to the Blosum 62 matrix (Pink = aliphatic/hydrophobic aa I, L, V, A, M; orange = aromatic aa F, W, Y; blue = positive aa K, R, H; red = negative aa D, E; green = hydrophilic aa S, T, N, Q; purple = conformationally special aa P, G; yellow = C). Four variable domains, distinguished by the gap insertions in the alignment, are interspersed between 5 highly conserved domains of the OmpA protein.

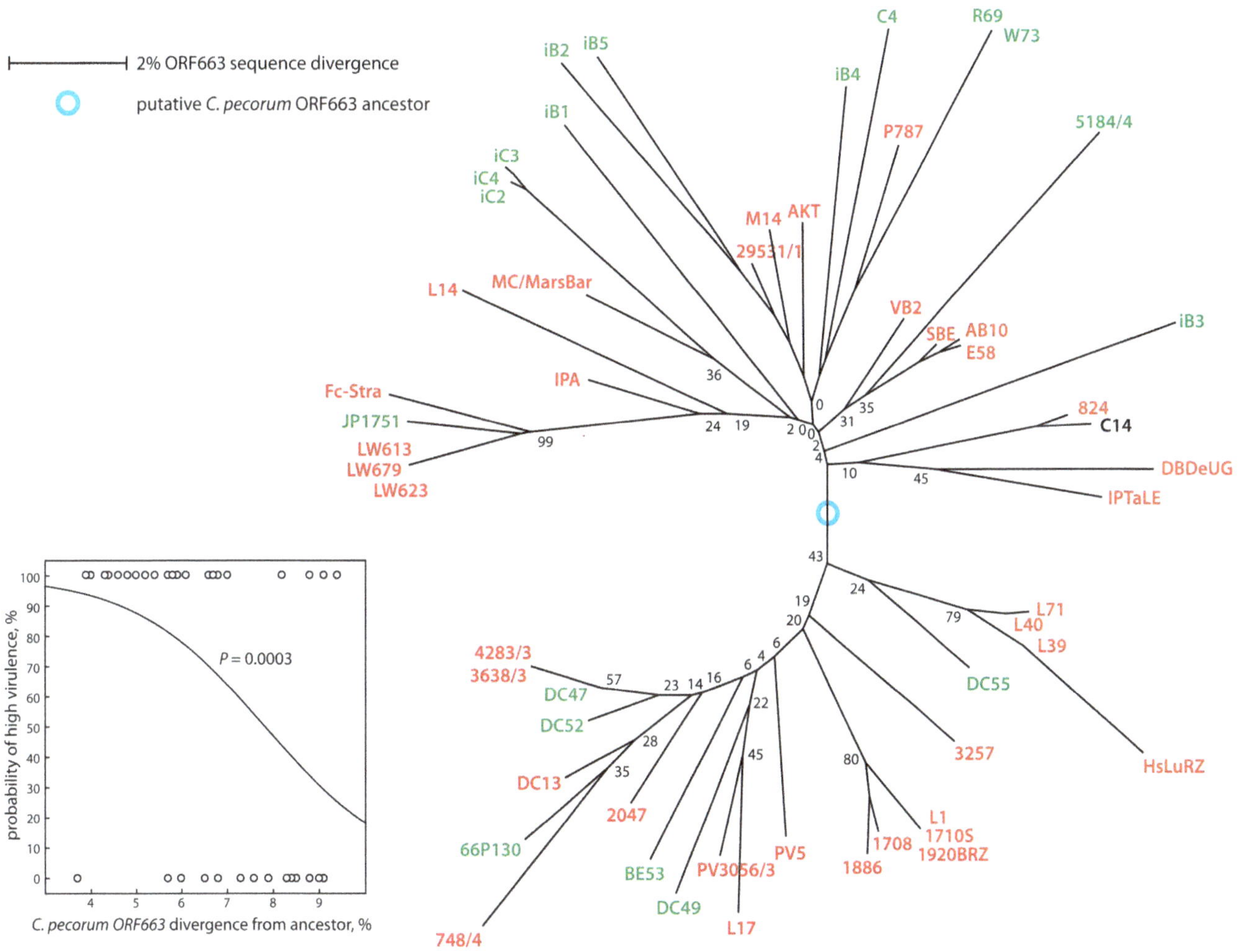

Figure 9. Evolutionary distance from the putative ORF663 ancestor correlates to virulence of *C. pecorum* strains. A neighbor-joining phylogram (not shown) was constructed of the conserved 5′ portion of ORF663 of all available *C. pecorum* sequences, and an outgroup composed of the ORF663 homologs found only in *C. abortus*, *C. psittaci*, *C. caviae*, and *C. pneumoniae*. In this unrooted phylogram based on the sequence alignment of the 3′ ORF663 fragment available for all 57 *C. pecorum* strains in this study, the putative ancestral ORF663 was assumed at the connection of this outgroup (blue circle). Indicators of bootstrap support, branch lengths, and strain virulence correspond to Figure 5. **Inset**: The relationship between ORF663 evolutionary distance from the putative ancestor and the probability of high virulence of *C. pecorum* strains was determined by logistic regression analysis. Long evolutionary distance is highly significantly correlated to low probability of high virulence.

Table 2. Mean pairwise sequence divergence between all 57 *C. pecorum* strains analyzed in this study, and between strains isolated from only healthy or from only diseased hosts.

% Pairwise Divergence	Healthy*	Diseased	All
ompA	13.23^{Ab} (12.57–13.88)	11.49^{C} (11.10–11.87)	12.49 (12.26–12.72)
incA	12.39^{Ab} (11.24–13.55)	9.65^{C} (9.16–10.14)	10.85 (10.51–11.20)
ORF663	12.93^{AB} (12.38–13.48)	10.09^{C} (9.81–10.36)	11.13 (10.94–11.31)

*Mean, 95% confidence interval.
^{A}Mean of Diseased highly significantly different at $P \leq 10^{-5}$.
^{b}Mean of All significantly different at $P<0.05$.
^{B}Mean of All significantly different at $P=10^{-9}$.
^{C}Mean of All significantly different at $P \leq 10^{-4}$.

Figure S2 *C. pecorum* IncA protein alignment. IncA peptide sequences of all 57 analyzed *C. pecorum* strains are shown. The complete IncA protein was used for alignment, and all available full-length sequences are shown in addition to the sequences encoded by the PCR fragment available for all strains. Amino acids 180 through C-terminal amino acid 326 of strain E58 correspond to the PCR fragment sequence used for phylogenetic reconstruction. Background Zappo colors correspond to alignment quality according to the Blosum 62 matrix. A highly conserved N-terminal region of approximately 275 amino acids is followed by a hypervariable region of inserted coding tandem repeats followed by a short conserved C-terminus of the IncA protein.

Figure S3 *C. pecorum* ORF663 protein alignment. ORF663 peptide sequences of all 57 analyzed *C. pecorum* strains are shown. The complete ORF5663 protein was used for alignment, and all available full-length sequences are shown in addition to the sequences encoded by the PCR fragment available for all strains. Amino acids 142-262 of strain E58 correspond to the PCR fragment sequence used for phylogenetic reconstruction. Background Zappo colors correspond to alignment quality according to the Blosum 62 matrix. A highly conserved N-terminal region of 154 or 165 amino acids is followed by a hypervariable region of inserted coding tandem repeats followed by short or long variants of a conserved C-terminus of the ORF663 protein.

Author Contributions

Conceived and designed the experiments: KM AR. Performed the experiments: KM. Analyzed the data: BK KSR KM AR. Contributed reagents/materials/analysis tools: KM AR BK KSR SM KS. Wrote the paper: BK KSR KM AR.

References

1. Kuo CC, Stephens RS, Bavoil PM, Kaltenboeck B (2011) Genus I. Chlamydia Jones, Rake and Stearns 1945, 55AL. In: Krieg NR, Staley JT, Brown DR, Hedlund BP, Paster BJ, et al. editors. Bergey's manual of systematic bacteriology. New York. Bergey's Manual Trust. pp. 846–865.
2. Kaltenboeck B., Kousoulas KG, Storz J (1993) Structures of and allelic diversity and relationships among the major outer membrane protein (*ompA*) genes of the four chlamydial species. J Bacteriol 175: 487–502.
3. Kaltenboeck B, Storz J (1992) Biological properties and genetic analysis of the *ompA* locus in chlamydiae isolated from swine. Am J Vet Res 53: 1482–1487.
4. Greco G, Corrente M, Buonavoglia D, Campanile G, Di Palo R, et al. (2008) Epizootic abortion related to infections by *Chlamydophila abortus* and *Chlamydophila pecorum* in water buffalo (*Bubalus bubalis*). Theriogenology 69: 1061–1069.
5. Jackson M, White N, Giffard P, Timms P (1999) Epizootiology of *Chlamydia* infections in two free-range koala populations. Vet Microbiol 65: 255–264.
6. Warren K, Swan R, Bodetti T, Friend T, Hill S, et al. (2005) Ocular chlamydiales infections of western barred bandicoots (*Perameles bougainville*) in Western Australia. J Zoo Wildl Med 36: 100–102.
7. Di Francesco A, Baldelli R, Donati M, Cotti C, Bassi P, et al. (2013) Evidence for *Chlamydiaceae* and *Parachlamydiaceae* in a wild boar (*Sus scrofa*) population in Italy. Vet Ital 49: 119–122.
8. Lenzko H, Moog U, Henning K, Lederbach R, Diller R, et al. (2011) High frequency of chlamydial co-infections in clinically healthy sheep flocks. BMC Vet Res 7: 29.
9. Godin AC, Björkman C, Englund S, Johansson KE, Niskanen R, et al. (2008) Investigation of *Chlamydophila* spp. in dairy cows with reproductive disorders. Acta Vet Scand 50: 39–44.
10. Reinhold P, Sachse K, Kaltenboeck B (2011) *Chlamydiaceae* in cattle: commensals, trigger organism, or pathogens? Vet J 189: 257–267.
11. Mohamad KY, Rodolakis A (2010) Recent advances in the understanding of *Chlamydophila pecorum* infections, sixteen years after it was named as the fourth species of the *Chlamydiaceae* family. Vet Res 41: 27.
12. Kauffold J, Melzer F, Henning K, Schulze K, Leiding C, et al. (2006) Prevalence of chlamydiae in boars and semen used for artificial insemination. Theriogenology 65: 1750–1758.
13. Wan C, Loader J, Hanger J, Beagley K, Timms P, et al. (2011) Using quantitative polymerase chain reaction to correlate *Chlamydia pecorum* infectious load with ocular, urinary and reproductive tract disease in the koala (*Phascolarctos cinereus*). Aust Vet J 89: 409–412.
14. Reinhold P, Jaeger J, Liebler-Tenorio E, Berndt A, Bachmann R, et al. (2008) Impact of latent infections with *Chlamydophila* species in young cattle. Vet J 175: 202–211.
15. Poudel A, Elsasser TH, Rahman KhS, Chowdhury EU, Kaltenboeck B (2012) Asymptomatic endemic *Chlamydia pecorum* infections reduce growth rates in calves by up to 48 percent. PLoS One 7: e44961.
16. Sachse K, Laroucau K, Vorimore F, Magnino S, Feige J, et al. (2009) DNA microarray-based genotyping of *Chlamydophila psittaci* strains from culture and clinical samples. Vet Microbiol 135: 22–30.
17. Pannekoek Y, Morelli G, Kusecek B, Morre SA, Ossewaarde JM, et al. (2008) Multi locus sequence typing of *Chlamydiales*: clonal groupings within the obligate intracellular bacteria *Chlamydia trachomatis*. BMC Microbiol 8: 42.
18. Laroucau K, Thierry S, Vorimore F, Blanco K, Kaleta E, et al. (2008) High resolution typing of *Chlamydophila psittaci* by multilocus VNTR analysis (MLVA). Infect Genet Evol 8: 171–181.
19. van Duynhoven YT, Ossewaarde JM, Derksen-Nawrocki RP, van der Meijden WI, van de Laar MJ (1998) *Chlamydia trachomatis* genotypes: correlation with clinical manifestations of infection and patient's characteristics. Clin Infect Dis 26: 314–322.
20. Gao X, Chen XS, Yin YP, Zhong MY, Shi MQ, et al. (2007) Distribution study of *Chlamydia trachomatis* serovars among high-risk women in China performed using PCR-restriction fragment length polymorphism genotyping. J Clin Microbiol 45: 1185–1189.
21. Vanrompay D, Ducatelle R, Haesebrouck F (1995) *Chlamydia psittaci* infections: a review with emphasis on avian chlamydiosis. Vet Microbiol 45: 93–119.
22. Denamur E, Sayada C, Souriau A, Orfila J, Rodolakis A, et al. (1991) Restriction pattern of the major outer-membrane protein gene provides evidence for a homogeneous invasive group among ruminant isolates of *Chlamydia psittaci*. J Gen Microbiol 137: 2525–2530.
23. Salinas J, Souriau A, De Sa C, Andersen AA, Rodolakis A (1996) Serotype 2-specific antigens from ruminant strains of *Chlamydia pecorum* detected by monoclonal antibodies. Comp Immunol Microbiol Infect Dis 19: 155–161.
24. Yousef Mohamad K, Rekiki A, Myers G, Bavoil PM, Rodolakis A (2008) Identification and characterisation of coding tandem repeat variants in *incA* gene of *Chlamydophila pecorum*. Vet Res 39: 56.

25. Mohamad KY, Roche SM, Myers G, Bavoil PM, Laroucau K, et al. (2008) Preliminary phylogenetic identification of virulent *Chlamydophila pecorum* strains. Infect Genet Evol 8: 764–771.
26. Kaltenboeck B, Heinen E, Schneider R, Wittenbrink MM, Schmeer N (2009) OmpA and antigenic diversity of bovine *Chlamydophila pecorum* strains. Vet Microbiol 135: 175–180.
27. Marsh J, Kollipara A, Timms P, Polkinghorne A (2011) Novel molecular markers of *Chlamydia pecorum* genetic diversity in the koala (*Phascolarctos cinereus*). BMC Microbiol 11: 77.
28. Jelocnik M, Frentiu FD, Timms P, Polkinghorne A (2013) Multilocus Sequence Analysis Provides Insights into Molecular Epidemiology of *Chlamydia pecorum* Infections in Australian Sheep, Cattle, and Koalas. J Clin Microbiol 51: 2625–2632.
29. Jee JB, DeGraves FJ, Kim TY, Kaltenboeck B (2004) High Prevalence of Natural *Chlamydophila* Species Infection in Calves. J Clin Microbiol. 42: 5664–5672.
30. Polkinghorne A (2014) Personal communication.
31. Mojica S, Huot Creasy H, Daugherty S, Read TD, Kim T, et al. (2011) Genome sequence of the obligate intracellular animal pathogen *Chlamydia pecorum* E58. J Bacteriol 193: 3690.
32. Sait M, Livingstone M, Clark EM, Wheelhouse N, Spalding L, et al. (2014) Genome sequencing and comparative analysis of three *Chlamydia pecorum* strains associated with different pathogenic outcomes. BMC Genomics 15: 23.
33. Mohamad KY, Rekiki A, Berri M, Rodolakis A (2010) Recombinant 35-kDa inclusion membrane protein IncA as a candidate antigen for serodiagnosis of *Chlamydophila pecorum*. Vet Microbiol 143: 424–428.
34. Coil DA, Vandersmissen L, Ginevra C, Jarraud S, Lammertyn E, et al. (2008) Intragenic tandem repeat variation between *Legionella pneumophila* strains. BMC Microbiol 8: 218.
35. Hood DW, Deadman ME, Jennings MP, Bisercic M, Fleischmann RD, et al. (1996) DNA repeats identify novel virulence genes in *Haemophilus influenzae*. Proc Natl Acad Sci U S A 93: 11121–11125.
36. Sachse K, Helbig JH, Lysnyansky I, Grajetzki C, Müller W, et al. (2000) Epitope mapping of immunogenic and adhesive structures in repetitive domains of *Mycoplasma bovis* variable surface lipoproteins. Infect Immun 68: 680–687.
37. Storz J, Collier JR, Eugster AK, Altera KP (1971) Intestinal bacterial changes in *Chlamydia*-induced primary enteritis of newborn calves. Ann NY Acad Sci 176: 162–175.
38. Rodolakis A (1976) Abortive infection of mice inoculated intraperitoneally with *Chlamydia ovis*. Ann Rech Vet 7: 195–205.
39. Benson G (1999) Tandem repeats finder: a program to analyze DNA sequences. Nucleic Acids Research 27: 573–580. Available: http://tandem.bu.edu.
40. Tamura K, Stecher G, Peterson D, Filipski A, Kumar S (2013) MEGA6: Molecular Evolutionary Genetics Analysis version 6.0. Mol Biol Evol 30: 2725–2729.
41. Waterhouse AM, Procter JB, Martin DMA, Clamp M, Barton GJ (2009) Jalview Version 2 - a multiple sequence alignment editor and analysis workbench. Bioinformatics 25: 1189–1191. doi:10.1093/bioinformatics/btp033.
42. Norton WL, Storz J (1967) Observations on sheep with polyarthritis produced by an agent of the psittacosis-lymphogranuloma venereum-trachoma group. Arthritis Rheum 10: 1–12.
43. Rodolakis A, Bernard F, Lantier F (1989) Mouse models for evaluation of virulence of *Chlamydia psittaci* isolated from ruminants. Res Vet Sci 46: 34–39.

The Dynamics of a Metapopulation: Changes in Life-History Traits in Resident Herring that Co-Occur with Oceanic Herring during Spawning

Arne Johannessen[1]*, Georg Skaret[2], Lise Langård[1,2], Aril Slotte[2], Åse Husebø[2], Anders Fernö[1,2]

1 Department of Biology, University of Bergen, Bergen, Norway, **2** Institute of Marine Research, Bergen, Norway

Abstract

Different populations of Atlantic herring are regarded as forming a metapopulation, but we know little about the dynamics of the connectivity and degree of interbreeding between the populations. Based on data from three periods between 1962 and 2011, we identified the presence of two components of herring in a small semi-enclosed coastal marine ecosystem based on different somatic growth patterns and mean vertebrae sum (VS). The two components were interpreted as belonging to a resident herring population and the migratory, oceanic Norwegian spring spawning (NSS) herring population, and they co-occurred during spawning. In the 1960s, resident herring characterized by slow growth and low VS co-occurred with rapid growth, high VS oceanic NSS herring. Similar slow-growing resident and rapid-growing NSS herring were found in the 1970–80s, but both populations now had low VS suggesting similar origins. Finally, in the 2000s both populations showed rapid growth. The changes coincided with the NSS herring going from a state of high abundance and oceanic distribution to a collapse in the late 1960s that resulted in a coastal distribution closer to resident herring populations, before full recovery and resumption of the migratory, oceanic pattern in the 1990s. During all three periods, NSS herring were only present in the local system up to an age of about five years, but the synchronous spawning of the populations supports mixed spawning and interbreeding. During the investigation period both longevity, length at age (growth) and length-at-first maturity increased markedly for the resident herring, which then became more similar to the NSS herring. Genetic and/or cultural factors are believed to be the main causes of the observed changes in life history traits, although some effect of changes in environmental factors cannot be excluded. Our study suggests that relationships among populations in a metapopulation can be highly dynamic.

Editor: Athanassios C. Tsikliras, Aristotle University of Thessaloniki, Greece

Funding: The work was funded by the University of Bergen and the Institute of Marine Research. The funders had no role in study design, data collection and analysis, decision to publish, or preparation of the manuscript

Competing Interests: The authors have declared that no competing interests exist.

* Email: Arne.Johannessen@bio.uib.no

Introduction

A metapopulation is defined as a set of populations with variable but moderate interbreeding [1,2]. Atlantic herring is composed of several populations characterized by different life-history traits [3–5], but the gene flow between the different populations is believed to be sufficient to define them as belonging to the same metapopulation [6]. Several studies have shown that different herring populations occur together during the feeding season [7–9], but spatial overlap during the spawning period is obviously a prerequisite for interbreeding in a metapopulation.

In addition to the geographically widespread, migratory, oceanic Norwegian spring spawning (NSS) herring, a number of spatially segregated local populations of Atlantic herring exist along the Norwegian coast [10–14]. Local populations are highly dependent upon local processes, but are also influenced by external replenishment [15]. The semi-enclosed coastal marine ecosystem in Lindåspollene in south-western Norway harbours a resident herring population (LP herring) [12]. Herring of a different origin have been observed in the system [12], and Johannessen et al. [16] even reported that the same pre-spawning school contained both LP herring and another component with a similar maturation pattern.

To identify different components of herring with similar external appearance is challenging. Molecular methods are typically employed to obtain a "snapshot" of the contemporary genetic composition, but cannot for instance be used on lost historic material. The otolith appearance was utilised to separate different components in an earlier study in Lindåspollene [16], but a closer examination has shown that this analysis does not allow for correct classification of young herring. The growth pattern and the vertebrae sum (VS) have traditionally been used to separate different populations of herring [3,5,17]. VS are primarily determined by environmental factors during embryogenesis [18] and should therefore make it possible to distinguish the geographical origin of different components by statistical comparisons of average VS. The local populations, including that in Lindåspollene, have generally lower VS than NSS herring [14,17,19]. They have also evolved different life-history traits than the oceanic NSS herring and are generally characterized by slower growth, a shorter life span and higher relative fecundity [11,12,14,17,19–21].

There are several related factors that could influence connectivity among populations; the population size could influence the migration patterns that in turn could determine to what extent different populations interact. Changes in distribution range and use of spawning locations have been reported for several herring stocks in the North Atlantic following fisheries-induced population collapses [22]. The spawning stock of NSS herring dropped from more than 10 million tonnes in the 1940–50s to near extinction in the early 1970s [23] and rose again to full recovery in the late 1980s [24]. During this period the migration pattern changed from oceanic to coastal and back to oceanic [25]. In the 1950s, NSS herring had feeding and wintering grounds in the Norwegian Sea, and spawned off the western Norwegian coast. After the stock collapse, the remaining herring stayed along the coast throughout the year [26,27]. After 1986, feeding migrations into the Norwegian Sea resumed [27], and spawning and overwintering areas gradually expanded northwards [26]. Hence, after the collapse, but prior to the recovery, the majority of NSS herring grew up along the coast and not in the traditional nursery areas in the Barents Sea [25]. Prior to the collapse, NSS larvae hatched from offshore banks and drifted into the fjords [25,26], whereas the demersal spawning occurred further inshore after the collapse [25–28]. This distributional shift brought NSS herring closer to the resident populations along the coast [25], possibly increasing the likelihood of interbreeding. In addition, this opened up for transmission of culturally mediated migration patterns [29–32].

The earlier study of the population structure in Lindåspollene [16] was primarily based on data from only one year, and historical samples are crucial to an understanding of metapopulation dynamics [33]. There exists a long time series of data (1962–2011) on growth pattern and VS of herring in Lindåspollene in a period extending from pre-collapse to post-recovery of NSS herring with the potential to separate LP herring and any NSS herring occurring in Lindåspollene in a period of extremely variable abundance and distribution of the NSS herring.

We investigated the overlap in time and space between LP herring and immigrants from the migratory NSS herring as well as possible changes in life history traits in LP herring. We evaluated the potential for interbreeding by examining the co-occurrence of LP and NSS herring and the degree of synchronization of the maturation stages. We present the most plausible explanations for our observations, although other interpretations cannot be excluded.

Materials and Methods

Ethics statement

Most of the fish caught were dead when collected from the gillnets, with the few fish still alive being killed by a blow to the head and bending the neck to confirm death. Permission to catch wild fish and killing them by this method was given by the Directorate of Fisheries (Fiskeridirektoratet, Reguleringsseksjonen) and the County Governor of Hordaland.

The Lindåspollene ecosystem

Lindåspollene is a small (~7 km^2) semi-enclosed ecosystem in south-western Norway comprising three 60–90 m deep basins (see [12]), with a main sill (7.5 m wide, 3.5 m deep) connecting to the outside fjord. The resident LP herring generally occur in two of the basins and play a key role in the system [12]. They are largely protected from harvesting [34].

The environment within the basins differs from that of the outside fjord (440 m deep with a sill at 20 m), in particular with regard to oxygen levels [12,35]. There is low zooplankton biomass in the basins and a dominance of small copepods compared to the outside fjord [36,37,38–40], which may be due to generally poor oxygen conditions and limited inflow due to the shallow sill [41]. There are large annual variations in temperature but no long-term trends over the course of the study period [12,35,42]. No data are available for an adequate comparison of other environmental conditions in the course of the study period including abundance of potential herring prey. In the outside fjord, unlike in other studied Norwegian fjords, mesopelagic fish are lacking, and crown jellyfish, *Periphylla periphylla*, occur in high concentrations [43,44]. As jellyfish are inefficient consumers of large zooplankton [44], this may have resulted in good feeding conditions for herring that migrated out of Lindåspollene. The development of the abundance of jellyfish and mesopelagic fish over our study period is not well documented. Fishermen observed jellyfish in varying amounts as early as in the late 1940s, and the first scientific study in the outside fjord reported high concentration of jellyfish in the early 1970s [45,43].

Biological sampling

Gillnet samples of herring were obtained from 1962 to 2011 in the three periods 1962–1964 (P1), 1970–1982 (P2) and 2005–2011 (P3) (see Table 1), where P1 coincides with the period prior to the eventual collapse of the NSS herring (year classes (YCs) represented: 1950–60s), P2 with the period during the collapse (YCs represented: 1960–1970s) and P3 with the period after the recovery (YCs represented: 1990s–2000s). Length and weight were recorded for nearly all fish and age for the majority of the fish. Meristic analyses (vertebrae sum, VS) were conducted throughout the period but with a low effort in the 1970s.

Herring were sampled with 25 m-long by 4 m-deep gillnets of mesh size 26 mm. The exceptions were 1979–1980 and 2005–2011 when mesh sizes ranged from 21 to 31 mm and 19–36 mm, respectively. Samples were taken throughout the year but mainly in February-April. The nets were set at the surface mainly within 1000 m from the overwintering and spawning areas of herring [42], based on local knowledge of fish distribution and/or acoustic recordings. The nets were usually set in series of three at 17:00-19:00 local time and hauled the next day (09:00-12:00). When catches were large, subsamples of up to 200 herring were analysed (see Table 1).

Herring were measured for total length (L, to the nearest 0.5 cm below, from the anterior head to the end of the caudal fin in natural position), total body weight (W, to the nearest gram), age, maturity stage and VS onboard the vessel, or were frozen immediately after capture and later measured in the laboratory. Age was determined by experienced age readers from scales until 1982 and thereafter mainly from otoliths. Eight maturity stages were distinguished based on macroscopic visual inspection of the gonads; 1–2, immature; 3–5 maturing or pre-spawning; 6, spawning/running; 7, spent; and 8, resting stage [46]. VS, the total number of vertebrae in the backbone including the urostyle, were recorded by experienced readers [46,47]. Fulton's condition factor K [48] was calculated on the basis of autumn samples (September-November) after the feeding period as: $K = 100 \cdot W/L^3$ for fish in the length range 27–31 cm, which was common for all periods.

Data analyses

We compared basic life history parameters (age, length, age and length at first maturity, and condition factor) of herring between the three sampling periods. In cases where the life history parameters were normally distributed (length, weight and condition factor), a regular ANOVA with a Tukey HSD post-hoc test

Table 1. Summary information of herring samples from Lindåspollene 1962–2011, with sampling month, total number of fish analysed (Total n), number of gillnet settings (n settings; referring to single gillnets from 2008 onwards and series of attached gillnets prior to 2008), number of fish sampled for age determination (Age) and VS.

Year	Month	Total n	n settings	Age	VS
1962	11	350	3	304	346
1963	3, 9–12	425	8	402	410
1964	1–7	809	14	747	478
1970	10–11	129	3	125	128
1971	1–11	978	13	925	566
1972	1,3–5, 8, 10–11	569	8	526	146
1973	1–3, 7, 9–10	266	8	258	50
1974	1–4	405	7	396	–
1975	3, 4	150	2	147	–
1977	3, 4, 9	751	9	652	–
1978	3, 4	340	7	331	–
1979	4, 9, 12	309	4	294	–
1980	3-5, 9–11	554	8	464	231
1981	3, 9	356	4	351	251
1982	1, 3	331	3	324	324
2005	2, 9	292	3	284	180
2006	2, 3, 10	351	4	325	89
2007	1–4	336	14	329	329
2008	1–4, 10–11	749	60	729	10
2009	1–4, 9	1035	75	875	92
2010	2–4, 8–10	1564	72	673	668
2011	4, 9	910	21	870	846

was employed to compare periods. In the case of age, the non-parametric Kruskal-Wallis test was used. The herring sampled during the three periods contained YC from five different decades (1950s, 1960s, 1970s, 1990s and 2000s) for which growth measured as length-at-age and VS could be compared independently of annual variability in sampling effort. In order to investigate length-at-age patterns at the start and end of the study period, the von Bertalanffy growth function (VBGF) [49] was fitted to length-at-age data from the YCs of the 1950 and 2000 decades using the FSA package in R. All statistics were performed using R version 3.0.0 (R Development Core Team 2013, http://www.rproject.org).

Results

Year class strength and longevity

In the two first sampling periods P1 and P2 (1960s and 1970–80s) the samples consisted of herring aged only up to about 10 years (Fig. 1). There were then some predominant YCs, for instance the 1969 YC that was found in the samples from age 2 (1971) to age 12 (1981). In the third sampling period P3 (from 2005) no single, predominant YC could be clearly followed over time, and from 2007 the age range became wider, and older herring (up to 20 years) were more frequent.

Variations in life history traits over time

There were significant increases in age, length, weight and condition factor from P1 to P3 ($p<0.001$, Table 2). Notably, mean age and weight approximately doubled from the 1960s to the 2000s.

Age at first maturity did not change over the 50 years (Fig. 2A); in all periods the herring started to mature at age 2 and close to 100% had matured by age 3. In accordance with the increasing size over time, the length at first maturity increased from a situation where close to 100% were maturing at the length of 22–25 cm in the 1960s and 1970s to a situation where less than 20% were maturing at this length in the 2000s (Fig. 2B).

Figure 3 provides the length-at-age as a proxy for growth for the YCs of the 1950s, 1960s, 1970s, 1990s and 2000s. There was an increased length-at-age with decadal period (ANOVA of age groups 6–9, age groups represented in data from all decades, $p<0.001$, Fig. 4), with the main increase taking place between the 1970s and the 1990s. Different mesh sizes were used in the 1960s and the 2000s, but net selectivity did not explain the observed rise in length in the 2000s. When the lengths of 2–12 year-old herring (age groups present in the catches from both periods) only caught with 26 mm mesh size were compared between the two sampling periods, there was still a marked difference in length (1960s: mean 25.9 cm$\pm$2.2, 2000s: 30.9 cm$\pm$2.6 ($p<0.001$).

Occurrence of different herring components

A visual inspection of the length-at-age plots in Figure 3 strongly suggest that two growth modes were present in young herring (<6 years) in the decades 1950–1970. In contrast, at ages ≥6 years the fast growing mode was hardly observed. In order to compare length-at-age at the start and end of the study period, two growth

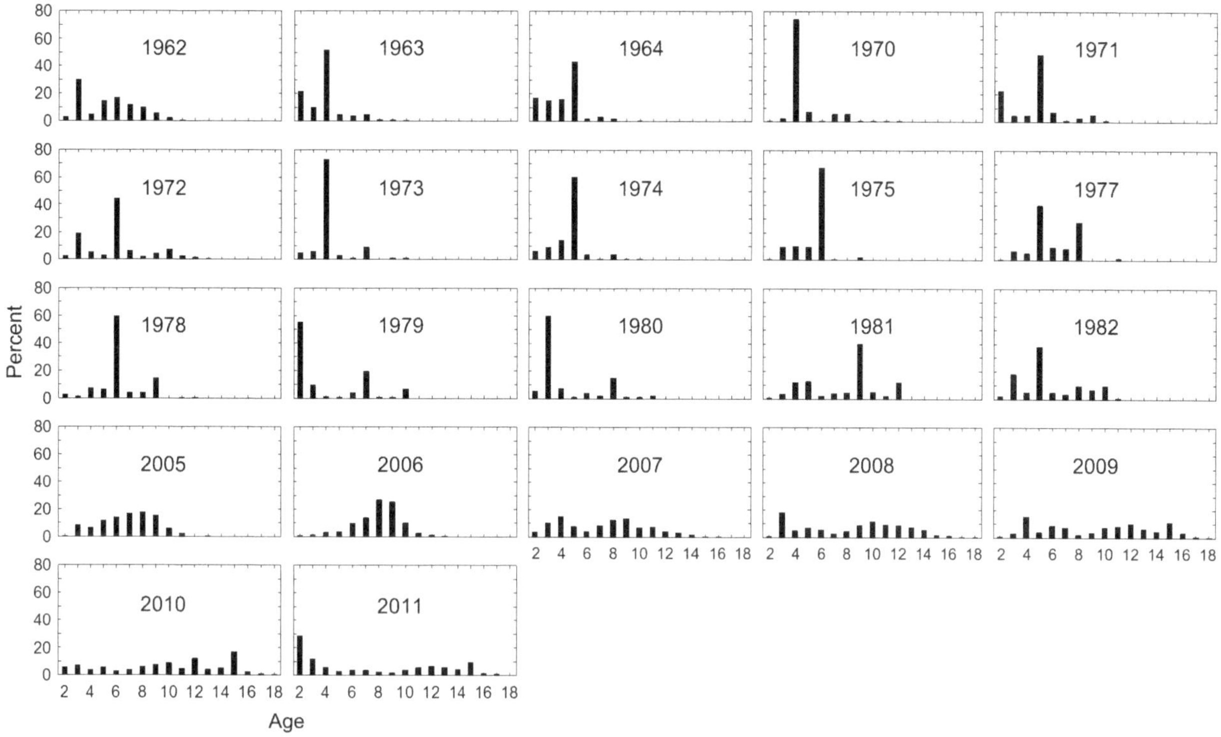

Figure 1. Annual age distribution. Percentages (%) by number in catches of all herring sampled 1962–2011.

curves were fitted with the VBGF (see Materials and Methods). The first was based on mean length-at-age for each age group in the 1950s YCs (red superimposed curve in Fig. 3), but excluding the rapid-growing mode of the 2–5 year-olds. The second was based on mean length-at-age for each age group in the 2000s YCs (green curve). The two growth curves clearly differed. The green curve of the 2000s had a steeper increase (k = 0.49 versus 0.22), indicating a faster growth rate of young fish, and reached the asymptote at a higher level ($L_{\infty 2000YCs}$ = 329 versus $L_{\infty 1950YCs}$ = 310) than the red curve of the 1950s, indicating increased maximum length. The red slow-growing curve fitted to the 1950s YCs corresponds to a slow growth pattern expected of resident herring, while the green curve fitted to the 2000s YCs is more similar to the NSS growth pattern. Notably, the young herring in the 1950s–1970s with rapid growth displayed growth patterns that fitted better with growth of the 2000s YCs (Fig. 3).

The connectivity between the two components changed with fish age. The proportion of fast growers decreased with age for the young herring from 0.59 for 2-year-olds to 0.12 for 5-year-olds (Table 3). Furthermore, the decrease was steeper during P1 than P2, indicating that the rapid-growing fish remained for longer in the system during P2.

In order to determine whether there was any difference in VS between rapid- and slow-growing fish indicating different areas of origin, VS was compared between fish with length-at-age closest to the VBGF fitted to the 1950s (slow-growers, red curve in Fig. 3) and the 2000s (fast-growers, green curve in Fig. 3). Comparisons were made for young (2–5 years old) herring for all decades for which data were available (not the 1990s). Figure 5 shows that fast-

Table 2. Age, length, weight and condition factor (CF) of herring (≥2 year old) in the different sampling periods.

Parameters	P1 (1960s)	P2 (1970–80s)	P3 (2000s)
Age	4.4±1.7 (2–13)	5.4±2.3 (2–15)	8.4±4.2 (2–20)
n (Age)	1453	4789	4137
Length (cm)	25.9±2.2 (19.5–33.5)	27.2±2.7 (17.5–35.0)	31.3±2.4 (22.0–44.7)
n (Length)	1584	5059	5152
Weight (g)	134±37 (55–275)	166±53 (35–415)	256±62 (81–602)
n (Weight)	1579	4797	5119
CF	0.78±0.08 (0.51–1.22)	0.86±0.11 (0.35–1.23)	0.91±0.13 (0.28–1.22)
n (CF)	264	595	146

Values are given as mean ± SD (min-max). n indicates the number of fish sampled in the different parameters. For CF, values refer to herring in the common length range 27–31 cm for September - November.

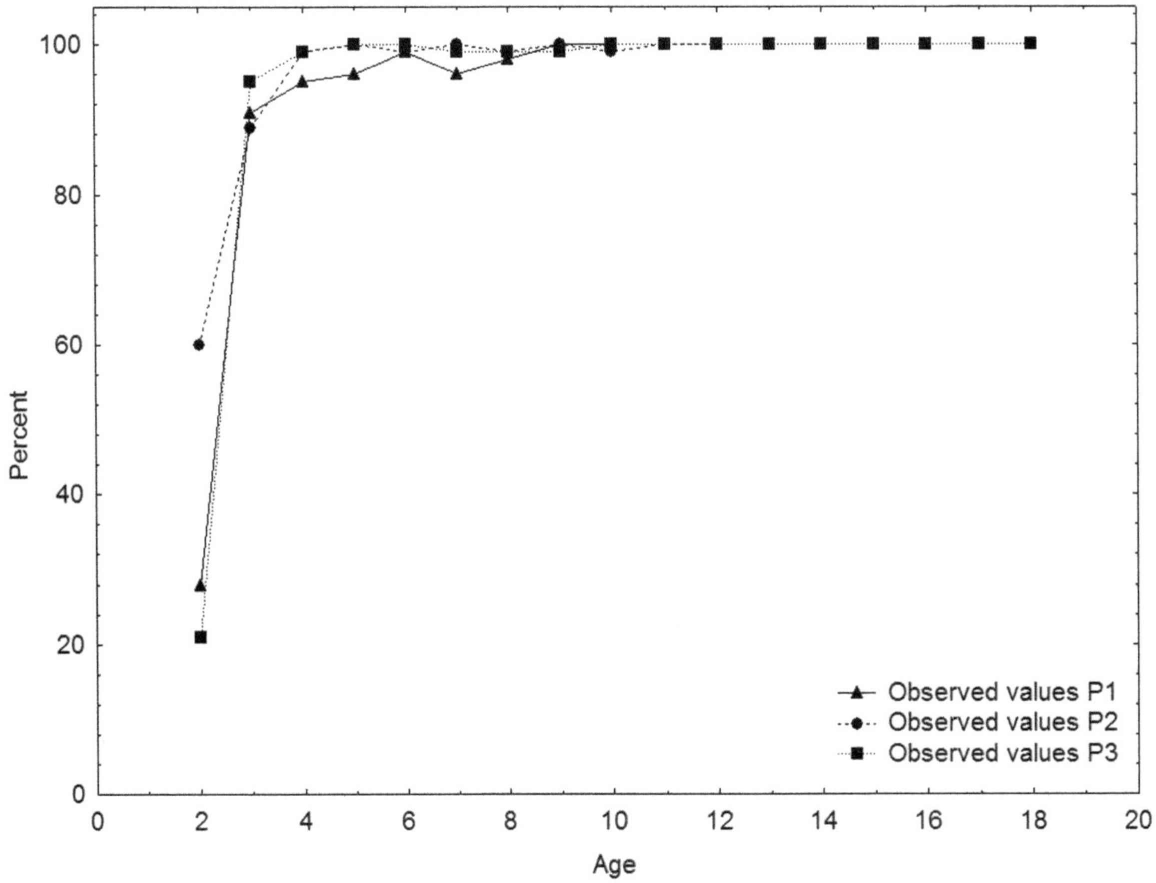
Percent
Age
Observed values P1
Observed values P2
Observed values P3

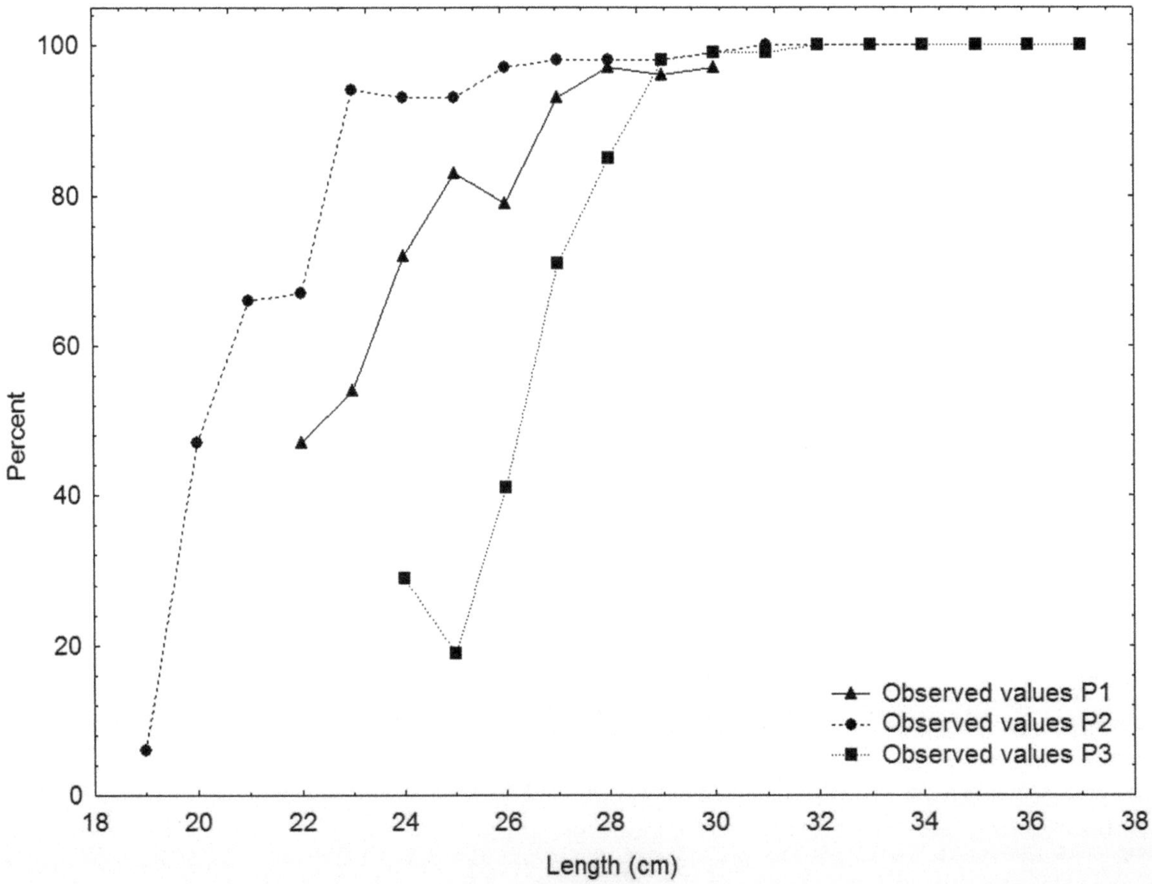
Percent
Length (cm)
Observed values P1
Observed values P2
Observed values P3

Figure 2. Maturity oogives. Percentages (%) of herring in stages 3–8 at (A) age and (B) length for periods P1, P2 and P3.

growing young herring in the 1950 and 1960 YCs had higher VS (t-test, p<0.001) than slow-growing fish of the same YCs. This indicates that these two groups of young herring originated from different areas and thus represent different populations, presumably NSS and LP herring. On the other hand, for young herring of the 1970 YCs, the VS of the rapid and slow growing components overlapped (t-test, p = 0.93), indicating that they originated from the same area. The 2000 YCs showed the same tendency as the 1950 and 1960 YCs, with higher VS for fast-growers than slow-growers (p = 0.08), but the presence of two growth modes was not clear among the young herring for the 2000 YCs so this method of comparing different growth modes is less adequate. For the older fish (6–11 years old) VS was in general similar to VS of the slow-growing young herring (about 56.6, Fig. 5) indicating a dominance of resident herring of local origin.

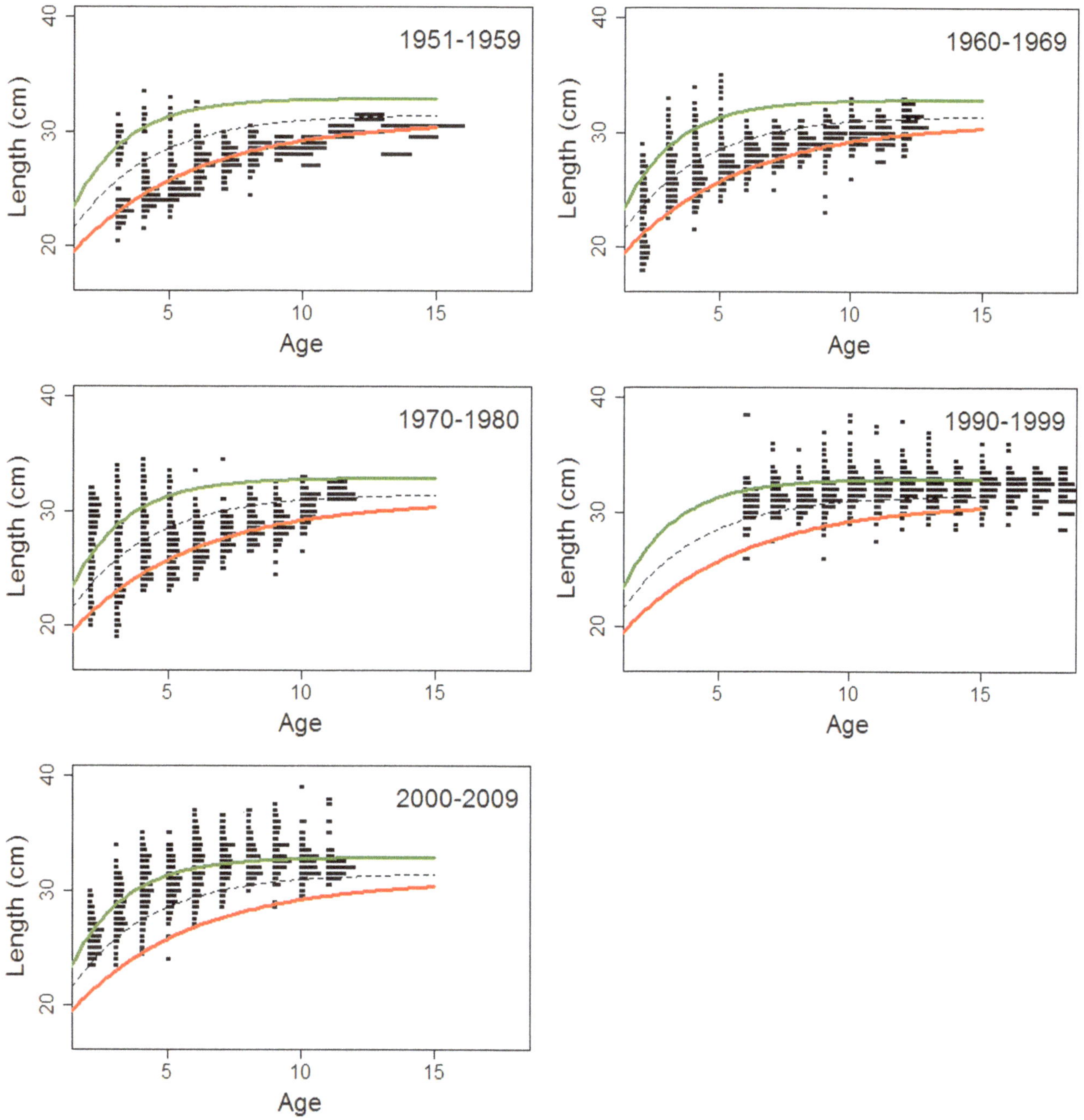

Figure 3. Length-at-age. Histograms averaged for herring YCs of the 1950s, 1960s, 1970s, 1990s and 2000s. A von Bertalanffy growth curve is fitted for the 1950s (red) and for the 2000s (green), and the dotted curve marks the midpoint between the two fitted curves.

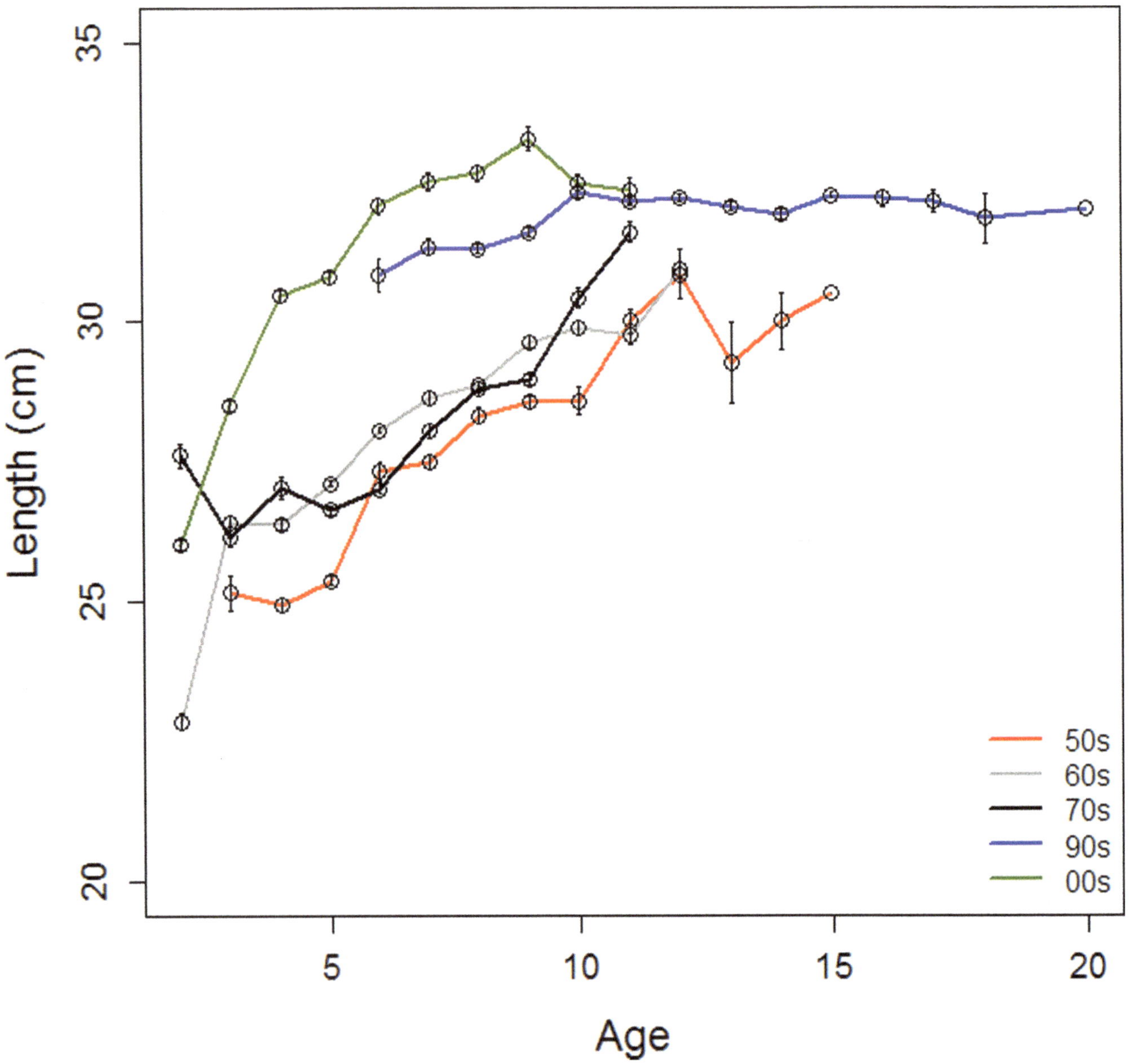

Figure 4. Average length-at-age. Mean ± SE of YCs of the decades 1950s to 2000s.

Table 3. Proportion of rapid growers for herring in age groups 2–5, in the sampling periods P1 (1960s) and P2 (1970–1980s).

Age	Prop fast total	Prop fast P1	N	Prop fast P2	N
2	0.59	0.74	223	0.52	476
3	0.52	0.56	245	0.50	634
4	0.26	0.19	342	0.30	582
5	0.12	0.04	391	0.14	1199

The two growth modes were not easily distinguishable during the sampling period P3.

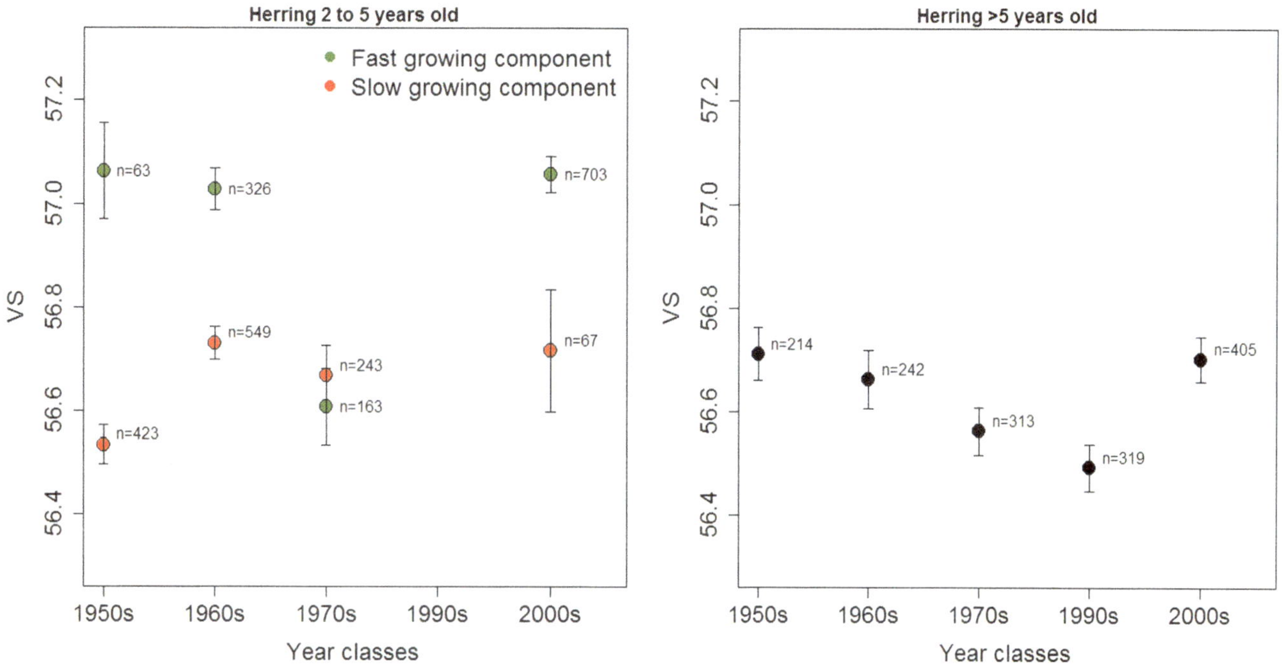

Figure 5. VS of young (2–5 years old) and older (6–11 years old) herring. Mean VS ± SE of YCs of the decades 1950s to 2000s. Individuals with length-at-age closest to the 1950s slow-growing curve shown in red and the ones closest to the 2000s fast-growing curve shown in green. Sample size is given for the different components.

Overlap in spawning time between different components of herring

A prerequisite for the existence of a metapopulation is that there is an overlap in timing of spawning between populations. To investigate the overlap between the two components identified, all fish in maturity stage 6 (spawning) were extracted from the data from P1 and P2 (Fig. 6, P3 excluded due to less certain discrimination of the components). Most samples contained herring from both components. Since the proportion of the rapid-growing component was usually lower, and most samples therefore only contained a limited number of this component, an additional comparison was made, including only stations at which at least 10 fish from both components were present in stage 6. In the four cases that fulfilled these criteria, the overlap was close to complete. The findings strongly suggest that spawning among the components was synchronized.

A small part of the data set could be used to resolve the temporal development of maturation on a finer scale. Between November 1963 and April 1964, the herring were sampled throughout the maturation period. Figure 7A shows that YCs 1959, 1960 and 1961 (3–5 years old spawners where two components should be present) had similar maturation patterns, with a large majority of the fish from all YCs in stage 6 ('spawning') or 7 ('spent') in April. There was no falling trend in length during the maturation period (Fig. 7B), as would have been expected if the fast-growing NSS herring left before spawning. The finding that the 1959 YC, despite being the oldest, generally had a smaller length and lower VS than the other YCs (Figure 7B, C) supports the idea that fast-growing NSS herring disappear from the area at a certain age.

Discussion

This is the first documentation of a long-term dynamic relationship between populations in a pelagic fish species during reproduction. We have found that a component of Atlantic herring inferred to belong to the resident LP herring population co-occurred in its coastal fjord home range with a herring component assumed to belong to the much larger population of Norwegian spring spawning (NSS) herring. The components remained together during spawning, with synchronised maturation states. Over a period of 50 years the resident herring underwent substantial changes in several life-history traits, including length-at-age, length at first maturity and longevity. The observed connectivity and mixed spawning between the components may be the drivers of the changes in life-history traits in the initially slow growing and relatively short-lived resident herring.

Herring components in Lindåspollene

VS are primarily determined by the environmental conditions during the embryonic stages [18,50,51] and thus have the potential to distinguish components of the herring that display different distribution and migration patterns. A central assumption in the interpretation of the data was that herring of coastal origin have lower VS than oceanic herring. A low VS and small size are characteristics of resident herring populations on the Norwegian coast [14,19], and the component with slow growth and low VS in Lindåspollene in the 1960s and the 1970–1980s was thus classified as belonging to the resident Lindås (LP) herring population. Oceanic NSS herring are characterized by rapid growth and high VS ([10,11,17,21], and young fish with these characteristics in the 1960s (P1) and 2000s (P3) were thus classified as belonging to the oceanic NSS population.

Classifying the herring components with different characteristics than the ones described above, is not a straightforward process. In

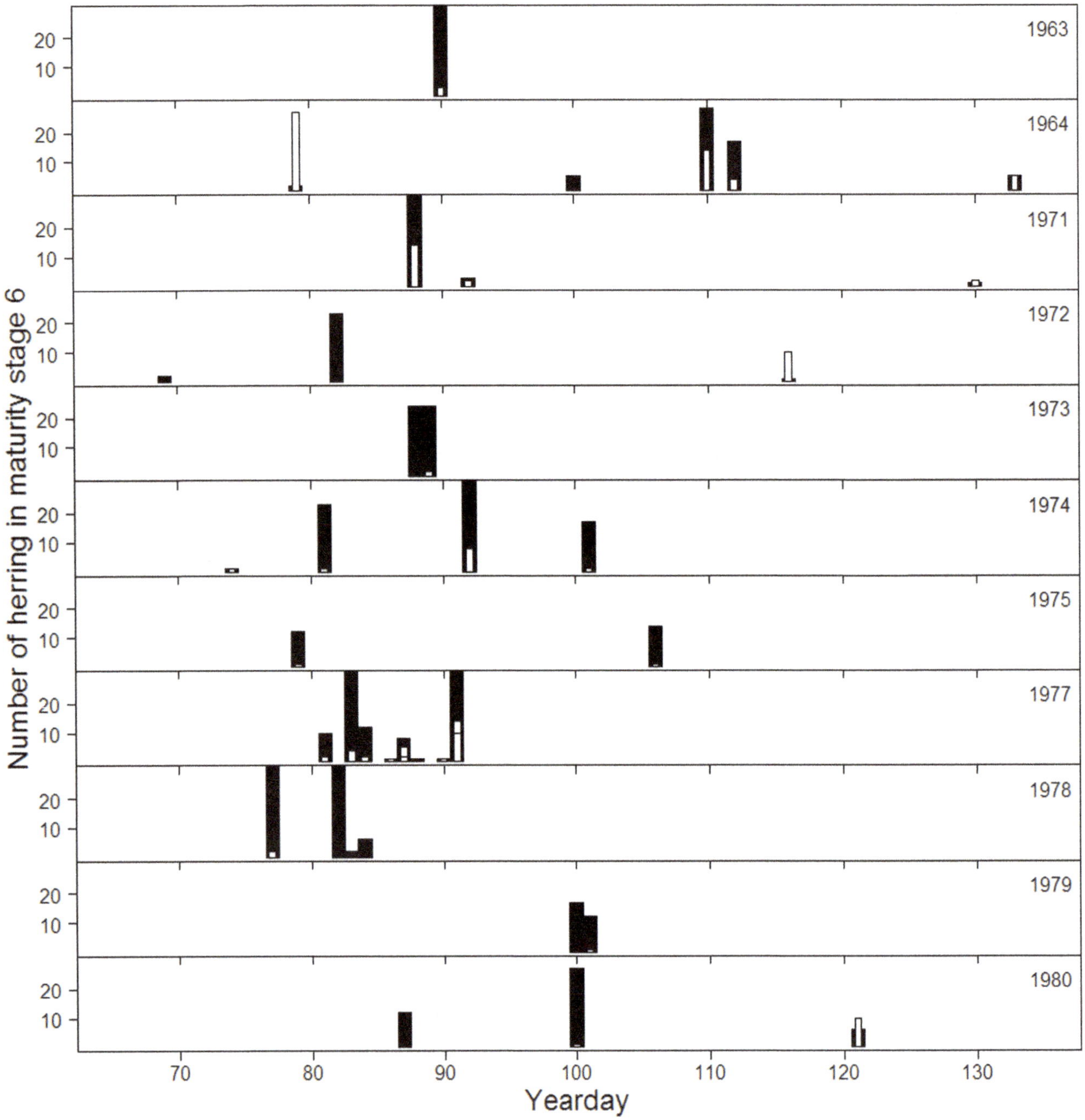

Figure 6. Timing of spawning in fast and slow growers. Number of fast (white) and slow growers (black) in stage 6 (spawning stage) as a function of yearday, all stations in P1 and P2 are included. For the sake of clarity, there is a cut-off on the y-axis at 30 individuals.

the 1970–1980s (P2) the rapid-growth component also had low VS. The population of NSS herring had by then nearly collapsed, and the remaining NSS herring stayed close to the coast throughout the year [23,52,53]. Spawning occurred more inshore than on the traditional spawning grounds on the large banks, with the vast majority of the progeny hatching in coastal areas [25]. The low VS suggests that NSS herring then utilised similar spawning areas as the resident populations [54]. We therefore assume that this component is NSS herring of coastal origin. In the 2000s, when the population of NSS herring had recovered, VS again increased in rapidly growing young fish (Figure 5). The decrease and subsequent increase in VS in young fish are in synchrony with the decline and rise of the oceanic NSS herring. In older fish, VS were generally low, supporting the notion that relatively few old oceanic NSS herring were present. Interestingly, the slow-growing component of the young fish with low VS had now been strongly reduced and a fast-growing component with low VS dominated. This component is interpreted as resident fast-growing herring. The greatest increase in length-at-age took place from the 1970s to 1990s YCs. Most herring in Lindåspollene in the 2000s (P3) were thus presumably rapid-growing oceanic NSS herring and resident herring (Figures 3 and 5). Any remaining old NSS herring of coastal origin could not be distinguished from fast-growing resident herring, but genetic analyses have demonstrated

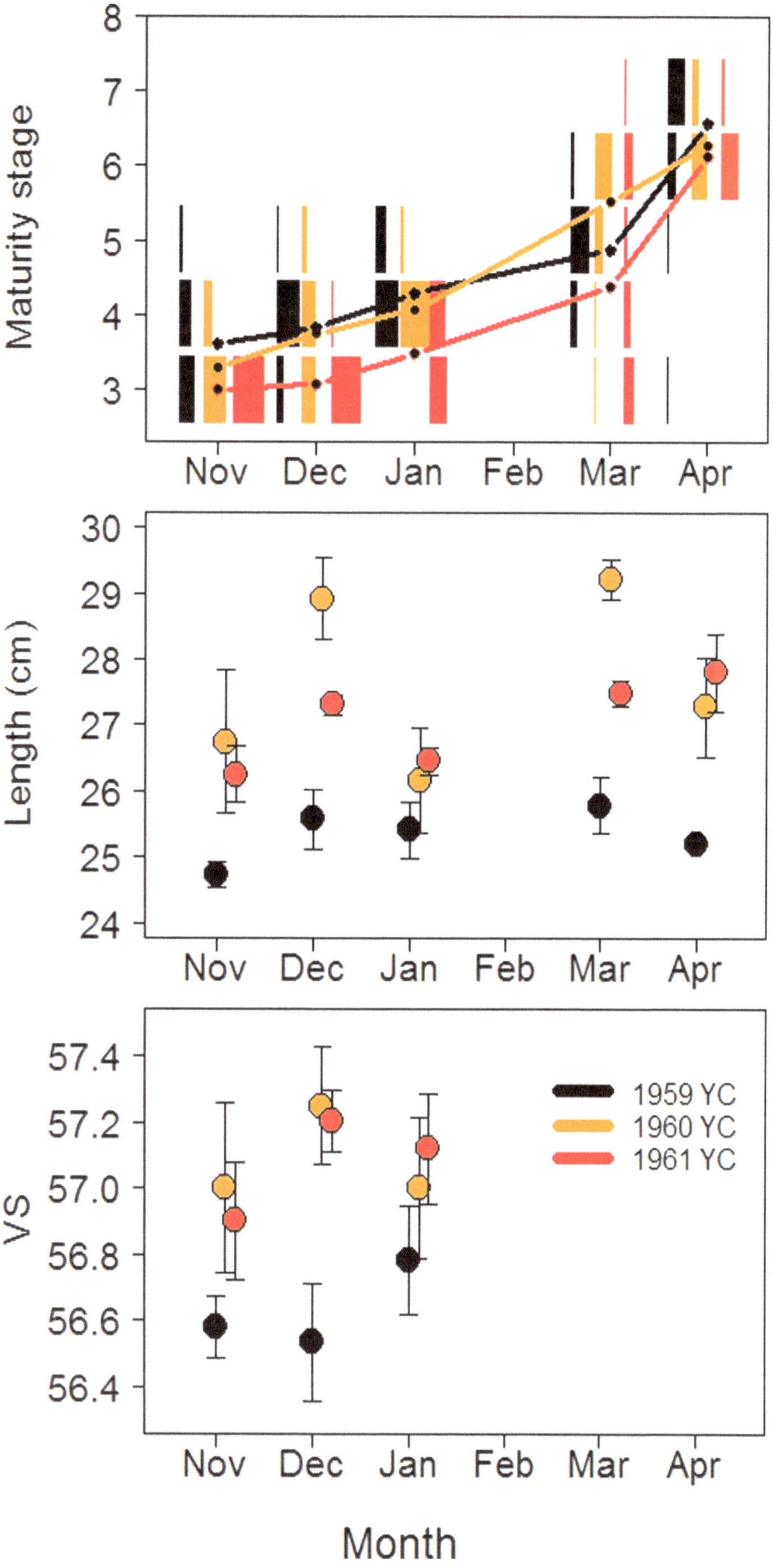
Maturity stage
8
7
6
5
4
3
Nov Dec Jan Feb Mar Apr
Length (cm)
30
29
28
27
26
25
24
Nov Dec Jan Feb Mar Apr
VS
57.4
57.2
57.0
56.8
56.6
56.4
1959 YC
1960 YC
1961 YC
Nov Dec Jan Feb Mar Apr
Month

Figure 7. A-C. Monthly mean values of maturity stage, length and VS of herring. Values are given for the period November 1963 through April 1964. Relative monthly proportions of maturity stages are given for each of the 1959–61 YCs.

that herring in Lindåspollene in the 2000s differ from NSS herring (Christophe Pampoulie, Marine Research Institute Iceland, unpublished observations). In the future, genetic analyses can be used to investigate the rate and extent of genetic change due to interbreeding.

The proportion of the component classified as NSS herring in Lindåspollene diminished markedly with age, with the proportion of five-year-old NSS herring only a fifth of the proportion of two-year-olds (Table 3). The NSS herring thus seem to enter the system as young or at the larval stage and spawn one or a few times with the resident herring. But why do they then leave the area? Size may be a threshold trait that determines migratory tactics [55], and herring above a certain size may migrate to more productive habitats, as has been observed in salmonids [56,57]. In the 2000s, resident herring of similar size as NSS herring remained in the system, but a genetic influence (see [58]) and different previous experience could result in different migration thresholds in LP and NSS herring. The decrease in proportion of NSS herring with age was strongest in P1, suggesting that NSS herring stayed longer in the system during P2 than P1. The chance of interbreeding between the resident and NSS herring could thus be higher during P2.

Interbreeding between NSS and LP herring

So do fish classified as NSS herring actually spawn with fish classified as LP herring? In the 1963–1964 data, no decrease in size during the course of the maturation period was observed, and the larger NSS herring thus seem to stay in Lindåspollene throughout spawning. Most importantly, young NSS and LP herring at maturation stage 6 (spawning), were caught together in the same gillnet settings. This stage of maturity typically lasts only for a day or so (Olav Kjesbu, Institute of Marine Research, pers. comm.). In addition, echosounder recordings and data from acoustic tags during spawning in Lindåspollene provide no evidence that herring split into groups according to origin [59,60]. Altogether, the evidence strongly indicates that LP herring and NSS herring interbreed in Lindåspollene.

Changes in life-history traits in LP herring

We have clearly demonstrated that several life-history traits in LP herring changed over time, with the mean age and weight of this component rising by almost 100% from the 1960s (P1) to the 2000s (P3), but it is not straightforward to explain these changes. There is no information available to suggest that changes have taken place in relevant environmental conditions in Lindåspollene. The marked changes in life-history traits could be a consequence of phenotypic plasticity, genetic responses or a combination of the two [61]. Herring is a flexible species with a high level of adaptability as a basic trait [62]. The genetic structure of life-history traits and high genetic diversity may provide a flexible species with large phenotypic plasticity and/or a large potential for rapid evolutionary changes [62,63].

The heritability of life-history traits is quite large [64], and interbreeding with NSS herring might well have changed genetically determined life-history traits in LP herring from slow growth and short life-span to faster growth and a longer life. Fish have limited energetic resources and there are trade-offs between somatic growth, survival and reproduction [65–67]. Rapid growth is crucial for the highly migrating NSS herring as the energy costs of swimming decrease as size increases [29]. For the more stationary LP herring large size is probably less critical. The trade-off between growth and reproduction is generally strong in herring [68,69]. LP herring should thus be expected to have invested more in reproduction when they were small early in the study period, and unpublished observations from the 1970s in fact indicate a high reproductive effort (RE). We would therefore expect that the RE of LP herring in the 2000s was similar to NSS herring, as both LP and NSS herring then had fast growth. However, RE was found to be considerably higher in LP than NSS herring also in the 2000s [20]. What is genetically determined is not a set value but a reaction norm [70], and by migrating to more productive outside waters (see below), LP herring could obtain more resources with low energy costs and thereby combine rapid growth and longevity with high RE.

McQuinn [6] argued that the structure of herring populations is of a behavioural rather than genetic nature. Although there are few data on movements of LP and NSS herring in and out of Lindåspollene, it is reasonable to assume that a change in feeding migration patterns was involved in the increased growth of LP herring. Migration patterns in herring are presumably maintained by local traditions [29–32], and pre-spawning LP herring have aggregated in a particular area in Lindåspollene for several decades [12,59,71]. By social transmission of migration patterns (see also [72]), NSS herring may have triggered a change in LP herring to migrating to more productive outside waters, leading in time to increased growth. As there is a genetic basis for migratory behaviour [73,74], interbreeding with migratory NSS herring may also have increased the tendency of LP herring to migrate. All in all, it is reasonable to assume that interactions between genetic factors that influence both growth potential and migratory tendency, and changes in migration traditions, are involved in the change of life- traits in LP herring.

Conclusions and Perspectives

We have presented new evidence that Atlantic herring form a metapopulation, and for the first time have found evidence of strong temporal dynamics between two herring populations. However, although the data material is unique and our explanations are plausible, we recognise that some questions remain to be answered. The application of VS is useful in the lack of more sophisticated identification methods such as genetic analyses, but due to the large individual variation in VS also within a single herring component, relatively large samples would be needed to obtain a robust basis for comparison. Nor is it clear whether most NSS herring initially enter Lindåspollene at the larval stage or later, and there are few data on movements of LP and NSS herring in and out of Lindåspollene. We have generally strong evidence that NSS herring leave Lindåspollene after a certain age and size, but this was more difficult to study in the 2000s with our methodology as LP and NSS herring were similar in size. This demonstrates the limitations of using VS and growth patterns to distinguish different components.

When NSS herring underwent a distributional shift after the stock collapse, the connectivity with resident herring could have led to changes in several life-history traits of the indigenous herring by interbreeding and/or a culturally mediated migratory shift. A metapopulation of herring could consist of populations that differ genetically but that also possess cultural diversity, such as different migration patterns, as an additional isolating mechanism. But migration routes can change. Historical records indicate that NSS herring have undergone large fluctuations in abundance during the past 500 years ([75] and references therein) causing changes in migration patterns [23]. Our finding that LP herring became more similar to NSS herring over the study period may suggest a cyclical pattern with genetic differentiation of the resident population during periods of weak connectivity resulting in slower growth and shorter life span alternating with periods of strong connectivity that reduce these differences.

Changes in migration patterns may be associated with a fishery-induced collapse [26]. This study suggests that anthropogenic influences such as a fishery should not be considered in isolation for a single herring population. Exploitation of neighbouring populations may alter the composition of the metapopulation in ways that could challenge the integrity of the resident population, and high rates of connectivity between populations could decrease the productivity and stability of a population [76]. Further studies of dynamic interchange between different populations of marine species and the relative importance of genetic and cultural factors behind the population structure are warranted, both from a basic ecological and evolutionary perspective and as a basis for sound management.

Acknowledgments

We appreciate discussions with colleagues at IMR and the University of Bergen, particularly Olav Kjesbu, Arild Folkvord and Mikko Heino for commenting on an earlier version of the manuscript. Two reviewers also made valuable suggestions for improvements. We would like to thank KO Fjeldstad, R Sevrinsen and K Øyjordsbakken on board the research vessel "Hans Brattstrøm" for their collaboration and practical assistance. We are very grateful to R Bergfjord for sampling and data collection, and for providing equipment and working/storage room. We are also indebted to J Røttingen, J de Lange, A-L Johnsen, A Haugsdal, V Lokøy and F Midtøy for their assistance with biological sampling. University of Bergen is thanked for funding open access of the article.

Author Contributions

Analyzed the data: AJ GS LL AS ÅH AF. Wrote the paper: AJ GS LL AS AF.

References

1. Levins R (1969) Some demographic and genetic consequences of environmental heterogeneity for biological control. Bull Entomol Soc Am 15: 237–240.
2. Levins R (1970) Extinction. In: Desternhaber M, editor, Some Mathematical Problems in Biology, American Mathematical Society, Providence. pp. 70–107.
3. Parrish BB, Saville RE (1965) The biology of the northeast Atlantic herring populations. Oceanogr Mar Biol Ann Rev 3: 323–373.
4. Iles TD, Sinclair M (1982) Atlantic Herring: Stock Discreteness and Abundance. Science 215: 627–633.
5. Rosenberg R, Palmén LE (1981) Composition of herring stocks in the Skagerrak-Kattegat and the relation of these stocks with those of the North Sea and adjacent waters. Fish Res 1: 83–104.
6. McQuinn I (1997a) Metapopulations in Atlantic herring. Rev Fish Biol Fish 7: 297–239.
7. Ruzzante DE, Mariani S, Bekkevold D, André C, Mosegaard H, et al. (2006) Biocomplexity in a highly migratory marine fish, Atlantic herring. Proc Roy Soc Ser B 273: 1459–1464.
8. Husebø Å, Slotte A, Clausen LAW, Mosegaard H (2005) Mixing of populations or year class twinning in Norwegian spring spawning herring? Mar Freshw Res 56: 763–772.
9. Stephenson RL, Melvin GD, Power MJ (2009) Population integrity and connectivity in Northwest Atlantic herring: a review of assumptions and evidence. ICES J Mar Sci 66: 1733–1739.
10. Runnström S (1941) Racial analysis of the herring in Norwegian waters. FiskDir Skr Ser HavUnders 6(7): 5–10.
11. Aasen O (1952) The Lusterfjord herring and its environment. Reports on Norwegian Fishery and Marine Investigations, 10.
12. Lie U, Dahl O, Østvedt OJ (1978) Aspects of the life history of the local herring stock in Lindåspollene, western Norway. FiskDir Skr Ser HavUnders 16: 369–404.
13. Jørstad KE, Pedersen SA (1986) Discrimination of herring populations in a northern Norwegian fjord: genetic and biological aspects. ICES CM 1986/H:63.
14. Hognestad PT (1994) The Lake Rossfjord herring (*Clupea harengus* L.) and its environment. ICES J Mar Sci 51: 281–292.
15. Kritzer JP, Sale PF (2004) Metapopulation ecology in the sea: from Levins' model to marine ecology and fisheries science. Fish Fish 5: 131–140.
16. Johannessen A, Nøttestad L, Fernö A, Langård L, Skaret G (2009) Two components of Northeast Atlantic herring within the same school during spawning: support for the existence of a metapopulation? ICES J Mar Sci 66: 1840–1748.
17. Sørensen ØB (2012) Comparative biology and population dynamics between Trondheimsfjord herring and Norwegian spring spawning herring, implications for management. Master's thesis, University of Bergen, Norway.
18. Pavlov DA, Shardin A (1998) Development of variation in the number of myomeres and vertebrae in the White Sea herring, *Clupea pallasi marisalbi*, under the influence of temperature. J Ichthyology 38: 251–261.
19. Jørstad KE, Nævdal G (1981) Significance of population genetics on management of herring stocks. ICES CM1981/H:64.
20. Silva F, Slotte A, Johannessen A, Kennedy J, Kjesbu OA (2013) Strategies for partition between body growth and reproductive investment in migratory and stationary populations of spring-spawning Atlantic herring (*Clupea harengus* L.) Fish Res 138: 71–79.
21. Eggers F (2013) Metapopulation dynamics in Atlantic herring (*Clupea harengus* L.) along the coast of southern Norway and in the local area of Landvikvannet. Master's Thesis, University of Bergen, Bergen, Norway.
22. Overholtz WJ, Friedland KD (2002) Recovery of the Gulf of Maine-Georges Bank Atlantic herring (*Clupea harengus*) complex: perspectives based on bottom trawl survey data. Fish Bull 100(3): 593–608.
23. Dragesund O, Hamre J, Ulltang Ø (1980) Biology and population dynamics of the Norwegian spring spawning herring. Rapp P.-v Réun Cons Int Explor Mer 177: 43–71.
24. Hay DE, Toresen R, Stephenson R, Thompson M, Claytor R, et al. (2001) Taking stock: an inventory and review of world herring stocks in 2000. In: Funk F, Blackburn J, Hay D, Paul AJ, Stephenson R, Toresen R, Witherell D, editors, Herring: Expectations for a New Millennium. University of Alaska Sea Grant, AK-SG-01-04, Fairbanks, Alaska, pp 381–454.
25. Holst JC, Slotte A (1998) Effects of juvenile nursery on geographic spawning distribution in Norwegian spring-spawning herring (*Clupea harengus* L.). ICES J Mar Sci 55: 987–996.
26. Dragesund O, Johannessen A, Ulltang Ø (1997) Variation in migration and abundance of Norwegian spring spawning herring (*Clupea harengus* L.). Sarsia 82: 97–105.
27. Røttingen I (1992) Recent migration routes of Norwegian spring spawning herring. ICES CM 1991/H:18.
28. Hamre J (1990) Life history and exploitation of the Norwegian spring-spawning herring. In: Monstad T, editor, Biology and fisheries of the Norwegian spring-spawning herring and blue whiting in the northeast Atlantic. Proceedings of the 4th Soviet-Norwegian Symposium, 12–16 June 1989, IMR, Norway, pp. 5–39.
29. Slotte A (1999) Differential utilization of energy during wintering and spawning migration in Norwegian spring-spawning herring. J Fish Biol 54: 338–355.
30. Corten A (2002) The role of "conservatism" in herring migrations. Rev Fish Bio Fish 11: 339–361.
31. Huse G, Fernö A, Holst JC (2010) Establishment of new wintering areas in herring co-occurs with peaks in the 'first time/repeat spawner' ratio. Mar Ecol Prog Ser 409: 189–198.
32. Petitgas JK, Secor DH, McQuinn I, Huse G, Lo N (2010) Stock collapses and their recovery: mechanisms that establish and maintain life-cycle closure in space and time. ICES J Mar Sci 67: 1841–1848.
33. Harma C, Brophy D, Minto C, Clarke M (2012) The rise and fall of autumn-spawning herring (*Clupea harengus* L.) in the Celtic Sea between 1959 and 2009: Temporal trends in spawning component diversity. Fish Res 121–122: 31–42.
34. Langård L, Nøttestad L, Johannessen A, Fernö A, Øvredal JT, et al. (2006) How and why acoustic detectability and catchability of herring change with individual motivation and physiological state in a variable environment: a multi-scale study on a local herring population in southwestern Norway. ICES CM. 2006/O:09.
35. Dahl O, Østvedt OJ, Lie U (1973) An introduction to a study of the marine ecosystem and the local herring stock in Lindåspollene. FiskDir Skr Ser HavUnders 16: 148–158.
36. Dahl O, Lie U (1981) Lindåspollene – et naturlig forskningsakvarium. Naturen nr. 3 (in Norwegian).
37. Ellingsen E (1973) Kvalitative og kvantitative zooplanktonundersøkelser i Lindåspollene mars 1971 – april 1972. Master's thesis, University of Bergen, Norway (in Norwegian).
38. Aksnes DL, Magnesen T (1983) Distribution, development, and production of *Calanus finmarchicus* (Gunnerus) in Lindåspollene, western Norway, 1979. Sarsia 68: 195–208.
39. Salvanes AGV, Aksnes D, Fosså JH, Giske J (1995) Simulated carrying capacities of fish in Norwegian fjords. Fish Oceanogr 4 (1): 17–32.

40. Ohman MD, Eiane K, Durbin EG, Runge JA, Hirche HJ (2004) A comparative study of *Calanus finmarchicus* mortality patterns at five localities in the North Atlantic. ICES J Mar Sci 61: 687–697.
41. Aksnes DL, Aure J, Kaartvedt S, Magnesen T, Richard J (1989) Significance of advection for the carrying capacities of fjord populations. Mar Ecol Prog Ser 50: 263–274.
42. Langård L, Slotte A, Skaret G, Johannessen A (2014) Thermal stratification influences maturation and timing of spawning in a local *Clupea harengus* population. J Fish Biol 84: 1202–1209.
43. Fosså JH (1992) Mass occurrence of *Periphylla periphylla* (*Scypozoa*, Coronatae) in a Norwegian fjord. Sarsia 77: 237–251.
44. Bagøien E, Kaartvedt S, Aksnes D, Eiane K (2001). Vertical distribution and mortality of overwintering *Calanus*. Limnol Oceanogr 46 (6): 1494–1510.
45. Johannessen P (1980) Resipientundersøkelse av enkelte fjordavsnitt i Lindås kommune med hovedvekt lagt på bunnforhold og bunndyr. Institutt for marinbiologi-Universitetet i Bergen 39pp (in Norwegian)
46. Mjanger H, Hestenes K, Svendsen BV, de Lange Wenneck T (2011) Håndbok for prøvetaking av fisk og krepsdyr. V. 3.16 (in Norwegian).
47. Burd AC (1985) Recent changes in the central and southern North Sea herring stocks. Can J Fish Aquat Sci 42 (Suppl. 1): 192–206.
48. Ricker WE (1975) Computation and interpretation of biological statistics of fish populations. Bull Fish Res Board Can 191: 1–382.
49. Bertalanffy L von (1934) Untersuchungen über die Gesetzlichkeit des Wachstums. Wilhelm Roux' Arkiv für Entwicklungsmechanik der Organismen 131: 613–652.
50. Tåning V (1952) Experimental study of meristic characters in fishes. Biological Reviews 27: 169–193.
51. Lindsey CC (1988) Factors controlling meristic variation. In: Hoar WS, Randall DJ, editors, Fish Physiology 11 (B). Academic Press, New York. pp. 197–274.
52. Hamre (1994) Biodiversity and exploitation of the main fish stocks in the Norwegian – Barents Sea ecosystem. Biodiv Cons 3: 473–492.
53. Dragesund O, Østvedt OJ, Toresen R (2008) Norwegian spring-spawning herring: history of fisheries, biology and stock assessment. In: Nakken O editor, Norwegian spring-spawning herring & Northeast Arctic cod. 100 Years of Research and Management. Tapir Academic Press, Trondheim pp. 41–82.
54. Metcalfe JD (2006) Fish population structuring in the North Sea: understanding processes and mechanisms from studies of the movements of adults. J Fish Biol 69(Suppl C): 46–65.
55. Sahashi G, Morita K (2013) Migration costs drive convergence of threshold traits for migratory tactics. Proc R Soc B 280 (1773): 20132539.
56. Jonsson B, Jonsson N (1993) Partial migration: niche shift versus sexual maturation in fishes. Rev Fish Biol Fish 3: 348–365.
57. Wysujack K, Greenberg LA, Bergman E, Olsson IC (2009) The role of the environment in partial migration: food availability affects the adoption of a migratory tactic in brown trout *Salmo trutta*. Ecology of Freshwater Fish 18: 52–59.
58. Theriault V, Garant D, Bernatchez L, Dobson JJ (2007) Heritability of life-history tactics and genetic correlation with body size in a natural population of brook charr (*Salvelinus fontinalis*). J Evol Biol 20: 2266–2277.
59. Langård L, Fatnes OA, Johannessen A, Skaret G, Axelsen BE, et al. (2014) State-dependent spatial and intra-school dynamics in pre-spawning herring *Clupea harengus* in a semi-enclosed ecosystem. Mar Ecol Prog Ser 501: 251–263.
60. Langård L (2013) Metapopulation dynamics and behavior of Atlantic herring (*Clupea harengus*) in a small semi-enclosed ecosystem. PhD thesis, University of Bergen, Norway.
61. Geffen AJ (2009) Advances in herring biology: from simple to complex, coping with plasticity and adaptability. ICES J Mar Sci 66: 1688–1695.
62. Barrett RDH, Schulter D (2008) Adaptation from standing genetic variation. TREE 23 (1): 38–44.
63. Lee CE, Gelembiuk GW (2008) Evolutionary origins of invasive populations. Evol Appl 1: 427–448.
64. Roff DA (2000) Trade-offs between growth and reproduction: an analysis of the quantitative genetic evidence. J Evol Biol 13: 434–445.
65. Roff DA (1983) An allocation model of growth and reproduction in fish. Can J Fish Aquat Sci 40: 1395–1404.
66. Stearns SC (1992) The Evolution of Life Histories. Oxford University Press, New York.
67. Burton OJ, Phillips BL, Travis JMJ (2010) Trade-offs and the evolution of life-histories during range expansion. Ecol Lett 13: 1210–1220.
68. Runnström S (1936) A study on the life history and migrations of the Norwegian spring-spawning herring based on the analysis of the winter rings and summer zones of the scale. FiskDir Skr Ser HavUnders 5(2): 1–103.
69. Engelhard G, Dieckmann U, Godø OR (2003) Age at maturation predicted from routine scale measurements in Norwegian spring-spawning herring (*Clupea harengus*) using discriminant and neural network analyses. ICES J Mar Sci 60: 304–313.
70. Engelhard GH, Heino M (2004) Maturity changes in Norwegian spring-spawning herring (*Clupea harengus*): compensatory or evolutionary response? Mar Ecol Prog Ser 272: 245–256.
71. Aksland M (1983) Acoustic abundance estimation of the spawning component of the local herring stock in Lindaaspollene, western Norway. FiskDir Skr Ser HavUnder 17: 297–334.
72. McQuinn I (1997b) Year-class twinning in sympatric seasonal spawning populations of Atlantic herring, *Clupea harengus*. Fish Bull 95: 126–136.
73. Kitano J, Ishikawa A, Kume M, Mori S (2012) Physiological and genetic basis for variation in migratory behavior in the three-spined stickleback, *Gasterosteus aculeatus*. Ichthyol Res 59: 293–303.
74. Hemmer-Hansen J, Nielsen EE, Therkildsen NO, Taylor MI, Ogden R, et al. (2013) A genomic island linked to ecotype divergence in Atlantic cod. Mol Ecol 22: 2653–2667.
75. Stephenson RL (2002) Stock structure and management structure: an ongoing challenge for ICES. ICES Mar Sci Symp 215: 305–314.
76. Alheit J, Hagen E (1997) Long-term climate forcing of European herring and sardine populations. Fish Oceanogr 6 (2): 130–139.

Tempo and Mode of Gene Duplication in Mammalian Ribosomal Protein Evolution

Asav P. Dharia[9], **Ajay Obla**[9], **Matthew D. Gajdosik, Amanda Simon, Craig E. Nelson***

University of Connecticut Department of Molecular and Cell Biology, Storrs, Connecticut, United States of America

Abstract

Gene duplication has been widely recognized as a major driver of evolutionary change and organismal complexity through the generation of multi-gene families. Therefore, understanding the forces that govern the evolution of gene families through the retention or loss of duplicated genes is fundamentally important in our efforts to study genome evolution. Previous work from our lab has shown that ribosomal protein (RP) genes constitute one of the largest classes of conserved duplicated genes in mammals. This result was surprising due to the fact that ribosomal protein genes evolve slowly and transcript levels are very tightly regulated. In our present study, we identified and characterized all RP duplicates in eight mammalian genomes in order to investigate the tempo and mode of ribosomal protein family evolution. We show that a sizable number of duplicates are transcriptionally active and are very highly conserved. Furthermore, we conclude that existing gene duplication models do not readily account for the preservation of a very large number of intact retroduplicated ribosomal protein (RT-RP) genes observed in mammalian genomes. We suggest that selection against dominant-negative mutations may underlie the unexpected retention and conservation of duplicated RP genes, and may shape the fate of newly duplicated genes, regardless of duplication mechanism.

Editor: Marc Robinson-Rechavi, University of Lausanne, Switzerland

Funding: This work was supported by National Science Foundation (NSF) Grant Award Number: 1024803 (http://nsf.gov/awardsearch/showAward?AWD_ID=1024803&HistoricalAwards=false). Author CEN was the principal investigator on the NSF grant who received the funding. The funders had no role in study design, data collection and analysis, decision to publish, or preparation of the manuscript.

Competing Interests: The authors have declared that no competing interests exist.

* Email: craig.nelson@uconn.edu

[9] These authors contributed equally to this work.

Introduction

Gene Duplication and Genome Evolution

In 1970, Susumu Ohno hypothesized that gene duplication provided the raw material required for the diversification of gene function. It is now appreciated that gene duplication and loss is a dynamic process that has given rise to many large gene families critical to the evolution of complex organisms [1]. Recent data reveal that lineage-specific expansion and contraction of gene families is more rapid than previously appreciated, and is responsible for major differences in gene family size between closely related mammalian genomes [2]. These differences are likely to have made major contributions to the divergence of mammalian lineages and to human evolution [3,4]. For this reason, understanding the forces that lead to the retention or loss of duplicated genes in complex genomes is fundamental to understanding genome evolution, and particularly to the evolution of complex organisms.

Scales, Mechanisms, and Outcomes of Gene Duplication

Duplications occur at all genomic scales, from a single nucleotide to the entire genome, and vary greatly in frequency. Depending on the nature of the duplication, these events can have a positive, negative or neutral effect upon an organism. For example, duplications involving a single gene or set of genes can be associated with enrichment for essential functions, while large scale duplications can be associated with important evolutionary transitions, major leaps in development, and the adaptive radiation of species [5,6]. Many physical mechanisms can give rise to duplication events: (1) whole-genome duplication (WGD), (2) tandem duplication (i.e., unequal crossing-over), (3) duplicative transposition, and (4) retrotransposition. WGD, tandem duplication, and duplicative transposition are DNA-mediated events, while retrotransposition is the Reverse Transcriptase (RT)-mediated insertion of a cDNA into the host genome. A WGD is believed to have occurred in yeast [7], and several have been inferred in the teleost lineage [8]; while the last WGD believed to have occurred in the mammalian lineage took place before the emergence of modern mammals [9,10]. Compared to these very large scale and rare events, duplicative transpositions and tandem duplications are likely to drive much of the duplication and loss giving rise to complex gene families. These are DNA-mediated processes that preserve varying amounts of the source gene's intron-exon structure [4]. Often, depending on the scale of the duplication, varying amounts of the intergenic regulatory DNA flanking the duplicated gene is also transferred, increasing the likelihood of a functional duplicate being created [11]. In contrast, retrotransposition is a process whereby a spliced mRNA transcript is reverse-transcribed into DNA and randomly re-integrated into the genome, creating a copy lacking introns and the promoter and enhancer elements of the source gene [12]. Such retrogenes have traditionally been regarded as non-functional. However, recent studies have shown that rampant retrotransposition can create genes that function as protein-coding genes [13] and small RNA's [14], and which can have dramatic phenotypic consequences [15]. Recent interest in retrotransposition is highlighted by the identification of several functional retrogenes, such as *Fgf4* and *c1orf37-dup* in mammals [16,17], and suggests that retroduplicaton may be a more important force in the evolution of complex gene families than has been widely appreciated.

Retroduplication in the Mammalian Genome

Previously, we reported that conserved retroduplicates are widespread in mammals, representing half of all gene duplicates under purifying selective pressure [18,19]. In addition, we noted that individual gene families have a strong tendency to evolve via DNA-mediated or RNA-mediated duplication, but not both. Developmentally important classes of genes, such as transcription factors, which often require large amounts of regulatory information to function properly, tend to evolve through DNA-mediated events. However, gene families involved in metabolic processes, such as protein synthesis, evolve primarily through RNA-mediated duplication. In fact, we reported that ribosomal protein (RP) genes are the largest class of conserved, retroduplicated genes in mammals [18]. While it is not surprising that the highly abundant ribosomal protein transcripts appear to be more frequently captured by retroviral reverse transcriptase than less abundant transcripts, it is intriguing that the slowly-evolving, highly-conserved ribosomal proteins have hundreds of intact duplicates in the genome.

Known Examples of Retrotransposed Duplicates

One of the most prominent examples of a RNA-mediated duplicate is Fgf4, a retrogene associated with the breed-defining chondrodysplasia in domestic dogs [15]. A human-specific example is the C1orf37-duplicate, derived through retrotransposition after divergence of human from chimp expressed selectively in certain human tissues, such as brain. It is suggested to encode a novel transmembrane protein [20]. Similar examples include TRMT12 retrogene [21], IMP3 gene [22,23] and other such retrogenes (see [24–28]). The majority of the aforementioned RT genes follow the convention that most retrogenes are in a state of "relaxed" selection. The molecular evolution of retrogenes is selectively neutral, allowing them to freely mutate, giving them a chance to be inactivated or positively selected, while parental genes remain subjected to purifying selection [29,30].

The Mammalian Ribosome and Ribosomal Proteins

The ribosome is an ancient molecular machine that is responsible for production of protein in all living cells. The mammalian ribosome consists of 79 RPs and four rRNAs. RPs play a central role in protein synthesis, are expressed at high levels, and evolve very slowly [31] showing strong conservation across the three domains of life [32]. Proper ribosomal biogenesis requires equimolar production of all RPs and rRNAs [33,34]. These transcriptional regulatory constraints have been extensively elucidated in various studies [35,36], along with the evidence that different ribosomal protein promoters exhibit equipotent strength [37]. Additionally, a strict copy number constraint is also observed as the majority of full length RP's have been shown to be single copy genes [38].

Due to the necessity of protein synthesis in any living cell, and the complexity of ribosome structure and assembly, it is perhaps unsurprising that mutations in ribosomal genes almost inevitably lead to pathological conditions such as Minutes in Drosophila [39] and Diamond-Blackfan anemia (DBA) in humans [40,41]. Despite, or perhaps because of, their stringent conservation, the evolution of RPs in vertebrates is relatively understudied.

Known Examples of Mammalian Ribosomal Protein Duplicates

One of the most recently published DNA-mediated ribosomal protein (DD-RP) gene duplicates is RPL22L1, paralog of ribosomal protein RPL22. These mouse paralogs play essential, distinct, and antagonistic roles in hematopoietic development [42]. Another known rodent-specific RT-RP duplicate is Rps23rg1, a gene originating from a retrotransposition of s23 mRNA that encodes proteins that decrease Alzheimer's β-amyloid level and tau phosphorylation [43]. There is also evidence for a ubiquitously expressed RT-RP duplicate, Rpl36al, and testis-specific RPL10L duplicate that have been implicated in compensation for the reduced dosage of X-linked RP genes [44]. As mentioned in a previous section, RPL3L is a DD-RP duplicate that has been observed to be highly expressed in a group of tissues where parent RPL3 has very little expression, exhibiting a potential functional role [45].

Focus of this study

In this study, we address several questions about gene duplication and the evolution of RPs. Many research groups have studied the dynamics of gene duplication in RPs in non-mammalian systems such as yeast [46]. In addition, some research has focused on individual human ribosomal proteins and their duplicates (eg: RPS19 and RPL7a) or comparative analyses between mammals [47–50]. To date, however, no study has encompassed all 79 mammalian RPs in a large set of mammalian genomes. In order to fill this gap and more thoroughly annotate RP gene duplication events during mammalian evolution, we created a pipeline that utilizes local synteny and conserved intron content to (1) identify each duplicated RP gene in eight mammalian species, (2) place each duplication event within the mammalian phylogeny, (3) discriminate between RNA- and DNA-mediated duplications, (4) estimate the degree of purifying selective pressure exerted on every duplicated RP gene, and (5) determine whether each duplicated RP gene copy exhibits evidence of expression.

Materials and Methods

Ribosomal Dataset

Seventy-six ribosomal protein (RP) sequences from nine species [human, chimp, monkey, mouse, rat, dog, cow, opossum, and chicken (outgroup)] were manually collected from Ensembl 62 [51]. Three RPs were excluded due to annotation issues. When a single gene encoded multiple transcripts, the longest was used. These protein sequences served as seed sequences, or input, to the pipeline (Figure 1).

Extraction of Gene Family Members

RP seed sequences were submitted to tBLASTn against donor genomes to capture as many putative duplicates of the seed gene as possible. Each resulting putative duplicate was processed using Pseudopipe [52] to determine the mechanism of duplication (DNA- or RNA-mediated) and the fate of the duplicate (intact or pseudogene). The default Pseudopipe filters for tBLASTn hits (E-

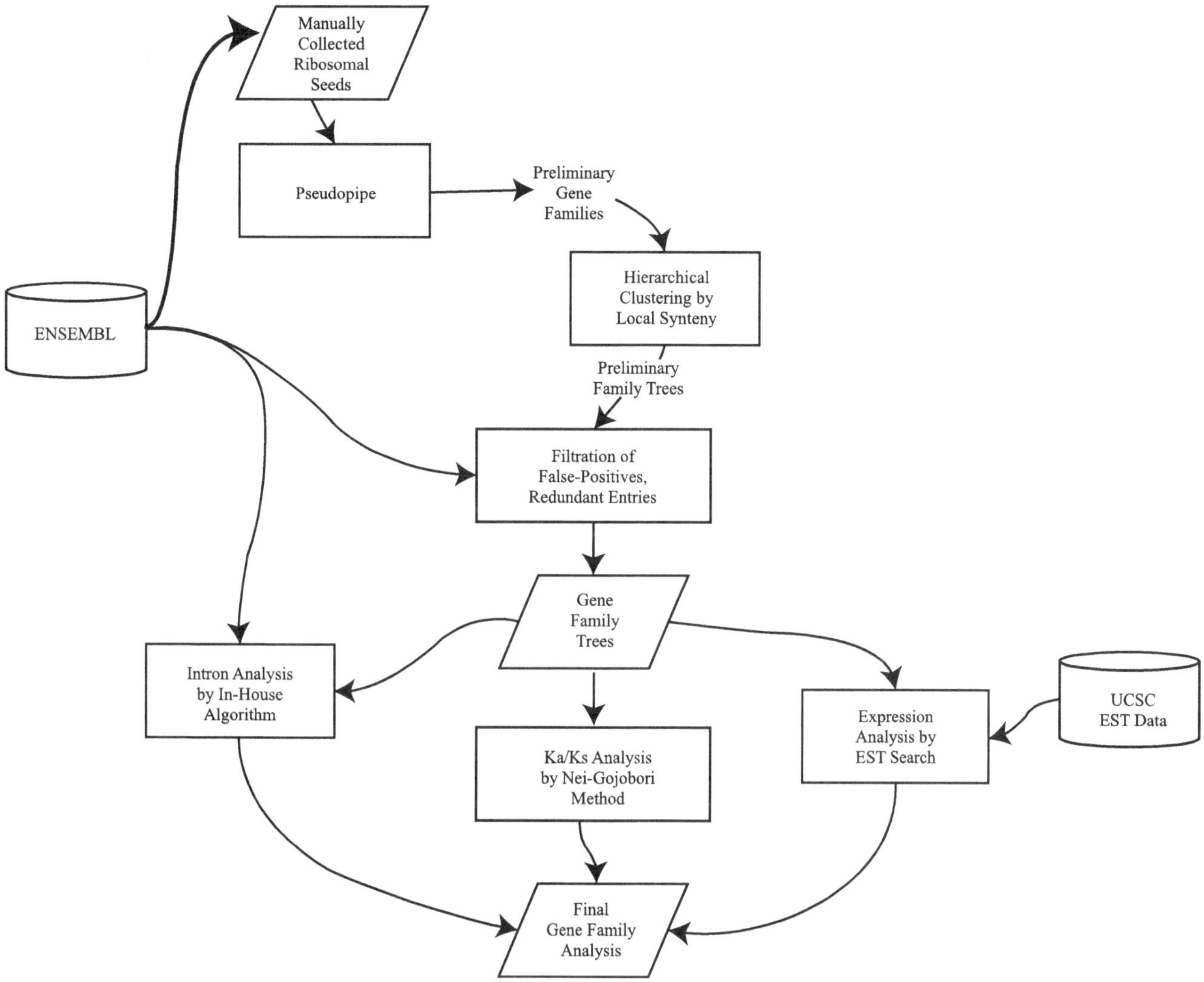

Figure 1. Pipeline for ribosomal protein family analyses. Protein sequence for all parental ribosomal proteins were collected manually from Ensembl62. These were input to tBLASTn against whole genomes to capture all putative duplicates. The resulting duplicates were processed by Pseudopipe to determine the mechanism of duplication (DNA or RNA) and the fate of the duplicate (intact or pseudogenized). We then utilized our in-house pipeline steps of hierarchical clustering by local synteny[3] in order to build our gene family trees after filtering false-positives and redundant entries. Final gene family analyses were conducted in 2 steps: 1) calculating the selective pressures on all gene duplicates using the Nei-Gojobori method against the species- and family-specific seed protein via an exon-based reconstruction, and 2) checking for expression signatures via EST analyses using the UCSC genome browser EST track for both human and mouse.

value cutoff $\leq 10^{-4}$ and identity and identity $\geq 40\%$) were used to define putative duplicates. Ambiguous duplicates, where the duplication mechanism was not confirmed, were resolved using an intron comparison algorithm [18], which compares intron/exon structure within a group while accounting for exon fusions and large insertions in exonic regions. These methods generated a set of RP superfamilies that consist of both protein-coding genes and related pseudogenes.

Identification of Duplications and Phylogenetic Analysis

Orthologous and paralogous relationships were determined using local synteny and a hierarchical clustering algorithm described in Jun et al, 2008 [18,19]. A local synteny score was assigned to all gene pairs based on the homology of genes (three upstream and three downstream) neighboring the two query genes. Pairwise synteny measures were obtained for all members of a gene family. The output generated based on these scores was used to construct phylogenetic trees in Newick format, representing the history of duplication in each family. Parsimony [53] was used to assign each inferred duplication event to a specific branch of the species tree [54,55]. 'Tube'-style phylogenetic trees for 74 mammalian RP genes were used to illustrate the history of DNA/RNA-mediated duplications across various evolutionary time periods (ancient vs. lineage specific) (See Appendix S1 for all trees).

Conservation and EST Analyses

Using exon-based reconstruction and the Nei Gojobori method, Ka/Ks ratios for all members of a gene family were calculated against the seed proteins. The putative exon-intron structures of duplicates were generated with an in-house algorithm, using these seed proteins. Results were then filtered based on p-values (<0.1) and the fraction of the parental gene represented by each duplicate ($>65\%$). Pairwise distances using ClustalW were also calculated as an added metric to evaluate sequence identity and account for all nucleotide level substitutions. Additionally, we also determined branch-wise omega values for 28 ribosomal protein families with following parameters, model = 2 & Nsite = 0, using codeML in PAML 4.7 [56]. In order to confirm the selective pressures, standard codon models M0, M1a, M2a were fitted to the data set with codeML. We used likelihood ratio tests (LRT) to determine the relative fit of the hierarchically nested models. Log likelihood ratio test statistic is $2\Delta\ell = 2(\ell 1-\ell 0)$, where $\ell 1$ is the log-likelihood of the model corresponding to the alternative hypothesis and $\ell 0$ represents the log-likelihood corresponding to the model used as null hypothesis. These values were compared with a chi-squared distribution in which the difference between the number of parameters of both models provides the degrees of freedom (df) [57,58]. Log likelihood values and parameter estimates are detailed in the results section and supplementary material.

In order to determine if duplicates were actively transcribed, human and mouse expressed sequence tags (EST) were mined from the UCSC genome browser EST. ESTs that mapped to multiple locations that showed less than 95% identity or 95% fraction length were discarded. Additionally, EST presence & absence calls were also made using data mined from Bgee database for annotated duplicates in our dataset [59].

Results

76 Ribosomal Protein Family Member Analyses

The first step of our pipeline identified all detectable duplicates of RP genes in eight mammalian genomes. RP families included 14,552 gene duplicates in the eight genomes analyzed: human, chimp, monkey, mouse, rat, dog, cow, and opossum (Figure 2A). Although data in figure 2A include duplicates with shared ancestry, the counts for each species represent the number of duplicate genes present in each extant species. To determine if sequencing coverage had a significant impact on our detection of RP gene duplicates, we compared the depth of sequence coverage in each species' genome to the number of duplications recovered in that species. We found no significant correlation between the number of duplications and genome coverage (Pearson's $r = -0.353$, $p = 0.391$, Figure 2B, S2). We also tested for bias in duplication types in each species and found no species-specific bias in duplication mechanisms. As we found significant association between species ($p = 6.07e\text{-}17$, two-way chi square test, Figure2B), all species were grouped for subsequent analyses.

Next we assessed the fate of each duplicate. Of the 14,552 duplication events detected, only 28 of these gene duplications are DNA-mediated (DD) events; the remainder (99.8%) are RNA-mediated (RT) duplications. Approximately 88% of RNA-mediated duplications are pseudogenes (12,800 duplicates), while 12% are intact (1724 duplicates, Figure 2C). A data table listing all RP gene duplicates recovered by our pipeline can be found in Table S1. We also examined every ribosomal protein gene's duplication history and evolutionary trajectory in the context of the encompassing species tree. One example of the resulting information is shown in Figure 3, for the ribosomal protein gene RPL36A. All 74 ribosomal protein gene family history trees are attached in supplementary material. Hereafter, all the intact RNA-mediated ribosomal protein gene duplicates will be referred to as RT-RPs, intact DNA-mediated copies as DD-RPs and RNA-mediated pseudogenes as RΨ-RPs. Leveraging previously published data by Jun et al; we observed a clear overrepresentation of RT-RPs among 8,872 gene families analyzed (Figure S1).

The Fate of Ribosomal Protein Duplications over time

In the second step of our pipeline, we determined the probable location of each RP duplication event in evolutionary history of these eight species, and distinguished between RNA- and DNA-mediated duplication events (Figure 3; [18]). Based on our methodology, Figure 4 clearly shows that the majority of detectable duplications have occurred during recent mammalian evolution: 100 million years ago (MYA) or more recently. However, a significant number of duplications date between 100–300 MYA. The majority of RP gene duplications older than 90 MYA result in RNA-mediated pseudogenes (RΨ-RPs) (190), though some events (25) are intact RNA-mediated duplications (RT-RPs), and a very small number (4) are linked to intact DNA-mediated duplicates (DD-RPs) [data not shown for DD-RPS due to small sample size]. It is important to note that many of the more ancient duplications detected represent incomplete clades; therefore we infer a considerable amount of gene loss. However, our inability to detect these genes may also be due to loss of synteny or other limitations of our pipeline.

The majority of duplicates (N = 13,588) observed in our dataset are young (91 MYA or younger). However, a few RT-RPs and DD-RPs have been conserved in all (or most) of the eight mammalian species analyzed (see the base of the tree in Figure 4).

Analysis of Selective Pressure Acting on All Ribosomal Gene Duplicates

To gain insight into the forces shaping the fate of these RP gene duplicates, nonsynonymous/synonymous substitution rates were evaluated using pair-wise and branch-wise methods (see Methods section and [13,60,61]. For the pair-wise method, we observe that DD-RP dups and RT-RP dups have mean Ka/Ks values of 0.166 (95% CI 0.083, 0.248) and 0.295 (95% CI 0.285, 0.305) suggesting that they are under strong purifying selective pressure. RΨ-RP's were under relatively less purifying selective pressure with a mean value of 0.455 (95%CI 0.453, 0.458) (Figure 5A). In order to avoid false positives with Ka/Ks >1, we did not include cases that had a very low Ka and Ks values. Calculation of pairwise DNA sequence distances reveals that the mean sequence distance for DD-RP duplicates was 0.091 (95% CI 0.062, 0.118), for RT-RP duplicates was 0.0059 (95% CI 0.062, 0.118) and for RΨ-RPs was 0.172 (95% CI 0.169, 0.173). This corroborates the evidence from the Ka/Ks analysis suggesting that these sequences are under strong selective pressure (Figure S3). Next we compared selective pressures on all RT-RP duplicates of various ages in each lineage. (DD-RPs were not included in this analysis due to the very small dataset.) Box-Whisker plots (Figure 5B and 5C) showed that RT-RP duplicates at all speciation nodes, irrespective of their age or lineage, are under strong selective pressures, as determined by Ka/Ks values. However, chimp (Pt) values seem to be an exception, likely due to a small sample size (Figure 5B). The trends appeared similar for all RΨ-RPs as the median Ka/Ks values are similar (~0.45) for all ages. Additionally, we also provide scatterplots for all speciation nodes to confirm the strong selection on all RP duplicates irrespective of age (see figures S6 and S7). While pairwise Ka/Ks calculations are computationally rapid and provide a good screen for selective pressure, especially within a gene family, for added support we wanted to cross-check our estimates of selective pressure using branch-specific omega values. To do this we used PAML to calculate branch-specific omega values for a sub-sample of 28 RP gene families (see Table S2). An example RP gene tree with all PAML branch-specific omega values is shown in Figure S4. Using this approach we obtained Ka/Ks values for RT-RP duplicates (mean = 0.162, CI 0.137, 0.188), and for the RΨ-RPs (mean = 0.357, CI 0.347, 0.367). As

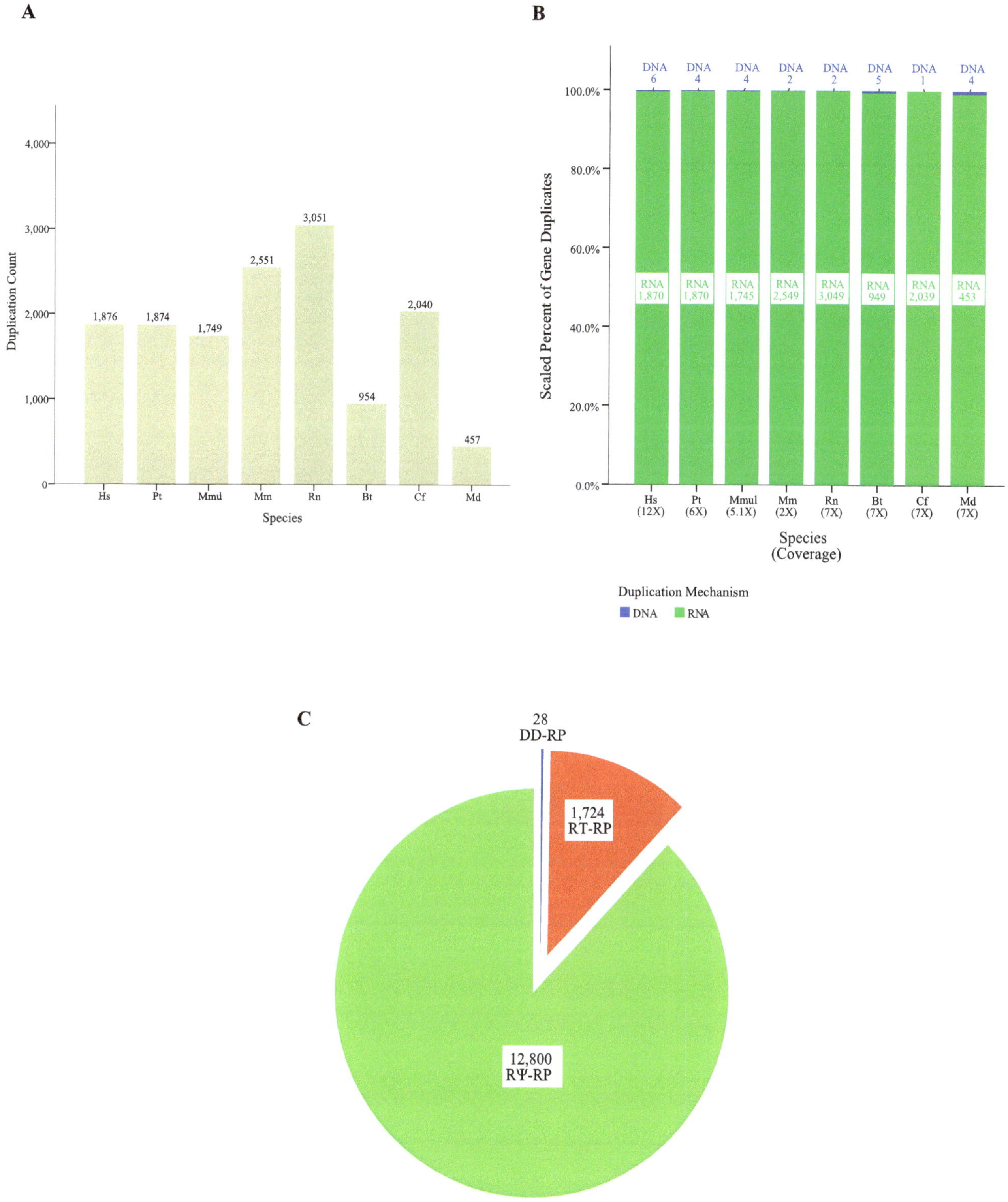

Figure 2. RP gene duplicates in 8 mammalian genomes. A) Distribution of duplication events in 8 mammalian genomes. B) Assessment of coverage or species-specific bias in ribosomal protein gene duplicates. C) Representation of DNA and RNA-mediated duplications in RP gene families. Abbreviations: Hs, *Homo sapiens* (human); Pt, *Pan troglodytes* (chimpanzee); Mmul, *Macaca mulatta* (Rhesus macaque); Mm, *Mus musculus* (house mouse); Rn, *Rattus norvegicus* (Norway rat); Bt, *Bos taurus* (cattle); Cf, *Canis familiaris* (dog); Md, *Monodelphis domestica* (gray short-tailed opossum); Gg, *gallus gallus* (chicken).

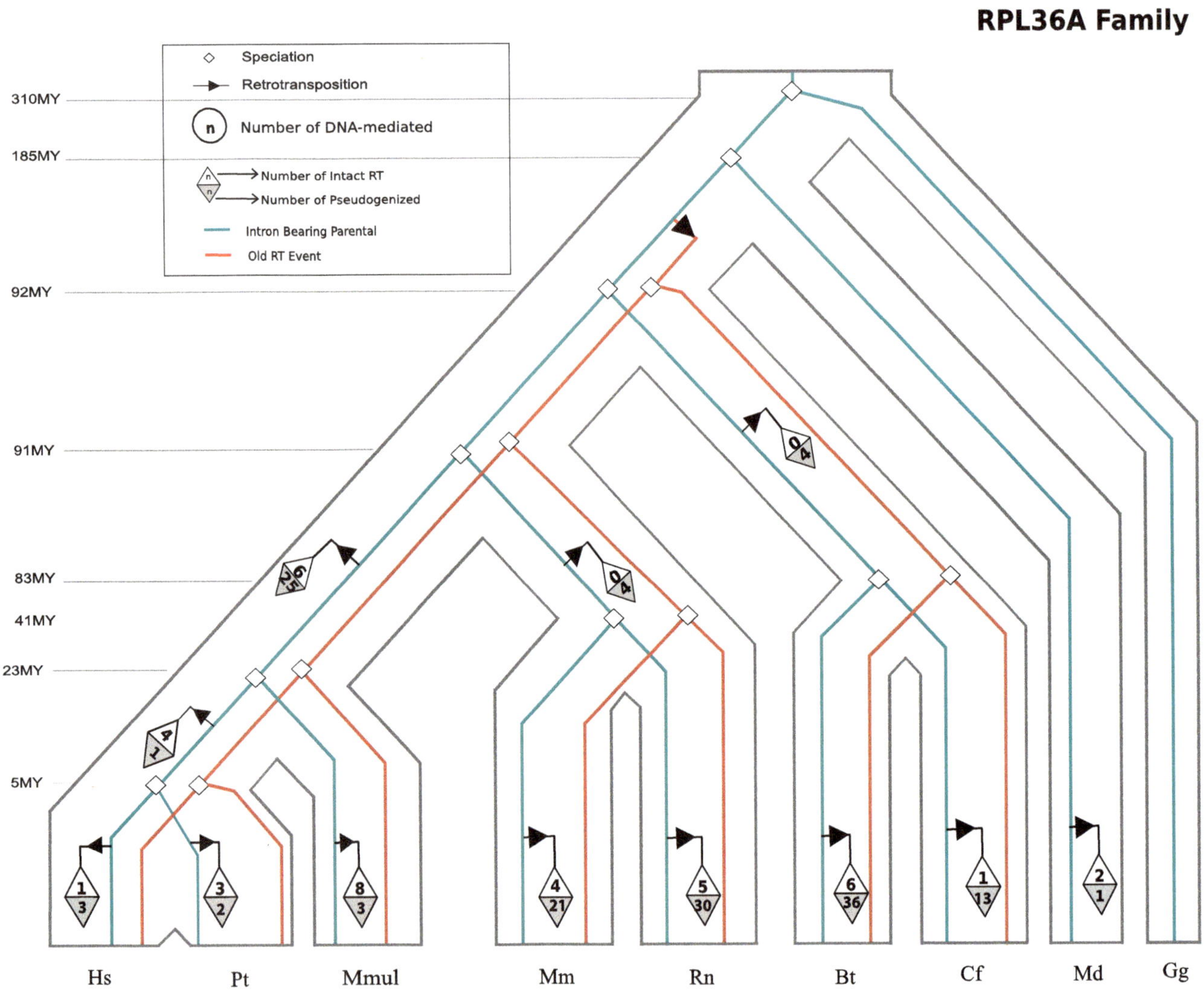

Figure 3. Example of the inferred evolutionary history for duplications of the ribosomal protein gene Rpl36a. Grey outlined tube tree represents the species tree that includes 8 mammals and chicken. Parental intron-bearing gene (in blue). RT-RPs (clear triangles), RΨ-RPs (grey triangles). An RT-RP duplicate generated from one of these events, Rpl36al (in red, at the base of the mammalian lineage on the branch between LCA with opossum and the other mammals) is conserved in all descendent species. All the 74 ribosomal protein gene family history trees are attached as supplementary material.

previously mentioned, to avoid false positives with Ka/Ks >>1, we excluded cases with very low Ka and Ks values. Both pairwise and PAML-based estimation methods confirm the strong purifying selective pressure acting on RT-RPs (Ka/Ks <0.3) and a slightly lower pressure on RΨ-RPs (Ka/Ks <0.5). As evolutionary pressure is often time dependent, we also plotted Ka against Ks estimated by both pair-wise (Figure 6A) and branch-wise (Figure 6B) methods. As expected the branch-wise method estimates higher divergence, as seen by the large distribution of Ks values (Figure 6B) compared to pairwise method. The influence of strong purifying selection over time is readily observed in Ka values for both methods as the data points of RP-RTs are compressed near the origin relative to RΨ-RPs, which have a much wider distribution (Figure 6).

In order to further confirm the nature of the selective pressure acting on our RP-RTs and RΨ-RPs, we also used different codon-substitution models developed by Nielsen and Yang [62] and Yang et al. [58]. Random-site models M0, M1a and M2a which assume variation in ω among sites but not among lineages were fitted to our data. The models used, parameter estimates and log-likelihood values are shown in Table S3. Table 1 shows the results of the LRT tests for these models. We applied the simplest of site-based models M0 [30], which assume a uniform ω ratio for all codons, to four random ribosomal genes namely Rps16, Rps18, Rpl14 and Rpl28. The estimated single ω value for each of these trees ranges from 0.22 to 0.35 (table S3). These values can be interpreted as an average of all lineages in the tree and over all sites in the protein. The low ω range obtained indicates a strong action of purifying selection in the evolution of ribosomal gene duplicates studied.

To test if branch-specific omegas are statistically justified, we compared Model M1a (nearly neutral), which constrains Ka/Ks ≤1 but not positive selection (Ka/Ks >1) and M2a which allows for positive selective pressure. This comparison leads us to reject the nearly neutral model as seen in table 1. Our final comparison was model M1a vs. M0 with a fix omega = 1 and we find out that model M1a fits our data better (p-value <0.0001). These results

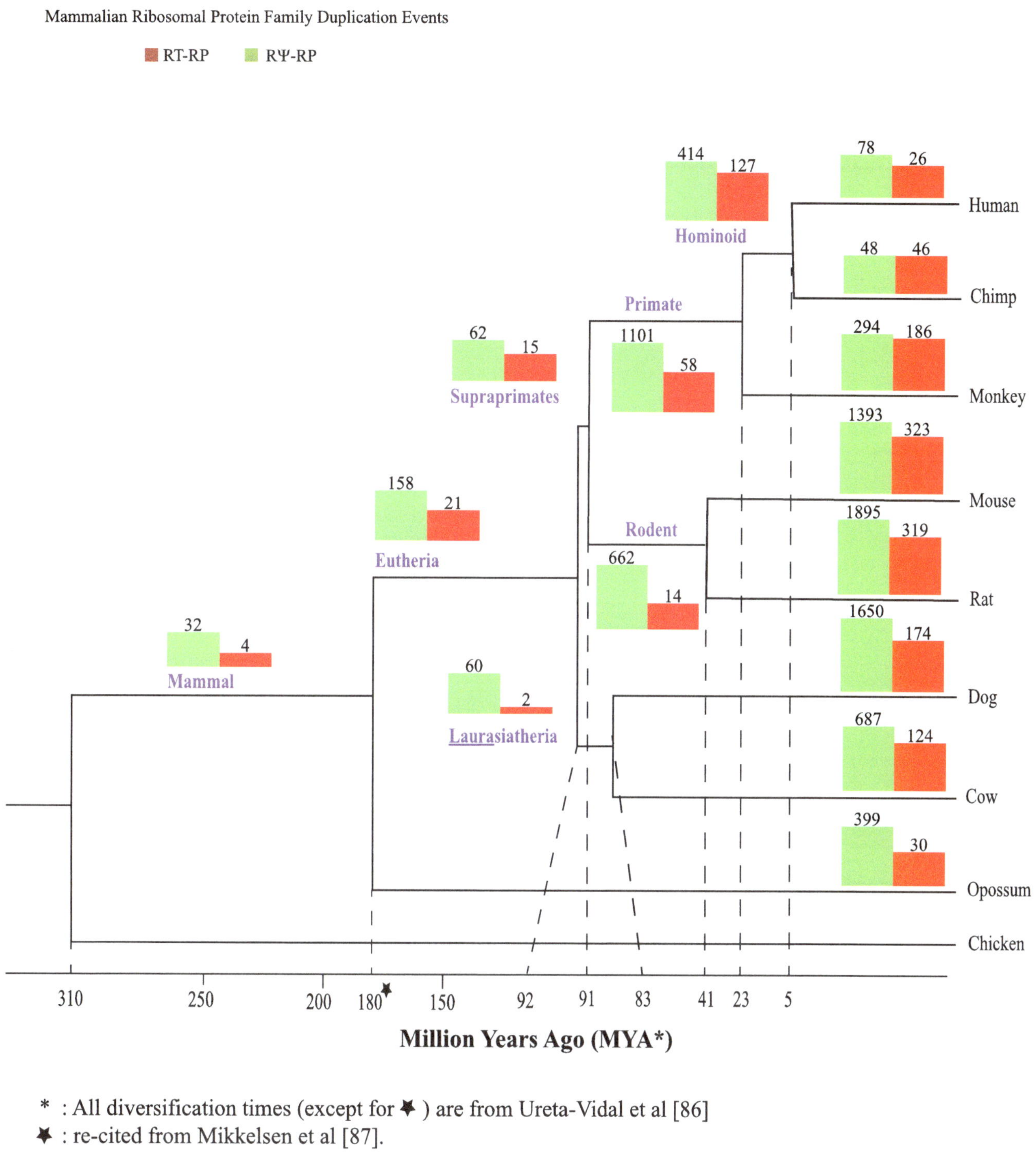

Figure 4. Ribosomal Protein Family duplication events based on age. All RP gene duplication events are displayed for 8 mammalian species. The bar charts at all speciation nodes show events classified by fate of duplication. The duplication counts on the bar charts are log normalized. RT-RPs are shown in red and RΨ-RPs in green. DD-RPs are not shown due to a very small sample size. The numbers above the bar charts represent the total number of gene duplication events at that speciation node. Age is marked at the bottom of the tree in millions of years (age estimates from [55,103]).

confirm that purifying selection is the predominant force acting in the evolution of ribosomal protein genes. Hence it further validates the Ka/Ks values obtained from both pair-wise and branch-wise methods.

EST Analysis for human and mouse RP duplicate genes

EST data for human and mouse were mined in the final step of our analysis pipeline. Using very stringent constraints (see Methods), we found evidence of expression for approximately 8% of all human and mouse duplicates. It should be noted that in

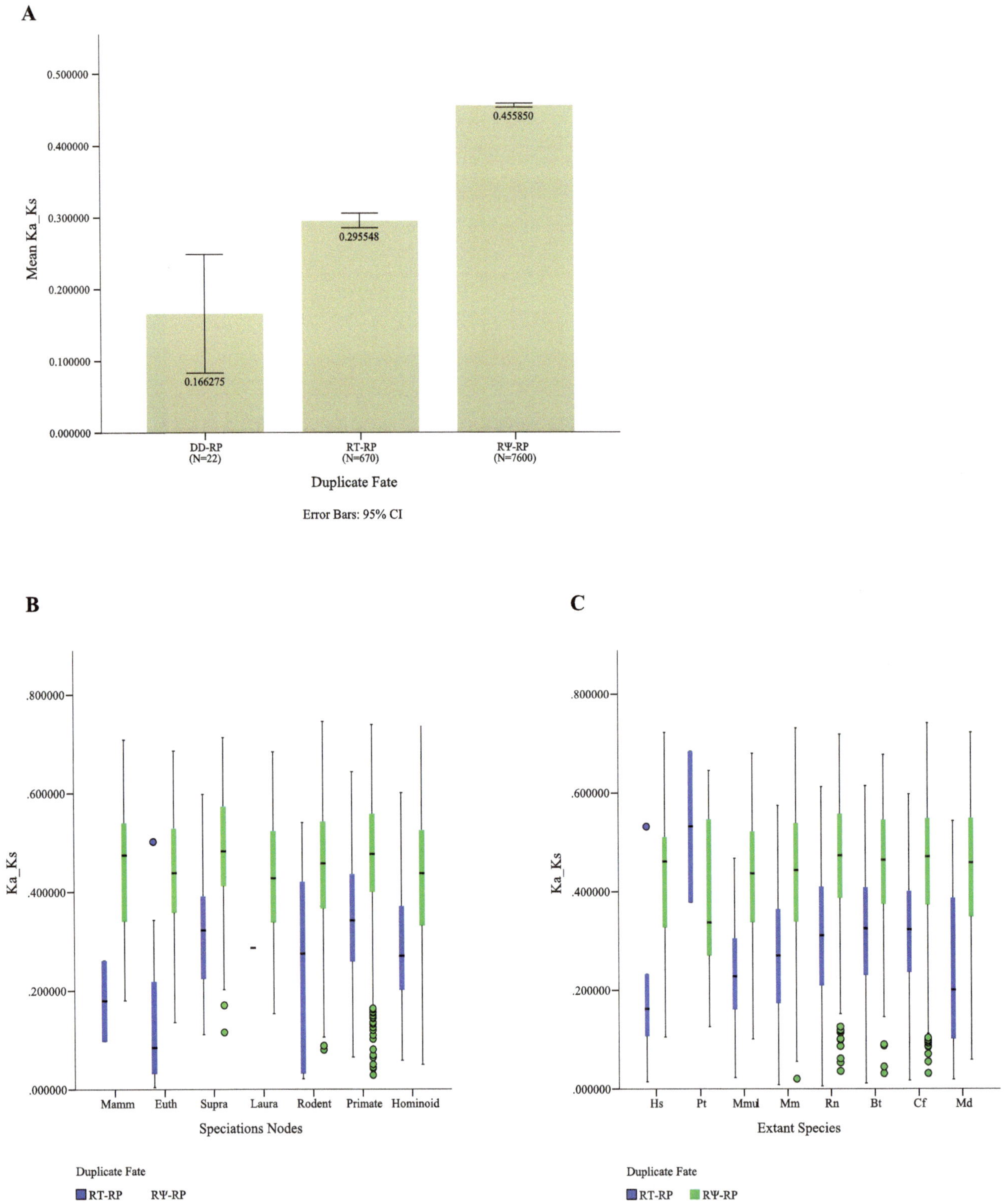

Figure 5. Selective Pressures on Ribosomal Protein Gene Duplicates. A) Mean Ka/Ks ratios were calculated for all classes (DD-RPs, RT-RPs and RΨ-RPs) of RP gene duplicates using the Nei Gojobori method. Results were then filtered based on p-values (<0.1) and the fraction of the parental gene represented by each duplicate (>65%). Error bars represent 95% confidence interval. B) Box and whisker plots for RT-RPs (blue) and RΨ-RPs (green) were generated for inner speciation nodes and C) Extant Species. DD-RPs were not included in the analyses due to small sample size (N = 3).

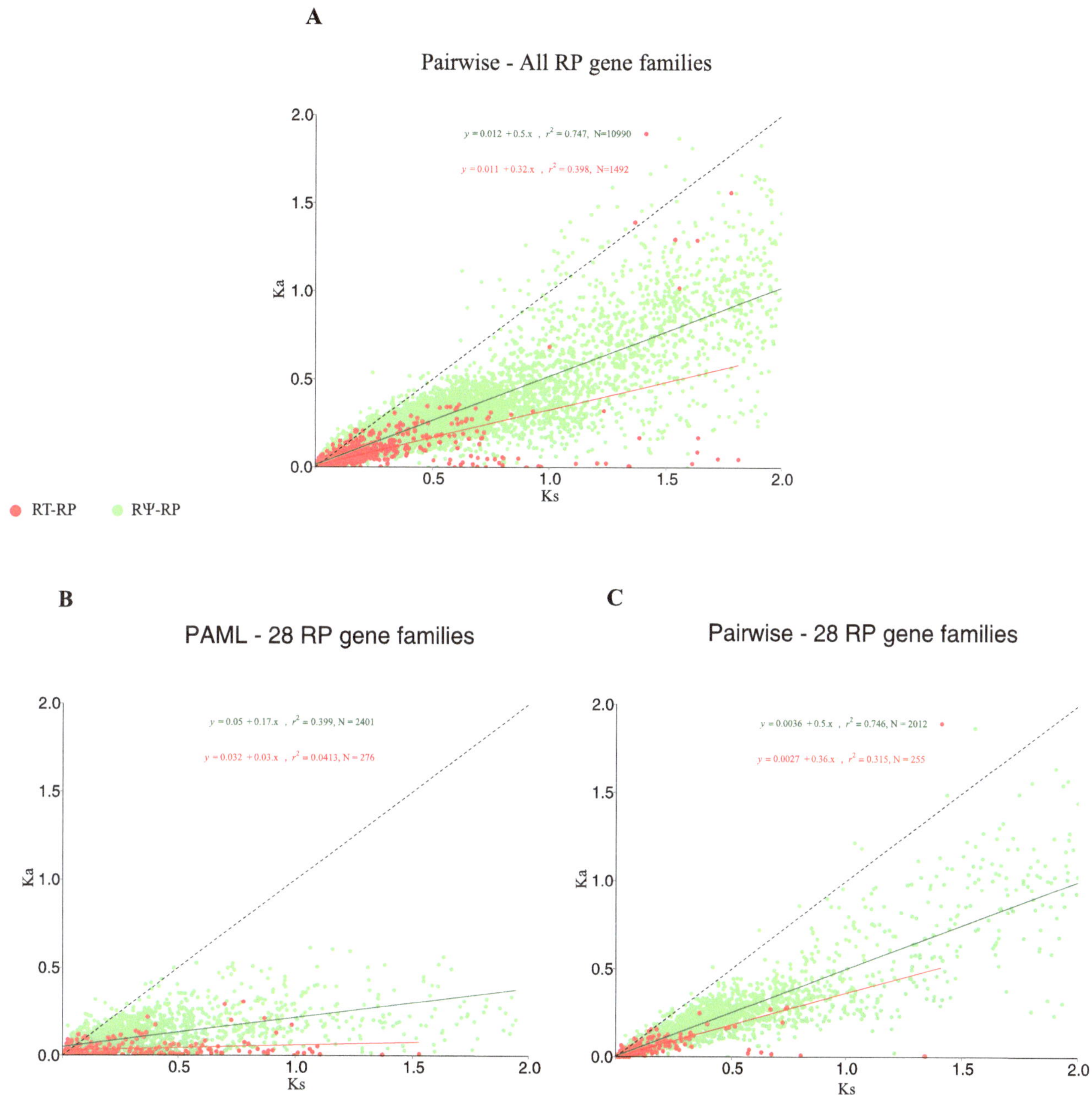

Figure 6. Scatterplots for pair-wise and branch-wise Ka against Ks values show that both methods capture the strong selective pressure acting on the RP gene duplicates. Plot of Ka against Ks for RP families with branch-wise and pair-wise methods. A) Distribution for all 76 RP families using the pair-wise selective pressure calculation method. Red dots represent RT-RPs and green dots represents RΨ-RPs. The black solid line represents Ka = Ks and the red & green line are the best line of fits for the distribution of RT-RPs and RΨ-RPs respectively. B) Distribution for 28 RP families analyzed by codeml program in PAML. C) Distribution of the aforementioned 28 families from PAML analysis using the pair-wise method.

order to avoid false positives resulting from the strong sequence similarity between parent genes and duplicates, a large number of EST matches were filtered out, suggesting that our estimates of active transcription are likely underestimates. The majority of EST data results from duplicates arising along younger portions of the mammalian lineage (younger primate or rodent lineage or the mouse, hominoid and human branches (Figure 7A)). For example, 320 out of 335 ESTs are either primate-specific or mouse-specific.

Finally, we compared the selective pressure on all RT-RP duplicate genes with evidence of expression (in the form of EST matches) to those without matching EST data. DD-RPs were not analyzed due to the small sample size (N = 3). Expressed RT-RP duplicates exhibit significantly higher levels of purifying selection than their non-transcribed counterparts (mean Ka/Ks value of 0.12 (95% CI 0.09, 0.15) compared to 0.24 (95% CI 0.22, 0.26) respectively (Figure 7B)). However, no similar difference in selective pressure is observed between expressed RΨ-RPs, whose mean Ka/Ks value is 0.36 (95% CI 0.31, 0.41), and their non-transcribed counterparts (mean Ka/Ks value of 0.35 (95% CI 0.34, 0.36)) (Figure 7B). Similarly, pairwise sequence distances for

Table 1. Likelihood Ratio Test statistics (LRT) for random site models.

Genes		M1a[a] vs. M2a[b]	M1a[a] vs. fix_omega = 1[b]
RPL28	df	2	2
	$2\Delta\ell$	−150.153	1209.854
	P-Value	1	<0.001
RPL14	df	2	2
	$2\Delta\ell$	−70.308	2043.084
	P-Value	1	<0.001
RPS16	df	2	2
	$2\Delta\ell$	1024.072	2410.42
	P-Value	0	<0.001
RPS18	df	2	2
	$2\Delta\ell$	−1269.931	1332.272
	P-Value	1	<0.001

[a]Alternative model;
[b]null model; $2\Delta\ell = 2(\ell_1 - \ell_0)$, df degrees of freedom.

these duplicated genes show that expressed RT-RP duplicates (mean pairwise distance of 0.02 (95% CI 0.01, 0.03)) have diverged less than intact non-transcribed duplicates (pairwise distance of 0.06 (95% CI 0.05, 0.07)) (Figure 7C). However, just as in the Ka/Ks analysis above, expressed RΨ-RPs have diverged less than non-transcribed pseudogenized duplicates (0.09 (95% CI 0.06, 0.13) compared to 0.16 (95% CI 0.15, 0.17) (Figure 7C). It is interesting to note that mean pairwise sequence distances are lower for pseudogenes with ESTs, suggesting purifying selective pressure prior to the pseudogenization event.

Discussion

Here we provide a near-comprehensive study of ribosomal protein gene sequence evolution, duplication, and loss in eight mammalian species. We find that these highly-conserved and highly-expressed genes are, not unexpectedly, frequently duplicated by retrotransposition, and comprise the largest such class of genes in mammalian genomes. It is quite clear that RNA-mediated RP duplicates (14,524 out of 14,552 events) dominate RP gene families. There is rare evidence of an old DNA duplicate, RPL3L (see Figure S5 for family tree) that has been retained for function (see [36,63]). However, the presence of only a very few such old duplicates and a complete absence of recent DD-RP duplicates, implies selection against the retention of DNA-mediated RP duplications. Negative selection against DNA-duplicates combined with the abundance of ribosomal protein gene mRNAs, and the observation that reverse transcription and transposition are more efficient on short GC-poor sequences like the ribosomal mRNAs [64,65], likely explain the almost complete dominance of retroduplication events in the evolution of the mammalian ribosomal protein genes.

Less expectedly, we also find that many of these retrotransposed RP duplicates are under strong purifying selective pressure (N = 1,724), and that this pressure is greatest amongst transcribed RP retroduplicates, regardless of whether these duplicates have been pseudogenized or retain intact coding regions. As gene duplicates are often found to be under relaxed selective pressures [66–68], the strength of selective pressure we observe across RT-RP duplicates was unexpected. It was not immediately obvious to us why so many duplicates are under selective pressure when the parental ribosomal genes exist almost exclusively in single copy, when DNA-mediated duplications appear to be selected against, where RP transcript levels are tightly regulated for optimal fitness, and the duplications are occurring over a timeframe where ribosomal evolution is thought to be almost stationary. Indeed, we expected RT-RP duplicates to be evolving neutrally for exactly these reasons.

The precise combination of forces enabling the retention of duplicated genes in complex genomes leading to the formation of gene families has been a subject of much study [69,70]. Several interesting studies have focused on the fate of ribosomal protein duplicates in non-mammalian lineages. RP duplicate fate after WGD events have been closely studied in yeasts and plants [71,72]. RP duplicates have been shown to be retained to maintain gene dosage after WGD [6,72–74]. But these retention events are not expected to affect the relative stoichiometry between RPs. However, the primary mode of duplication observed in the present study is RNA-mediated, small-scale duplications, which could result in severe stoichiometric imbalance. Additionally, it has been implied that RP duplicates after WGD's can be selected for defined functions like increasing levels of gene expression and divergence of gene function [75]. But evidence for this is not readily apparent in mammalian RT-RPs.

Population genetics suggests that duplicates should be lost long before adaptive forces can fix them in the population [76]. Many models have been forwarded that attempt to explain this apparent paradox and provide scenarios within which duplicated genes will be retained at the levels observed in many genomes (for an excellent review see [77]). In an attempt to understand the origin of the widespread selective pressure we observe on mammalian ribosomal protein retroduplicates, we focus this discussion on the ability of current models to account for this phenomenon.

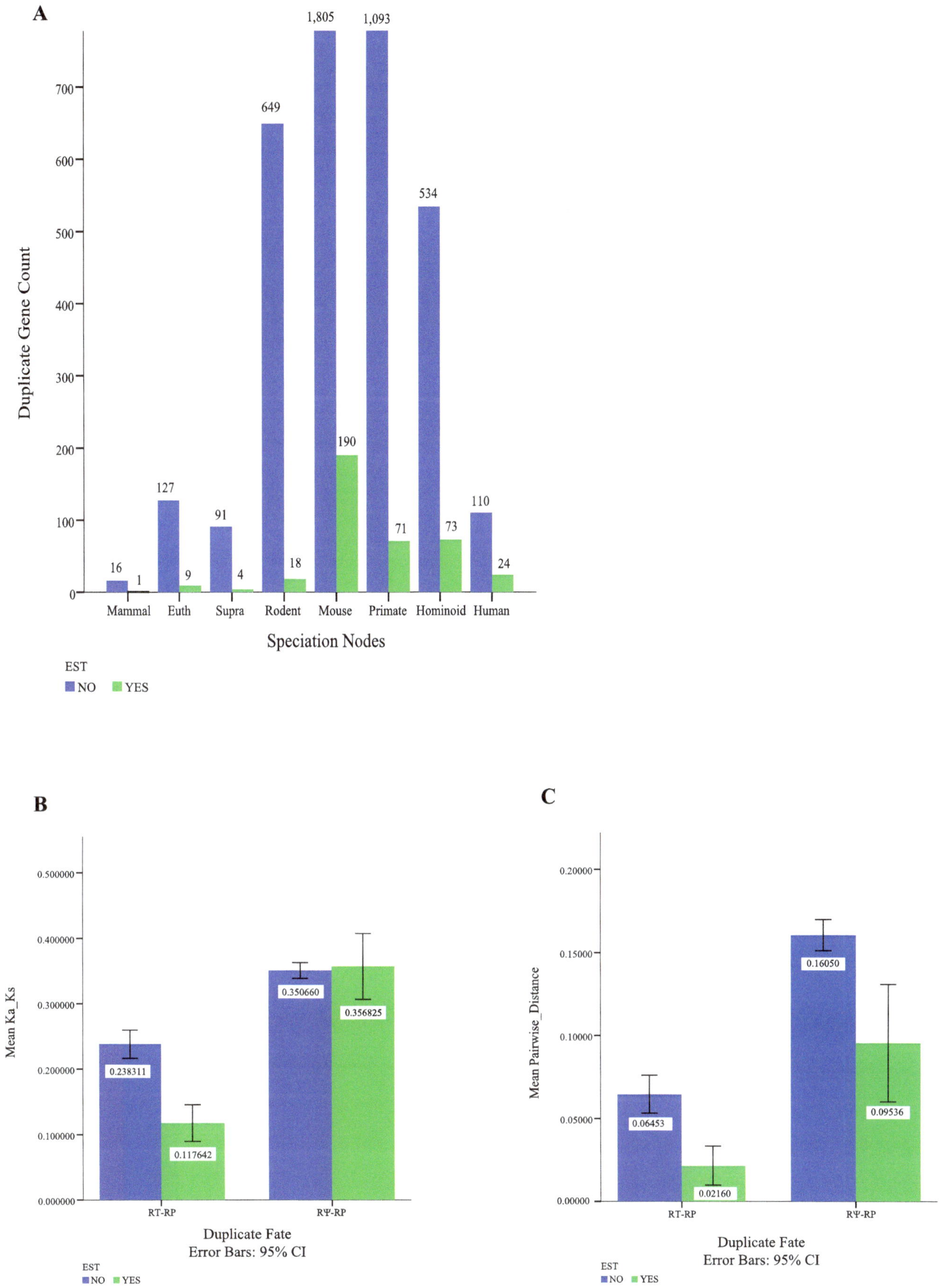
A
Duplicate Gene Count
Speciation Nodes
Mammal
Euth
Supra
Rodent
Mouse
Primate
Hominoid
Human
16
1
127
9
91
4
649
18
1,805
190
1,093
71
534
73
110
24
EST
NO
YES
B
Mean Ka_Ks
0.238311
0.117642
0.350660
0.356825
RT-RP
RΨ-RP
Duplicate Fate
Error Bars: 95% CI
C
Mean Pairwise_Distance
0.06453
0.02160
0.16050
0.09536
Duplicate Fate
Error Bars: 95% CI

Figure 7. Human/Mouse EST Counts and evolutionary selective pressure. A) Counts of human and mouse genes with EST (green) versus NO EST (blue) at all speciation nodes were calculated. B) Ka/Ks values were calculated for RT-RPs and RΨ-RPs for gene duplicates with (green) and without expression (blue). C) Pairwise distances for RP-RTs and RΨ-RPs with and without expression.

Can existing models account for observed number of conserved RP-dups?

Neofunctionalization. After assessing the current literature for existing retrogenes and ribosomal gene duplicates, we wanted to evaluate the gene duplication models. Gene duplication models for neofunctionalization, namely, the Dyhkhuizen-Hartl model, the Adaptation model, and the Adaptive Radiation model, predict that the rate of evolution after gene duplication will be accelerated in the duplicated copy and constrained in the original gene [67,78]. However, these models fail to account for thousands of ribosomal retrogenes in our dataset which demonstrates that rather than experiencing neutral selection, the new copy is under stringent purifying selection. Moreover, while some extra-ribosomal functions for divergent RP duplicates has been observed these events appear to be very rare [79]. Therefore, neofunctionalization models appear unlikely to account for the very large number of conserved ribosomal protein gene retroduplicates in mammalian genomes.

Subfunctionalization. Subfunctionalization, and its most cited model, DDC does appear to account for the retention of some number of gene duplicates [80]. DDC postulates that the genetic drift and accumulation of mutations will cause the loss of specific subfunctions from each copy of the duplicated genes. Once one copy has lost an essential function, selection on that function in the other duplicate will be reasserted. Eventually the two copies preserve largely non-overlapping complementary functions and both must be maintained by selection [80]. This division of function can result from changes in the regulatory regions or the coding regions of duplicated genes, and is most often envisioned as a driving force for the divergence of gene expression [for example see [74]].

However, DDC seems an improbable model for retention of the ribosomal protein retrogenes due to the fact that rather than appearing to drift, the coding regions of these duplicates are under strong purifying selective pressure. Large numbers of degenerative mutations in the coding regions are not observed until after pseudogenization. Also, because RT-RP duplicates do not carry any regulatory information, the most likely scenario for DDC, the evolution of complementary regulatory regions is unlikely. In addition, EST signatures retrieved from our pipeline and a review of existing literature [81] suggests that retroduplicates typically have a much narrower expression profile compared to the ubiquitous expression patterning of their parents, while the parents never seem to lose ubiquitous expression, as would be expected under DDC. Hence, division of function in such a manner seems improbable for ubiquitously expressed ribosomal parental genes. Other subfunctionalization models like EAC [82], and specialization and gene sharing [83] require neutral selection on the duplicate copy [84] and are not consistent with the purifying selective pressure we observe.

Gene conservation. Gene conservation is another outcome that can be used to explain the retention of retrogenes. The primary gene conservation model that has been employed to explain gene retention is the dosage model, which posits that gene duplicates are retained in order to produce more of the same gene product [69]. In comparison, the dosage compensation model states that the gene duplicates can compensate for the activity of the source gene [85].

The RP genes are under strict transcriptional regulatory control to maintain equimolar ratio of ribosomal constituents [86–88], and changes in ribosomal protein levels, including overexpression, are often highly deleterious [89]. This point is confirmed by DeSmet et al. 2013 paper [90] as they suggest that retention of small scale duplications (SSDs) will result in the stoichiometric imbalance among protein complexes and that the dosage balance hypothesis would work for a WGD as relative ratios among subunits can be flawlessly maintained, which would not be the case with SSDs. Similar conclusions were drawn for SSDs, suggesting that they would be selected against in a highly connected protein network [91]. This suggests that retroduplications that alter gene dosage would be selected against, not favored. Another very important piece of evidence that argues strongly against the retention of RP-RT duplicates by dosage is the study conducted by Kittler et al and Gilsdorf et al. [92,93]. In this study 34 ribosomal retrogenes (highly conserved old, new, intact and pseudogenized candidates) were knocked-down with no detectable phenotypic defects. However, knock-down of each of 70 parental RP genes had drastic phenotypic defects on the cells, with no evidence of retrogenes compensating for the loss of parental gene products (data obtained and analyzed from [94]). Previous work done on *Paramecium tetraurelia* [95,96] discusses about dosage compensation affecting the short term retention rate of duplicate genes after WGD's, while maintaining stoichiometry. While they correctly predict selection against the retention of non-balanced duplicates, they do not predict the knock-down results obtained in mammalian RT-RPs discussed earlier.

Retention of RT-RPs cannot be readily explained. Due to the fact that RT-RPs are very abundant, are highly conserved, lack parental regulatory regions, and because changes in ribosomal gene dosage are strongly selected against, the retention of the RT-RP duplicates is not readily explained using current models of Ohno's three trajectories of dosage, subfunctionalization, and neofunctionalization [1]. Because of this, identifying the forces leading to the retention of these thousands of highly conserved, expressed duplicates in mammalian genomes is likely to require a new model for the retention of gene duplicates.

Dominant-Negative effects acting to preserve RT-RP duplicates. One factor not fully explored in most existing models for the retention of retroduplicated genes is the potential for dominant-negative effects of missense mutations on cellular processes. Mutant proteins can act in a dominant-negative fashion in a wide variety of ways [97–99], and this mechanism could account for the strong purifying selective pressure we observe on duplicated genes. Importantly, a dominant-negative mechanism does not require complementation, neofunctionalization, subfunctionalization, ubiquitous expression, or a selective advantage for the new copy. The acquisition of dominant-negative mutations in duplicates may represent a threat to the viability of an organism *via* expression alone. Thus, these gene copies will remain under purifying selection until they are inactivated (pseudogenized or transcriptionally silenced).

We suggest that dominant negative phenotypes may exert an immediate and strong purifying selective pressure upon any duplicated gene, with this pressure varying directly with the potential for the gene product to act in a dominant negative fashion [99]. The pairwise Ka/Ks values of less than 1 that we observe may be the average of a regime of intense purifying selective pressure, followed by drift after pseudogenization. In a recent study of flowering plants, De Smet et al., postulate a very similar idea that dominant negative model constrains genes to be maintained as single copies to avoid non-specific interactions [90]. Strongly conserved multiprotein complexes like the ribosome are the most commonly observed context for dominant negative phenotypes, but dominant negative phenotypes are not restricted to such multi-protein complexes, in fact, they are widespread [98]. Because selection against dominant negative alleles acts immediately upon newly duplicated genes, and serves to maintain gene products in a very restricted portion of protein conformational space, it likely facilitates the retention of duplicates by many of the models described above by increasing the half-life of functional alleles in the population and the exploration of the small local region of allowable variation in protein conformation. Liberles and coworkers have proposed similar models in the context of negative pleiotropy [91,100–102], and have reached parallel conclusions on the impact of these bottlenecks in sequence space during evolution. To gain support for these models of gene family evolution, it will be important to functionally test the predictions of these models in experimentally tractable systems.

Supporting Information

Figure S1 Observed frequencies for RNA-mediated duplicates are much higher than expected frequencies in RP families. Observed frequencies and expected frequencies (in brackets) shown for each speciation node for 5 mammalian genomes. The values were generated using data table created in Jun et al [68]. Observed frequencies for RP-RTs were derived from ribosomal families (Number of RP-RTs/Total Number of Duplicates in RPs) and expected frequencies for intact retro-duplicates were derived from 8872 non-ribosomal gene families (Number of RTs in non-rp gene families/Total Number of duplicates in non-rp gene families). All diversification times are from Ureta-Vidal et al [54].

Figure S2 No Species-specific bias seen based on duplicate fates in ribosomal protein gene families. Distribution of duplicates (annotated by fate) across 8 mammalian species. DD-RPs, RT-RPs and RΨ-RPs are shown in blue, green and yellow respectively.

Figure S3 High sequence conservation observed across all duplicate fates in terms of pairwise distances. DD-RPs and RT-RPs were seen to be under comparatively stronger conservation than RΨ-RPs. Using pairwise distances instead of Ka/Ks ratios as seen in figure 4. Error bars represent 95% Confidence Interval.

Figure S4 Gene Tree for RPL10A showing PAML branch specific omega values leading up to a clade. An abridged Gene tree of RPL10A generated by parsimony-based syntenic method (see Methods). The branch specific omega values are listed at each node in purple. Ka/Ks values are represented at all leaves in green. The RT-RP duplicates and their omega values are highlighted in red.

Figure S5 Reconstructed evolutionary history for ribosomal protein gene RPL3. Tube Tree showing RPL3L which is a DD-RP that can be seen persistent in all mammals since its origination (green line). Refer to figure 3 legend for tree annotation.

Figure S6 Scatterplots for pair-wise Ka against Ks values show a strong selective pressure acting on the RP gene duplicates in older clades. Panel of Ka against Ks graphs for RP duplicates using pair-wise method. Each panel represents different clades for 8 mammalian genomes we studied. Red dots represent RT-RPs and green dots represents RΨ-RPs. The black dashed line represents Ka = Ks and the red & green line are the best line of fits for the distribution of RT-RPs and RΨ-RPs respectively.

Figure S7 Scatterplots for pair-wise Ka against Ks values show variable levels of purifying selection acting on the RP gene duplicates across all lineage specific clades. Each panel represents lineage specific clade of 8 mammalian genomes. Refer to figure S6 for more information on the graph. Refer to figure 2 for abbreviations.

Table S1 Table representing all our RP gene duplicates in 8 mammalian genomes and outgroup chicken. The clade_num represents the syntenic relationships between gene duplicates.

Table S2 Table for all RP families that have PAML branch-specific Ka and Ks values. The clade_num represents the syntenic relationships between duplicates.

Table S3 Log-likelihood and parameter estimates generated from random-site models for RP genes. P = number of free parameters for each model, l = log-likelihood value for each model.

Appendix S1 74 RP gene trees with all annotated duplication events.

Acknowledgments

We thank Dr. Jin Jun (University of Connecticut and Max-Plank Institute Leipzig, Germany) for immense help with setting up the project and valuable suggestions during the inception of the work presented in the manuscript. We thank Dr. Victoria Robinson (University of Connecticut) for valuable suggestions in the manuscript regarding ribosomal protein biology.

Author Contributions

Conceived and designed the experiments: CEN APD. Performed the experiments: APD AO MDG. Analyzed the data: APD CEN AO AS. Contributed reagents/materials/analysis tools: CEN MDG AO APD. Wrote the paper: APD AO CEN.

References

1. Ohno S (n.d.) Evolution by gene duplication. Allen and Unwin, London.
2. Demuth J, Bie T, Stajich J, Cristianini N (2006) The evolution of mammalian gene families. PloS one.
3. Bailey J, Gu Z, Clark R, Reinert K, Samonte R (2002) Recent segmental duplications in the human genome. Science (New York, N.Y.).

4. Bailey JA, Eichler EE (2006) Primate segmental duplications: crucibles of evolution, diversity and disease. Nature reviews. Genetics 7: 552–64. doi:10.1038/nrg1895.
5. Maere S, Bodt S, Raes J (2005) Modeling gene and genome duplications in eukaryotes. Proceedings of the.
6. Hakes L, Pinney JW, Lovell SC, Oliver SG, Robertson DL (2007) All duplicates are not equal: the difference between small-scale and genome duplication. Genome biology 8: R209. doi:10.1186/gb-2007-8-10-r209.
7. Wolfe K, Shields D (1997) Molecular evidence for an ancient duplication of the entire yeast genome. Nature.
8. Meyer A, Schartl M (1999) Gene and genome duplications in vertebrates: the one-to-four (-to-eight in fish) rule and the evolution of novel gene functions. Current opinion in cell biology.
9. Panopoulou G, Hennig S, Groth D, Krause A (2003) New evidence for genome-wide duplications at the origin of vertebrates using an amphioxus gene set and completed animal genomes. Genome.
10. Dehal P, Boore J (2005) Two rounds of whole genome duplication in the ancestral vertebrate. PLoS biology.
11. Zhang L, Lu H, Chung W, Yang J, Li W (2005) Patterns of segmental duplication in the human genome. Molecular biology and evolution.
12. Brosius J, Gould SJ (1992) On "genomenclature": a comprehensive (and respectful) taxonomy for pseudogenes and other "junk DNA". Proceedings of the National Academy of Sciences of the United States of America 89: 10706–10.
13. Harrison PM, Zheng D, Zhang Z, Carriero N, Gerstein M (2005) Transcribed processed pseudogenes in the human genome: an intermediate form of expressed retrosequence lacking protein-coding ability. Nucleic Acids Res 33: 2374–83. doi:10.1093/nar/gki531.
14. Watanabe T, Totoki Y, Toyoda A, Kaneda M (2008) Endogenous siRNAs from naturally formed dsRNAs regulate transcripts in mouse oocytes. Nature.
15. Parker H, VonHoldt B, Quignon P, Margulies E (2009) An expressed fgf4 retrogene is associated with breed-defining chondrodysplasia in domestic dogs. Science (New York, N.Y.).
16. Parker H, VonHoldt B, Quignon P, Margulies E (2009) An expressed fgf4 retrogene is associated with breed-defining chondrodysplasia in domestic dogs. Science (New York, N.Y.).
17. Yu H, Jiang H, Zhou Q, Yang J, Cun Y (2006) Origination and evolution of a human-specific transmembrane protein gene, c1orf37-dup. Human molecular.
18. Jun J, M II, Mandoiu II, oiu, Nelson CE (2009) Identification of mammalian orthologs using local synteny. BMC Genomics 10: 630. doi:10.1186/1471-2164-10-630.
19. Jun J, Ryvkin P, Hemphill E, Nelson C (2009) Duplication mechanism and disruptions in flanking regions determine the fate of Mammalian gene duplicates. Journal of computational biology: a journal of computational molecular cell biology 16: 1253–66. doi:10.1089/cmb.2009.0074.
20. Yu H, Jiang H, Zhou Q, Yang J, Cun Y (2006) Origination and evolution of a human-specific transmembrane protein gene, c1orf37-dup. Human molecular.
21. Rodriguez V, Chen Y, Elkahloun A, Dutra A, Pak E, et al. (2007) Chromosome 8 BAC array comparative genomic hybridization and expression analysis identify amplification and overexpression of TRMT12 in breast cancer. Genes, Chromosomes and Cancer 46: 694–707. doi:10.1002/gcc.20454.
22. Jiang Z, Chu P, Woda B, Liu Q, Balaji K (2008) Combination of quantitative IMP3 and tumor stage: a new system to predict metastasis for patients with localized renal cell carcinomas. Clinical Cancer.
23. Jiang Z, Lohse C, Chu P, Wu C, Woda B (2008) Oncofetal protein IMP3. Cancer.
24. Tan J, Dunn J, Jaeken J (1996) Mutations in the MGAT2 gene controlling complex N-glycan synthesis cause carbohydrate-deficient glycoprotein syndrome type II, an autosomal recessive disease. American journal of human.
25. Rosengren A, Jokubka R, Tojjar D, Granhall C (2010) Overexpression of alpha2A-adrenergic receptors contributes to type 2 diabetes. Science.
26. Roman T, Polanczyk G, Zeni C, Genro J (2005) Further evidence of the involvement of alpha-2A-adrenergic receptor gene (ADRA2A) in inattentive dimensional scores of attention-deficit/hyperactivity disorder. Molecular.
27. Carnemolla A, Fossale E, Agostoni E (2009) Rrs1 is involved in endoplasmic reticulum stress response in Huntington disease. Journal of Biological.
28. Almeida A, Murakami Y, Layton D, Hillmen P (2006) Hypomorphic promoter mutation in PIGM causes inherited glycosylphosphatidylinositol deficiency. Nature medicine.
29. Yu Z, Morais D, Ivanga M, Harrison P (2007) Analysis of the role of retrotransposition in gene evolution in vertebrates. BMC Bioinformatics.
30. Wang P (2004) X chromosomes, retrogenes and their role in male reproduction. Trends in Endocrinology & Metabolism.
31. Hori H, Higo K, Osawa S (1977) The rates of evolution in some ribosomal components. Journal of Molecular Evolution.
32. Wool IG, Chan YL, Gluck A (n.d.) Structure and evolution of mammalian ribosomal proteins. Biochem Cell Biol 73: 933–47.
33. Wool IG, Chan Y, Gluck A (n.d.)Structure and evolution of mammalian ribosomal proteins. Biochem Cell Biol 73: 933–47.
34. Mager W (2013) Control of ribosomal protein gene expression. Biochimica et biophysica acta 949: 1–15.
35. Fried M, Colombo P, Colombo MF (1992) Functional elements of the ribosomal protein L7a (rpL7a) gene promoter region and their conservation between mammals and birds. Nucleic acids research 20: 3367.
36. Thorrez L, Deun K, Tranchevent L (2008) Using ribosomal protein genes as reference: a tale of caution. PloS one: 1–8.
37. Hariharan N, Kelley, Perry R (1989) Equipotent mouse ribosomal protein promoters have a similar architecture that includes internal sequence elements. Genes & development.
38. Kenmochi N, Kawaguchi T, Rozen S, Davis E (1998) A map of 75 human ribosomal protein genes. Genome.
39. Marygold S, Coelho C, Leevers S (2005) Genetic analysis of RpL38 and RpL5, two minute genes located in the centric heterochromatin of chromosome 2 of Drosophila melanogaster. Genetics.
40. Willig T, Draptchinskaia N, Dianzani I (1999) Mutations in ribosomal protein S19 gene and Diamond Blackfan anemia: wide variations in phenotypic expression. Blood.
41. Draptchinskaia N, Gustavsson P, Andersson B, Pettersson M, Willig TN, et al. (1999) The gene encoding ribosomal protein S19 is mutated in Diamond-Blackfan anaemia. Nature genetics 21: 169–75. doi:10.1038/5951.
42. Zhang Y, Duc A, Rao S, Sun X, Bilbee A (2013) Control of Hematopoietic Stem Cell Emergence by Antagonistic Functions of Ribosomal Protein Paralogs. Developmental cell.
43. Huang X, Chen Y, Li W, Cohen S (2010) The Rps23rg gene family originated through retroposition of the ribosomal protein s23 mRNA and encodes proteins that decrease Alzheimer's β-amyloid level and tau. Human molecular.
44. Uechi T, Maeda N, Tanaka T (2002) Functional second genes generated by retrotransposition of the X-linked ribosomal protein genes. Nucleic Acids Res.
45. Thorrez L, Deun K, Tranchevent L (2008) Using ribosomal protein genes as reference: a tale of caution. PloS one: 1–8.
46. Wapinski I, Pfeffer A, Friedman N, Regev A (2007) Automatic genome-wide reconstruction of phylogenetic gene trees. Bioinformatics.
47. Flygare J, Aspesi A, Bailey J, Miyake K (2007) Human RPS19, the gene mutated in Diamond-Blackfan anemia, encodes a ribosomal protein required for the maturation of 40S ribosomal subunits: 1–8.
48. Fried M, Colombo P, Colombo M (1992) Functional elements of the ribosomal protein L7a (rpL7a) gene promoter region and their conservation between mammals and birds. Nucleic acids research 20: 3367.
49. Zhang Z, Harrison P, Gerstein M (2002) Identification and analysis of over 2000 ribosomal protein pseudogenes in the human genome. Genome Res.
50. Balasubramanian S, Zheng D, Liu Y, Fang G, Frankish A, et al. (2009) Comparative analysis of processed ribosomal protein pseudogenes in four mammalian genomes. Genome Biol 10: R2. doi:10.1186/gb-2009-10-1-r2.
51. Flicek P, Amode M, Barrell D, Beal K, Brent S, et al. (n.d.) Ensembl 2011. Nucleic acids research 39: D800-6.
52. Zhang Z, Carriero N, Zheng D, Karro J (2006) PseudoPipe: an automated pseudogene identification pipeline. Bioinformatics (Oxford, England).
53. Goodman M, Czelusniak J, Moore GW, Romero-Herrera AE, Matsuda G (1979) Fitting the Gene Lineage into its Species Lineage, a Parsimony Strategy Illustrated by Cladograms Constructed from Globin Sequences. Systematic Biology 28: 132–163. doi:10.1093/sysbio/28.2.132.
54. Ureta-Vidal A, Ettwiller L, Birney E (2003) Comparative genomics: genome-wide analysis in metazoan eukaryotes. Nature Reviews Genetics.
55. Mikkelsen T, Wakefield M, Aken B, Amemiya C (2007) Genome of the marsupial Monodelphis domestica reveals innovation in non-coding sequences. Nature.
56. Yang Z (2007) PAML 4: Phylogenetic Analysis by Maximum Likelihood. Molecular biology and evolution 24: 1586–1591.
57. Farfán M, Miñana-Galbis D, Fusté MC, Lorén JG (2009) Divergent evolution and purifying selection of the flaA gene sequences in Aeromonas. Biology Direct 2009 4: 23 4: 23. doi:10.1186/1745-6150-4-23.
58. Yang Z, Swanson WJ (2002) Codon-Substitution Models to Detect Adaptive Evolution that Account for Heterogeneous Selective Pressures Among Site Classes. Molecular biology and evolution 19: 49–57.
59. Bastian F, Parmentier G, Roux J, Moretti S, Laudet V, et al. (n.d.) Bgee: Integrating and Comparing Heterogeneous Transcriptome Data Among Species: 124–131. doi:10.1007/978-3-540-69828-9_12.
60. Khachane AN, Harrison PM (2009) Assessing the genomic evidence for conserved transcribed pseudogenes under selection. BMC Genomics 10: 435. doi:10.1186/1471-2164-10-435.
61. Zheng D, Zhang Z, Harrison PM, Karro J, Carriero N, et al. (2005) Integrated Pseudogene Annotation for Human Chromosome 22: Evidence for Transcription. Journal of Molecular Biology 349: 27–45. doi:10.1016/j.jmb.2005.02.072.
62. Nielsen R, Yang Z (1998) Likelihood Models for Detecting Positively Selected Amino Acid Sites and Applications to the HIV-1 Envelope Gene. Genetics 148: 929–936.
63. Raay TJ, Connors TD, Klinger KW, L GM, Landes GM, et al. (1996) A Novel Ribosomal Protein L3-like Gene (RPL3L) Maps to the Autosomal Dominant Polycystic Kidney Disease Gene Region. Genomics 37: 172–176. doi:10.1006/geno.1996.0538.
64. Gonçalves I, Goncalves I, Duret L, Mouchiroud D (n.d.) Nature and Structure of Human Genes that Generate Retropseudogenes. Genome research 10: 672–678. doi:10.1101/gr.10.5.672.
65. Zhang Z, Harrison P, Gerstein M (2002) Identification and analysis of over 2000 ribosomal protein pseudogenes in the human genome. Genome Res.
66. Han MV, Hahn MW (2009) Identifying parent-daughter relationships among duplicated genes. Pacific Symposium on Biocomputing. Pacific Symposium on Biocomputing: 114–25.

67. Hahn MW (2009) Distinguishing among evolutionary models for the maintenance of gene duplicates. J Hered 100: 605–17. doi:10.1093/jhered/esp047.
68. Yu Z, Morais D, Ivanga M, Harrison P (2007) Analysis of the role of retrotransposition in gene evolution in vertebrates. BMC Bioinformatics.
69. Ohno S (n.d.) Evolution by gene duplication. Allen and Unwin, London.
70. Jun J, Ryvkin P, Hemphill E, M I, Mandoiu I, et al. (2009) The birth of new genes by RNA- and DNA-mediated duplication during mammalian evolution. Journal of computational biology: a journal of computational molecular cell biology 16: 1429–44. doi:10.1089/cmb.2009.0073.
71. Blanc G, Wolfe K (2004) Functional divergence of duplicated genes formed by polyploidy during Arabidopsis evolution. The Plant Cell Online.
72. Evangelisti AM, Conant GC (2010) Nonrandom Survival of Gene Conversions among Yeast Ribosomal Proteins Duplicated through Genome Doubling. Genome Biology and Evolution 2: 826–834. doi:10.1093/gbe/evq067.
73. Freeling M (2009) Bias in Plant Gene Content Following Different Sorts of Duplication: Tandem, Whole-Genome, Segmental, or by Transposition. Annual Review of Plant Biology 60: 433–453. doi:10.1146/annurev.arplant.043008.092122.
74. Amoutzias GD, He Y, Gordon J, Mossialos D, Oliver SG, et al. (2010) Posttranslational regulation impacts the fate of duplicated genes. Proceedings of the National Academy of Sciences. Vol. 107. pp. 2967–2971. doi:10.1073/pnas.0911603107.
75. Seoighe C, Wolfe KH (1999) Yeast genome evolution in the post-genome era. Current Opinion in Microbiology 2: 548–554. doi:10.1016/S1369-5274(99)00015-6.
76. Zhang J (2003) Evolution by gene duplication: an update. Trends in ecology & evolution.
77. Hahn M (2009) Distinguishing among evolutionary models for the maintenance of gene duplicates. J Hered 100: 605–17.
78. Innan H, Kondrashov F (2010) The evolution of gene duplications: classifying and distinguishing between models. Nature reviews. Genetics 11: 97–108. doi:10.1038/nrg2689.
79. Xue S, Barna M (2012) Specialized ribosomes: a new frontier in gene regulation and organismal biology. Nature Reviews Molecular Cell Biology, Published online: 23 May 2012; | doi:10.1038/nrm3359 13: 355–369. doi:doi:10.1038/nrm3359.
80. Force A, Lynch M, Pickett FB, Amores A, Yan YL, et al. (1999) Preservation of duplicate genes by complementary, degenerative mutations. Genetics 151: 1531–45.
81. Cusack B, Wolfe K (2007) Not born equal: increased rate asymmetry in relocated and retrotransposed rodent gene duplicates. Molecular biology and evolution.
82. Marais D, Rausher M (2008) Escape from adaptive conflict after duplication in an anthocyanin pathway gene. Nature.
83. Conrad B, Antonarakis SE (2007) Gene Duplication: A Drive for Phenotypic Diversity and Cause of Human Disease. Annual Review of Genomics and Human Genetics 8: 17–35. doi:10.1146/annurev.genom.8.021307.110233.
84. Innan H, Kondrashov F (2010) The evolution of gene duplications: classifying and distinguishing between models. Nature reviews. Genetics 11: 97–108.
85. Zhang J (2003) Evolution by gene duplication: an update. Trends in ecology & evolution.
86. Dennis PP, Nomura M (1974) Stringent control of ribosomal protein gene expression in Escherichia coli. Proceedings of the National Academy of Sciences of the United States of America 71: 3819–23.
87. Mager W (2013) Control of ribosomal protein gene expression. Biochimica et biophysica acta 949: 1–15.
88. Zhao Y, McIntosh K, Rudra D (2006) Fine-structure analysis of ribosomal protein gene transcription. and cellular biology.
89. Marygold S, Roote J, Reuter G, Lambertsson A (2007) The ribosomal protein genes and Minute loci of Drosophila melanogaster. Genome.
90. de Smet R, Adams KL, Klaas V, Vandepoele K, epoele, et al. (2013) Convergent gene loss following gene and genome duplications creates single-copy families in flowering plants. PNAS 110: 2898–2903. doi:10.1073/pnas.1300127110.
91. Konrad A, Teufel AI, Grahnen JA, Liberles DA (2011) Toward a General Model for the Evolutionary Dynamics of Gene Duplicates. Genome Biology and Evolution 3: 1197–1209. doi:10.1093/gbe/evr093.
92. Gilsdorf M, Horn T, Arziman Z, Pelz O, Kiner E, et al. (2010) GenomeRNAi: a database for cell-based RNAi phenotypes. 2009 update. Nucleic acids research 38: D448–52.
93. Kittler R, Putz G, Pelletier L, Poser I, Heninger A (2004) An endoribonuclease-prepared siRNA screen in human cells identifies genes essential for cell division. Nature.
94. Gilsdorf M, Horn T, Arziman Z, Pelz O, Kiner E, et al. (2010) GenomeRNAi: a database for cell-based RNAi phenotypes. 2009 update. Nucleic acids research 38: D448–52. doi:10.1093/nar/gkp1038.
95. Hughes T, Ekman D, Ardawatia H, Elofsson A (2007) Evaluating dosage compensation as a cause of duplicate gene retention in Paramecium tetraurelia. Genome.
96. Aury J, Jaillon O, Duret L, Noel B, Jubin C, et al. (2006) Global trends of whole-genome duplications revealed by the ciliate Paramecium tetraurelia. Nature 444: 171–178. doi:10.1038/nature05230.
97. Veitia RA (2010) A generalized model of gene dosage and dominant negative effects in macromolecular complexes. The FASEB Journal 24: 994–1002. doi:10.1096/fj.09-146969.
98. Veitia R (2007) Exploring the molecular etiology of dominant-negative mutations. The Plant Cell Online.
99. Gibson TJ, Spring J (1998) Genetic redundancy in vertebrates: polyploidy and persistence of genes encoding multidomain proteins. Trends in genetics: TIG 14: 46–9; discussion 49–50.
100. Liberles DA, Teufel AI, Liu L, Stadler T (2013) On the Need for Mechanistic Models in Computational Genomics and Metagenomics. Genome Biology and Evolution 5: 2008–2018. doi:10.1093/gbe/evt151.
101. Liberles DA, Tisdell MDM, Grahnen JA (2011) Binding constraints on the evolution of enzymes and signalling proteins: the important role of negative pleiotropy. Proceedings of the Royal Society B: Biological Sciences 278: 1930–1935. doi:10.1098/rspb.2010.2637.
102. Hughes T, Da Liberles (2007) The pattern of evolution of smaller-scale gene duplicates in mammalian genomes is more consistent with neo-than subfunctionalisation. Journal of Molecular Evolution.
103. Ureta-Vidal A, Ettwiller L, Birney E (2003) Comparative genomics: genome-wide analysis in metazoan eukaryotes. Nature Reviews Genetics.

6

Molecular Evolution of the Primate α-/θ-Defensin Multigene Family

Dong-Qiang Cheng[1,2], Ying Li[1,3], Jing-Fei Huang[1,4]*

1 State Key Laboratory of Genetic Resources and Evolution, Kunming Institute of Zoology, Chinese Academy of Sciences, Kunming, China, **2** Kunming College of Life Science, University of Chinese Academy of Sciences, Beijing, China, **3** Institute of Animal Genetics and Breeding, College of Animal Science and Technology, Sichuan Agricultural University, Yaan, China, **4** Kunming Institute of Zoology-Chinese University of Hongkong Joint Research Center for Bio-resources and Human Disease Mechanisms, Kunming, China

Abstract

The primate α-/θ-defensin multigene family encodes versatile endogenous cationic and amphipathic peptides that have broad-spectrum antibacterial, antifungal and antiviral activity. Although previous studies have reported that α-/θ-defensin (*DEFA/DEFT*) genes are under birth-and-death evolution with frequent duplication and rapid evolution, the phylogenetic relationships of the primate *DEFA/DEFT* genes; the genetic bases for the existence of similar antimicrobial spectra among closely related species; and the evolutionary processes involved in the emergence of cyclic θ-defensins in Old World monkeys and their subsequent loss of function in humans, chimpanzees and gorillas require further investigation. In this study, the *DEFA/DEFT* gene repertoires from primate and treeshrew were collected, followed by detailed phylogenetic, sequence and structure, selection pressure and comparative genomics analyses. All treeshrew, prosimian and simian *DEFA/DEFT* genes are grouped into two major clades, which are tissue-specific for enteric and myeloid defensins in simians. The simian enteric and myeloid α-defensins are classified into six functional gene clusters with diverged sequences, variable structures, altered functional constraints and different selection pressures, which likely reflect the antimicrobial spectra among closely related species. Species-specific duplication or pseudogenization within each simian cluster implies that the antimicrobial spectrum is ever-shifting, most likely challenged by the ever-changing pathogen environment. The *DEFT* evolved from the myeloid *DEFA8*. The prosegment of θ-defensin is detected with adaptive changes coevolving with the new protein fold of mature peptide, coincident with the importance of the prosegment for the correct folding of the mature peptide. Lastly, a less-is-hitchhiking hypothesis was proposed as a possible explanation for the expansion of pseudogene *DEFTP* and the loss of functional *DEFT*, where the gain or loss of the hitchhiker is determined by its adjacent driver gene during the birth-and-death evolutionary process.

Editor: Alexander M. Cole, University of Central Florida College of Medicine, United States of America

Funding: This work was supported by the National Natural Science Foundation of China (Grant No. 31123005) and Chinese Academy of Sciences (Grant No. 2007211311091) for JFH, and the "100-Talent Program" in Sichuan and National Natural Science Foundation of China (Grant No. 30600066) for YL. The funders had no role in study design, data collection and analysis, decision to publish, or preparation of the manuscript.

Competing Interests: The authors have declared that no competing interests exist.

* E-mail: huangjf@mail.kiz.ac.cn

These authors contributed equally to this work.

Introduction

Host immune systems evolve various defensive weapons against pathogens during the pathogen-host arms race. Defensins are small antimicrobial peptides that act as both effectors and mediators of host immunity. These endogenous cationic and amphipathic peptides have broad-spectrum antibacterial, antifungal and antiviral activity [1–6]. They also modulate the innate and adaptive immune systems by promoting or suppressing the proinflammatory responses during microbial infection [7–9]. In mammals, there are three structural subfamilies of defensins—α-, β- and θ-defensins—that differ by their tridisulfide motifs. The β-defensins are the oldest members and are found in most classes of vertebrates [10–13]. The α-defensins, which originate from the β-defensins, are younger members. The α-defensins are found in marsupials and most mammals but were lost in some Laurasiatheria species, such as cattle and dogs [14–17]. The θ-defensins are the only mammalian cyclic peptides currently known and originate from α-defensins through a nonsense mutation in the mature peptide. The θ-defensins are found in Old World monkeys and certain apes but underwent a loss of function in humans, chimpanzees and gorillas due to a nonsense mutation in the signal peptide [18,19].

Primate α-/θ-defensin (*DEFA/DEFT*) genes are located as multigene clusters on chromosome regions that are homologous with human chromosome 8p23 [16,20]. Like other multigene families, *DEFA/DEFT* genes can be constitutively expressed at high levels to produce variant functional proteins. In humans, there are six functional and tissue-specific α-defensin peptides. Human HNP1–HNP4, encoded by the *DEFA1*, *DEFA3* and *DEFA4* genes, are primarily expressed in neutrophils [1]. The *DEFA1* and *DEFA3* genes are genetic variants that encode proteins with a single amino acid difference and are also referred to as *DEFA1A3* genes. Unlike the single-copy *DEFA4*, the *DEFA1A3* genes have copy number polymorphisms [21–23]. In contrast, human HD5 and HD6, encoded by *DEFA5* and *DEFA6*, respectively, are expressed primarily in Paneth cells of the small intestine and play important roles in intestinal host defense and homeostasis [24–26]. Apart from the functional genes, there are many defensin pseudogenes—*DEFAP* and *DEFTP*—in humans [19,27]. In macaques, the new member θ-defensin encoded by *DEFT* is primarily expressed in the bone marrow and leukocytes [28].

Defensins are synthesized as pre-pro-defensins containing a signal peptide, a prosegment and a mature peptide. The prosegment, which serves as an intramolecular chaperone, assists in the correct disulfide pairing and proper folding of the mature peptide [29] and keeps the mature peptide inactive until it is cleaved by various proteolytic enzymes [30–32]. The mature peptides are cationic and amphipathic, which are important properties for inducing the depolarization and permeabilization of the microbial membrane [2,33]. The α-defensin monomer has a three-stranded antiparallel β-sheet structure with three intramolecular disulfide pairs linked as Cys1-Cys6, Cys2-Cys4 and Cys3-Cys5. Two monomers form an amphipathic dimer, which is stabilized by hydrophobic interactions and intermolecular hydrogen bonds between residues 18 and 20 (HNP4 numbering) in the second β-sheet [33,34]. The dimerization of α-defensins, in addition to their cationic and amphipathic character, is also important for their antimicrobial ability [35–37]. In contrast to the structure of α-defensins, the θ-defensins form a cyclic octadecapeptide through the posttranslational head-to-tail ligation of two nonapeptides and harbor three intermolecular disulfide pairs [18]. Recently, synthetic defensins have been studied and are being developed as potential antimicrobial peptide drugs [38–41].

Because of the frequent duplication and rapid evolution of primate α-/θ-defensins, the nomenclature and phylogenetic relationships among this multigene family are still ambiguous. Moreover, there is no clear phylogenetic classification related to the expression pattern or the confounding antimicrobial function of these α-defensins, although many functional studies indicate that α-defensins are effective microbicidal peptides against a wide variety of microorganisms. Previous studies have demonstrated that the α-/θ-defensin multigene family, like many other multigene families, is subject to birth-and-death evolutionary process with frequent gene duplication, pseudogenization and significant positive selection [42–44]. However, the molecular evolution of the undocumented antimicrobial spectra that are composed of functionally divergent α-/θ-defensins in humans and closely related primates should be further explored. In this study, the phylogenetic classification, sequence divergence and structural diversification of the primate α-/θ-defensins were investigated using molecular evolution and molecular dynamics analyses. Furthermore, the evolutionary processes involved in the emergence of cyclic θ-defensins and their subsequent loss of function in humans, chimpanzees and gorillas require investigation. Loss of function is a major driving force for phenotypic change and can be advantageous, deleterious or tolerated, as explained by the hypotheses of less-is-more, less-is-less and less-is-nothing, respectively [45]. Because of the frequent gene duplication and functional redundancy within multigene families, losing certain members is usually "tolerable". However, the θ-defensins have significant antiviral activity that is higher than that of α-defensins and β-defensins, including anti-HIV-1 activity [39,46]. The loss of functional θ-defensin seems to be "deleterious". Therefore, the expansion of pseudogene *DEFTP* and the loss of functional *DEFT* in humans, chimpanzees and gorillas cannot be well explained by the three hypotheses noted above. The evolutionary fate of θ-defensin genes may be influenced by other factors, which should be further investigated. In this study, a less-is-hitchhiking hypothesis was proposed to explain this process, using comparative genomics analyses.

Materials and Methods

Sequence identification

To identify sequences for the primate α-/θ-defensin gene family, a literature search and exhaustive genome blast searches were carried out. First, primate *DEFA/DEFT* sequences were collected from the literature and defined as queries. Second, sequence-based searches of both BlastN and BLAT were performed using these queries, and these BlastN and BLAT searches were repeated using the newly identified genes as queries until no additional genes could be identified. Last, all identified genes were manually checked for functional genes or pseudogenes. Primate *DEFA/DEFT* DNA sequences or completely assembled genomes from the NCBI database or the UCSC Genome Browser were used, including human (*Homo sapiens*; assembly: GRCh37.p5; March 2011), chimpanzee (*Pan troglodytes*; assembly: Pan_troglodytes-2.1.4; March 2011), gorilla (*Gorilla gorilla*; assembly: gorGOR3.1; October 2011), Sumatran orangutan (*Pongo abelii*; assembly: AC206038.3; August 2008), northern white-cheeked gibbon (*Nomascus leucogenys*; assembly: Nleu1.0; October 2011), siamang (*Symphalangus syndactylus*; sequences: AY128121.1 and AY128122.1), macaque (*Macaca mulatta*; assembly: rheMac2; January 2006), crab-eating macaque (*Macaca fascicularis*; assemblies: AEHL01390702.1 and CAEC01613298.1; October 2011), olive baboon (*Papio anubis*; assembly: AC116559.30; September 2010), pig-tailed macaque (*Macaca nemestrina*; AY128123.1), eastern black-and-white colobus (*Colobus guereza kikuyuensis*; AY128124), marmoset (*Callithrix jacchus*; assembly: calJac3; June 2007), Bolivian squirrel monkey (*Saimiri boliviensis boliviensis*; assembly: saiBol1; November 2011), gray mouse lemur (*Microcebus murinus*; assembly: ABDC01305195.1; July 2007), bush baby (*Otolemur garnettii*; assembly: AAQR03188273.1; March 2011), and Philippine tarsier (*Tarsius syrichta*; assembly: ABRT010372935.1; September 2008). The treeshrews (northern treeshrew *Tupaia belangeri*; assembly: AAPY01804345.1; June 2006; Chinese treeshrew *Tupaia belangeri chinensis*; assembly: ALAR1000000; February 2013), which are close relatives of primates, were also used. The exon-intron structures were determined by GeneWise. Sequences that encoded α-/θ-defensins containing the six conserved cysteines (for *DEFA*) and the three conserved cysteines (for *DEFT*) were identified as functional genes. Sequences with frame-shift (insertion or deletion) mutations or premature stop codons were considered pseudogenes. A list of identified sequences is detailed in Table S1.

Molecular evolutionary analyses

To understand the evolutionary processes of the primate α-/θ-defensin gene family, molecular evolutionary analyses and comparative genomics analyses were performed. Coding regions of the *DEFA/DEFT* functional genes were aligned codon-to-codon using the MUSCLE program as implemented in the MEGA6 software [47], and then corresponding regions of the pseudogenes were manually matched to the multiple sequence alignment of the functional genes.

Phylogenetic trees were built using either the more conserved signal-prosegment region or the entire coding region. To obtain reliable phylogenetic relationships, three different tree-building methods were combined. Bayesian inference (BI) phylogenetic trees were constructed using the BEAST software [48]. Neighbor-joining (NJ) and maximum likelihood (ML) phylogenetic trees were constructed using the MEGA6 software [47]. The BI trees were computed using the general time-reversible (GTR) nucleotide substitution model, and the site heterogeneity model was assumed with a discrete gamma distribution for among-site rate variation and a proportion of invariant sites (G + I (4 categories)). Two independent Markov chain Monte Carlo (MCMC) runs were conducted for 10,000,000 steps, with a sampling frequency of every 1000 steps, a UPGMA starting tree and other default parameters. When the effective sample sizes (ESSs) of all quantities were larger than 200, the MCMC was considered converged, and the results were analyzed with 20% burn-in. The best nucleotide substitution patterns for ML trees were selected based on the analyses of best-fit models in MEGA6. The ML trees were computed using the Kimura 2-parameter substitution model (K2) with a bootstrap test (1000 replicates), and the rate variation model was allowed with a discrete gamma distribution (4 categories). The evolutionary distances for NJ trees were computed using the Kimura 2-parameter method with a bootstrap test (1000 replicates), and the rate variation among sites was modeled with a gamma distribution (gamma parameter =2.6 for the signal-prosegment region and 1.4 for the entire coding region).

The phylogenetic trees of BI, NJ and ML were combined using the TreeGraph2 software [49]. The BI trees were selected as background trees, and the clades or clusters of similar topologies in the BI, NJ and ML trees were labeled with posterior probabilities (BI) and bootstrap support values (NJ and ML).

The *DEFA/DEFT* genes from the human, chimpanzee, orangutan, macaque and marmoset genomes were used for synteny analysis by mapping their loci onto chromosomes in scale.

Functional divergences (type I and type II) between clusters were tested using the Gu99 and Gu2013 probabilistic models implemented in the DIVERGE 2.0 software [50,51]. Pairwise coefficients ($\theta_{ij}\pm$SE) and likelihood ratio statistics (LRT) for simian enteric or myeloid defensin clusters were analyzed using the bootstrapped Gu99 probabilistic method. Functional distance analysis was performed based on the Wang and Gu 2001 method [52]. The type I functional distance (d_F) between clusters was defined as $d_F=-\ln(1-\theta_{ij})$. For two clusters A and B, $d_F(A,B)=b_F(A)+b_F(B)$; thus, the functional branch length, b_F, for each cluster can be calculated using a standard least-squares method. A large b_F value for a gene cluster indicates that the evolutionary conservation may be shifted at many sites. The effective number of sites (n_e) related to functional divergences was estimated following the Gu2013 protocol. Type I sites, typically generated by type I functional divergence, are conserved in one gene cluster but highly variable in the other following gene duplication, whereas type II sites, which are primarily caused by type II functional divergence, refer to radical changes of conserved amino acids at the same site in both gene clusters.

To investigate the selection pressure, maximum likelihood estimations of positive selection were performed using the site-specific models from the CODEML program implemented in the PAML 4.5 package [53]. Parameters for the models M0 (one ratio), M1a (neutral), M2a (selection), M7 (beta), M8 (beta and ω (equivalent to Ka/Ks)) and M8a (beta and $\omega = 1$) were calculated. The M0 model assumes a uniform selective pressure among sites. The M1a model assumes a variable selective pressure but no positive selection. The M2a model assumes a variable selective pressure with positive selection. The M7 model assumes a beta-distributed variable selective pressure. The M8 model assumes a beta-distributed variable selective pressure plus positive selection. The M8a assumes a beta distributed variable selective pressure without positive selection. Three likelihood ratio tests (LRTs) were compared using the following paired models: M1a-M2a, M7-M8 and M8a-M8.

To investigate changes in the selection pressure along the *DEFA* and *DEFT* sequences, amino acid identities and average pairwise Ka/Ks ratios were computed using a sliding window method with a window size of 10 residues and a step size of 5 residues. Within each window, the numbers of nonsynonymous and synonymous substitutions per site (denoted as Ka and Ks, respectively) were calculated using the Nei and Gojobori (1986) method incorporated into the KaKs_Calculator software [54]. The values of Ka/Ks were reanalyzed using a bootstrap test of 1000 replicates to assess the effect of potential biases from a few genes in the gene set and to test the statistical significance of the difference between the average Ka/Ks ratio and 1. Suppose there were N original sequences, the average Ka/Ks from a random resampling of N sequences was calculated, and the resampling process was repeated 1000 times. The histogram for the bootstrap estimates of average Ka/Ks showed a normal distribution, and the parameters of the normal distribution were calculated. The null hypothesis (H0) was that there was no difference between Ka/Ks and 1, and the alternative hypothesis (H1) was that there was a difference between Ka/Ks and 1. If the average Ka/Ks for the original sequences was larger than 1, the H0 was defined as "Ka/Ks≤1" and the H1 was defined as "Ka/Ks>1." If the average Ka/Ks for the original sequences was smaller than 1, the H0 was defined as "Ka/Ks≥1" and the H1 was defined as "Ka/Ks<1." The p-value was calculated based on the parameters of the normal distribution.

Sequence logos were generated from the web server WebLogo [55].

Molecular dynamics simulations

Structures of the α- and θ-defensins were searched using BlastP against the Protein Data Bank (PDB) database. One representative structure for each simian α-defensin cluster was analyzed, including human HNP1 (PDB code: 3GNY) [56] representing the DEFA1 cluster, human HNP4 (PDB code: 1ZMM) for the DEFA4 cluster, human HD5 (PDB code: 1ZMP) for the DEFA5 cluster and human HD6 (PDB code: 1ZMQ) for the DEFA6 cluster [34]. No native structures were solved for the DEFA8 and DEFA9 clusters; thus, homology models of the rhesus macaque DEFA8 and marmoset DEFA9a were built using the MODELER automodel method [57] implemented in the Discovery Studio 3.1 software. The corresponding template structures selected for this modeling were HNP1 (41.9% sequence identity with DEFA8) and HD5 (39.4% sequence identity with DEFA9). Monomers were aligned into dimers by superimposing four intermolecular hydrogen bonds between the backbone residues 18 and 20 (HNP4 numbering) through a customized TCL script running in the VMD 1.9 software [58].

Molecular dynamics (MD) simulations were performed following the method described below. The simulation system was parameterized by applying the standard AMBER force field ff99SB for bio-organic systems using the leap module implemented in the AmberTools 12 package [59]. The preparative processes included adding hydrogen atoms and disulfide bonds, neutralizing with counter ions (Cl-) and solvating in a periodic box with TIP3P water to at least a 10-Å distance around the protein. Molecular dynamics simulations were then performed using the NAMD 2.7 software [60]. The entire simulation process was composed of the following steps: minimization, heating, equilibration and production. Three rounds of energy minimization were performed by releasing the restraints in a stepwise fashion. In the first round, all protein atoms were constrained for 50,000 steps with 2.0 kcal/(mol•Å^2) restraints. In the second round, only atoms of the backbone were restrained for 50,000 steps with 2.0 kcal/(mol•Å^2)

restraints. In the third round, all atoms were relaxed for 100,000 steps. Short-range electrostatic and Van der Waals interactions were truncated at 9 Å, whereas long-range electrostatic forces were computed using the particle-mesh Ewald (PME) summation method. Next, the system was gently annealed from 0 to 310 K using a Langevin thermostat with a coupling coefficient of 5.0 ps^{-1} and employing a force constant with 2.0 kcal/(mol•$Å^2$) restraints. All subsequent steps were carried out in the isobaric-isothermal NPT ensemble with a target pressure of 1 atm. Five rounds of equilibration (100 ps, each at 310 K) were performed with decreasing restraint weights from 2.0, 1.5, 1.0, 0.5 to 0.0 kcal/(mol•$Å^2$). By releasing all the restraints, the system was again equilibrated for 500 ps. After the final equilibration step, the production phase of MD was run without any restraints for a total of 16 ns.

Tertiary structures and electronic properties were analyzed after performing MD simulations. The flexibility of the dimer was evaluated using two distances and a dihedral angle involving four highly flexible residues. The distance between the C^{α} atoms of residue 22 in each dimer (HNP4 numbering) was defined as A22C^{α}-B22C^{α}. The distance between the C^{α} atoms of residue 11 in each dimer was defined as A11C^{α}-B11C^{α}. The dihedral angle was defined as A11C^{α}-A22C^{α}-B22C^{α}-B11C^{α}. Distances and dihedral angles were extracted using a customized TCL script program.

The electrostatic potential was mapped by averaging the last five frames of each MD trajectory using the APBS plugin [61] implemented in the VMD software.

Results

Identification, nomenclature and phylogeny of the primate α-/θ-defensin multigene family

The primate α-/θ-defensin gene repertoires, which were identified from the published literature [19,42,62] and exhaustive genomic BlastN and BLAT searches, were manually checked. A total of 144 *DEFA/DEFT* sequences from primates were identified, including 105 functional genes and 39 pseudogenes. Moreover, 10 *DEFA* sequences from two species of treeshrews were collected and served as closely related outgroups for the primate sequences (Table S1). Almost one-third of the primate α-/θ-defensin genes have lost their functional protein-coding ability because of nonsense or frame-shift mutations that have occurred during the process of rapid birth-and-death evolution. The translated amino acid sequences of functional genes were aligned with MUSCLE (Figure S1). Most of the protein sequences are computationally translated from genome sequences rather than being demonstrated to be produced in vivo. The multiple sequence alignment shows that the mature peptide of α-defensins has a motif with six conserved cysteine residues, represented as C-x-C-x(3,4)-C-x(9)-C-x(6,9)-C, whereas the mature peptide of θ-defensins has a nonapeptide motif with three conserved cysteine residues, represented as x-C-x-C-x(4)-C, with the exception of the Eastern black-and-white colobus θ-defensin (*cgue_DEFT*), which has the motif x-C-x-C-x(8)-C.

To explore the evolutionary relationships among these defensin genes, phylogenetic trees were constructed based on nucleotide sequences from either the more conserved signal-prosegment region or the entire coding region, and the phylogenetic trees inferred by the BI, NJ and ML methods were combined (Figure 1). All sequences are classified into major clades, which are referred to as the treeshrew, prosimian and simian clades based on the species included. In all the trees from the signal-prosegment region (Figures 1A, S2 and S3), the simian defensins are classified into two major clades of tissue-specific expression profiles and are thus named simian myeloid α-defensins and simian enteric α-defensins. In all the trees from the entire coding region (Figures 1B, S4 and S5), all sequences are separated into two groups, with group I containing the simian enteric α-defensins and the treeshrew clade 2 and group II containing the simian myeloid α-defensins and the treeshrew clade 1, which indicates a possible early duplication and divergence before the split of treeshrews and primates. However, for the prosimian clades, phylogenetic incongruence was observed. The tree inferred from the signal-prosegment region has two major prosimian clades (Figure 1A), whereas the tree inferred from the entire coding region divides the prosimian clade 2 into four groups (Figure 1B). Because incongruence between phylogenetic trees may result from the independence of evolutionary changes at different sites and the homogeneity of the substitution process [63], the phylogenetic incongruence was further assessed by reconstructing the phylogenetic tree after removing the homogeneity sites. Some amino acid sites in the mature peptide were identified as homogeneity sites of convergent or parallel evolution (Figure S6). When those sites were removed, the prosimian α-defensins grouped together, suggesting that these convergent or parallel evolution sites have resulted in long-branch attraction such that distant clades grouped together.

The simian myeloid and enteric α-defensins are classified into seven clusters, which are named following the nomenclature used for human *DEFA/DEFT* genes. The simian myeloid α-defensins contain the *DEFA1/DEFA3*, *DEFA8/DEFA10/DEFT*, *DEFA4* and *DEFA7* clusters, and the simian enteric α-defensins contain the *DEFA9/DEFA11*, *DEFA6* and *DEFA5* clusters. The *DEFA7* cluster contains only pseudogenes and is found only in Old World monkeys and hominoids. The other six gene clusters, each containing at least one functional gene, are functional gene clusters. All six functional gene clusters include species from New World monkeys, Old World monkeys and hominoids, suggesting that the duplication and diversification of these clusters occurred before the species differentiation of simians. The six functional gene clusters, which diverged prior to the simian differentiation, provide the genetic bases for the similar antimicrobial spectra in closely related simian species.

The *DEFA9/DEFA11* cluster shows different topologies among the phylogenetic trees (Figures S2 and S4). In the NJ and ML trees from the signal-prosegment region, the pseudogene *DEFA11P* and the *DEFA9* from Old World monkeys and hominoids group into one cluster, and the *DEFA9* from New Old monkeys is the outgroup. However, in the BI tree from signal-prosegment region and the BI/NJ/ML trees from the entire coding region, *DEFA11P* and *DEFA9* form separate clusters. To further clarify the relationship between *DEFA9* and *DEFA11P*, the sequence alignment from the complete gene including the intron was examined, and different patterns of sequence similarity were observed before and after a clear boundary in the genes. Therefore, phylogenetic trees were inferred based on either the sequences before the boundary (part I) or the sequences after the boundary (part II), using *DEFA6* as the outgroup (Figure S7). The trees based on part I show that *DEFA11P* is duplicated from *DEFA9* after the split of New World and Old World monkeys, whereas the trees based on part II show that *DEFA11P* is the outgroup of all *DEFA9*. The different topologies indicate that the two parts of *DEFA11P* might be derived from different ancestors. The *DEFA8* from Old World monkeys and hominoids also shows different phylogenetic positions among the phylogenetic trees (Figures S2 and S4). In the NJ tree from the signal-prosegment region, the *DEFA8* from Old World monkeys and hominoids clusters together with *DEFA10/DEFT*. However, in all the other

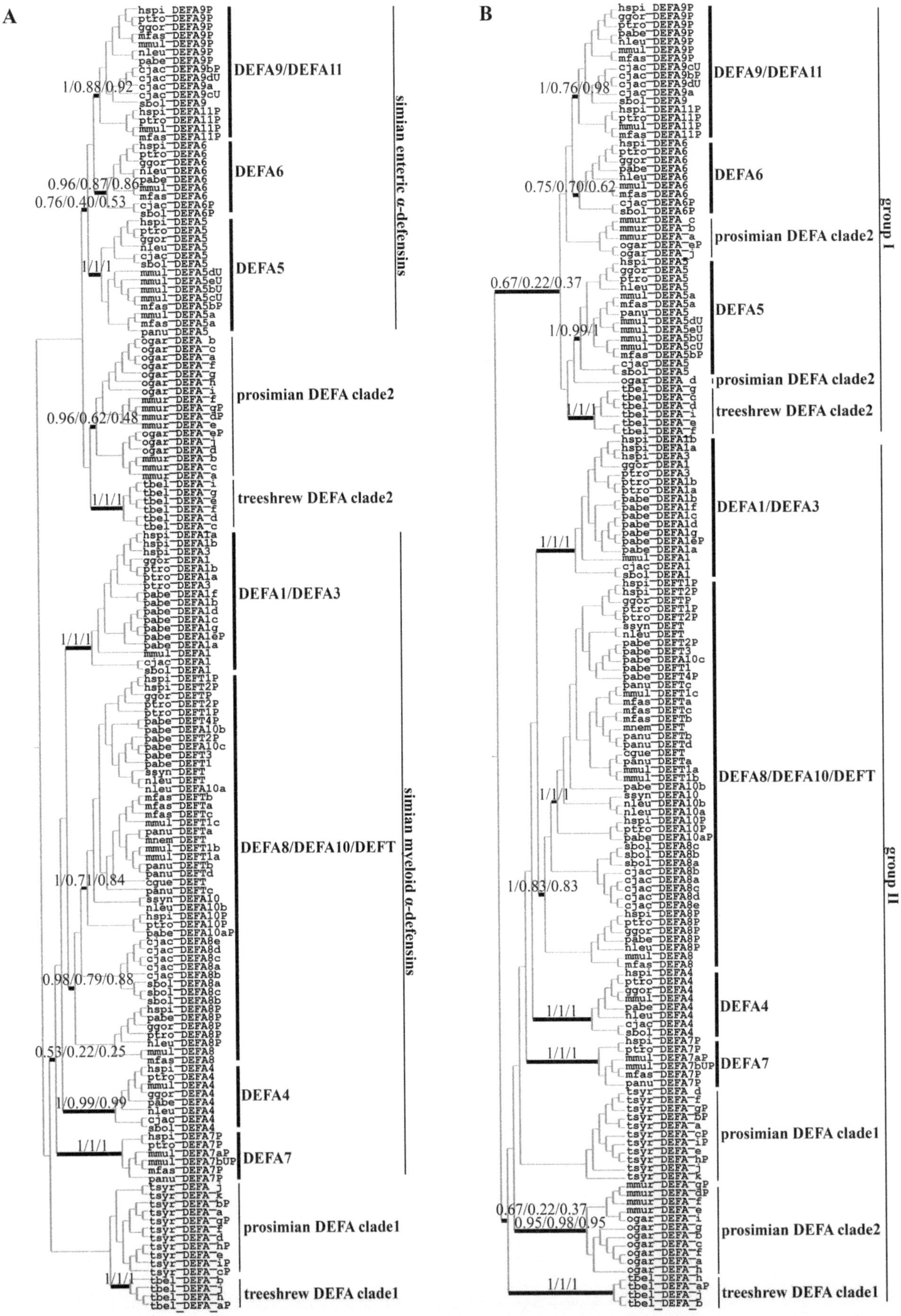
A
DEFA9/DEFA11
DEFA6
DEFA5
prosimian DEFA clade2
treeshrew DEFA clade2
DEFA1/DEFA3
DEFA8/DEFA10/DEFT
DEFA4
DEFA7
prosimian DEFA clade1
treeshrew DEFA clade1
simian enteric α-defensins
simian myeloid α-defensins
1/0.88/0.92
0.96/0.87/0.86
0.76/0.40/0.53
1/1/1
0.96/0.62/0.48
1/1/1
1/1/1
1/0.71/0.84
0.98/0.79/0.88
0.53/0.22/0.25
1/0.99/0.99
1/1/1
1/1/1
B
DEFA9/DEFA11
DEFA6
prosimian DEFA clade2
DEFA5
prosimian DEFA clade2
treeshrew DEFA clade2
DEFA1/DEFA3
DEFA8/DEFA10/DEFT
DEFA4
DEFA7
prosimian DEFA clade1
prosimian DEFA clade2
treeshrew DEFA clade1
group I
group II
1/0.76/0.98
0.75/0.70/0.62
0.67/0.22/0.37
1/0.99/1
1/1/1
1/1/1
1/1/1
1/0.83/0.83
1/1/1
1/1/1
0.67/0.22/0.37
0.95/0.98/0.95
1/1/1

Figure 1. Phylogenetic trees of the α-/θ-defensin (*DEFA/DEFT*) genes in primates and treeshrews. A: Phylogenetic tree of primate and treeshrew *DEFA/DEFT* genes based only on the signal-prosegment region. The BI tree is selected as the background tree. The major clades or clusters having similar topologies from all three tree-building methods (BI, NJ and ML) are combined and labeled with the BI posterior probabilities and the bootstrap support values from the NJ and ML analyses. Primate *DEFA/DEFT* genes are clustered into prosimian and simian clades. The simian clades are classified and named following the nomenclature used for the human *DEFA/DEFT* genes. The "P" in node labels denotes a pseudogene, and the "U" indicates a gene with an ambiguous locus from the species in the synteny map in Figure 7. B: Phylogenetic tree of primate and treeshrew *DEFA/DEFT* genes based the entire coding region. The BI tree is selected as the background tree. The treeshrew *DEFA* genes are the outgroups of the two separate clades.

trees, the *DEFA8* from Old World monkeys and hominoids is the outgroup in the *DEFA8/DEFA10/DEFT* cluster. Similarly, different patterns of sequence similarity were observed before and after a clear boundary in the genes of this cluster. Phylogenetic trees suggest that part I and part II of the *DEFA8* from Old World monkeys and hominoids might be derived from different ancestors (Figure S8). The complicated phylogenetic relationships within both the *DEFA9/DEFA11* cluster and the *DEFA8/DEFA10/DEFT* cluster imply that gene conversion or interlocus recombination might have occurred during the evolutionary process.

Unlike the *DEFA* genes, which are present in all primates, *DEFT* genes are present in only Old World monkeys and certain hominoids. Both the tree inferred from the entire coding region and the tree inferred from the signal-prosegment region show that *DEFT* groups together with *DEFA10*. *DEFA10*, which is present only in hominoids, is the basal clade in the *DEFA10/DEFT* cluster (Figure 1). The relationship between *DEFA10* and *DEFT* and the origin of *DEFT* were further analyzed using their introns. The BI/NJ/ML trees from the introns support that *DEFT* and *DEFA10* are grouped into separate clusters, and both clusters are derived from the duplication of *DEFA8* (Figure S9).

Sequence divergence and structural variation of simian α-defensins

After identifying different clusters of primate α-/θ-defensins, the pattern of natural selection operating on each cluster was examined using the site-specific models of heterogeneous selection pressure among sites. Likelihood ratio tests (LRTs) were carried out between models M1a (neutral) and M2a (selection), M7 (beta) and M8 (beta and ω (equivalent to Ka/Ks)), and M8a (beta and $\omega = 1$) and M8 [53]. The models M2a and M8, which are significantly favored over the other models, indicate positive selection, and the positively selected sites with posterior probabilities greater than 0.95 are listed (Table 1 and S2). Positive selection is not significantly detected for every cluster. For the two prosimian clades and the simian *DEFT*, *DEFA1* (including *DEFA3*), *DEFA5* and *DEFA8* (including *DEFA10*) clusters, positively selected sites are detected and are located primarily in the mature peptide. However, the simian *DEFA4*, *DEFA6* and *DEFA9* clusters are under purifying selection, and no positively selected site is detected. The different selection patterns among these clusters reflect their variant functional constraints. Therefore, the sequence divergence and structural variations among the six clusters (excluding the *DEFA7* pseudogene cluster) were further investigated to understand their functional diversity, which most likely constitutes their antimicrobial spectra.

During the evolutionary process following gene duplication, due to functional divergence, the evolutionary rate of duplicated genes will initially increase and may subsequently shift with altered functional constraints (type I functional divergence) or remain at the original rate with no altered functional constraints (type II functional divergence) [50,51,64]. Type I and type II functional divergence patterns among simian myeloid and enteric α-defensins were examined using the Gu99 and Gu2013 methods (Figure 2). Statistical analysis suggests that clusters of both simian myeloid α-defensins and simian enteric α-defensins have significantly diverged (Figure 2A). The pairwise coefficients of type I functional divergence (θ_{ij}) are all significantly greater than 0 ($p<0.05$), except for the comparison between *DEFA9* and *DEFA6* ($p = 0.07$). The type I functional distance for each pair of clusters was estimated to measure whether one cluster has more shifted evolutionary rates than the other cluster following gene duplication [52]. The type I functional distance (d_F) ranges from 0.78 to 1.40 for the simian myeloid α-defensins (with the functional branch length of *DEFA1*, *DEFA4* and *DEFA8* being 0.94, 0.32 and 0.46, respectively) and from 0.69 to 3.93 for the simian enteric α-defensins (with the functional branch length of *DEFA5*, *DEFA6* and *DEFA9* being 3.41, 0.53 and 0.17, respectively). Type I functional distances among enteric α-defensins are much higher than among myeloid ones, except for the distance between *DEFA9* and *DEFA6*. The enteric α-defensins have a higher level of functional divergence, which reflects a more challenging microbial environment in the gut. Critical amino acid sites responsible for the type I and type II functional divergences (hereafter referred to as type I and type II sites, respectively) were detected (Figure 2B and C). For simian myeloid α-defensin clusters, the type I sites of *DEFA1* are located in the mature peptide, whereas the type I sites of *DEFA8* and *DEFA4* are located in the signal-prosegment (Figure 2D). The different locations of the type I sites among simian myeloid α-defensin clusters suggest a specific altered functional constraint on the mature peptide of *DEFA1*. For simian enteric α-defensin clusters, the type I sites of *DEFA5* are located in the signal-prosegment, whereas the type I sites of *DEFA9* and *DEFA6* are primarily located in the mature peptide. This result indicates a specific altered functional constraint on the signal-prosegment region of *DEFA5*. Therefore, different clusters of α-defensins have undergone various functional constraints and have likely been specialized in their antimicrobial spectra. No site corresponding to the type I or type II functional divergence was detected between simian myeloid α-defensins and simian enteric α-defensins because of the high level of divergence within both clades.

The dimerization and the cationic and amphipathic properties of the α-defensin structures are important for their antimicrobial activities [2,33,35–37]; hence, these characters of the representative structures from six clusters were analyzed. The α-defensin dimers were superimposed based on the backbone residues 18 and 20, which form four pairs of intermolecular hydrogen bonds (Figure 3A and B). These superimposed dimers include the myeloid α-defensins DEFA1 (human HNP1), DEFA4 (human HNP4) and DEFA8 (modeled rhesus macaque DEFA8) and the enteric α-defensins DEFA5 (human HD5), DEFA6 (human HD6) and DEFA9 (modeled marmoset DEFA9a). The β-sheets overlap well, whereas the two loops connecting β1-β2 strands and β2-β3 strands show structural variation. Residues 11 and 22 represent the most distal amino acids on the two flexible loops. Thus, the distances of $A22C^{\alpha}$-$B22C^{\alpha}$ and $A11C^{\alpha}$-$B11C^{\alpha}$, as well as the dihedral angle of $A11C^{\alpha}$-$A22C^{\alpha}$-$B22C^{\alpha}$-$B11C^{\alpha}$, were monitored in molecular dynamics simulations (see Materials and Methods) to assess their structural variation (Figure 3B). The distance of

Table 1. Log-likelihood values and positively selected sites for primate *DEFA/DEFT* genes under site models.

Cluster	Models (2ΔlnL)	Positively Selected Sites	P value
DEFT (n = 16)	M1a versus M2a (12.57)	67I 69R	<0.001
	M7 versus M8 (12.60)	17Q 50D 67I **69R**	0.002
	M8a versus M8 (12.58)		<0.001
DEFA1 (n = 16)	M1a versus M2a (22.42)	29V **67Y 74I**	<0.001
	M7 versus M8 (23.33)	29V **67Y 74I** 86Q	<0.001
	M8a versus M8 (22.40)		<0.001
DEFA4 (n = 8)	M1a versus M2a (3.02)		0.082
	M7 versus M8 (3.32)		0.190
	M8a versus M8 (3.02)		0.082
DEFA8 (n = 15)	M1a versus M2a (119.67)	**65R 67I 70R** 71G **72I 75L 76L 79R 80Y 82S 84A 85F 87G 92I**	<0.001
	M7 versus M8 (121.57)	**65R 67I 70R** 71G **72I 75L 76L 79R 80Y 82S 84A 85F 87G 92I**	<0.001
	M8a versus M8 (119.66)		<0.001
DEFA5 (n = 13)	M1a versus M2a (59.78)	**68T 70R** 72A **73T 74R** 77L **80V 82E 83I 84S**	<0.001
	M7 versus M8 (63.85)	25R 62A **68T 70R** 72A **73T 74R** 77L **80V 82E 83I 84S**	<0.001
	M8a versus M8 (59.72)		<0.001
DEFA6 (n = 7)	M1a versus M2a (3.33)		0.068
	M7 versus M8 (3.35)		0.187
	M8a versus M8 (3.32)		0.068
DEFA9 (n = 4)	M1a versus M2a (2.51)		0.113
	M7 versus M8 (2.52)		0.284
	M8a versus M8 (2.50)		0.114
prosimian DEFA clade1 (n = 6)	M1a versus M2a (16.11)	69I 79R **83V** 87R	<0.001
	M7 versus M8 (16.84)	66H 69I 79R **83V** 87R	<0.001
	M8a versus M8 (16.32)		<0.001
prosimian DEFA clade2 (n = 14)	M1a versus M2a (76.36)	**66R 73G** 79Y **87F 90L**	<0.001
	M7 versus M8 (79.40)	37T **66R** 71R **73G** 74F 78T **79Y 87F 90L**	<0.001
	M8a versus M8 (72.84)		<0.001

Notes: positively selected sites with significance at the 99% level are bold, whereas the remaining sites have significance at the 95% level. The likelihood ratio tests are analyzed by comparing the following pairs of models: M1a-M2a, M7-M8 and M8a-M8.

A22C$^{\alpha}$-B22C$^{\alpha}$ is almost equal among the six dimers (Figure 3C), suggesting that the dimers are not separated into monomers and remain stable during the 16-nanosecond molecular dynamics simulation. The distance of A11C$^{\alpha}$-B11C$^{\alpha}$ ranges from 30 Å to 37 Å, and the dihedral angle of A11C$^{\alpha}$-A22C$^{\alpha}$-B22C$^{\alpha}$-B11C$^{\alpha}$ ranges from 75° to 120° (Figure 3C and D), showing significant variations in these dimer structures and providing a structural foundation for their different antimicrobial activities. In addition to dimerization, the surface electrostatic potentials of simian enteric and myeloid α-defensins create distinct cationic and amphipathic patterns (Figure 3E), which likely affect their membrane depolarization and permeabilization abilities through electrostatic and hydrophobic interactions with microbial membranes.

Sequence profiles and structural features of θ-defensins

The θ-defensins are cyclic octadecapeptides formed by the head-to-tail ligation of two nonapeptides that can be homodimeric or heterodimeric to increase their diversity [62,65]. The nonapeptide motif of θ-defensins is presented as RC[IVLF]C[RTGVL][RL]G[VFI]C (Figure 4A), which shows both a conserved cysteine pattern and a conserved hydrophilic/hydrophobic pattern. The conserved cysteines at positions 2, 4 and 9 of the consensus motif form the tridisulfide ladder that has been proven to be important for structural stability but not for antimicrobial activity [66]. Positions 1, 5 and 6 are mostly cationic and hydrophilic arginine residues, whereas positions 3, 7 and 8 are mostly hydrophobic residues. The side chains of arginine residues and the tridisulfide ladder are on opposite sides of the cyclic backbone plane and thus form a polarized structure. The conserved hydrophilic/hydrophobic pattern generates the cationic and amphipathic properties, which might play a major role in the membrane depolarization and permeabilization of target cells or enveloped viruses. The cationic and amphipathic properties of θ-defensins, like α-defensins, have been reported to be important for antimicrobial activities by disrupting microbial membrane structures [2,33].

Sites 3I and 5R in the nonapeptides (corresponding to 67I and 69R in the full sequence) were found to be under significant positive selection (Table 1). The surface electronic potentials for two types of θ-defensin homodimers, $(RCICRRGVC)_2$ that is under positive selection and $(RCVCTRGVC)_2$ that is not under

A

	Functional Divergence Gene Clusters (i,j)	Type I θ_{ij}±SE	Type I LRT	Type I p value	Type I d_F	Type I n_e	Type II θ_{ij}±SE	Type II n_e
simian myeloid α-defensins	DEFA1, DEFA8	0.74±0.15	24.71	<0.0001	1.40	8	0.20±0.17	1
	DEFA1, DEFA4	0.72±0.29	6.53	0.01	1.26	2	0.09±0.15	-
	DEFA8, DEFA4	0.54±0.17	10.30	0.001	0.78	3	0.34±0.16	4
simian enteric α-defensins	DEFA6, DEFA5	0.98±0.25	15.11	<0.0001	3.93	7	0.13±0.16	-
	DEFA9, DEFA5	0.97±0.21	22.22	<0.0001	3.57	8	0.22±0.15	2
	DEFA9, DEFA6	0.50±0.29	3.22	0.07	0.69	1	0.20±0.11	4

B

	position (k)	Type I: 69 74 78 81 91 5 17 32 36 37	Type II: 20 31 35 50 52
DEFA1	hspi_DEFA1a	I A R T F A A P A A	E A I E L
	hspi_DEFA1b	I A R T F A A P A A	E A I E L
	hspi_DEFA3	I A R T F A A P A A	E A I E L
	ptro_DEFA1a	I A R T F A A P P A	E A I E L
	ptro_DEFA1b	I A R T F A A P P A	E A I E L
	ptro_DEFA3	I A R T F A A P P A	E A I K L
	ggor_DEFA1	I A R T F A A P P A	E A I E L
	pabe_DEFA1a	I A R T F A A Q P A	E A I E L
	pabe_DEFA1b	I A R T F A A P P A	E A I E L
	pabe_DEFA1c	I A R T F A A Q P A	E A I K L
	pabe_DEFA1d	I A R T F A A Q P A	E A I K L
	pabe_DEFA1f	I A R T F A A P P A	E A I E L
	pabe_DEFA1g	I A R T F A A Q P A	E A I E L
	mmul_DEFA1	I A R T F V A Q P T	E A I E L
	cjac_DEFA1	I A R T F I A K L A	G A I E L
	sbol_DEFA1	I A R T F I A K S A	G A I E L
DEFA8	pabe_DEFA10b	R L R S I A A Q G A	E A P E A
	pabe_DEFA10c	R F H S I A A Q G A	E A P E A
	nleu_DEFA10a	R A R T A A A Q G A	E A P E A
	nleu_DEFA10b	R A R T A A A Q G A	E A T E A
	ssyn_DEFA10	T L H S I A A Q G A	E A T E A
	mmul_DEFA8	R P R T Y A A Q R A	E A P E A
	mfas_DEFA8	R P R T Y A A Q R A	E A P E A
	cjac_DEFA8a	P S D A L A A Q G A	E A P E A
	cjac_DEFA8b	L S L F V A A Q G A	E A P Q A
	cjac_DEFA8c	L S D T L A A Q G A	E A P E A
	cjac_DEFA8d	R A D A L A A Q G A	E A P E A
	cjac_DEFA8e	R A Q I F A A Q G A	E A P E A
	sbol_DEFA8a	Q F P T F A A Q G A	E A P E A
	sbol_DEFA8b	F S E F C A A Q G A	E A P E A
	sbol_DEFA8c	V S Q T Y A A Q G A	E A P E A
DEFA4	hspi_DEFA4	L R R N Y A V Q G P	G G R K S
	ptro_DEFA4	L R R N Y A V Q G P	G G R K S
	ggor_DEFA4	L R R N Y A V Q G P	G G R K S
	pabe_DEFA4	L R R N Y A V Q G A	G G P K S
	nleu_DEFA4	L R R N Y T A Q G A	G G R K S
	mmul_DEFA4	L R R N Y A V Q G P	G G R K S
	cjac_DEFA4	L P R S Y A A Q G V	G G P K S
	sbol_DEFA4	L P S S Y A A Q G V	G G H K S

C

	position (k)	Type I: 8 10 18 22 23 31 51 71 74 83 69	Type II: 31 51 5 35 47 77
DEFA5	hspi_DEFA5	A I Q L Q T L R T E T	T L A S A S
	ptro_DEFA5	A I Q L Q T L H I E I	T L A S A S
	ggor_DEFA5	A I Q L Q T L P N E T	T L T S A S
	nleu_DEFA5	A I Q L Q T L L A Y S	T L A S A S
	mmul_DEFA5a	A I Q L Q T L G R I I	T L A P E S
	mmul_DEFA5bU	A I Q L Q T L R T K N	T L A P E F
	mmul_DEFA5cU	A I Q L Q T L R T V T	T L T P E F
	mmul_DEFA5dU	A I Q L Q T L R R R I	T L T P E S
	mmul_DEFA5eU	A I Q L Q T L R R N F	T L T P E S
	mfas_DEFA5a	A I Q L Q T L H R I I	T L A P E S
	panu_DEFA5	A I Q L Q T L R R I I	T L A P E S
	cjac_DEFA5	A I Q L Q T L C Y T T	T L A P A S
	sbol_DEFA5	A I Q L Q T L C Y T S	T L A S A S
DEFA6	hspi_DEFA6	T V K L Q A S S S T R	A S T R A Y
	ptro_DEFA6	T V K L Q A S S S T R	A S T R A Y
	ggor_DEFA6	T V K L Q A S S S T R	A S T R A Y
	pabe_DEFA6	T V K L Q T S S S T R	T S T R A Y
	nleu_DEFA6	A V K L Q T S S S T R	T S T R A Y
	mmul_DEFA6	A I Q L Q A N S S T R	A N T R A Y
	mfas_DEFA6	A I Q L Q A N S S T G	A N T R A Y
DEFA9	cjac_DEFA9a	A I Q S A A S V L A I	A S A P T S
	cjac_DEFA9cU	A I Q F L A S V L A I	A S A P T S
	cjac_DEFA9dU	A I Q S P A S V L A I	A S A P T S
	sbol_DEFA9	A I Q T L A S V L S I	A S A P T S

D

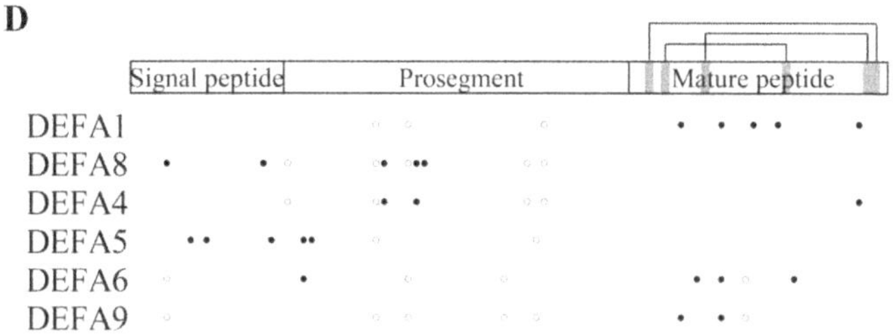

Figure 2. Type I and type II functional divergences among simian enteric or myeloid α-defensin clusters. A: Statistical analyses of the type I and type II functional divergences. Clusters within both simian enteric α-defensins and simian myeloid α-defensins have diverged. Pairwise coefficients (θ_{ij}±SE) and likelihood ratio statistics (LRT) are analyzed using the bootstrapped Gu99 method. Type I functional distance (d_F) between clusters is calculated as $d_F = -\ln(1-\theta_{ij})$. The effective number of functional divergence related sites (n_e) is estimated following the Gu2013 method. B: Critical amino acid sites responsible for type I and type II functional divergences from simian myeloid α-defensin clusters. These sites are identified based on the n_e cutoff after ranking posterior probabilities. C: Critical amino acid sites responsible for type I and type II functional divergences from simian enteric α-defensin clusters. D: The diagram of the sites responsible for type I and type II functional divergences in different simian clusters. Closed circles represent sites responsible for type I functional divergence, and opened circles denote sites responsible for type II functional divergence.

positive selection (Figure 4B), were analyzed based on molecular dynamics simulations. The surface electronic potentials exhibit differences that may affect the polarization of θ-defensin, suggesting that the positively selected sites are related to the antimicrobial function of θ-defensin.

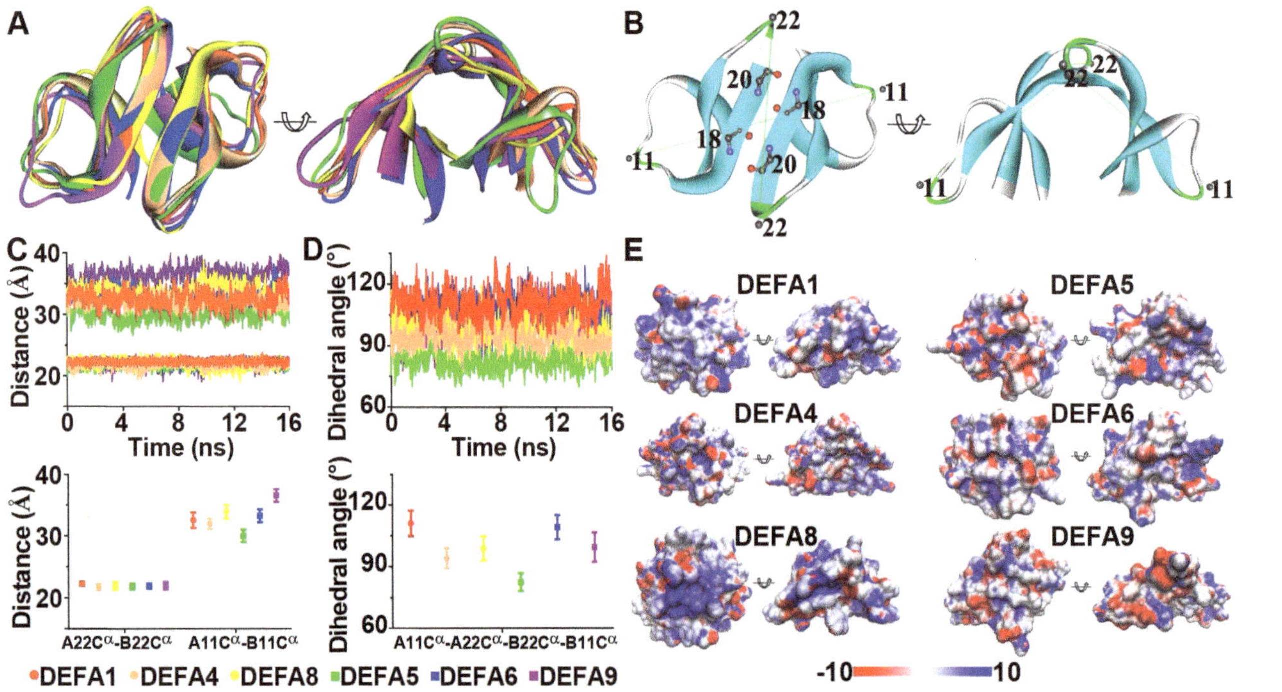

Figure 3. Structural flexibility and surface electrostatic potential for the simian myeloid and enteric α-defensin dimers. A: Superimposed α-defensin dimers for the myeloid DEFA1, DEFA4 and DEFA8 proteins as well as the enteric DEFA5, DEFA6 and DEFA9 proteins. B: Diagram illustrating the intermolecular hydrogen bonds, the two distances (A22C$^{\alpha}$-B22C$^{\alpha}$ and A11C$^{\alpha}$-B11C$^{\alpha}$) and the dihedral angle (A11C$^{\alpha}$-A22C$^{\alpha}$-B22C$^{\alpha}$-B11C$^{\alpha}$). C: Plots of the A22C$^{\alpha}$-B22C$^{\alpha}$ and A11C$^{\alpha}$-B11C$^{\alpha}$ distances during the 16 nanoseconds of molecular dynamics simulations (upper panel) and plots of the average A22C$^{\alpha}$-B22C$^{\alpha}$ and A11C$^{\alpha}$-B11C$^{\alpha}$ distances (lower panel). Similar A22C$^{\alpha}$-B22C$^{\alpha}$ distances indicate that the dimer structures are well maintained during simulation processes, whereas diverse A11C$^{\alpha}$-B11C$^{\alpha}$ distances reflect the flexibility of the dimers. Error bars represent standard deviations. D: Plots of the dihedral angle A11C$^{\alpha}$-A22C$^{\alpha}$-B22C$^{\alpha}$-B11C$^{\alpha}$. The variant dihedral angles among different clusters of myeloid and enteric α-defensins indicate different dimer topologies. E: Surface electrostatic potential generated using the smoothed trajectory from the last five frames of molecular dynamics simulations. The electrostatic potential (±10 kT/e) is colored red (−) or blue (+). The left and right views of each structure are the same as in panel A.

Adaptive changes in the prosegment coevolving with the new protein fold of the mature peptide

The θ-defensins are the youngest members of the defensin family. A premature stop codon mutation in the mature peptide gives birth to the θ-defensin, which contains a new protein fold that is different from that of α-defensins. The selection patterns along the length of *DEFA* and *DEFT* were examined to check whether there were any adaptive changes in the signal-prosegment region that are related to the new protein fold. For this purpose, a sliding window method for the amino acid identity and the Ka/Ks was used to analyze the selection pressures on different regions of all *DEFA*, *DEFT* and separate *DEFA* clusters (Figure 5, Table S3). Furthermore, bootstrap tests for the Ka/Ks were performed to assess the effect of potential biases from a few genes in the gene set and to test the significance of the difference between the average Ka/Ks ratio and 1.

The amino acid identities decrease from the signal peptide and prosegment to the mature peptide for both *DEFA* and *DEFT* (Figure 5A and B). The signal peptide is the most conserved region, whereas the mature peptide is highly divergent. This result also supported our rationale for inferring a phylogenetic tree using the more conserved signal-prosegment region (Figure 1A).

The Ka/Ks ratios are not homogeneous among the sliding windows for *DEFA*, *DEFT* and separate *DEFA* clusters, indicating that varying selection pressures act on different fragments. The signal peptides are tested to be under purifying selection (Ka/Ks $<$ 1) for both *DEFA* and *DEFT*. When the *DEFA* clusters are analyzed separately, only the *DEFA5* cluster (sites 6–15) and the prosimian *DEFA* clade 1 (sites 1–10) are detected with positive selection in the signal peptides. However, the mature peptides are under positive selection (Ka/Ks$>$1) for both *DEFA* and *DEFT*. For separate simian *DEFA* clusters, the mature peptides of *DEFA5* and *DEFA8* are under strong positive selection. The mature peptides of the prosimian *DEFA* clade 1 and clade 2 are also detected with positive selection. The positive selection acting on the mature peptides is likely due to the strong selection pressure through their direct interactions with variable microbes.

The prosegments of the *DEFA* and the *DEFT* are under different selection pressures. For *DEFA*, the fragment 56–65 in the prosegment is under positive selection (Ka/Ks = 1.46, bootstrap test Ka/Ks significantly $>$1, $P<0.01$) (Figure 5A), whereas the same fragment from separate simian *DEFA* clusters (Figure 5C) is not detected with positive selection. The fragment 56–65 is a crucial region that harbors several posttranslational cleavage sites [31,67,68]. The positive selection acting on this cleavage region of *DEFA* may be explained by the diverse splicing mechanisms among different *DEFA* clusters, which are conserved within each cluster. The prosimian clade 1 and clade 2 are also detected to be under positive selection in the cleavage region, indicating that both clades contain divergent α-defensins.

For *DEFT*, the fragment 41–50 in the prosegment is under strong positive selection (Ka/Ks = 1.68, Ka/Ks significantly $>$1, $P<0.05$), compared to the corresponding fragment of *DEFA* (Ka/Ks significantly $<$1, $P<0.01$) and the closely related *DEFA8* (Ka/

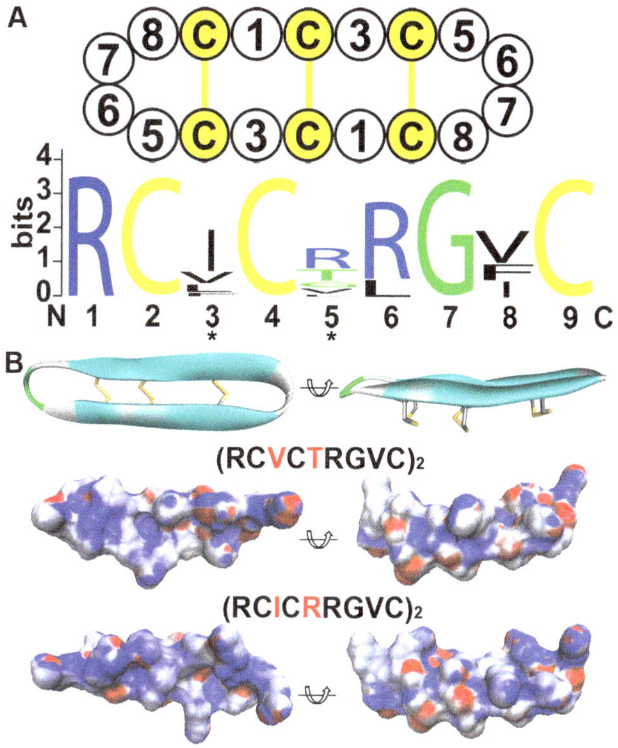

Figure 4. Sequence motifs and electrostatic features of primate cyclic θ-defensins. A: The θ-defensin octadecapeptide is formed by the head-to-tail ligation of two nonapeptides. The sequence logo for the nonapeptide is generated using all the primate θ-defensins. Sites 3I and 5R are under positive selection and are marked with asterisks. B: Surface electrostatic potential of the θ-defensins without or with the positively selected sites (3I and 5R).

Ks = 1.11, Ka/Ks not significantly >1, P = 0.16). However, the prosegment of *DEFA* is detected with positive selection in another region, namely, the fragment 26–35 (Ka/Ks = 1.26, Ka/Ks significantly >1, P<0.01). The distinct positive selection in the prosegments of *DEFA* and *DEFT* indicates a specific change in the prosegment of *DEFT*. Because the prosegment is important for the correct folding of defensins [29], it is most likely that the prosegment of θ-defensins has undergone adaptive changes that are related to the new protein fold. The diagram in Figure 6 presents our proposed hypothesis for this evolutionary pattern and illustrates the adaptive evolution of the prosegment along with the new protein fold of the mature peptide.

Genetic hitchhiking of the pseudogene *DEFTP* and loss of functional θ-defensin in humans and chimpanzees

To understand the evolutionary dynamics of the α-/θ-defensin multigene family, the comparative synteny map from five simian species, including humans, chimpanzees, orangutans, macaques and marmosets, was analyzed and was combined with the phylogenetic results. The primate *DEFA/DEFT* genes originate from tandem duplication and are located in the regions homologous to the human chromosome 8p23 (Figure 7A). Conserved synteny relationships are observed among the five simian species. The *DEFA5* orthologs are located at the 5′ terminus, and the *DEFA9*, *DEFA8*, *DEFA4* and *DEFA6* orthologs are located at the 3′ terminus in a 5′ to 3′ order. The simian myeloid and enteric *DEFA* genes, which are clustered into distinct clades in both phylogenetic trees (Figure 1), are not located as two clearly separated blocks in the synteny map. Loci of the genes from the two major clades (simian enteric and myeloid defensins) are intermingled, implying that duplication of a segment containing two or more genes has occurred. Orthologs from some species have mutated into pseudogenes, which include *DEFA6P* in the marmoset; *DEFA9P* in the macaque, orangutan, chimpanzee and human; and *DEFA8P* in the orangutan, chimpanzee and human. In the phylogenetic trees, genes from the same species group together in each simian cluster (Figure 1), indicating species-specific duplication, which is a common phenomenon during the birth-and-death evolution of multigene families. In the synteny map (Figure 7A), species-specific duplication is not homogeneous across this region, with the middle section having more species-specific duplications. Based on the current versions of assembled genomes, the marmoset has multiple duplications of *DEFA8* and *DEFA9*, and the macaque has multiple duplications of *DEFT*. In the orangutan, chimpanzee and human, there are several species-specific duplications of *DEFA1* and *DEFA10/DEFT*.

A nonsense mutation at codon 17 of *DEFT* results in its loss of function in orangutans, chimpanzees and humans (Figure 7B), and these *DEFTP* pseudogenes contain the *hspi_DEFT1P*, *hspi_DEFT2P*, *ptro_DEFT1P*, *ptro_DEFT2P* and *pabe_DEFT4P* (excluding the *pabe_DEFT2P* pseudogene which is generated by frame-shift mutations). In humans, chimpanzees and gorillas, the *DEFT* has been lost, and only *DEFTP* pseudogenes have been maintained. Our results demonstrate that θ-defensin had a conserved structural motif and was under positive selection. Therefore, the driving force for the expansion of the pseudogene *DEFTP* and for the loss of functionally important θ-defensins in humans and chimpanzees were investigated. A less-is-hitchhiking hypothesis was proposed based on the synteny map and phylogenetic relationships. The synteny map suggests that the loci of the *DEFA1* and the *DEFT/DEFA10* are linked and have been duplicated together in orangutans, chimpanzees and humans (Figure 7A). The genes of the *DEFA1* cluster are all functional and are under positive selection; thus, we infer that each box in the synteny map contains a driver from the *DEFA1* cluster and a hitchhiker from the *DEFT/DEFA10* cluster. The orangutan has multiple copies of functional *DEFT* as well as the pseudogene *pabe_DEFT4P* containing the nonsense mutation at site 17 (Figure 7A and B). This result suggests that, in the ancestor of the orangutan, chimpanzee and human, the selection pressure on the multiple-copied *DEFT* was relaxed and the pseudogene *DEFTP* (containing the nonsense mutation at site 17) emerged. Both *DEFT* and *DEFTP* are linked to *DEFA1* and can expand, but only *DEFTP* (containing the nonsense mutation at site 17) is maintained, whereas *DEFT* is lost in the chimpanzees and humans (Figure 7A). The retention of *DEFTP* and loss of *DEFT* might be independent of their own fitness and depends on the fate of the driver. The *pabe_DEFT4P* is linked to the *pabe_DEFA1g*, which harbors a significantly positively selected site whose expansion might be advantageous: 74I, adjacent to the third cysteine. This 74I is not present in other *pabe_DEFA1* (*a, b, c, d and f*) sequences (Table 1, Figure S1). Therefore, the *DEFTP* is linked to a *DEFA1*, whose expansion might be advantageous.

The less-is-hitchhiking hypothesis depicts the evolutionary process of the expansion of the pseudogene *DEFTP* and the loss of functional θ-defensin in the humans and chimpanzees (Figure 8). With the expansion of the gene (driver) under positive selection, the adjacent linked gene (hitchhiker) is duplicated in the process of segmental duplication. The pseudogenization of the hitchhiker occurs in an ancestral species following the duplication process. During the birth-and-death evolution, the hitchhiker that gains more copies may be either the pseudogene or the functional gene,

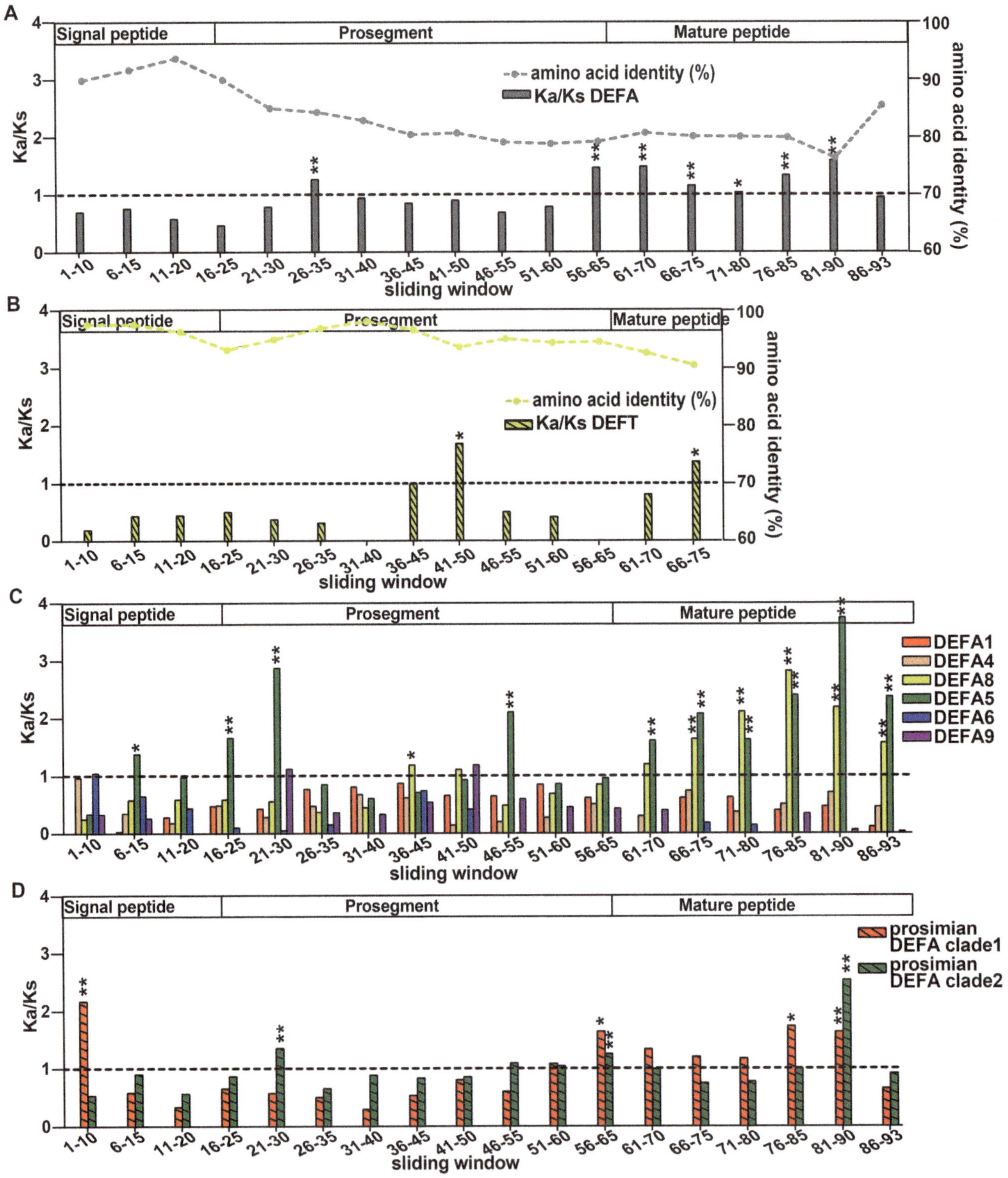

Figure 5. Sliding window analyses of the Ka/Ks for all *DEFA*, *DEFT* and separate *DEFA* clusters. A: Sliding window analysis of the amino acid identity and Ka/Ks for all *DEFA*. The sequence conservation decreases from the signal peptide and prosegment to the mature peptide. The selection pressures acting on different windows are different as indicated by the Ka/Ks ratios. Fragment 26–35 in the prosegment region, fragment 56–65 in the cleavage region and most fragments in the mature peptide are under positive selection (Ka/Ks>1). The significance of Ka/Ks>1 is tested by a bootstrap method and is indicated by * ($P<0.05$) or ** ($P<0.01$). B: Sliding window analysis of the amino acid identity and Ka/Ks for *DEFT*. The sequence conservation also decreases from the signal peptide and prosegment to the mature peptide. The selection pressure on different regions is also variable. Compared to the same region of all *DEFA* and that of *DEFA8*, fragment 41–50 in the prosegment of *DEFT* is under strong positive selection. C: Sliding window analysis of Ka/Ks for each simian *DEFA* cluster. Only the *DEFA5* and *DEFA8* clusters contain fragments that are under positive selection. D: Sliding window analysis of Ka/Ks for the two prosimian clades. The fragment 56–65 in the cleavage region is also under positive selection for both prosimian clades, consistent with that of all *DEFA*.

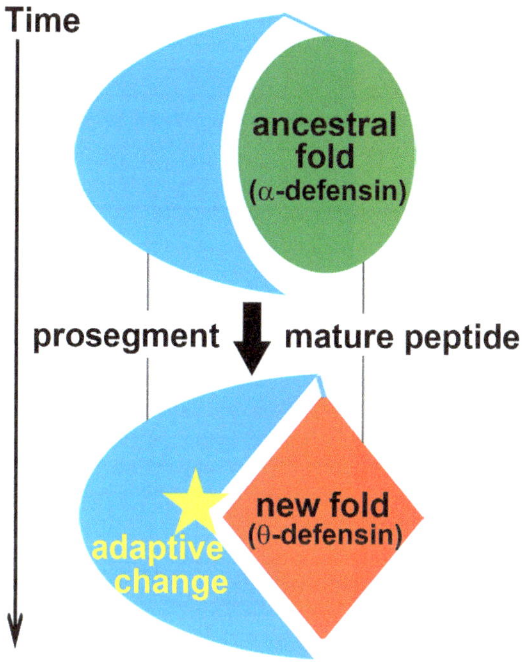

Figure 6. Coevolution of the prosegment with the new protein fold of the mature peptide. The ancestral protein fold refers to the α-defensin mature peptide, and the new protein fold refers to the θ-defensin mature peptide. Because the prosegment is important for the correct folding of the mature peptide, the prosegment coevolves with the mature peptide through adaptive changes.

which is determined by the fate of its adjacent driver. The former scenario is the less-is-hitchhiking hypothesis. The less-is-more, less-is-less and less-is-nothing hypotheses depend on the phenotypic change of the loss-of-function variant, which can be advantageous, deleterious or tolerated, respectively. However, the less-is-hitchhiking hypothesis provides another explanation for the driving force of the loss-of-function variant expansion that does not involve phenotypic consequences. The fate of the loss-of-function hitchhiker is determined by the surrounding genomic environment of a driver gene (i.e., an adjacent gene under positive selection). The loss-of-function hitchhiker can be advantageous, deleterious or tolerated.

Discussion

In this study, we collected the primate *DEFA/DEFT* gene repertoires and performed extensive analyses of the primate α-/θ-defensin multigene family. Through systematic phylogenetic analyses, a detailed classification and nomenclature of the primate α-/θ-defensins was provided. The classification of simian α-/θ-defensins based on phylogenetic trees is related to their expression patterns in myeloid and enteric tissues. Further phylogenetic classification of simian α-defensins into six functional gene clusters corresponds to their functional divergence. In a previous study, this multigene family was classified into three classes based on neighbor-joining (NJ) and maximum parsimony (MP) trees using genes from simians that included human, chimpanzee, orangutan, macaque and marmoset sequences as well as one mouse α-defensin sequence as the outgroup [42]. A recent study focused on the phylogenetic relationships of the α-, β- and θ-defensins using *DEFA/DEFT* sequences from 10 primate species and two human *DEFB* sequences as the outgroup based on the NJ method [69]. In

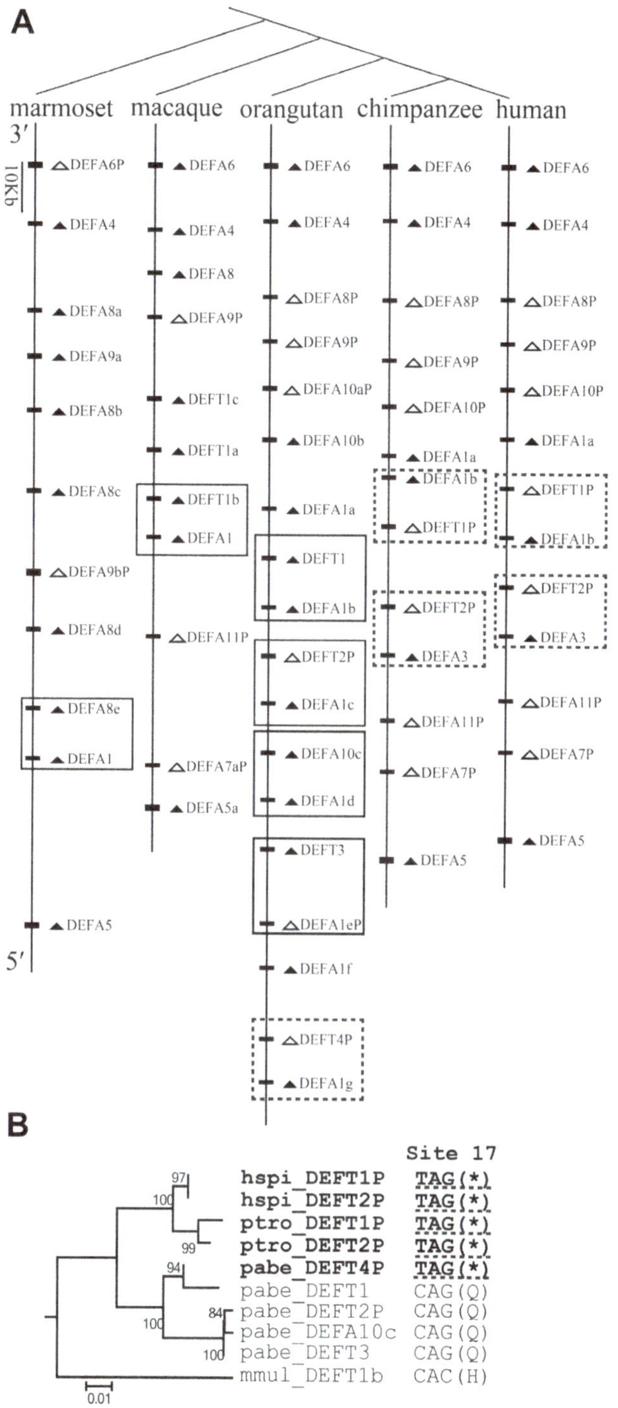

Figure 7. Hitchhiking of *DEFT/DEFTP* during the birth-and-death evolution of the primate *DEFA/DEFT* multigene family. A: Comparative synteny map of the *DEFA/DEFT* gene loci in humans, chimpanzees, orangutans, macaques and marmosets. The boxes highlight the two genes that are linked and duplicate together, including a driver from the *DEFA1* cluster and a hitchhiker from the *DEFT/DEFA10* cluster. Arrowheads indicate transcriptional orientation. Pseudogenes are in white, and functional genes are in black. The dashed-line box includes the pseudogene *DEFTP* containing the nonsense mutation at site 17. B: The NJ tree based on the introns of the boxed *DEFT/DEFTP* from humans (*hspi*), chimpanzees (*ptro*), orangutans (*pabe*) and macaques (*mmul*). The nonsense mutations (*) at codon site 17 of the pseudogenes *hspi_DEFT1P*, *hspi_DEFT2P* *ptro_DEFT1P*, *ptro_DEFT2P* and *pabe_DEFT4P* are underlined with

dashed lines, suggesting that these *DEFTP* pseudogenes are derived from a common ancestor.

our study, we identified 144 primate *DEFA/DEFT* sequences from 16 species of simians and prosimians as well as 10 *DEFA* sequences from two treeshrews, which are closely related to primates. We used sequence alignments of both the more conserved signal-prosegment region and the entire coding region to more precisely reconstruct α-/θ-defensin orthologous relationships and to uncover their potential functional divergence.

The primate *DEFA/DEFT* sequences that are used for this study were collected from the public database, which may contain errors or uncertainty of the sequencing and assembly. One of such ambiguous sequences might be the *pabe_DEFA10c* sequence, which is highly similar to the *pabe_DEFT* sequences and is phylogenetically far from the *DEFA10* cluster (Figure 1A and S8). This sequence is named *DEFA10* only because site 77 is not a premature stop codon. Thus, *pabe_DEFA10c* was excluded in the phylogenetic analysis based on the introns from *DEFA8/DEFA10/DEFT* (Figure S9). For the analysis of positive selection using PAML, sequencing and assembly errors may cause bias. Our results of PAML analysis for separate clusters show that most of the positively selected sites are located in the mature peptide, similar to the results of a previous study that analyzed all the *DEFA* sequences together [42]. For the sliding window analysis of Ka/Ks, we used a bootstrap method to eliminate the effect of potential bias from the sequencing and assembly error in a few genes. For the synteny analysis, there are sequences (*cjac_DEFA9cU* and *cjac_DEFA9dU*) that are not mapped onto the chromosomes of the marmoset. In macaques, several sequences (*mmul_7bUP*, *mmul_-DEFA5bU*, *mmul_DEFA5cU*, *mmul_DEFA5dU* and *mmul_DEFA5eU*) are mapped onto a chromosome region that is quite far from the defensin gene locus presented in the synteny map. The human, chimpanzee, orangutan, macaque and marmoset genomes were selected for the synteny analysis because of their more complete

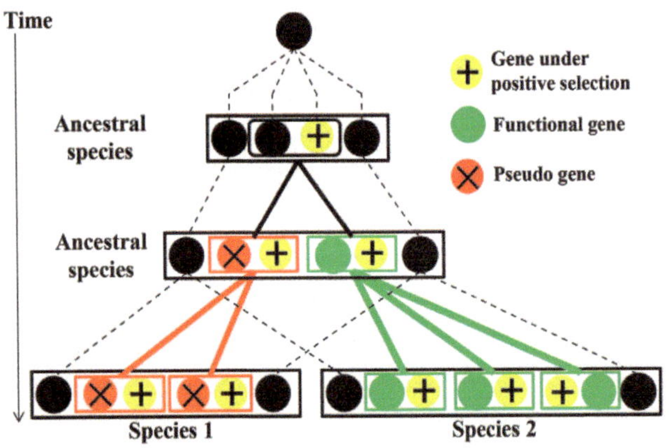

Figure 8. The less-is-hitchhiking hypothesis during the birth-and-death evolution of multigene families. With the expansion of the gene (driver) under positive selection, the adjacent linked gene (hitchhiker) is duplicated in the process of segmental duplication. Pseudogenization of the hitchhiker occurs in an ancestral species following the duplication process. During the process of birth-and-death evolution, the hitchhiker that gains more copies can be either the pseudogene or the functional gene with the other one being lost, and the gain or loss of the hitchhiker is determined by the fitness of its adjacent driver. The expansion of the pseudogene and loss of the functional gene scenario is defined as the less-is-hitchhiking hypothesis.

assemblies in the genomic region harboring the *DEFA1* and *DEFA8/DEFA10/DEFT* gene cluster, whereas the gorilla and gibbon genomes were not used due to the poor assembly in this region.

The functional shifting of simian α-defensins, which is implied by their sequence divergence, structural variation and selection pattern differences among the six functional gene clusters, most likely corresponds to differentiated antimicrobial activity. However, orthologs within the same gene cluster from different species are still under functional constraints, most likely due to their similar antimicrobial mechanisms and challenges from similar types of pathogens in closely related species. Based on these results, we speculate that these defensin clusters constitute a wide range of antimicrobial spectra and that each cluster occupies one part of the spectra. Within each cluster, there is species-specific cluster duplication or pseudogenization following the species diversifications of simians, which implies that the antimicrobial spectra are continuously changing. At the same time, deconstructing the undocumented antimicrobial spectra of α-/θ-defensins is not feasible, even for the most closely related species, because of the countless types of pathogens and the frequent duplication and divergence of defensins. Many experimental studies have shown that members of this family have different antimicrobial activities and are effective against different groups of microbes. For example, the antibacterial activity and specificity of the six human α-defensins against Gram-positive and Gram-negative bacteria are different [70]. For human myeloid defensins, HNP1-3 defensins are of greater potencies than HNP4 against Gram-positive bacteria such as *S. aureus*, whereas HNP4 is of greater potency against Gram-negative bacteria such as *E. coli* and *E. aerogenes*. For human enteric defensins, HD5 is effective against both Gram-negative and Gram-positive bacteria, whereas HD6 shows significantly lower antimicrobial activity against Gram-negative and Gram-positive bacteria than HD5. The myeloid α-/θ-defensins and the enteric HD5 have also been reported to have inhibitory potential against enveloped and non-enveloped viruses, such as the herpes simplex virus [71], the human immunodeficiency virus [72,73], papillomaviruses [74], adenoviruses [75], polyomaviruses [76], the SARS coronavirus [77], influenza viruses [78,79] and others. In addition to its role in host defense, HD5 also has a homeostatic role in establishing and maintaining the intestinal microbiota [80]. Overall, these experimental studies indicate that the variable antimicrobial abilities of mammalian α-/θ-defensins compose an antimicrobial spectrum. Defensins with continuously shifting antimicrobial activity, together with other versatile antimicrobial factors, evolve to act as the first line of defense of the innate immune system and play important roles in the early host defense. Because of the therapeutic potential of antimicrobial peptides, synthetic defensins are being developed as peptide drugs, such as the fusion inhibitors. Exploring the natural defensin repertories, especially in species closely related to humans, will help us understand the stability and plasticity of α-/θ-defensins and aid in designing potent peptide drugs.

The sliding window analysis of *DEFT* (Figure 5B) suggests that fragment 41–50 in the prosegment and fragment 66–75 in the mature peptide are under positive selection. These two fragments contain the positively selected sites detected by PAML (50D, 67I and 69R, corresponding to the sliding window sites 49, 66 and 68, respectively). Other amino acid changes that may contribute to the high Ka/Ks value are 41G/E, 45S/A and 48R/W in fragment 41–50 as well as 69R/L, 71V/I/F and 73R/Q in fragment 66–75 (Figure S1). To be spliced into an 18-residue mature θ-defensin, the three C-terminal residues must be removed during the maturation process [18]. The three C-terminal residues were not found to be under positive selection using the PAML software. However, the three C-terminal residues of θ-defensin show a pattern of RLL or QLL that is quite different from the α-defensins (Figure S1), which might be related to their involvement in the maturation process of θ-defensins.

Multigene families play important roles in the immune system, the sensory system, development and other processes. These families can express gene products at high levels, such as highly conserved histones and nuclear ribosomal RNAs, and produce proteins with diverse functional spectra, such as α-defensins, major histocompatibility complex proteins, immunoglobulins, chemoreceptors and olfactory receptors. The evolutionary processes of multigene families have been explained by birth-and-death evolutionary models [43]. Frequent gene conversion, interlocus recombination, gene duplication and pseudogenization are involved in the evolutionary processes of multigene families. Strong purifying selection or positive selection can also act on multigene families to conserve gene function or give rise to new adaptive phenotypes. The expansion of genes under selection through segmental duplications may impact the fate of the adjacent genes linked with them, which can also affect the evolution of multigene families. Our less-is-hitchhiking hypothesis depicts this phenomenon that the retention of pseudogenes and the loss of functional gene are determined by the fate of the adjacent gene during the birth-and-death evolutionary process. We believe that the expansion of the *DEFTP* pseudogene and the loss of functional θ-defensins in humans and chimpanzees is a representative case for the less-is-hitchhiking hypothesis based on the following criteria. First, the driver and the hitchhiker belong to phylogenetically separate clusters. Second, the driver and the hitchhiker are genetically linked and duplicate together on the chromosome. Third, positive selection acts on the functional drivers. Lastly, the pseudogene hitchhiker expands and the functional hitchhiker has been lost, as determined by the fitness of the driver.

Recently, it has been demonstrated that genetic hitchhiking is pervasive and the mutational cohort that includes both the driver and the hitchhikers drives adaptation [81]. Although these hitchhikers mostly refer to point mutations, there are few reports of an entire gene being a hitchhiker, with one exception reported for the yellow monkey flower (*Mimulus guttatus*), in which a copper tolerance locus under selection and its tightly linked hybrid incompatibility locus spread to fixation in a copper mine population by genetic hitchhiking [82,83]. Future research on the evolution of genetic hitchhiking involving two or more closely linked genes from both case studies and whole-genome comparisons will help to uncover the adaptation of complex traits from linked genes and to understand the genetic and evolutionary basis of certain disease-related traits during the hitchhiking processes.

Supporting Information

Figure S1 Multiple sequence alignment of the primate α-/θ-defensins. Most of the protein sequences are computationally translated from genome sequences rather than being demonstrated to be produced in vivo. The mature peptide of α-defensins has a motif with six conserved cysteine residues, represented as C-x-C-x(3,4)-C-x(9)-C-x(6,9)-C. A nonsense mutation at position 77 (Q77*) generates a θ-defensin precursor with three conserved cysteines. The mature peptide of θ-defensins has a nonapeptide motif with three conserved cysteine residues, represented as x-C-x-C-x(4)-C, with the exception of the Eastern black-and-white colobus θ-defensin (*cgue_DEFT*), which has the motif x-C-x-C-x(8)-C. The nonsense mutation at position 17 (Q17*) results in the loss of the downstream protein-coding ability of the θ-defensins. The initial methionines (M) and conserved cysteines (C) in the alignment are highlighted in blue. The translational ends (*) of the θ-defensin are highlighted in magenta. The blocks highlight the sliding windows of fragments 41–50 and 65–75 of *DEFT*, which are detected with Ka/Ks >1 ($P<0.05$) in Figure 5. The question marks indicate ambiguous amino acids because of the single nucleotide polymorphisms from the genome sequences. The genus and species for each abbreviation can be found in Table S1.

Figure S2 Phylogenetic trees of primate and treeshrew DEFA/DEFT genes based on the signal-prosegment region. The trees are built using the (A) Bayesian inference (BI), (B) neighbor-joining (NJ) and (C) maximum likelihood (ML) methods. The BI tree is labeled with posterior probabilities. The NJ and ML trees are labeled with bootstrap support values. All three trees are drawn to scale, with branch lengths proportional to the estimated evolutionary distances. "P" in node labels denotes a pseudogene.

Figure S3 Phylogenetic tree of primate and treeshrew DEFA/DEFT genes based on the signal-prosegment region. The BI tree is selected as the background tree. The major clades or clusters having similar topologies from all three tree-building methods (BI, NJ and ML) are combined and labeled with the BI posterior probabilities and the bootstrap support values from the NJ and ML analyses.

Figure S4 Phylogenetic trees of primate and treeshrew *DEFA/DEFT* genes based on the entire coding region. The trees are built using the BI (A), NJ (B) and ML (C) methods. The BI tree is labeled with posterior probabilities. The NJ and ML trees are labeled with bootstrap support values. All three trees are drawn to scale, with branch lengths proportional to the estimated evolutionary distances.

Figure S5 Phylogenetic tree of primate and treeshrew *DEFA/DEFT* genes based on the entire coding region. The BI tree is selected as the background tree. The major clades or clusters having similar topologies from all three tree-building methods (BI, NJ and ML) are combined and labeled with the BI posterior probabilities and the bootstrap support values from the NJ and ML analyses.

Figure S6 Phylogenetic incongruence is caused by the long-branch attraction of homogeneity sites. The homogeneity sites under convergent or parallel evolution in the mature peptide (62, 63, 64, 65, 68, 71, 77, 81, 88, 89, 92 and 93) are highlighted in different colors. These homogeneity sites can cause phylogenetic incongruence between the trees constructed using the entire coding region versus the signal-prosegment region. The phylogenetic tree on the left is inferred based on the amino acid sequences of the entire coding region using NJ method without removing the homogeneity sites, whereas the tree on the right is built using NJ method after removing the homogeneity sites. When the long-branch attraction effect is eliminated, the sequences of the prosimian *DEFA* clade 2 group together.

Figure S7 The different phylogenetic relationships of *DEFA9/DEFA11* are determined by different parts of the

gene. A: Different patterns of sequence similarity are observed before and after a clear boundary in the *DEFA9/DEFA11* genes, namely part I and part II. Phylogenetic trees are separately inferred from part I (the tree on the left) and part II (the tree on the right), using the *DEFA6* cluster as the outgroup. The combined tree based on part I shows that *DEFA11* is duplicated from *DEFA9* after the split of New World and Old World monkeys. Whereas, in the combined tree of part II, *DEFA11* is the outgroup of all *DEFA9* sequences. The BI tree is computed using the GTR + G + I (4 categories) model. The NJ tree is computed using the K2 + G (shape parameter = 1.8 for part I and 3.3 for part II) model. The ML tree is computed using the K2 + G (4 categories) model. B: The BI/NJ/ML trees with branch lengths proportional to the estimated distances inferred from part I. C: The BI/NJ/ML trees inferred from part II.

Figure S8 The different phylogenetic positions of hominoid *DEFA8* are determined by different parts of the gene. A: Different patterns of sequence similarity are also observed before and after a clear boundary in the *DEFA8/DEFA10/DEFT* genes, which is different from that of *DEFA9/DEFA11* genes. Similarly, phylogenetic trees are separately inferred from part I (the tree on the left) and part II (the tree on the right), using the *DEFA1* cluster as the outgroup. In the combined tree inferred from part I, the *DEFA8* from Old World monkeys and hominoids clusters together with *DEFA10/DEFT*. Whereas in the tree inferred from part II, the *DEFA8* from Old World monkeys and hominoids is the outgroup in the *DEFA8/DEFA10/DEFT* cluster. The BI tree is computed using the GTR + G + I (4 categories) model. The NJ tree is computed using the K2 + G (shape parameter = 1.5 for part I and 2.0 for part II) model. The ML tree is computed using the K2 + G (4 categories) model. The sequence *pabe_DEFA10c* is clustered with *DEFT*, likely resulting from sequence assembly error. Thus, the sequence *pabe_DEFA10c* was excluded in following analysis in Figure S9. B: The BI/NJ/ML trees with branch lengths proportional to the estimated distances inferred from part I. C: The BI/NJ/ML trees inferred from part II.

Figure S9 Phylogenetic trees constructed using BI (A), NJ (B) and ML (C) methods based on the introns of *DEFA8/DEFA10/DEFT*. *DEFT* is found in Old World monkeys and hominoids, whereas *DEFA10* is found in hominoids only. The *DEFA10* and *DEFT* are clustered into independent groups and are both derived from the duplication of *DEFA8* before the divergence of Old World monkeys and hominoids. But the *DEFA10* is lost in the ancestor of Old World monkeys. The BI tree is computed using the GTR + G + I (4 categories) model. The NJ tree is computed using the K2 + G (shape parameter = 1.8) model. The ML tree is computed using the K2 + G (4 categories) model.

Table S1 The repertoires of α-/θ-defensin (*DEFA/DEFT*) genes from primates and treeshrews.

Table S2 Log-likelihood values and parameter estimates for primate DEFA/DEFT genes under site models. Notes: $\omega = dN/dS$; p, number of free parameters; positively selected sites with significance at the 99% level are bold, whereas the remaining sites have significance at the 95% level. The likelihood ratio tests are analyzed by comparing the following pairs of models: M1a-M2a, M7-M8 and M8a-M8.

Table S3 The average values of Ka, Ks and Ka/Ks from the sliding window analyses for the primate DEFA/DEFT genes and the P values from the bootstrap test. Notes: avg. Ka: average pairwise Ka value; avg. Ks: average pairwise Ks value; avg. Ka/Ks: average pairwise Ka/Ks ratio; Bold number: Ka/Ks significantly >1 ($P<0.05$); NA: Not available.

Author Contributions

Conceived and designed the experiments: DQC YL JFH. Performed the experiments: DQC. Analyzed the data: DQC YL. Contributed reagents/materials/analysis tools: DQC. Wrote the paper: DQC YL JFH.

References

1. Ganz T, Selsted ME, Szklarek D, Harwig SS, Daher K, et al. (1985) Defensins. Natural peptide antibiotics of human neutrophils. J Clin Invest 76: 1427–1435.
2. Ganz T (2003) Defensins: antimicrobial peptides of innate immunity. Nat Rev Immunol 3: 710–720.
3. Lehrer RI (2004) Primate defensins. Nat Rev Microbiol 2: 727–738.
4. Selsted ME, Ouellette AJ (2005) Mammalian defensins in the antimicrobial immune response. Nat Immunol 6: 551–557.
5. Klotman ME, Chang TL (2006) Defensins in innate antiviral immunity. Nat Rev Immunol 6: 447–456.
6. Lehrer RI, Lu W (2012) alpha-Defensins in human innate immunity. Immunol Rev 245: 84–112.
7. Yang D, de la Rosa G, Tewary P, Oppenheim JJ (2009) Alarmins link neutrophils and dendritic cells. Trends Immunol 30: 531–537.
8. Rehaume LM, Hancock RE (2008) Neutrophil-derived defensins as modulators of innate immune function. Crit Rev Immunol 28: 185–200.
9. Yang D, Biragyn A, Hoover DM, Lubkowski J, Oppenheim JJ (2004) Multiple roles of antimicrobial defensins, cathelicidins, and eosinophil-derived neurotoxin in host defense. Annu Rev Immunol 22: 181–215.
10. Maxwell AI, Morrison GM, Dorin JR (2003) Rapid sequence divergence in mammalian beta-defensins by adaptive evolution. Mol Immunol 40: 413–421.
11. Sugiarto H, Yu PL (2004) Avian antimicrobial peptides: the defense role of beta-defensins. Biochem Biophys Res Commun 323: 721–727.
12. Zou J, Mercier C, Koussounadis A, Secombes C (2007) Discovery of multiple beta-defensin like homologues in teleost fish. Mol Immunol 44: 638–647.
13. Dalla Valle L, Benato F, Maistro S, Quinzani S, Alibardi L (2012) Bioinformatic and molecular characterization of beta-defensins-like peptides isolated from the green lizard Anolis carolinensis. Dev Comp Immunol 36: 222–229.
14. Belov K, Sanderson CE, Deakin JE, Wong ES, Assange D, et al. (2007) Characterization of the opossum immune genome provides insights into the evolution of the mammalian immune system. Genome Res 17: 982–991.
15. Lynn DJ, Bradley DG (2007) Discovery of alpha-defensins in basal mammals. Dev Comp Immunol 31: 963–967.
16. Hollox EJ, Barber JC, Brookes AJ, Armour JA (2008) Defensins and the dynamic genome: what we can learn from structural variation at human chromosome band 8p23.1. Genome Res 18: 1686–1697.
17. Bruhn O, Paul S, Tetens J, Thaller G (2009) The repertoire of equine intestinal alpha-defensins. BMC Genomics 10: 631.
18. Tang YQ, Yuan J, Osapay G, Osapay K, Tran D, et al. (1999) A cyclic antimicrobial peptide produced in primate leukocytes by the ligation of two truncated alpha-defensins. Science 286: 498–502.
19. Nguyen TX, Cole AM, Lehrer RI (2003) Evolution of primate theta-defensins: a serpentine path to a sweet tooth. Peptides 24: 1647–1654.
20. Linzmeier R, Ho CH, Hoang BV, Ganz T (1999) A 450-kb contig of defensin genes on human chromosome 8p23. Gene 233: 205–211.
21. Aldred PM, Hollox EJ, Armour JA (2005) Copy number polymorphism and expression level variation of the human alpha-defensin genes DEFA1 and DEFA3. Hum Mol Genet 14: 2045–2052.
22. Linzmeier RM, Ganz T (2005) Human defensin gene copy number polymorphisms: comprehensive analysis of independent variation in alpha- and beta-defensin regions at 8p22-p23. Genomics 86: 423–430.
23. Mars WM, Patmasiriwat P, Maity T, Huff V, Weil MM, et al. (1995) Inheritance of unequal numbers of the genes encoding the human neutrophil defensins HP-1 and HP-3. J Biol Chem 270: 30371–30376.
24. Jones DE, Bevins CL (1993) Defensin-6 mRNA in human Paneth cells: implications for antimicrobial peptides in host defense of the human bowel. FEBS Lett 315: 187–192.

25. Porter EM, Liu L, Oren A, Anton PA, Ganz T (1997) Localization of human intestinal defensin 5 in Paneth cell granules. Infect Immun 65: 2389–2395.
26. Bevins CL, Salzman NH (2011) Paneth cells, antimicrobial peptides and maintenance of intestinal homeostasis. Nat Rev Microbiol 9: 356–368.
27. Patil A, Hughes AL, Zhang G (2004) Rapid evolution and diversification of mammalian alpha-defensins as revealed by comparative analysis of rodent and primate genes. Physiol Genomics 20: 1–11.
28. Tang YQ, Yuan J, Miller CJ, Selsted ME (1999) Isolation, characterization, cDNA cloning, and antimicrobial properties of two distinct subfamilies of alpha-defensins from rhesus macaque leukocytes. Infect Immun 67: 6139–6144.
29. Wu Z, Li X, Ericksen B, de Leeuw E, Zou G, et al. (2007) Impact of pro segments on the folding and function of human neutrophil alpha-defensins. J Mol Biol 368: 537–549.
30. Valore EV, Martin E, Harwig SS, Ganz T (1996) Intramolecular inhibition of human defensin HNP-1 by its propiece. J Clin Invest 97: 1624–1629.
31. Tongaonkar P, Golji AE, Tran P, Ouellette AJ, Selsted ME (2012) High fidelity processing and activation of the human alpha-defensin HNP1 precursor by neutrophil elastase and proteinase 3. PLoS One 7: e32469.
32. Glenthoj A, Glenthoj AJ, Borregaard N (2013) ProHNPs are the principal alpha-defensins of human plasma. Eur J Clin Invest 43: 836–843.
33. Hill CP, Yee J, Selsted ME, Eisenberg D (1991) Crystal structure of defensin HNP-3, an amphiphilic dimer: mechanisms of membrane permeabilization. Science 251: 1481–1485.
34. Szyk A, Wu Z, Tucker K, Yang D, Lu W, et al. (2006) Crystal structures of human alpha-defensins HNP4, HD5, and HD6. Protein Sci 15: 2749–2760.
35. Zhang Y, Lu W, Hong M (2010) The membrane-bound structure and topology of a human alpha-defensin indicate a dimer pore mechanism for membrane disruption. Biochemistry 49: 9770–9782.
36. Pazgier M, Wei G, Ericksen B, Jung G, Wu Z, et al. (2012) Sometimes it takes two to tango: contributions of dimerization to functions of human alpha-defensin HNP1 peptide. J Biol Chem 287: 8944–8953.
37. Rajabi M, Ericksen B, Wu X, de Leeuw E, Zhao L, et al. (2012) Functional determinants of human enteric alpha-defensin HD5: crucial role for hydrophobicity at dimer interface. J Biol Chem 287: 21615–21627.
38. Henriques ST, Craik DJ (2010) Cyclotides as templates in drug design. Drug Discov Today 15: 57–64.
39. Lehrer RI, Cole AM, Selsted ME (2012) theta-Defensins: cyclic peptides with endless potential. J Biol Chem 287: 27014–27019.
40. Jarczak J, Kosciuczuk EM, Lisowski P, Strzalkowska N, Jozwik A, et al. (2013) Defensins: natural component of human innate immunity. Hum Immunol 74: 1069–1079.
41. Wood MP, Cole AL, Ruchala P, Waring AJ, Rohan LC, et al. (2013) A compensatory mutation provides resistance to disparate HIV fusion inhibitor peptides and enhances membrane fusion. PLoS One 8: e55478.
42. Das S, Nikolaidis N, Goto H, McCallister C, Li J, et al. (2010) Comparative genomics and evolution of the alpha-defensin multigene family in primates. Mol Biol Evol 27: 2333–2343.
43. Nei M, Rooney AP (2005) Concerted and birth-and-death evolution of multigene families. Annu Rev Genet 39: 121–152.
44. Lynn DJ, Lloyd AT, Fares MA, O'Farrelly C (2004) Evidence of positively selected sites in mammalian alpha-defensins. Mol Biol Evol 21: 819–827.
45. Quintana-Murci L (2012) Genetics. Gene losses in the human genome. Science 335: 806–807.
46. Seidel A, Ye Y, de Armas LR, Soto M, Yarosh W, et al. (2010) Cyclic and acyclic defensins inhibit human immunodeficiency virus type-1 replication by different mechanisms. PLoS One 5: e9737.
47. Tamura K, Stecher G, Peterson D, Filipski A, Kumar S (2013) MEGA6: Molecular Evolutionary Genetics Analysis version 6.0. Mol Biol Evol 30: 2725–2729.
48. Drummond AJ, Suchard MA, Xie D, Rambaut A (2012) Bayesian phylogenetics with BEAUti and the BEAST 1.7. Mol Biol Evol 29: 1969–1973.
49. Stover BC, Muller KF (2010) TreeGraph 2: combining and visualizing evidence from different phylogenetic analyses. BMC Bioinformatics 11: 7.
50. Gu X, Zou Y, Su Z, Huang W, Zhou Z, et al. (2013) An update of DIVERGE software for functional divergence analysis of protein family. Mol Biol Evol 30: 1713–1719.
51. Gu X (1999) Statistical methods for testing functional divergence after gene duplication. Mol Biol Evol 16: 1664–1674.
52. Wang Y, Gu X (2001) Functional divergence in the caspase gene family and altered functional constraints: statistical analysis and prediction. Genetics 158: 1311–1320.
53. Yang Z (2007) PAML 4: phylogenetic analysis by maximum likelihood. Mol Biol Evol 24: 1586–1591.
54. Zhang Z, Li J, Zhao XQ, Wang J, Wong GK, et al. (2006) KaKs_Calculator: calculating Ka and Ks through model selection and model averaging. Genomics Proteomics Bioinformatics 4: 259–263.
55. Crooks GE, Hon G, Chandonia JM, Brenner SE (2004) WebLogo: a sequence logo generator. Genome Res 14: 1188–1190.
56. Wei G, de Leeuw E, Pazgier M, Yuan W, Zou G, et al. (2009) Through the looking glass, mechanistic insights from enantiomeric human defensins. J Biol Chem 284: 29180–29192.
57. Eswar N, Eramian D, Webb B, Shen MY, Sali A (2008) Protein structure modeling with MODELLER. Methods Mol Biol 426: 145–159.
58. Humphrey W, Dalke A, Schulten K (1996) VMD: visual molecular dynamics. J Mol Graph 14: 33–38, 27–38.
59. Case DA, Cheatham TE, 3rd, Darden T, Gohlke H, Luo R, et al. (2005) The Amber biomolecular simulation programs. J Comput Chem 26: 1668–1688.
60. Phillips JC, Braun R, Wang W, Gumbart J, Tajkhorshid E, et al. (2005) Scalable molecular dynamics with NAMD. J Comput Chem 26: 1781–1802.
61. Baker NA, Sept D, Joseph S, Holst MJ, McCammon JA (2001) Electrostatics of nanosystems: application to microtubules and the ribosome. Proc Natl Acad Sci U S A 98: 10037–10041.
62. Garcia AE, Osapay G, Tran PA, Yuan J, Selsted ME (2008) Isolation, synthesis, and antimicrobial activities of naturally occurring theta-defensin isoforms from baboon leukocytes. Infect Immun 76: 5883–5891.
63. Delsuc F, Brinkmann H, Philippe H (2005) Phylogenomics and the reconstruction of the tree of life. Nat Rev Genet 6: 361–375.
64. Gu X (2001) Maximum-likelihood approach for gene family evolution under functional divergence. Mol Biol Evol 18: 453–464.
65. Leonova L, Kokryakov VN, Aleshina G, Hong T, Nguyen T, et al. (2001) Circular minidefensins and posttranslational generation of molecular diversity. J Leukoc Biol 70: 461–464.
66. Conibear AC, Rosengren KJ, Daly NL, Henriques ST, Craik DJ (2013) The cyclic cystine ladder in theta-defensins is important for structure and stability, but not antibacterial activity. J Biol Chem 288: 10830–10840.
67. Ghosh D, Porter E, Shen B, Lee SK, Wilk D, et al. (2002) Paneth cell trypsin is the processing enzyme for human defensin-5. Nat Immunol 3: 583–590.
68. Kamdar K, Maemoto A, Qu X, Young SK, Ouellette AJ (2008) In vitro activation of the rhesus macaque myeloid alpha-defensin precursor proRMAD-4 by neutrophil serine proteinases. J Biol Chem 283: 32361–32368.
69. Li D, Zhang L, Yin H, Xu H, Trask JS, et al. (2014) Evolution of primate alpha and theta defensins revealed by analysis of genomes. Mol Biol Rep.
70. Ericksen B, Wu Z, Lu W, Lehrer RI (2005) Antibacterial activity and specificity of the six human {alpha}-defensins. Antimicrob Agents Chemother 49: 269–275.
71. Yasin B, Wang W, Pang M, Cheshenko N, Hong T, et al. (2004) Theta defensins protect cells from infection by herpes simplex virus by inhibiting viral adhesion and entry. J Virol 78: 5147–5156.
72. Cole AM, Hong T, Boo LM, Nguyen T, Zhao C, et al. (2002) Retrocyclin: a primate peptide that protects cells from infection by T- and M-tropic strains of HIV-1. Proc Natl Acad Sci U S A 99: 1813–1818.
73. Zhang L, Yu W, He T, Yu J, Caffrey RE, et al. (2002) Contribution of human alpha-defensin 1, 2, and 3 to the anti-HIV-1 activity of CD8 antiviral factor. Science 298: 995–1000.
74. Buck CB, Day PM, Thompson CD, Lubkowski J, Lu W, et al. (2006) Human alpha-defensins block papillomavirus infection. Proc Natl Acad Sci U S A 103: 1516–1521.
75. Smith JG, Nemerow GR (2008) Mechanism of adenovirus neutralization by Human alpha-defensins. Cell Host Microbe 3: 11–19.
76. Dugan AS, Maginnis MS, Jordan JA, Gasparovic ML, Manley K, et al. (2008) Human alpha-defensins inhibit BK virus infection by aggregating virions and blocking binding to host cells. J Biol Chem 283: 31125–31132.
77. Wohlford-Lenane CL, Meyerholz DK, Perlman S, Zhou H, Tran D, et al. (2009) Rhesus theta-defensin prevents death in a mouse model of severe acute respiratory syndrome coronavirus pulmonary disease. J Virol 83: 11385–11390.
78. Leikina E, Delanoe-Ayari H, Melikov K, Cho MS, Chen A, et al. (2005) Carbohydrate-binding molecules inhibit viral fusion and entry by crosslinking membrane glycoproteins. Nat Immunol 6: 995–1001.
79. Salvatore M, Garcia-Sastre A, Ruchala P, Lehrer RI, Chang T, et al. (2007) alpha-Defensin inhibits influenza virus replication by cell-mediated mechanism(s). J Infect Dis 196: 835–843.
80. Salzman NH, Hung K, Haribhai D, Chu H, Karlsson-Sjoberg J, et al. (2010) Enteric defensins are essential regulators of intestinal microbial ecology. Nat Immunol 11: 76–83.
81. Lang GI, Rice DP, Hickman MJ, Sodergren E, Weinstock GM, et al. (2013) Pervasive genetic hitchhiking and clonal interference in forty evolving yeast populations. Nature 500: 571–574.
82. Presgraves DC (2013) Hitchhiking to speciation. PLoS Biol 11: e1001498.
83. Wright KM, Lloyd D, Lowry DB, Macnair MR, Willis JH (2013) Indirect evolution of hybrid lethality due to linkage with selected locus in Mimulus guttatus. PLoS Biol 11: e1001497.

7

On the Evolution of *Yeti*, a *Drosophila melanogaster* Heterochromatin Gene

Roberta Moschetti[1], Emanuele Celauro[2], Fulvio Cruciani[2], Ruggiero Caizzi[1], Patrizio Dimitri[2]*

1 Dipartimento di Biologia, Università degli Studi di Bari, Bari, Italy, **2** Dipartimento di Biologia e Biotecnologie "Charles Darwin" and Istituto Pasteur Fondazione Cenci-Bolognetti, Sapienza Università di Roma, Roma, Italy

Abstract

Constitutive heterochromatin is a ubiquitous and still unveiled component of eukaryotic genomes, within which it comprises large portions. Although constitutive heterochromatin is generally considered to be transcriptionally silent, it contains a significant variety of sequences that are expressed, among which about 300 single-copy coding genes have been identified by genetic and genomic analyses in the last decades. Here, we report the results of the evolutionary analysis of *Yeti*, an essential gene of *Drosophila melanogaster* located in the deep pericentromeric region of chromosome 2R. By FISH, we showed that *Yeti* maintains a heterochromatin location in both *D. simulans* and *D. sechellia* species, closely related to *D. melanogaster*, while in the more distant species e.g., *D. pseudoobscura* and *D. virilis*, it is found within euchromatin, in the syntenic chromosome Muller C, that corresponds to the *2R* arm of *D. melanogaster* chromosome 2. Thus, over evolutionary time, *Yeti* has been resident on the same chromosomal element, but it progressively moved closer to the pericentric regions. Moreover, *in silico* reconstruction of the *Yeti* gene structure in 19 *Drosophila* species and in 5 non-drosophilid dipterans shows a rather stable organization during evolution. Accordingly, by PCR analysis and sequencing, we found that the single intron of *Yeti* does not undergo major intraspecies or interspecies size changes, unlike the introns of other essential *Drosophila* heterochromatin genes, such as *light* and *Dbp80*. This implicates diverse evolutionary forces in shaping the structural organization of genes found within heterochromatin. Finally, the results of d_S - d_N tests show that *Yeti* is under negative selection both in heterochromatin and euchromatin, and indicate that the change in genomic location did not affected significantly the molecular evolution of the gene. Together, the results of this work contribute to our understanding of the evolutionary dynamics of constitutive heterochromatin in the genomes of higher eukaryotes.

Editor: Igor V. Sharakhov, Virginia Tech, United States of America

Funding: This research was supported by grants from Ministero dell'Università e della Ricerca- Miur (Prin 2008), Istituto Pasteur-Fondazione Cenci-Bolognetti and Fondazione Roma terzo settore. The funders had no role in study design, data collection and analysis, decision to publish, or preparation of the manuscript.

Competing Interests: The authors have declared that no competing interests exist.

* Email: patrizio.dimitri@uniroma1.it

Introduction

Constitutive heterochromatin is commonly found in large blocks near centromeres and telomeres; it consists mostly of repetitive DNA sequences and maintains its characteristic organization on both homologous chromosomes. It is a ubiquitous component of eukaryotic genomes and, in many species, comprises large chromosomal portions, or even entire chromosomes. For example, about 30% of the *Drosophila* and human genomes, and up to 70–90% of certain nematode and plant genomes, are made up of constitutive heterochromatin [1,2,3], yet the reasons for its widespread occurrence are still unclear.

Heterochromatin was originally defined at cytological level as the chromosome portion that stains deeply at prophase and maintains a compact organization throughout all stages of the mitotic cell cycle [4]. Historically, distinctive antagonistic properties compared to the rest of the genome were identified: 1) strongly reduced level of meiotic recombination; 2) low gene density; 3) mosaic inactivation of the expression of euchromatic genes when moved nearby (position effect variegation, PEV); 4) late replication during S phase; 5) transcriptional inactivity; 6) enrichment in the so-called "junk" repetitive DNA, such as satellite sequences and truncated transposable element remnants.

Together, these properties led to the view of constitutive heterochromatin as a "desert" of genetic functions [5]. In the last three decades, however, studies primarily conducted in *Drosophila melanogaster* have shown that constitutive heterochromatin does in fact play roles in important cellular functions, such as chromosome organization and inheritance [6,7,8,9,10,11]. Although generally regarded as transcriptionally silent, constitutive heterochromatin has been found to contain actively transcribed genes [3]. For example, in *Drosophila melanogaster*, more than 40 genes essential for viability or fertility have been mapped to pericentric heterochromatin [12,13,14,15,16,17].

In the last decade, the release of *D. melanogaster* heterochromatin sequence by the Berkeley Drosophila Genome Project (http://www.fruitfly.org/) and Drosophila Heterochromatin Genome Project (http://www.dhgp.org/index_release_notes.html) has greatly facilitated studies of mapping, molecular organization and function of genes located in pericentromeric heterochromatin [18].

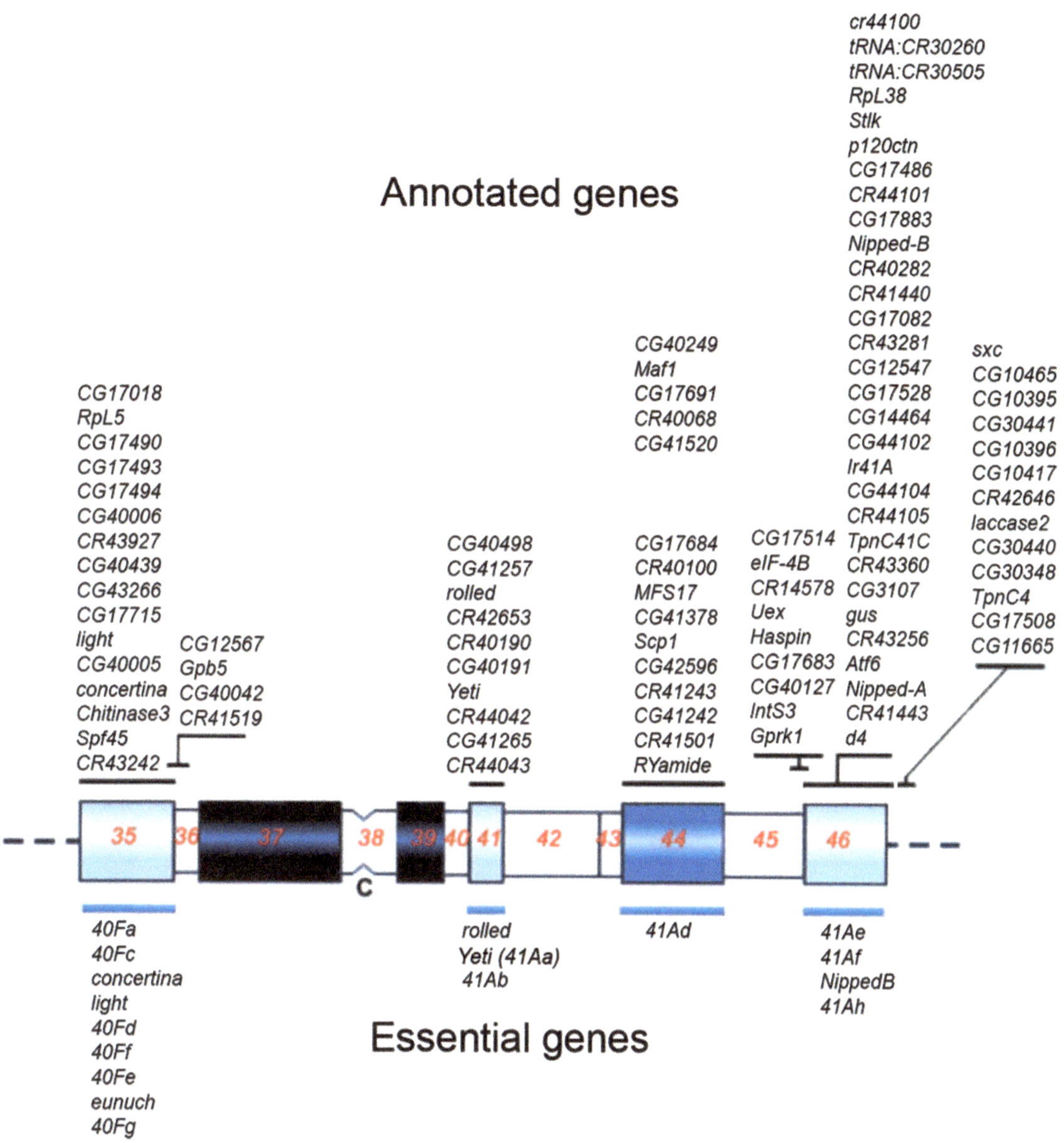

Figure 1. Cytogenetic mapping of heterochromatin genes of chromosome 2. The map was modified from that shown in previous papers [3,45]. The diagram shows the essential genes defined by mutational analyses (below) and annotated genes defined by the heterochromatin genome project (above). Shades of blue correspond to the intensity of DAPI staining, with the darkest blue blocks representing regions with strong fluorescence intensity and open blocks representing non fluorescent regions. The different cytological regions are numbered.

More recently, an improved whole genome shotgun assembly [19] has been produced, which includes 20.7 Mb of draft-quality heterochromatin sequence. In the last years, 15 Mb of this sequence have been further improved or completed [20] and a BAC-based physical map of 13 Mb of pericentric heterochromatin, together with the cytogenetic map that locates some 11 Mb to specific heterochromatin regions, have been constructed [20]. About 250 protein-coding genes were defined in the release 5.1 annotation of the currently sequenced heterochromatin DNA [21]. According to these results, the number of active genes in constitutive heterochromatin of *D. melanogaster* appears to be higher than defined by genetic analysis. Notably, these genes encode proteins involved in important cellular and developmental processes [3].

Further mapping of *D. melanogaster* heterochromatin was performed by comparative genomic hybridization [22]. The transcription profiles of mapped sequences by microarray analysis also revealed region-specific temporal patterns of transcription within heterochromatin during oogenesis and in early embryonic development.

Evolutionary studies have shown that *D. melanogaster* heterochromatin genes, such as *light* and others, originated from progenitors that were originally located within euchromatin in the drosophilid lineage [23,24]. Here we have focussed our study on the evolutionary origin of *Yeti*, an essential heterochromatin gene of *D. melanogaster*, which encodes a protein belonging to the evolutionarily conserved BCNT family of chromatin remodellers [25,26]. We report that *Yeti* locates in euchromatin in distant species, e.g. *D. pseudoobscura* and *D. virilis*, similarly to what has been found for *light* and other genes [23,24]. Moreover, we found that the *Yeti* gene structure remains rather stable during the evolution of *Drosophila* species. In particular, the second exon that

Table 1. List of the *Yeti* orthologs and their encoded proteins.

Species	Gene ID	Database Location	GenBank A.C. (position)	Uniprot reference	Amino acids
D.melanogaster	FBgn0128734	2Rh:1343403..1345119	NW_001848856.1	B4J7U2	241
D.simulans	FBgn0191193	3R:38,269..39,056 [+]	NT_167061.1	B4QUX4	241
D.sechellia	FBgn0166002	2h;scaffold_170:38,673..3,460 [−].	NW_001999858.1	B4IME9	241
D.yakuba	FBgn0236265	2h;v2_chr2h_random_005:668,463..673,172.	NW_002052891.1	B4IT75	215
D.erecta	FBgn0103193	scaffold 4929	NW_001956548.1	B3N457	236
D.eugracilis	not available	not available	KB464511.1 (19716..18949)	not available	235
D.biarmipes	not available	not available	KB462255.1 (18839..19580)	not available	229**
D.takahashi	not available	not available	KB460683.1 (11044..10283)	not available	235**
D.elegans	not available	not available	KB458177.1 (6204..6991)	not available	244**
D.ananassae	FBgn0088065	2R euchromatic region scaffold 13266	NW_001939294.1	B3MBR1[a]	272[a]
D.bipectinata	not available	not available	KB464371.1 (48012..48773)	not available	232
D.pseudoobscura	FBgn0245984	2R euchromatic region	NC_009006.2	B5E0W2	275
D.persimilis	FBgn0148669	not available	NW_001985955.1	B4GBJ4	275
D.miranda	not available	not available	CM001519.2 (366522..367496)	not available	275
D.willistoni	FBgn0212772	2R euchromatic arm	NW_002032340.1	B4MJ08	273
D.mojavensis	FBgn0143134	2R euchromatic arm	NW_001979114.1	B4KQ8	295
D.virilis	FBgn0209341	2R euchromatic arm	NW_002014420.1	B4LKP7	300
D.albomicans	not available	not available	JH859027.1 (5894805..5895706)	not available	279
D.grimshawi	FBgn0128734	2R euchromatic arm	NW_001961673.1	B4J7U2	285
C.quinquefasciatus	CPIJ018830	not available	NW_001888048.1	B0XH12	284
A.aegypti	AAEL007422	supercontig 1.255, euchromatic region 2p25 [46]	NW_001810963.1	Q0IF03	279
A.darlingi	not available	not available	ADMH02000690.1 (45710.. 44829)	not available	
A.gambiae	AGAP005152	2L:euchromatic region 21E	NT_078265.2	Q7PPY4	293
M.destructor	not available	not available	GL501532.1 (369211.. 368309)	not available	300

**ORF defective. Amino acids deducted from bestfit alignments.
[a)]Bestfit protein is 236 aminoacids long.

encodes the last 30 aminoacids of the conserved BCNT domain is invariably 91 bp-long. Finally, we found that the single intron of *Yeti* does not undergo major size changes in *D. melanogaster* and closely related species, unlike the introns of other essential *Drosophila* heterochromatin genes [27].

Results

Evolutionary repositioning of the *Yeti* gene from euchromatin to pericentric heterochromatin

The single-copy *Yeti* gene of *D. melanogaster* maps to the region h41 of chromosome 2R mitotic heterochromatin (Figure 1; Table 1), which corresponds to division 41A of salivary gland polytene chromosomes [25,26,28,29].

To characterize the chromosomal location of *Yeti* among *Drosophila* genus species, we performed fluorescent *in situ* hybridization (FISH) experiments on polytene chromosomes of *D. simulans, D. sechellia*, two sibling species of *D. melanogaster*, and on two distantly related species: *D. pseudoobscura*, belonging to the Sophophora subgenus and *D. virilis*, belonging to the *Drosophila* subgenus. These species cover nearly a 40 million years divergence time and thus represent a wide spectrum of the evolutionary history of *Yeti*.

To map *Yeti* in *D. simulans* and *D. sechellia*, we used the *D. melanogaster Yeti* cDNA probe (RE36623), while PCR species-specific probes were used in *D. pseudoobscura* and *D. virilis*. PCR probes were amplified over a less conserved region located outside the C-terminal BCNT coding domain of YETI protein (see Materials and Methods).

The results of this analysis are shown in Figure 2. In *D. melanogaster, D. simulans* and *D. sechellia* the *Yeti* cDNA probe produces a signal mapping to the base of division 41A, in the right arm of chromosome 2 (Figure 2A,B,C). Notably, the signals show a large diffuse structure very different from the sharp hybridization signals usually seen with euchromatic probes; such a morphology is a distinctive mark for sequences derived from partially polytenized heterochromatin regions [30,31]. Together, our FISH results indicate that *Yeti* maintains a heterochromatic location in *D. simulans* and *D. sechellia*. The FlyBase localization of *Yeti* in *D. sechellia* is in 2 h (scaffold_170:38,673..3,460; Table 1), in accord with our mapping results, while that in *D. simulans* in 3R (38,269..39,056; Table 1) is apparently conflicting and may reflect an assembly error in the *D. simulans* genome sequences (see discussion), as reported by Schaffer et al. [32]. In both *D. pseudoobscura and D. virilis* a single FISH signal was observed in the euchromatic arms of polytene chromosomes, in agreement with FlyBase (Figure 2D, E; Table 1). In *D. pseudoobscura*, the *Yeti PCR* probe produced a sharp signal that maps to region 63C in the proximal euchromatin of chromosome 3, while in *D. Virilis* the *Yeti* signal is found at region 53E, in the distal euchromatin of chromosome 5. Thus, independently of their genome localization (heterochromatin or euchromatin), in the analysed species *Yeti* lies

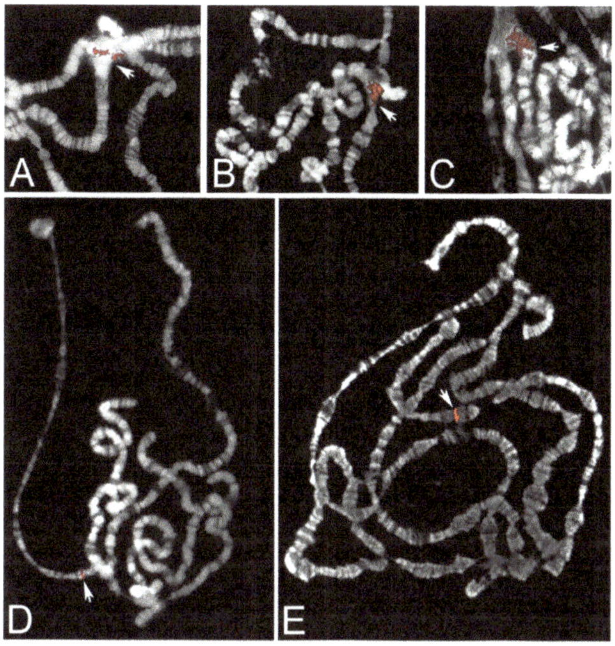

Figure 2. Examples of FISH mapping of *Yeti* probes to polytene chromosomes of *Drosophila* species. Salivary gland polytene chromosomes were stained with DAPI and pseudocolorated in blue; fluorescent signals were pseudocolorated in red. In *D. melanogaster* (A) and in the closely related *D. simulans* (B) and *D. sechellia* (C) species, the *Yeti* cDNA probe maps to *2Rh* at the base of polytene division 41. The large and diffuse morphology of the *Yeti* signal found in these species, reflects the disorganized and poorly banded structure of the heterochromatin in the chromocenter. The arrows point the base of *2Rh*. In *D. pseudobscura*, the hybridization signal of *Yeti* PCR probe maps to region 63C in the proximal euchromatin (D). In *D. virilis* the *Yeti* hybridization signal maps to region 53E, in the distal euchromatin of chromosome 5 (E).

in the syntenic chromosome Muller C, that corresponds to the *2R* arm of *D. melanogaster* chromosome 2.

In silico reconstruction of *Yeti* gene organization in different sequenced genomes

To study the evolutionary conservation of the *Yeti* gene organization, we have characterized the structure of the *Yeti* orthologs in 19 *Drosophila* species (*D. melanogaster, D. simulans, D. sechellia, D. yakuba, D. erecta, D. eugracilis, D. biarmipes, D takahashii, D. elegans, D. ananassae, D. bipectinata, D. pseudobscura, D. persimilis, D. miranda, D. willistoni, D. mojavensisi, D. virilis, D. albomicans,* and *D. grimshawi*) and in five non-drosophilid dipterans (*C. quinquefasciatus, A. Aegypti, A. darlingi, A. gambiae* and *M. destructor* pest crop).

All the *Yeti* DNA sequences were retrieved from FlyBase. In case of annotated genes, the YETI protein sequences were extracted from the Ortho DB database [33], where they are reported as orthologous sequences belonging to the BCNT family complex [26,34]. For the recently sequenced genomes, the *Yeti* DNA sequences were recovered by the TblastN procedure, using the *D. melanogaster* YETI protein sequence. The alignments of the retrieved *Yeti* orthologs are shown in Figure 3 and their coordinates are reported in Table 1.

We were able to recover a deducted complete protein sequence except for the recently added genomes of *D. biarmipes*, *D. takahashi* and *D. elegans*, where frame-shift mutations were found in the 5′ end of the gene, probably due to errors in sequencing that needs to be improved. However, in each case the 3′end of the gene, containing the BCNT domain-coding region, was detected (Figure 3). The protein sequence alignments show a strong conservation of the BCNT domain in the *Drosophila* genus and in non-drosophilid dipterans (Figure 4).

We next compared the reconstructed molecular organization of the *Yeti* gene in the above-mentioned species, to study whether it underwent substantial structural changes during evolution. The results of this analysis are shown in Figure 5. It appears that the structure of the *Yeti* orthologs, with two exons and one intron remained highly conserved during the divergence of the lineages in the *Drosophila* genus, the only exception being *D. willistoni*, where an additional intron of 75 bp is present. The gene size is identical among *D. melanogaster, D. simulans* and *D. sechellia*, the only detectable difference is represented by the intron that in *D. melanogaster* is 7 bp longer. In general, in the *Drosophila* genus both the first exon and the single intron undergo changes in size: the exon varies from 557 bp (*D. yakuba*) to 812 bp (*D. virilis*), while the intron spans from 51 bp (*D. biarmipes*) to 70 bp (*D. willistoni*). Notably, the size of the second exon, that encodes the last portion of the BCNT domain, shows a striking conservation in the *Drosophila* genus, being invariably 91 bp long. Finally, in the five genome species of non-drosophilid dipterans (*C. quinquefasciatus*, *A. Aegypti*, *A. darlingi*, *A. gambiae* and *M. destructor*) the organization of *Yeti* differs in that the intron disappears and a single exon is present.

Characterization of the *Yeti* intron in *Drosophila* species

Intraspecific and interspecific size polymorphisms of heterochromatin gene introns have been found in *Drosophila*, which are likely to be associated with *de novo* insertions of TE-related sequences [27]. We then asked whether the intron of *Yeti* is prone to TE insertions or to other gross changes in length. To answer this question, we PCR amplified a region of about 180 bp comprising the *Yeti* intron from genomic DNAs extracted from *D. melanogaster* and *D. simulans* strains. Most of the analyzed strains derived from geographically distant natural populations (see materials and methods for the complete list). We also included in the analysis a single strain of *D. sechellia* and *D. teissieri*. The rationale of these experiments is that if an insertion have targeted the intron, it would have in turn increased the expected size of the amplified region.

We analysed 25 wild type strains and 4 laboratory stocks of *D. melanogaster*, as well as 9 wild type strains of *D. simulans* (see materials and methods). As shown in Figure 6, PCR amplification of *Yeti* produces a prominent band of the expected size in all the different strains of *D. melanogaster* (Figure 6A) and of *D. simulans* (Figure 6B) and in both *D. sechellia* and *D. tessieri* strains (Figure 6C). Sequencing of the purified PCR products from Iso and Scansano (*D. melanogaster*), Chicharo and Death Valley (*D. simulans*), *D. sechellia* and *D. tessieri* confirmed that they correspond to the *Yeti* intron-containing region (Figure 7). In conclusion, the results of this analysis suggest that the small intron of *Yeti* does not frequently undergo significant increase in size, unlike other essential heterochromatin genes of *Drosophila* [27]. This conclusion is in agreement with the observation that the *Yeti* gene structure tends to be stable during the evolution of the *Drosophila* genus, with the single intron that retains its short size (Figure 5).

Yeti is under negative selection

We next asked whether *Yeti* has evolved under negative (purifying) or positive selection, and whether the change in

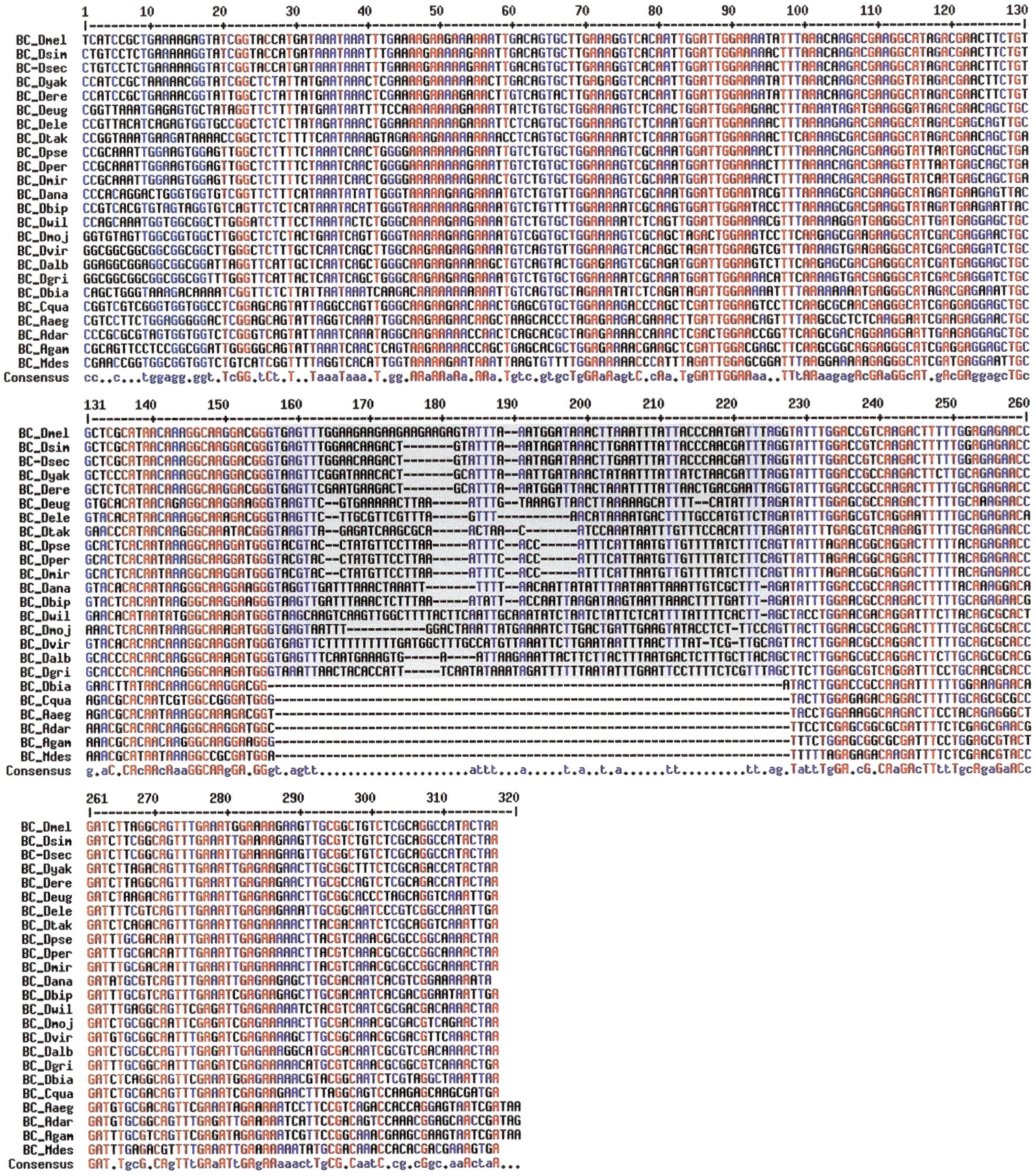

Figure 3. Alignment of the *Yeti* ortholog sequences encoding the BCNT-C domain. The grey area corresponds to the intron present in the *Drosophila* species.

location from euchromatin to heterochromatin may have affected the molecular evolution of the gene. To this aim we performed a d_S - d_N tests, a codon-based test of selection (for details see material and methods), using DNA sequences from five representative *Yeti* genes: three located in heterochromatin (*Dmel\Yeti, Dsec\Yeti and Dere\Yeti*) and two in euchromatin (*Dpse\Yeti* and *Dvir\Yeti*). The results suggest that *Yeti* is under purifying (negative) selection when present both in heterochromatin and euchromatin. Thus, the change in genomic location does not appear to have affected significantly the molecular evolution/function of *Yeti* (Table 2).

Discussion

In this paper, we have studied the evolutionary origin of *Yeti*, an essential gene of *Drosophila melanogaster* (Figure 1) located in the

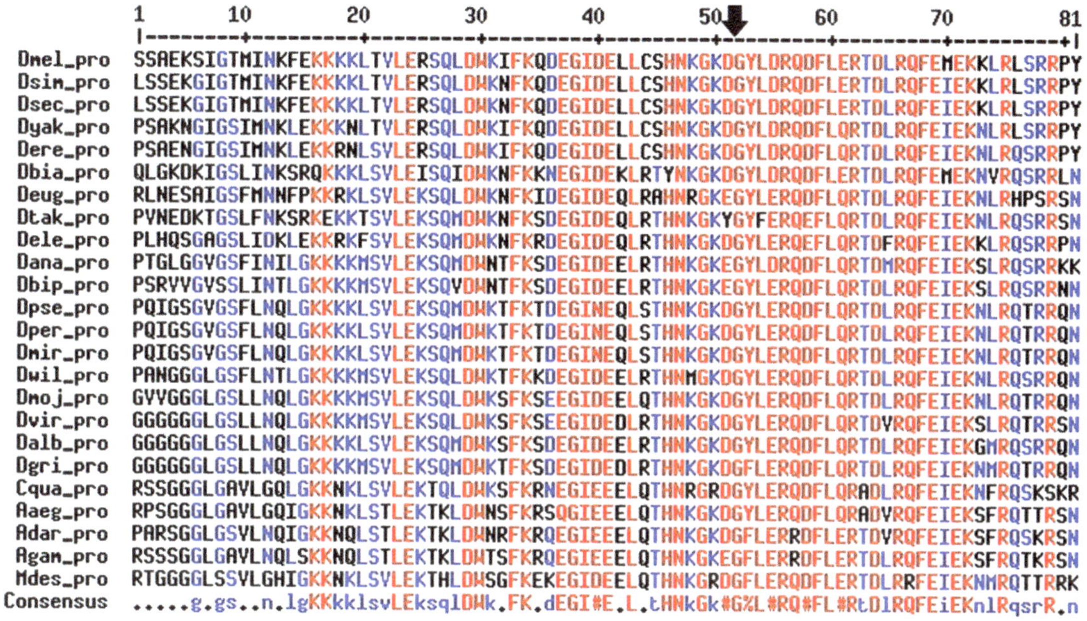

Figure 4. Alignment of the BCNT domain of YETI proteins among species. The arrow points the intron position in the corresponding coding region of *Drosophila* species.

deep heterochromatin of chromosome 2 [25,26,28,29] and found that it has evolved from a euchromatic ancestor in drosophilids.

Our FISH analysis shows that *Yeti* maintains a heterochromatin location in *2Rh*, at the base of division 41A, in both *D. simulans* and *D. sechellia* sibling species of *D. melanogaster* (Figure 2). The FlyBase localization of *Yeti* in *D. sechellia* scaffold_170:38,673..3,460), is in accord with our mapping results, while *D. simulans Yeti* is reported to map to the 3R arm (3,765..73,764; Table 1) in a 70 kb gene-poor genomic region. However, we are confident about the heterochromatin location of *Yeti* in *D. simulans* for the following reasons: First, as discussed in the result section, the FISH signal morphology produced by the *Yeti* probe is different from that usually seen with euchromatic probes and represents a distinctive mark for sequences derived from polytenized heterochromatin [30,31]. Second, our FISH mapping of *Yeti* in *D. simulans* is based on three reproducible experiments, each carried out on several polytene chromosome figures obtained from at least 10 larvae. Finally, the paucity of genes around *D. simulans Yeti* is *per se* highly suggestive of heterochromatin localization. Thus, it is possible that the apparent discrepancy between our data and FlyBase may reflect an assembly error that occurred in the *D. simulans* genome sequence assembly, as reported by Schaffer et al. [32].

Our FISH analysis show that in two distantly related species, *D. pseudoobscura* and *D. virilis*, *Yeti* is located in euchromatin (Figure 2). In *D. pseudobscura*, *Yeti* maps to chromosome 3 at polytene division 63C, while in *D. virilis* it is found in chromosome 5, at polytene division 53E. Interestingly, *Yeti* lies in the syntenic chromosome Muller C that corresponds to the *2R* arm of *D. melanogaster* chromosome 2. Together, the results of our analysis indicate that during the evolution of the *Drosophila* genus, *Yeti* has been resident on the same chromosomal element, but over time it progressively moved closer to the pericentric regions. Such movements would have occurred in about 40 million of years, the estimated divergence time between *D. melanogaster* and *D. virilis* (Figure 8). A similar evolutionary trend was reported for *light* and other neighboring genes in *2L* heterochromatin [23] and for other genes of chromosome 3 heterochromatin [24].

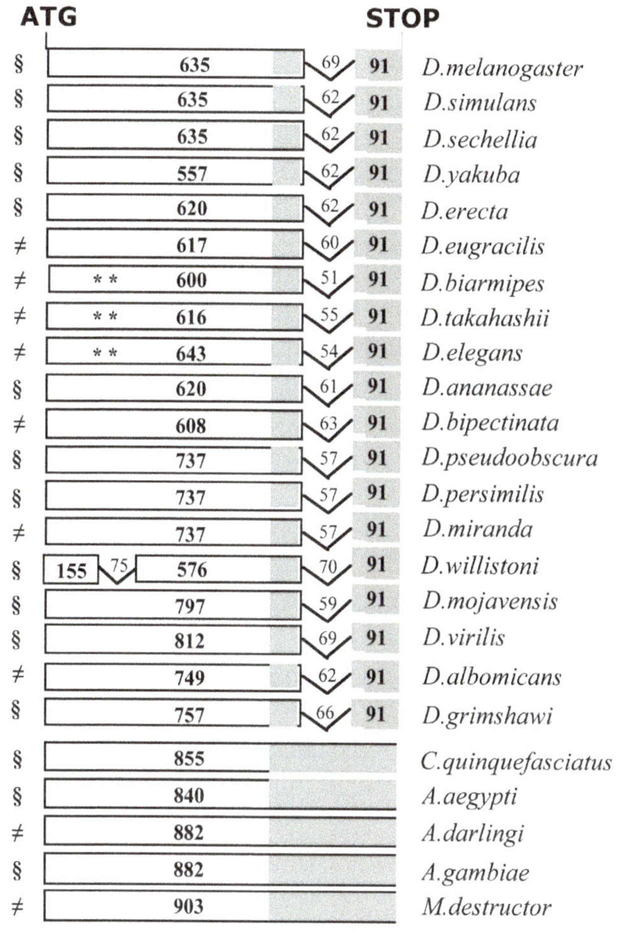

Figure 5. Comparison of the *Yeti* gene structure among sequenced genomes. Only the coding regions are showed. Exons are in boxes and numbers refer to nucleotides. Symbols: §, annotated genes; ≠, this study; **, defective ORF. The grey area at the 3′ end represents the conserved BCNT-C domain in the protein.

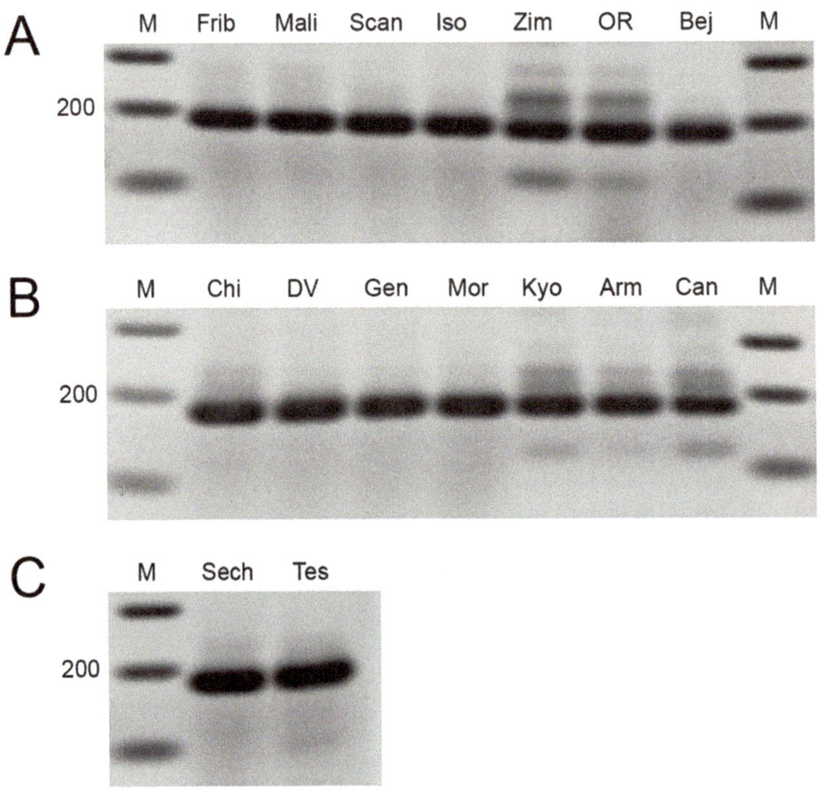

Figure 6. PCR amplification of the genomic region containing the *Yeti* intron in *Drosophila* species. A single PCR product of about 180 bp was found, in *D. melanogaster* (A) and *D. simulans*, (B) *D. sechellia* (C) and *D. tessieri* (C) related species. M = Marker; Frib = Friburgo; Mal = Mali; Scan = Scansano; Iso = y^1; cn^1 $bw^1 sp^1$ isogenic strain; OR = Oregon-R; Bej = Bejin; Chi = Chicharo; DV = Death Valley; Gen = Genoa; Mor = Moruya; Kyo = Kyogle; Arm = Armidale; Can = Canaries; Sech = *D. Sechellia*; tes = *D. tessieri*. Molecular weight is in bps.

A striking difference between heterochromatin and euchromatin genes lies in the generally larger size and complex molecular structure of the former compared to the latter. The example of the 'giant' Y-chromosome fertility factors of *D. melanogaster* mentioned above is paradigmatic in this respect [12]. Some of the essential heterochromatin genes of chromosomes 2 and 3 are also large due to the presence of large introns that harbour truncated TE copies (or TE "remnants") [27,35,36,37]. In this context, the *Yeti* gene of *D. melanogaster* with a 900 bp-long genomic region represents an exception [26]. The same is true for *RpL38*, *RpL5* and *RpL15*, three essential ribosomal protein-coding genes located in the heterochromatin of chromosomes 2 and 3, all of which are of relatively small size [24,26,37,38]. How might these observations be explained?

```
                    1        10        20        30        40        50        60        70        80        90
                    |--------+---------+---------+---------+---------+---------+---------+---------+---------|
     Yeti-gDNA/1-134                            TAAGATCGGTTCTCTCCAAAAAGTCTTGACGGTCCAAATACCTAAATCATTGGGTAATAAATTT
    Scansano-F/1-164 GTGCAGCATTCTTTA-TTTAAACTGCCTAGATCGGTTCTCTCCAAAA-GTCTTGACGGTCCAAATACCTAAATC-TTGGGTAATAAATTT
    Chicharo-F/1-154 GGGC-GCGTTCTTTTCTTTAAACTGCCGAGA-CGGTTCTCTCCAAAA-GTCTTGACGGTCCAAATACCTAAATCGTTGGGTAATAAATTC
Death-Valley-F/1-155  GGTCTCGTTCTTT-CTTTAAACTGCCGAGA-CGGTTCTCTCCAAAA-GTCTTGACGGTCCAAATACCTAAATCGTTGGGTAATAAATTC
   Sechellia-F/1-155   GGTGCGTTCTTT-CTTTAAACTGCCGAGA-CGGTTCTCTCCAAAA-GTCTTGACGGTCCAAATACCTAAATCGTTGGGTAATAAATTC
    Tessieri-F/1-155  GGTCTCGTTCTTT-CTTTAAACTGCCGAGA-CGGTTCTCTCCAAAA-GTCTTGACGGTCCAAATACCTAAATCGTTGGGTAATAAATTC
           Consensus ..g...c.ttcttt..tttaaactgcc.AGAtCGGTTCTCTCCAAAA.GTCTTGACGGTCCAAATACCTAAATC.TTGGGTAATAAATTt

                    91       100       110       120       130       140       150       160    167
                    |--------+---------+---------+---------+---------+---------+---------+------|
     Yeti-gDNA/1-134 AAGTTTATCCATTTAAATACTCTTCTTCTTCTTCTTCCAAACTCACCCGTCCTTGCCTTTGTTATGCGAG
    Scansano-F/1-164 AAGTTTATCCATTTAAATACTCTTCTTCTTCTTCTTCCAAACTCACCCGTCCTTGCCTTTGTTATGCGAGCACAAAA
    Chicharo-F/1-154 AAGTTTATCTATTTAAATACA-------GTCTTGTTCCAAACTCACCCGTCCTTGCCTTTGTTATGCGAGCACA
Death-Valley-F/1-155 AAGTTTATCTATTTAAATACA-------GTCTTGTTCCAAACTCACCCGTCCTTGCCTTTGTTATGCGAGCACAAA
   Sechellia-F/1-155 AAGTTTATCTATTTAAATACA-------GTCTTGTTCCAAACTCACCCGTCCTTGCCTTTGTTATGCGAGCACAAAA
    Tessieri-F/1-155 AAGTTTATCTATTTAAATACA-------GTCTTGTTCCAAACTCACCCGTCCTTGCCTTTGTTATGCGAGCACAAA
           Consensus AAGTTTATCcATTTAAATACtcttcttctTCTTcTTCCAAACTCACCCGTCCTTGCCTTTGTTATGCGAGcaca...
```

Figure 7. Sequencing of the purified PCR products from *Drosophila* species. Sequence alignments from Iso and Scansano (*D. melanogaster*), Chicharo and Death Valley (*D. Simulans*), *D. sechellia* and *D. tessieri*. Sequence analysis confirmed that they correspond to the *Yeti* intron containing region. The *Yeti* intron is shown in normal text, the flanking exons are in bold. The *D. simulans* and *D. sechellia* intron lacks a 7 bp stretch (see the gap), in agreement with the genome sequence data (see results and Figure 5). The *D. Tessieri* intron sequence is identical to that of *D. simulans* and *D. sechellia*.

Table 2. Codon-based test of selection.

Number of species compared	Number of codons[a]	d_S - d_N	P[b]
5	232	10.2	$<10^{-6}$
3 (heterochromatic sequences)	233	5.3	$<10^{-6}$
2 (euchromatic sequences)	275	8.7	$<10^{-6}$

[a]All positions containing gaps were eliminated.
[b]The probability (P) of rejecting the null hypothesis of strict-neutrality ($d_N = d_S$) in favor of the alternative hypothesis (purifying selection, dN<dS) is shown.

One may imagine that, during evolution, genes increased their size by becoming targets for reiterated transposable-element insertions in the intronic regions, depending on their time of residence in heterochromatin. This, however, does not seem to have been the case. In fact, the *light* and *Yeti* genes, although having likely resided in heterochromatin for a comparable evolutionary time (less than 30 million of years), underwent a different molecular architecture; the *light* gene structure dramatically changed during the evolutionary transition from euchromatin to heterochromatin, due to a remarkable increase in the size of introns targeted by TEs [23]; *Yeti* retained its original organization in all analysed species, with a short genomic region carrying a single short intron (Figure 5). In addition, by PCR analysis we found that the *Yeti* intron does not undergo significant interspecies or intraspecies changes of its physical size (Figure 6). Similarly, the *RpL15* gene shows a conserved structure among *Drosophila* species, independently from its genomic location [24].

How to explain the different behaviour of heterochromatin genes? In particular, there might have been a selective pressure to maintain some genes of short size (with few, short introns) despite of their genomic location, owing to their particular functional requirement: interestingly in that respect, highly expressed genes are known to harbour substantially shorter introns than genes that are expressed at low levels [39]. This may be the case of *Yeti* and *RpL38*, *RpL5* and *RpL15* heterochromatic genes of *Drosophila melanogaster*, which are all highly expressed and all have indeed short size and carry short introns [26,37,38]. *Yeti* itself encodes an important chromatin-remodeling factor required for development [26] and ribosomal protein coding genes are also essential for proper development. It is not unreasonable to speculate that these genes maintained the original structure, in spite of their transition to heterochromatin during massive chromosomal rearrangements that occurred over time, because of the requirement for their efficient expression during early development. In addition, it is possible that once in heterochromatin, a given sequence might be differentially targeted by transposable elements, with some sequences being more refractory than others. These observations suggest that the evolutionary forces that acted in shaping the structural organization of genes currently found in *D. melanogaster* heterochromatin are molecularly diverse.

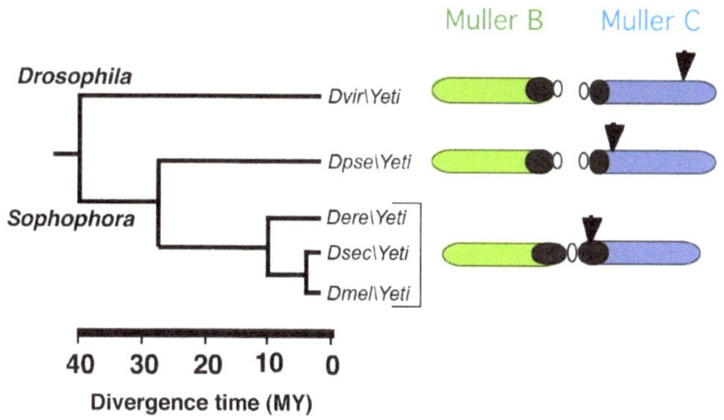

Figure 8. Evolutionary repositioning of the *Yeti* gene. Schematic representation of *Yeti* gene transition of from euchromatin to heterochromatin. The arrows point the chromosomal position of *Yeti*. It appears that *Yeti* has been resident on the Muller C chromosomal element, but over evolutionary time it progressively approached to pericentric heterochromatin and in *D. melanogaster* it is found in the deep portion of *2Rh*.

Finally, the results of our d_S - d_N tests, showing that *Yeti* is under negative selection both in heterochromatin and euchromatin (Table 2), are in accord with its evolutionary conserved function and suggest that the change in genomic location did not affected significantly the molecular evolution of the gene.

Together, these results contribute to expand our understanding of the molecular dynamics driving the evolution of the heterochromatin genome in higher eukaryotes.

Materials and Methods

Drosophila strains

Fly cultures were carried out at 25°C in standard cornmeal yeast medium. *D. melanogaster* strains derived from natural populations: Altamura I and Altamura II (South Italy), Beijing (Cina), California and California I (USA), Charolles (France), Charolles 1999 (France), Friburgo 1997 (Germany), HJ30 (France), Hobo (Grece) Israel-4 (Israel), Lanuvio (Italy), Luminy (France), Mali and Mali I (West Africa), Marrakesh (North Africa), Scansano (Central Italy), Vallecas (Spain), W15 (USA), W30 (USA), W90 (USA), W130 (USA), W135 (USA), Zimbabwe (Africa), Gaiano (North Italy). *D. melanogaster* laboratory strains (separated by comma): isogenized y^1; cn^1,bw^1,sp^1 strain (Iso) [40]. *l(2)LP2/SM1,Cy*. *Cy/Sp;Sb,Delta2-3,ry*^{506}*/TM6,ry*^{506}. *l(2)EMS-31/SM1,Cy*. *D. simulans* strains derived from natural populations: Armidale (New South Wales, Australia), Canaries (Atlantic Spain), Can River (Australia) Chantal (France), Chicharo (Portugal), Death Valley (USA), Genoa (Italy), Kyogle (Australia) and Moruya (New South Wales, Australia). *D. melanogaster* and *D. simulans* wild-type strains were derived from natural populations collected in the wild before the end of year 2000 and are gifts of Sergio Pimpinelli, Nikolaj Junakovic, Chantal Vaury and Pierre Capy.

Cytology and fluorescent in situ hybridization

Polytene chromosomes prepared according to Pardue [41] were stained with DAPI. The *D. melanogaster* RE36623 cDNA *Yeti* probe was labelled by nick-translation with Cy3-dCTP (Amersham). Species-specific PCR probes were used for FISH in D. pseudobscura and D. virilis (see below). FISH procedures were performed according to Dimitri [42]. Digital images were obtained using an Olympus epifluorescence microscope equipped with a cooled CCD camera. Gray scale images, obtained separately recording Cy3 and DAPI fluorescence by specific filters, were pseudo colored and merged for the final image using the Adobe Photoshop software.

Nucleic acid manipulation and sequence alignments

Genomic DNA extraction and PCR were performed according to the protocol of Berkeley Drosophila Genome Project (http://www.fruitfly.org/about/methods/index.html). To PCR amplify *Yeti* genomic fragments from *D. melanogaster, D. simulans* strains, *D. sechellia* and *D. tessieri* the following primers were used: F: 5′-TTAGTATGGCCTGCGAGACA-3′; R: 5′-TGTGCTCGCA-TAACAAAGGC-3′.

PCR cycle were: 40″ at 98°C, 40 x (10″ 98°C, 30″ 58°C, 8″ 72°C). Amplified fragments were gel purified and sequenced by Bio-Fab research s.r.l. The *Yeti* probes from *D. pseudoobscura* and *D. virilis* were generated by PCR over genomic DNA with the following primers:

Dpse-F 5′-GCGACGATGATAGCATCAAT-3′; Dpse-R 5′-GTGAGTGCTCAGCTGCTCAT-3′

Dvir-F 5′-AGCTAAACGTAGCACGCGTC-3′; Dvir-R 5′-TGTGTACGCAGATCCTCGTC-3′

PCR cycle were: 4′ at 94°C, 35 x (30″ 95°C, 45″ 60°C, 30″ 72°C). Amplified fragments were cloned in pGEM-T vector (Promega) and verified by DNA sequencing.

Multiple sequence alignment were performed by ClustalW procedure available at EMBL-EBI (http://www.ebi.ac.uk) or with the multialin interface at http://multalin.toulouse.inra.fr.

Testing for signatures of selection

A codon-based test of selection was conducted in MEGA 6.0 [43], using the Nei-Gojobori method [44]. The statistic test ($d_S - d_N$) is expected to be zero under the null hypothesis of neutrality. d_S and d_N are the numbers of synonymous and nonsynonymous substitutions per site, respectively. Three analyses were performed. The first analysis involved five coding sequences of the *Yeti* gene from five different Drosophila species (*D. melanogaster*, *D. simulans*, *D. erecta*, *D. pseudoobscura* and *D. virilis*). The other two analyses involved either the three Drosophila species where the *Yeti* gene is heterochromatic (*D. melanogaster*, *D. simulans* and *D. erecta*) or the two species where the *Yeti* gene is euchromatic (*D. pseudoobscura* and *D. virilis*).

Acknowledgments

We are grateful to Patrizia Lavia for critical reading of the manuscript. We also thank Sergio Pimpinelli, Chantal Vaury and Pierre Capy for gift of *Drosophila* strains.

Author Contributions

Conceived and designed the experiments: RC EC FC RM PD. Performed the experiments: RC EC FC RM PD. Analyzed the data: RC EC FC RM PD. Contributed reagents/materials/analysis tools: RC FC PD. Wrote the paper: PD.

References

1. Moritz KB, Roth GE (1976) Complexity of germline and somatic DNA in Ascaris. Nature 259: 55–57.
2. Peterson DG, Pearson WR, Stack SM (1998) Characterization of the tomato (Lycopersicon esculentum) genome using in vitro and in situ DNA reassociation. Genome 41: 346–356.
3. Dimitri P, Caizzi R, Giordano E, Carmela Accardo M, Lattanzi, et al. (2009) Constitutive heterochromatin: a surprising variety of expressed sequences. Chromosoma 118: 419–35.
4. Heitz E (1928) Das Heterochromatin der Moose. Jb Wiss Bot 69: 762–818.
5. John B (1988) The biology of heterochromatin. In Heterochromatin: Molecular and Structural aspects *ed. V. R. S.*, Cambridge University Press. p. 1–128.
6. Williams SM, Robbins LG (1992) Molecular genetic analysis of *Drosophila* rRNA arrays. Trends Genet 8: 335–340.
7. Dernburg AF, Sedat JW, Hawley RS (1996). Direct evidence of a role for heterochromatin in meiotic chromosome segregation. Cell 86: 135–146.
8. Elgin SCR (1996) Heterochromatin and gene regulation in *Drosophila*. Current Opinion in Genetics and Development 6: 193–200.
9. Karpen GH, Le MG, Le H (1996) Centric heterochromatin and the efficiency of achiasmate disjunction in *Drosophila* female meiosis. Science 273: 118–122.
10. Pardue ML (2007) Following the chromosome path to the garden of the genome. Annu Rev Cell Dev Biol. 2007; 23: 1–22.
11. Villasante A, Mendéz-Lago M, Abad JP, Montejo de Garcini E (2007) The birth of the centromere. Cell Cycle 6: 2872–2876.
12. Gatti M, Pimpinelli S (1992). Functional elements in *Drosophila melanogaster* heterochromatin. Annu Rev Genet 26: 239–275.
13. Hilliker AJ (1976) Genetic analysis of the centromeric heterochromatin of chromosome 2 of *Drosophila melanogaster*: deficiency mapping of EMS-induced lethal complementation groups. Genetics 83: 765–782.
14. Marchant GE, Holm DG (1988) Genetic analysis of the heterochromatin of chromosome 3 in *Drosophila melanogaster*. II. Vital loci identified through EMS mutagenesis. Genetics 120: 519–532.
15. Dimitri P (1991) Cytogenetic analysis of the second chromosome heterochromatin of *Drosophila melanogaster*. Genetics 127: 553–564.
16. Koryakov DE, Zhimulev IF, Dimitri P (2002) Cytogenetic analysis of the third chromosome heterochromatin of *Drosophila melanogaster*. Genetics 160: 509–517.
17. Fitzpatrick KA, Sinclair DA, Schulze SR, Syrzycka M, Honda BM (2005) A genetic and molecular profile of third chromosome centric heterochromatin in *Drosophila melanogaster*. Genome 48: 571–584.
18. Adams MD, Celniker SE, Holt RA, Evans CA, Gocayne JD, et al. (2000) The genome sequence of *Drosophila melanogaster*. Science 287: 2185–2195.
19. Hoskins RA, Smith CD, Carlson JW, Carvalho AB, Halpern A, et al. (2002) Heterochromatic sequences in a *Drosophila* whole-genome shotgun assembly. Genome Biology 3: research0085.1–0085.16.
20. Hoskins RA, Carlson JW, Kennedy C, Acevedo D, Evans-Holm M, et al. (2007) Sequence finishing and mapping of *Drosophila melanogaster* heterochromatin. Science 316: 1625–1628.
21. Smith CD, Shu S, Mungall CJ, Karpen GH (2007) The Release 5.1 annotation of *Drosophila melanogaster* heterochromatin. Science 316: 1586–1591.
22. He B, Caudy A, Parsons L, Rosebrock A, Pane A, et al. (2012) Mapping the pericentric heterochromatin by comparative genomic hybridization analysis and chromosome deletions in Drosophila melanogaster. Genome Res 22: 2507–2519.
23. Yasuhara JC, DeCrease CH, Wakimoto BT (2005) Evolution of heterochromatic genes of *Drosophila*. Proc Natl Acad Sci USA 102: 10958–10963.
24. Schulze SR, McAllister BF, Sinclair DA, Fitzpatrick KA, Marchetti M, et al. (2006) Heterochromatic genes in Drosophila: a comparative analysis of two genes. Genetics 173: 1433–1445.
25. Cenci G, Belloni G, Dimitri P (2003) 1(2)41Aa, a heterochromatic gene of Drosophila melanogaster, is required for mitotic and meiotic chromosome condensation. Genet Res 81: 15–24.
26. Messina G, Damia E, Fanti L, Atterrato MT, Celauro E, et al. (2014) Yeti, an essential Drosophila melanogaster gene, encodes a protein required for chromatin organization. J Cell Sci 127: 2577–2588. doi:10.1242/jcs.150243 Epub 2014 Mar 20.
27. Dimitri P, Junakovic N, Arcà B (2003) Colonization of heterochromatic genes by transposable elements in *Drosophila*. Mol Biol Evol 20: 503–512.
28. Dimitri P, Corradini N, Rossi F, Vernì F, Cenci G, et al. (2003) Vital genes in the heterochromatin of chromosomes 2 and 3 of Drosophila melanogaster. Genetica. 117: 209–215.
29. Rossi F, Moschetti R, Caizzi R, Corradini N, Dimitri P (2007) Cytogenetic and molecular characterization of heterochromatin gene models *in Drosophila melanogaster*. Genetics 175: 595–607.
30. Zhang P, Spradling AC (1995) The Drosophila salivary gland chromocenter contains highly polytenized subdomains of mitotic heterochromatin. Genetics. 1995 Feb; 139(2): 659–70.
31. Berghella L, Dimitri P (1996) The heterochromatic rolled gene of Drosophila melanogaster is extensively polytenized and transcriptionally active in the salivary gland chromocenter. Genetics. 1996 Sep; 144(1): 117–25.
32. Schaeffer SW, Bhutkar A, McAllister BF, Matsuda M, Matzkin LM, et al. (2008) Polytene chromosomal maps of 11 Drosophila species: the order of genomic scaffolds inferred from genetic and physical maps. Genetics 179: 1601–55.

33. Waterhouse RM, Zdobnov EM, Tegenfeldt F, Li J, Kriventseva EV (2011) OrthoDB: the hierarchical catalog of eukaryotic orthologs in 2011. Nucleic Acids Res. 39(Database issue): D283–8.
34. Iwashita S, Osada N, Itoh T, Sezaki M, Oshima K, et al. (2003). A transposable element-mediated gene divergence that directly produces a novel type bovine Bcnt protein including the endonuclease domain of RTE-1. Mol Biol Evol 20: 1556–1563.
35. Devlin RH, Bingham B, Wakimoto BT (1990a) The organization and expression of the *light gene*, a heterochromatic gene of *Drosophila melanogaster*. Genetics 125: 129–140.
36. Tulin A, Stewart D, Spradling AC (2002) The *Drosophila* heterochromatic gene encoding poly (ADP-ribose) polymerase (PARP) is required to modulate chromatin structure during development. Genes and Development 16: 2108–2119.
37. Schulze SR, Sinclair DA, Fitzpatrick KA, Honda BM (2005) A genetic and molecular characterization of two proximal heterochromatic genes on chromosome 3 of *Drosophila melanogaster*. Genetics 169: 2165–2177.
38. Marygold SJ, Coelho CM, Leevers SJ (2005) Genetic analysis of *RpL38* and *RpL5*, two minute genes located in the centric heterochromatin of chromosome 2 of *Drosophila melanogaster*. Genetics 169: 683–695.
39. Castillo-Davis CI, Mekhedov SL, Hartl DL, Koonin EV, Kondrashov FA (2002) Selection for short introns in highly expressed genes. Nat Genet 31: 415–418.
40. Brizuela BJ, Elfring L, Ballard J, Tamkun JW, Kennison JA (1994) Genetic analysis of the brahma gene of Drosophila melanogaster and polytene chromosome subdivisions 72AB. Genetics 137: 803–813.
41. Pardue ML (1986) In situ hybridization to DNA of chromosome and nuclei. In: Roberts DB (ed) Drosophila: a practical approach. IRL Press, Oxford, pp. 111–137.
42. Dimitri P (2004) Fluorescent in situ hybridization with transposable element probes to mitoticchromosomal heterochromatin of Drosophila. Methods Mol Biol 260: 29–39.
43. Tamura K, Stecher G, Peterson D, Filipski A, Kumar S (2013) MEGA6: Molecular Evolutionary Genetics Analysis version 6.0. Molecular Biology and Evolution 30: 2725–2729.
44. Nei M, Gojobori T (1986). Simple methods for estimating the numbers of synonymous and nonsynonymous nucleotide substitutions. Molecular Biology and Evolution 3: 418–426.
45. Coulthard AB, Alm C, Cealiac I, Sinclair DA, Honda BM, et al. (2010) Essential loci in centromeric heterochromatin of Drosophila melanogaster. I: the right arm of chromosome 2. Genetics 185: 479–495. doi:10.1534/genetics.110.117259 Epub 2010.
46. Timoshevskiy VA, Kinney NA, deBruyn BS, Mao C, Tu Z, et al. (2014) Genomic composition and evolution of Aedes aegypti chromosomes revealed by the analysis of physically mapped supercontigs. BMC Biol 12: 27. doi:10.1186/1741-7007-12-27.

8

The α1,6-Fucosyltransferase Gene (*fut8*) from the *Sf*9 Lepidopteran Insect Cell Line: Insights into *fut8* Evolution

Sylvie Juliant[1], Anne Harduin-Lepers[2], François Monjaret[1], Béatrice Catieau[2,3], Marie-Luce Violet[1], Pierre Cérutti[1], Annick Ozil[1], Martine Duonor-Cérutti[1]*

1 CNRS UPS3044 Baculovirus et Thérapie, Saint Christol Lèz Alès, France, **2** CNRS UMR8576, Unité de Glycobiologie Structurale et Fonctionnelle, Université Lille Nord de France, Lille1, Villeneuve d'Ascq, France, **3** Laboratoire Français du Fractionnement et des Biotechnologies de Lille, Lille, France

Abstract

The core alpha1,6-fucosyltransferase (FUT8) catalyzes the transfer of a fucosyl moiety from GDP-fucose to the innermost asparagine-linked *N*-acetylglucosamine residue of glycoproteins. In mammals, this glycosylation has an important function in many fundamental biological processes and although no essential role has been demonstrated yet in all animals, FUT8 amino acid (aa) sequence and FUT8 activity are very well conserved throughout the animal kingdom. We have cloned the cDNA and the complete gene encoding the FUT8 in the *Sf9* (*Spodoptera frugiperda*) lepidopteran cell line. As in most animal genomes, *fut8* is a single-copy gene organized in different exons. The open reading frame contains 12 exons, a characteristic that seems to be shared by all lepidopteran *fut8* genes. We chose to study the gene structure as a way to characterize the evolutionary relationships of the *fut8* genes in metazoans. Analysis of the intron-exon organization in 56 *fut8* orthologs allowed us to propose a model for *fut8* evolution in metazoans. The presence of a highly variable number of exons in metazoan *fut8* genes suggests a complex evolutionary history with many intron gain and loss events, particularly in arthropods, but not in chordata. Moreover, despite the high conservation of lepidoptera FUT8 sequences also in vertebrates and hymenoptera, the exon-intron organization of hymenoptera *fut8* genes is order-specific with no shared exons. This feature suggests that the observed intron losses and gains may be linked to evolutionary innovations, such as the appearance of new orders.

Editor: Andreas Hofmann, Griffith University, Australia

Funding: This work was supported by Centre National de la Recherche Scientifique to MD-C, AH-L, FM, MLV, Institut National de la Recherche Agronomique to PC, SJ, AO, Laboratoire Français du Fractionnement et des Biotechnologies de Lille to BC, and PPF Bioinformatique Lille1 to AH-L. The funders had no role in study design, data collection and analysis, decision to publish, or preparation of the manuscript.

Competing Interests: The authors have declared that no competing interests exist.

* Email: martine.cerutti@cnrs.fr

Introduction

Glycosylation of proteins is a key process. Indeed, congenital disorders of glycosylation lead to severe dysfunction and disability. Maturation of glycoproteins in the Golgi apparatus requires hundreds of enzymes (i.e., glycosyltransferases, glycosidases), known as Carbohydrate-Active Enzymes (CAZymes) [1], and also chaperones that act through complex protein-protein interactions.

Fucosylation is one of the most common post-translational modifications. Fucosylated glycans are involved in various biological processes, such as cell adhesion, growth factor receptor modulation, microbial and viral infections, cancer and atherosclerosis [for review 2, 3]. Several fucosyltransferases (FucTs) have been identified and classified in the CAZy glycosyltransferase (GT) families GT-11 (α1,2-FucTs), GT-10 (α1,3/4-FucTs), GT-23 (core α1,6-FucTs, known as fut8), GT-37 (α1,2-FucTs), GT-65 (O-FucTs pofut1 and fut12) and GT-68 (O-FucTs pofut2 and fut13). A previous phylogenetic analysis of vertebrates, invertebrates and bacterial FucTs genes highlighted their ancient and divergent evolution from one or two ancestral genes [4].

While α1,2- and α1,3/4-FucTs are implicated in terminal fucosylation (e.g., histo-blood group antigens) [5], core α1,6-FucT (FUT8) adds fucose to the innermost asparagine-linked *N*-acetylglucosamine (GlcNAc) of the chitobiose disaccharide-core unit of glycoproteins. This core fucosylation has an essential role in regulating the function of many glycoproteins, such as activation of growth factor receptors [6] and modulation of antibody-dependent cell-mediated cytotoxicity (ADCC) *in vitro* and *in vivo* [7]. In mammals, FUT8 physiological relevance has been demonstrated by genetic ablation of the *fut8* gene: 80% of these mice die three days after birth [8] and the survivors present severe growth retardation.

The crystal structure of human FUT8 has been resolved [9]. It is a type-II glycoprotein with a short N-terminal cytoplasmic tail, a transmembrane domain, a stem region and a C-terminal catalytic domain. The latter encompasses three peptidic consensus sequences (motif I, motif II and motif III) [4,9] located in the

Rossmann fold [10], a protein motif that binds to nucleotides. These motifs are conserved among α1,2-, α1,6- and *O*-FucTs, strongly suggesting their implication in the fucose transfer reaction [4,11–14]. Site-directed mutagenesis [9,12] confirmed this hypothesis demonstrating that most of the highly conserved aa in this region are essential for enzymatic activity. Four disulfide bonds may be important for protein folding and stability [9]. A SH3 domain is also present at the C-terminus in all known FUT8 sequences [15,16], but its role has not been elucidated yet.

Lepidopteran cells can perform most of the post-translational modifications, including *N*- and *O*-glycosylation. *N*-glycan processing [for review, 17] is comparable to what observed in mammalian cells during the early stages of the glycosylation pathway with the synthesis of GlcNAcβ1,2Manα1,3[Manα1,6]-Manβ1,4GlcNAcβ1,4[Fucα1,6]GlcNAc [18]. The glycan structures most frequently identified on glycoproteins synthesized in insect cells are fucosylated paucimannose structures (Man_3(Fuc)$GlcNAc_2$) and, to a lesser extent, oligomannose-type glycans [19]. The absence of complex oligosaccharides correlates with the absence of β1,4-galactosyltransferase I (β1,4GalTI) [20,21] and sialyltransferases [18]. Despite the *in silico* identification of α2,6-sialyltransferase (*st6gal*) gene sequences in some lepidopterans, such as *Bombyx mori* [22], no sialyltransferase activity has been detected in insect cell lines, such as *Sf*9 (*S. frugiperda*) cells [18,23]. In addition, the very low expression level of *N*-acetylglucosaminyltransferase I (GNT-I) [18,24,25] and GNT-II [26] and the presence of a Golgi-associated *N*-acetylglucosaminidase (FDL) [27,28] explain the formation of paucimannose structures. In contrast to mammals, two core-fucosylation events with α1,6- and α1,3-linkage have been reported in insects and in plants [29]. Analysis of fucosyltransferase activities in several lepidopteran cell lines showed distinct activities in each cell line. While core α1,3- and core α1,6-fucose are found on glycoproteins expressed in *Masmestra brassicae* [20,30,31] and *Trichoplusia ni* (High five) cell lines [32], only core α1,6-fucose was found in *B mori* and *Sf9* cells [20,30,33].

Although its essential role has not yet been demonstrated in all animals, FUT8 aa sequence and FUT8 enzymatic activity are well conserved throughout the animal kingdom as testified by several molecular cloning and functional studies in vertebrates and invertebrates [34,35,36,37,38]. In the present study, we report the molecular cloning and functional characterization of a cDNA encoding *fut*8 from the *Sf9* lepidopteran insect cell line. As in most animal genomes, *fut8* is a single-copy gene organized in several exons. These properties and the high conservation of amino acids in FUT8 catalytic domain were used to retrieve unique *fut8* orthologs from a large variety of metazoan genomes and to study their exon-intron structure evolution in several insect orders. The results of this analysis allow us to propose a model of *fut8* evolution throughout the animal kingdom. Furthermore, *fut8* evolutionary history could be used to measure the divergences among insect genomes highlighting the frequent intron losses and gains in arthropods.

Results

Cloning of *fut8* cDNA from *Sf9* cells

A partial cDNA sequence was obtained using a reverse transcription - polymerase chain reaction (RT-PCR) approach and degenerate primers (**Table S1**). This 860-bp (base pairs) sequence was compared to all non-redundant coding sequences (cds) in the GenBank database using the BlastX program [39] and was identified as belonging to a *fut8* gene, with a high identity score (51% aa identity with human FUT8). This sequence was then used to design new exact-match primers for 3′- and 5′-RACE. The full-length cDNA sequence of *Sf*9 *fut8* gene is available under the accession number KC538901.

Sequence analysis of this full-length cDNA sequence showed the presence of three in-frame ATG codons as potential initiation codons (**Figure 1**). Inspection of the sequences immediately flanking the initiation codon of genes expressed in *Sf*9 cells (n = 168) [40] allowed us to propose a consensus sequence (**Table 1**) for initiation of translation (most frequent for gene expression with high or low efficiency) that was in agreement with the start translation sites reported by Cavener and Ray [41] for invertebrates. Based on this consensus sequence, the third ATG codon (**Figure 1**, ATG3) seemed to have the most favorable environment as potential initiation codon (**Table 2**). Sequence analysis showed that this cDNA encodes a type II protein of 561-aa with a short cytoplasmic tail of 8-aa at the N-terminus followed by a 17-aa (residues 9–25) transmembrane domain. The Golgi luminal part contains the catalytic C-terminal domain with the three highly conserved motif I (residues 336–348), motif II (residues 381–394) and motif III (residues 429–455) consensus sequences [4,9]. All the critical aa residues (Arg-343, Asp-346, Lys-347, Glu-351, Tyr-360, Asp-387, Asp-431 and Ser-447) implicated in FUT8 enzymatic activity are also perfectly conserved in the *Sf9* Fut8 protein (**Figure 1**). The catalytic domain is preceded by a 78-aa proline-rich region (Pro-279 to Pro-336), like in mammalian FUT8 [15,35,36]. As for all known FUT8 sequences, *Sf9* FUT8 has a 60-aa SH3 domain at its C-terminus (residues 484–543) [15,16] (**Figure 1**). All cysteine residues involved in the formation of disulfide bridges in human FUT8, are also perfectly conserved, except Cys-472 in motif III that is replaced by a glycine residue (G-450) (**Figure 1 and Figure S1**). The amino acid identity of the *Sf9* FUT8 aa sequences was 86.81%, and 79.89% with the lepidopteran *Manduca sexta* and *Danaus Plexippus* FUT8. Identity with dipteran FUT8 sequences ranged between 50.28% (*Culex quinquefasciatus pipiens*) and 46.90% (*Drosophila melanogaster*) and was about 44% with vertebrate FUT8 aa sequences (44.24% with human and 44.42% with bovine *fut8*).

To confirm that this cDNA encoded *Sf9* FUT8, we evaluated the enzymatic activity of a recombinant protein produced by expressing the cloned cDNA in the baculovirus-*Sf9* cell expression system. As shown in **Table 3**, a characteristic fucosyltransferase activity was observed.

Restriction-enzyme digested genomic DNA from *Sf*9 cells hybridized to a single fragment, suggesting the presence of only one copy of the *fut8* gene in the *Sf*9 genome (**Figure 2A**). *Fut8* expression was then analyzed by Northern blotting using a specific mono-exonic (exon 3) *fut8* probe. Only one transcript of the expected size was detected (**Figure 2B**).

Phylogenomic analyses

We took advantage of the growing list of sequenced metazoan genomes to identify *in silico* several potential orthologs of the previously described FUT8 protein sequences. BLAST searches using the human FUT8 sequence and the previously described [4,9] fucosyl motifs I, II and III, as hallmarks for ortholog identification, allowed the identification of 96 highly conserved FUT8-related sequences from early metazoan genomes to protostome (mainly arthropods) and deuterostome genomes (**Table S2**). In most of these genomes, one single-copy *fut8* gene was identified with the notable exception of *Danio rerio, Xenopus laevis* and *Saccoglossus kowalevskii* in which two *fut8* paralogs (named fut8A and fut8B) were found. Multiple sequence alignments using ClustalW (**Figure 3**) highlighted the presence

```
AAG TTA TTA AGT TTT AAA AGA AGT CTT GCG TGG ATA TTA CCT GTC GTT GAT AAA AAG CCA GTT GCT GCT ACT AAA CAG ATA AAT GAC GAT

GCG GCA ACA GTG AGG AGT CGC TAC AAG GAT GCC GCC GCG ACA ATG AAA CAC TAC TCG AGT GGG CCC GCG ACA GTC AAT GAC TAC TTT ATG
                                                        1                                                             2
1/1                             10/30                                    20/60                                    30/90
ATG TAC CTG GCC AAA TGG AAG CGC GCC GCC GTC GTC TTG CTT TTC GTT TGG ATT GTC GTC ACC TAC CTG GTT ATA TCT CCA CTC AGA TGC
M   Y   L   A   K   W   K   R   A   A   V   V   L   L   F   V   W   I   V   V   T   Y   L   V   I   S   P   L   R   C
3
                                40/120                                   50/150                                   60/180
AAT GGA AAC CAA GAT GAG GTA GCA GAA TTC CAG GAG AGG TTA AAA AAG GTG TCG TCA CAA CTG GAG ACG TTG AGG CAG CAG CAT AGT AAT
N   G   N   Q   D   E   V   A   E   F   Q   E   R   L   K   K   V   S   S   Q   L   E   T   L   R   Q   Q   H   S   N

                                                                         80/240                                   90/270
CTC ATA TCA CAG ATA AAG AAG TCT TCT AGC CTC AAT GGT AAC CTG AAG GAC ATA GAC ACA GGG TCC CTG TTC CTA GAC GGG GTT CAG GGA
L   I   S   Q   I   K   K   S   S   S   L   N   G   N   L   K   D   I   D   T   G   S   L   F   L   D   G   V   Q   G

                                100/300                                  110/330                                  120/360
CCC ACA GAA GAC TAT GAG AAT CTC AGA AGG AGA ATA TAC TCC AAC ACC AAG GAG CTA TGG TAC TAT GTA CAC CAT GAG CTC ACT AAG GTC
P   T   E   D   Y   E   N   L   R   R   R   I   Y   S   N   T   K   E   L   W   Y   Y   V   H   H   E   L   T   K   V

                                130/390                                  140/420                                  150/450
ATA CAA GAA GAT AAC AAA GCG GAG AAG ATA CAG ACT ATA TTG GAC CAA ATT GGT GAA AGG AAA AGG TCG CTG CTC TCG GAC CAG CAA TCG
I   Q   E   D   N   K   A   E   K   I   Q   T   I   L   D   Q   I   G   E   R   K   R   S   L   L   S   D   Q   Q   S

                                160/480                                  170/510                                  180/540
CTA CCG GAG CTG GAC GGG TAC AAT GAG TGG AGG CAG GCG GAG GCA GCC AAC GTG AGC GAC CTC GTG CAA CGA CGA CTG CAC TAT CTA CAG
L   P   E   L   D   G   Y   N   E   W   R   Q   A   E   A   A   N   V   S   D   L   V   Q   R   R   L   H   Y   L   Q

                                190/570                                  200/600                                  210/630
AAC CCG CCC GAC TGC AAG GAC GCT AGG AAG CTC ATA TGT AAC TTG AAT AAG GGT TGC GGC TTC GGT TGC CAG CTA CAT CAC ATA GTA TAC
N   P   P   D   C   K   D   A   R   K   L   I   C   N   L   N   K   G   C   G   F   G   C   Q   L   H   H   I   V   Y
            C185(C204)*                     C193(C212)                  C199(C218)         C203(C222)
                                220/660                                  230/690                                  240/720
TGC CTC ATA TTC GCG TAC GCA ACG GAA CGC ACC CTG ATC CTC AAC TCC AAA GGC TGG AGG TAC AAC AAT AAG GGC TGG GAG TAT GTC TTC
C   L   I   F   A   Y   A   T   E   R   T   L   I   L   N   S   K   G   W   R   Y   N   N   K   G   W   E   Y   V   F
C211(C230)
                                250/750                                                                           270/810
ATG CCC ATA TCA GAC ACC TGC ACC AGT GCA TAC GAT GAC AAA GTC GTT ATG TGG CCA GTA ACG TAT GAC GCC AAG GTG TTA GCC CTG ACG
M   P   I   S   D   T   C   T   S   A   Y   D   D   K   V   V   M   W   P   V   T   Y   D   A   K   V   L   A   L   T
                        C247(C266)
                                280/840                                  290/870                                  300/900
TTC ATA GAC TCG ATA TCT CAG AAG CCC AAG TTC TTA CCG CTC GCC ATA CCT AAG GAT TTG GCA CAT CGG ATA ACG AGG TTT AAC GGT GAC
F   I   D   S   I   S   Q   K   P   K   F   L   P   L   A   I   P   K   D   L   A   H   R   I   T   R   F   N   G   D

                                310/930                                  320/960                                  330/990
CCG TCA TCA TGG TGG ATA GGC CAG ATG CTG AAA TAC ATT CTG AAG CCG CGC CAG GCG ATG CAG AAA GCT ATC AAT GAT ACT ATA GCT AAG
P   S   S   W   W   I   G   Q   M   L   K   Y   I   L   K   P   R   Q   A   M   Q   K   A   I   N   D   T   I   A   K

                                340/1020                                 350/1050                                 360/1080
ATG AAC TTC AAG AAA CCT ATC GTT GGG GTC CAC ATA AGG AGA ACG GAC AAG GTG GGC ACG GAG GCG GCG TTC CAC CAC ATC CAC GAG TAC
M   N   F   K   K   P   I   V   G   V   H   I   R   R   T   D   K   V   G   T   E   A   A   F   H   H   I   H   E   Y
                              Motif I   R343(R365)D346(D368)K347(K369) E351(E373)                           Y360(Y382)
                                370/1110                                 380/1140                                 390/1170
ATG GTG CAC GTG AAG GAC TAC TAC GAC CAG CTC GAG CTC ACC CGC CCC GTC GCC GTG CGC CGC GTC TAC CTG GCC ACT GAC GAC GCC AAC
M   V   H   V   K   D   Y   Y   D   Q   L   E   L   T   R   P   V   A   V   R   R   V   Y   L   A   T   D   D   A   N
                                                                                        Motif II  D387(D403)
                                400/1200                                 410/1230                                 420/1260
GTG TTG GAG GAC GCC CGC AGC AAG TAC CCG TCG TAC GAG TTC CTG GGC GAC GCG TCC ATC GCC AAG ACG GCG GCC ACG CAC CGC CGC TAC
V   L   E   D   A   R   S   K   Y   P   S   Y   E   F   L   G   D   A   S   I   A   K   T   A   A   T   H   R   R   Y

                                430/1290                                 440/1320                                 450/1350
ACG CCG CTG TCG CTG ACC GGC CTG CTG GTG GAT CTG CAT CTA CTC GCC ATG TGC GAC TAC CTC GTC TGC ACC TTC AGT AGC CAG GTC GGC
T   P   L   S   L   T   G   L   L   V   D   L   H   L   L   A   M   C   D   Y   L   V   C   T   F   S   S   Q   V   G
                                D431(D453)                         Motif III
                                                             C438(A460)        C443(C465)        S447(S469) G450(C472)
                                460/1380                                 470/1410                                 480/1440
AGA GTG GCG TAC GAG ATG ATG CAG ACG AAC CGC GTG GAC GCG TCC GAC AGC TTC TTC TCG CTG GAC GAC ATC TAC TAC TTC GGC GGA CAG
R   V   A   Y   E   M   M   Q   T   N   R   V   D   A   S   D   S   F   F   S   L   D   D   I   Y   Y   F   G   G   Q

                                490/1470                                                                          510/1530
AAC GCG CAC GAC CGG GTC GCC GTC ATG CCC AAC GAC GGC GCC TAC CAG GAT ATA TCC TTC CAG GTT GGC GAT TTA ATA GGC ATC GCG GGC
N   A   H   D   R   V   A   V   M   P   N   D   G   A   Y   Q   D   I   S   F   Q   V   G   D   L   I   G   I   A   G

                                520/1560                                 530/1590                                 540/1620
AAC CAC TGG AAC GGT TAC GGT CGA GGC ACC AAC AAA CGA ACA AAC ATG AAC GGA CTG ATA CCC TGG TAC AAG ACT GCG GAC CAC CTG GTC
N   H   W   N   G   Y   G   R   G   T   N   K   R   T   N   M   N   G   L   I   P   W   Y   K   T   A   D   H   L   V

                                550/1650                                 561/1686
CTC TAC CCA TTC CCG GAG TAC AAA CAG GTG CCT CTC TAC ACA GAC ACG CGA CAG AAA GAC TTA TAA ggA AGG CCC TAG TGC GGA GCT GCG
L   Y   P   F   P   E   Y   K   Q   V   P   L   Y   T   D   T   R   Q   K   D   L   -

GAC TAC CTA GCG GGT TTA CCG GGG CTC CGG CTC GAA AAG CAG GAG AAG GAA CGG GGT GGT TTT TAG TCA GTA AGA GTC TGA CAC TCC CTC

TCG CCT CGC CCA AGG CGG GAG AAG TCA TTG GAT GAC TTT TCC TCC TTA AAA AAA AAA AAA AAA
```

Figure 1. Nucleotide and deduced aa sequences of the *Sf*9 α1,6-fucosyltransferase cDNA. The three potential start codons are presented in boxes. ATG3 was considered as the start codon (see Table 2). A potential transmembrane domain (residues 9 to 25) is highlighted in grey. The conserved motifs I, II and III in the active site are underlined. The putative SH3 domain is in bold and underlined with a dashed line. The two potential *N*-glycosylation sites are underlined with a double line. Solid arrowheads indicate the position of introns. The numbers between brackets represent the position of critical aa conserved in human FUT8.

Table 1. Determination of the optimal sequence for initiation of translation in *Sf9* cells.

Sequences analyzed n = 168	**Position upstream of the start codon***										
	-10	**-9**	**-8**	**-7**	**-6**	**-5**	**-4**	**-3**	**-2**	**-1**	**AUG**
A	**43**	**36**	**27**	**35**	**27**	22	**39**	**88**	**64**	**57**	
U	26	**34**	**29**	**27**	**31**	**41**	11	0	13	6	
G	14	17	14	7	**28**	12	7	10	3	11	
C	17	13	**30**	**31**	14	25	**43**	2	20	26	
Consensus sequence	**A**	A/U	A/U/C	A/U/C	A/U/G	**U**	A/C	**A**	**A**	**A**	**AUG**

168 genes expressed in *Sf9* cells [40] were included in this analysis.
*Nucleotide flanking *Sf9* start codons, in percentage.

Table 2. Analysis of the sequences immediately flanking the three potential start codons (ATG1, ATG2 and ATG3 in Figure 1) identified in the *fut8* cDNA.

	Position upstream of the start codon										
	-10	**-9**	**-8**	**-7**	**-6**	**-5**	**-4**	**-3**	**-2**	**-1**	**AUG**
Consensus sequence[1]	**A**	**A/U**	**A/U/C**	**A/U/C**	**A/U/G**	**U**	**A/C**	**A**	**A**	**A**	**AUG**
Consensus sequence[2]	**A**	**U**	**A**	**A**	**A**	**U**	**A/C**	**A**	**A**	**C**	**AUG**
	C	G	**C**	**C**	**G**	C	G	**A**	C	**A**	**AUG (1)**
	U	G	**A**	**C**	**U**	A	**C**	U	U	U	**AUG (2)**
	C	**U**	**A**	**C**	**U**	**U**	U	**A**	U	G	**AUG (3)**

[1]Consensus sequence determined in this work (Table 1).
[2]Consensus sequence determined by Cavener and Ray (1991) [41] for invertebrates.

Table 3. Determination of α1-6-fucosyltransferase activity.

Sample	α1,6-fucosyltransferase activity
2D production	1,229+/−80
3D production	2,554+/−35
Negative control	62+/−1

Recombinant FLAG-tagged *Sf9* FUT8 was produced using the baculovirus-*Sf9* cell system and purified using anti-FLAG M2 affinity gel. The fucosyltransferase activity of the recombinant protein was tested with GDP-[^{14}C]-L-fucose, as substrate, after two days (2D) and three days (3D) of production. The transfer of [^{14}C]-fucose is expressed in cpm.

of the three conserved motifs (motifs I, II and III) and of a new "Cys-rich" peptide motif that is highly specific to FUT8 proteins.

We then assessed the evolutionary relationships of these animal sequences as described in the Materials and Methods section. To understand the orthology relationships of arthropods FUT8 sequences, we performed Maximum Likelihood (ML) (**Figure 4**) and Neighbor Joining (NJ) reconstructions (**Figure S2**). The recovered ML topology was very similar to the one generated by the NJ method, showing high conservation in vertebrates and in each insect order. The newly described *S. frugiperda fut8* sequence grouped with other early diverging lepidopteran FUT8 sequences. Both tree topologies tended to place nematode FUT8 sequences outside the protostome branch suggesting that these are rapidly evolving FUT8 sequences. We also investigated the extent of synteny and gene order conservation in the metazoan FUT8 sequences visualized at the Genomicus web site [42]. This preliminary analysis suggested microsynteny conservation in each insect order and no synteny conservation between insect orders (data not shown).

Intron-exon organization of the *fut8* genes

Twelve exons were identified in *Sf9 fut8*, ranging from 75 bp to 216 bp in size. The intron-exon boundaries of *Sf9 fut8* were in agreement with the consensus sequence for the splicing donor and acceptor sites (GT at position 1 and 2 at 5′ splice sites and AG at position -1 and -2 at 3′ splice sites) (not shown) [43,44]. Comparative analysis of *fut8* orthologs showed that the intron-exon organization was order-specific (**Figure 5**). The structure of lepidopteran *fut8* genes was overall well-conserved with only minor length disparities in exons 3, 9 and 12. Particularly, exon 3 contained 67 codons in *Heliconius melpomene*, 69 in *D. plexippus* and 71 codons in *S. frugiperda* and *B. mori*; exon 9 of *H. melpomene fut8* included 112 codons as a result of exon 9 and 10 fusion.

Despite the overall highly variable organization of *fut*8 genes (**Figure 5**), some exons (such as exons 4 and 8) that encode domains involved in protein folding and catalytic activity were shared among arthropod *fut8* genes. Exon 4 was also present in *Pediculus humanus* (phtiraptera) and *Daphnia pulex* (diplostraca) *fut8*, whereas exon 5 was found in *P. humanus, Tribolium castaneum* (coleoptera), *Acyrthosiphon pisum* and *D. pulex fut8*. Exon 8 was also found in Culicidae (e.g., *C. pipiens and Anopheles gambiae*). Lepidopteran exons 6 and 7 seemed to result from the splitting of an exon shared (2-78-2) by many arthropoda (phtiraptera, coleoptera, hemiptera and crustacea).

*Sf*9 *fut8* contained 11 introns (noted i*1*l to i*11*l), a much higher number compared to other insect *fut8* genes (between 2 and 9 introns) (**Table 4**). Diptera had the lowest number of introns: 3 in Drosophilidae and 2 in Culicidae. Strikingly, although all the essential sequence motifs required for the enzymatic activity were conserved in the hymenoptera *fut8* genes, the intron insertion sites identified in these genes (three in *Apis mellifera, Bombus impatiens, Megachile rotundata* and four in *Atta cephalotes, Solenopsis invicta, Camponotus floridanus, Harpagnathos saltator*) were hymenoptera-specific and were not shared by any of the other arthropod fut*8* genes analyzed here. In addition, the intron position i*3*h, which splits the hymenopteran exon 3 in apidae and megachilidae, appeared to be formicidae-specific (**Table 5, Figure 5**).

To understand the evolution of *fut8* genomic organization, we analyzed the repartition of intronic insertion sites in 56 complete orthologous *fut8* genes from organisms belonging to different animal phyla. As intron position analysis relies on consistent aa alignments, FUT8 sequences were aligned using Clustal Omega. The high aa identity observed in FUT8 catalytic domain (from exon 4 {E*4*l} to exon 11 {E*11*l}) allowed us to compare the intron-exon structure in this region. In contrast, the N-terminal region showed a very low level of aa conservation (**Figure S1**). As previously reported [44], when the insertion site was conserved, the intron phase was conserved as well. Several intron insertion sites were very well conserved (**Table 4 and Table 6**), for instance i*3*l, i*5*l, i*7*l and i*9*l in chordata, hemichordata (*S. kowalevskii*) and echinodermata (*Strongylocentrotus purpuratus*).

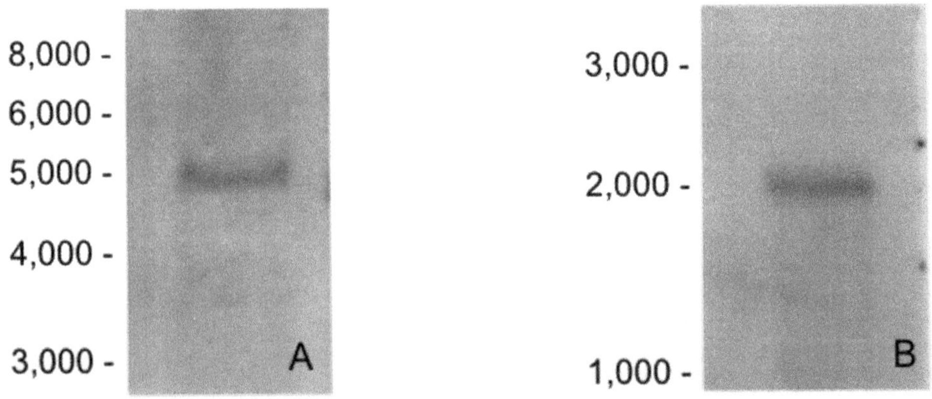

Figure 2. Southern and Northern blot analyses. A. Ten µg/lane of EcoRI-digested genomic DNA from *Sf9* cells was hybridized with a mono-exonic (exon 3, {E3l}) digoxigenin-labeled Fut8 probe prepared by PCR using the Forint10 and B24 primers (Table S1), as described in Materials and Methods. **B.** One µg/lane of total RNA was analyzed by Northern blotting. Hybridization was performed with a *fut8* anti-sense RNA probe (pos. 825 to pos. 1686 on the cDNA, Figure 1). Molecular weight markers in bp are indicated on the left of each panel.

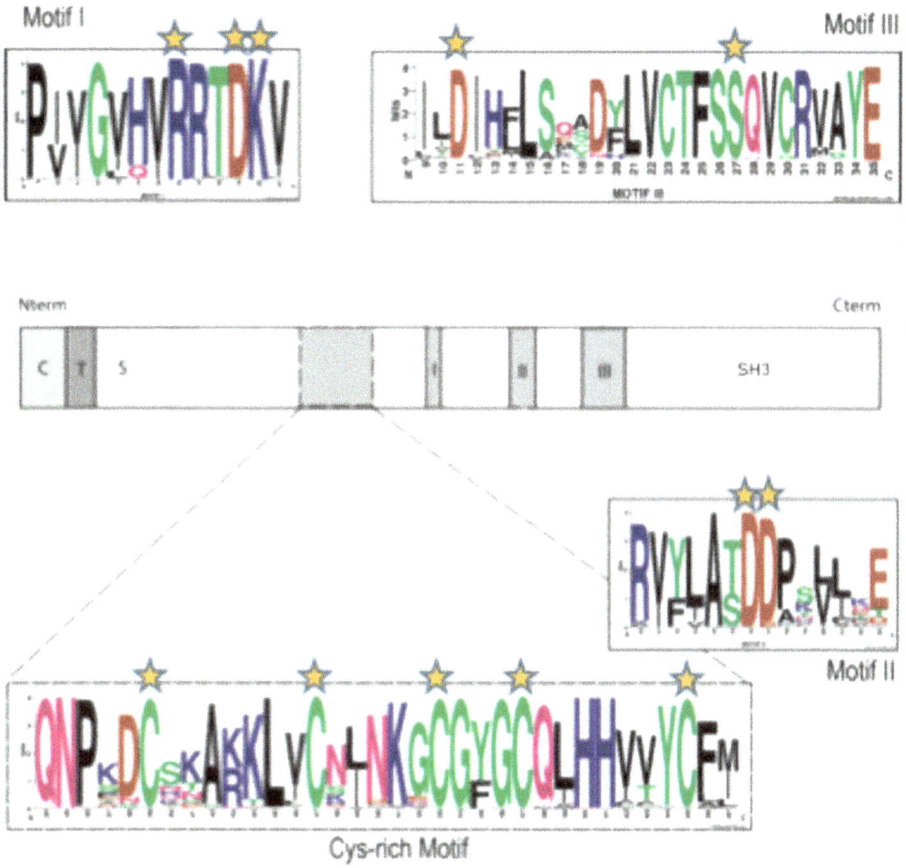

Figure 3. Conserved aa and motifs found in all the α1,6-fucosyltransferases sequences. Schematic representations of the FUT8 protein showing the cytoplasmic (C), transmembrane (T) and stem region (S) characteristic of α1,6-fucosyltransferases. The catalytic domain is in white and motifs I, II and III in grey. In addition, a region found only in α1,6-fucosyltransferase with conserved cysteine residues is indicated by dashed lines and was named "Cys-rich" domain. Conserved aa and those implicated in the enzymatic activity are highlighted with orange stars. The conserved peptide sequences used to generate the motif I, motif II and motif III sequence logos were extracted from multiple alignments of 96 α1,6-fucosyltransferase sequences identified in the databases (Table S2) and visualized at the Weblogos site at Berkeley, as described previously [63]. In the logos, aa are colored according to their chemical properties: polar aa (G, C, S, T, Y) are green, basic (K, R, H) are blue, acidic (D, E) are red, hydrophobic (A, V, L, I, P, W, F, M) are black and neutral polar aa (N, Q) are pink. The overall height of the stacks indicates the sequence conservation at a given position, while the height of the symbol within the stack indicates the relative frequency of each aa at that position. [69,70].

In nematoda, *fut8* gene was interrupted by 9 or 7 intronic sequences, but only one in rhabditida (*Caenorhabditis brenneri*) and three in trichocephalida (*Trichinella spiralis*) (*i.e.* i9l and i6l, i7l and i9l, respectively) were shared with *S. frugiperda fut8*. The intron sites i7l and i9l were also conserved in placozoa (*Trichoplax adhaerens*), cnidaria (*Hydra magnipapillata*) and annelida (*Capitella teleta*) (**Table 6**).

In the *A. pisum* (arthropoda) and chordata *fut8* genes, exon *1*-58-*2* and exon *1*-50-*0* were flanked by the conserved intron sites i7l and i9l (**Figure 5**). Comparison of these exons with the corresponding lepidopteran exon 8 (*1*-51-*0*) and exon 9 (*0*-58-*0*) led us to hypothesize that these two exons derived from a common ancestral exon (*1*-109-*0*) that was found in *T. castaneum*. Similarly, the exon pairs *1*-55-*0*/*0*-62-*1* (E4l and E5l) in lepidoptera and *1*-38-*0*/*0*-78-*1* in chordata (E4c and E5c), which were flanked by the conserved intron sites i3l and i5l, could derive from a common ancestral exon.

Discussion

FUT8 can be considered part of the GT-23 family within the CAZy classification, which groups enzymes acting on carbohydrates. This classification is based on the structural (presence of conserved peptide motifs) and mechanistic features (α1,6-fucosyltransferase) of these glycosyltransferases. GT-23 is a single-copy nuclear gene family, whereas most GT families are polygenic. *fut8* cDNAs have been cloned essentially from a few mammalian species (bovine, rat, pig, murine, and human, [15,34,35,37,45] and from two invertebrates, *Caenorhabditis elegans* and *Drosophila* [38]. In this work, we report the identification and molecular cloning of a *fut8* ortholog in the lepidoptera *S. frugiperda* and the analysis of the molecular function and genomic organization of this new lepidopteran gene.

All conserved motifs (I, II and III) and all essential aa residues implicated in FUT8 enzymatic activity were identified in the *Sf*9 FUT8 sequence. However, analysis of the cysteine residues involved in the disulfide bridges formation in most FUT8 proteins shows that one cysteine residue, which corresponds to Cys-472 in human FUT8, is missing in the lepidopteran protein. This change is accompanied by the presence of a new cysteine residue (Cys-438 in *S. frugiperda* FUT8) that is also conserved in many FUT8 sequences (tricocephalida, spirudida and ascaridia, *Ciona intestinalis*, *Nematostella vectensis* and in the hymenoptera FUT8 sequences analyzed in this study) (**Figure 1 and Figure S1**). The

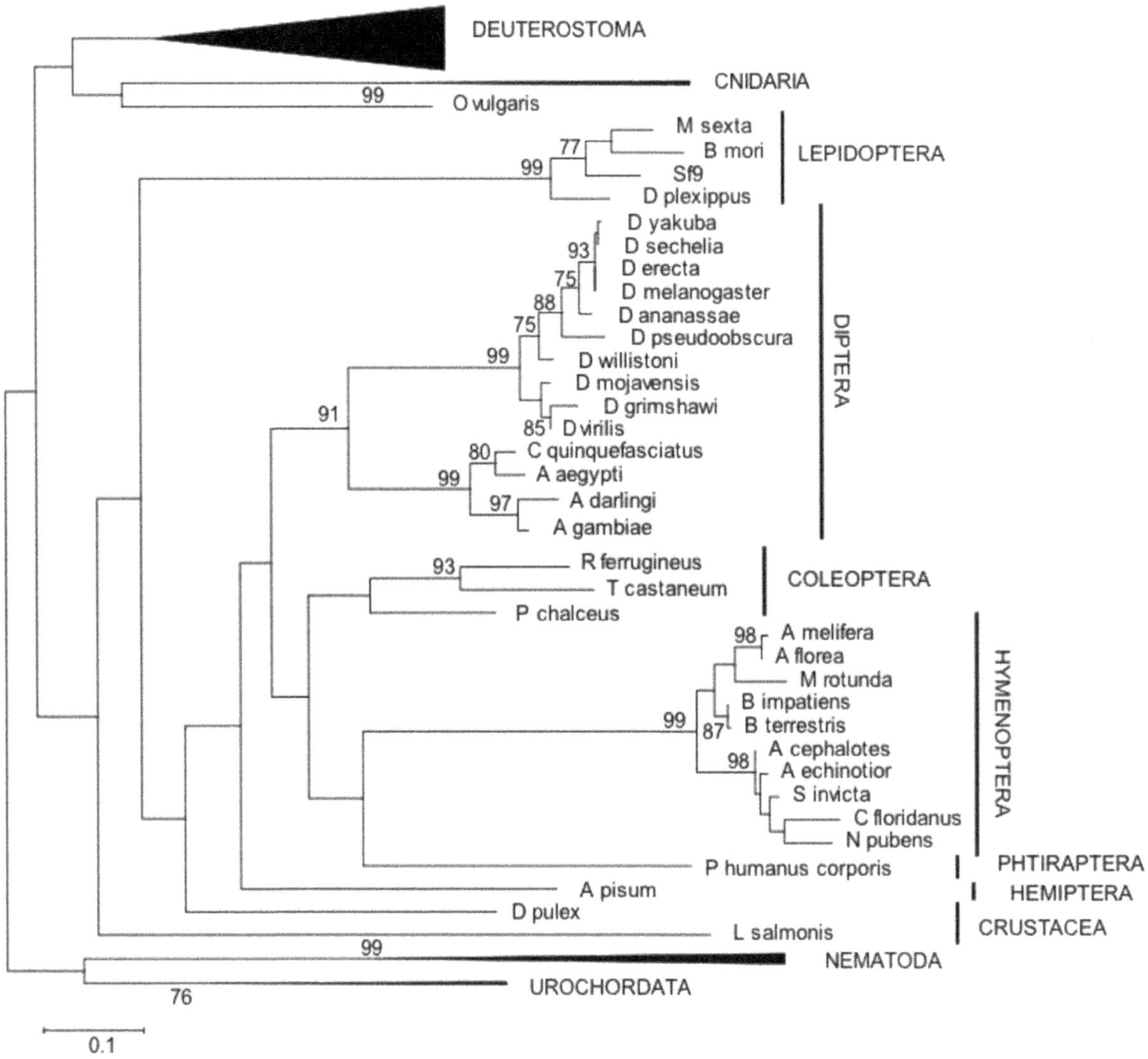

Figure 4. Molecular phylogenetic analysis using the Maximum Likelihood method. Evolutionary analyses were conducted in MEGA5 [67] and the evolutionary history was inferred using the Maximum Likelihood method based on the Whelan and Goldman model [72]. This analysis involved 92 FUT8 aa sequences and the final dataset contained 336 positions (42% of 787). The bootstrap consensus tree inferred from 1050 replicates is taken to represent the evolutionary history of the analyzed taxa. The percentage of trees (only those >75%) in which the associated taxa clustered together is shown next to the branches [73]. Initial tree(s) for the heuristic search were obtained automatically as follows. When the number of common sites was <100 or less than one fourth of the total number of sites, the maximum parsimony method was used; otherwise the BIONJ method with the MCL distance matrix was used.

high enzymatic activity measured with recombinant *Sf*9 FUT8 suggests that the disulfide bridge detected between Cys-465 and Cys-472 in most species can probably be equally established between Cys-465 with Cys-438 in *S. frugiperda* FUT8 without affecting its enzymatic activity or protein stability. Two potential *N*-glycosylation sites (Asn-167 and Asn-325) (**Figure 1**) are also present in the *Sf9* sequence, but these sites are not conserved in agreement with the fact that the enzyme activity is not dependent on the presence of *N*-glycans [15].

In an attempt to trace back the origin and evolutionary relationships of *fut8* genes, we identified *fut8* orthologs in a large panel of metazoan genomes and determined their gene organization.

Most GT families have significantly expanded in early vertebrates through whole genome duplications, and differential loss or retention of duplicated genes has contributed to the functional divergence of these GT families (e.g., sialyltransferases families) [46,47]. GT-23 represents a unique case of an evolutionary ancient GT family that did not diverge during metazoan evolution. We have chosen to study its gene structure as a way to characterize the evolutionary relationships of these genes.

Comparative analysis of the *fut8* gene in various animal genomes provided few insights into the evolution of this single-copy gene in metazoans. In contrast with genes encoding α1,3-, α1,3/4- and α1,2-FucTs, all the characterized *fut8* genes presented a common poly-exon organization. Metazoan *fut8* genes contained a highly variable number of exons (from 12 exons in lepidoptera to 2 or 3 in diptera and no introns in chilopoda *Lepeophtheirus salmonis*), suggesting a complex evolutionary history with many intron gain and loss events. Analysis of the structure and evolution of the *fut8* genes in arthropods showed that the high evolutionary rates found for instance in the coleoptera and diptera branches (see branch length differences

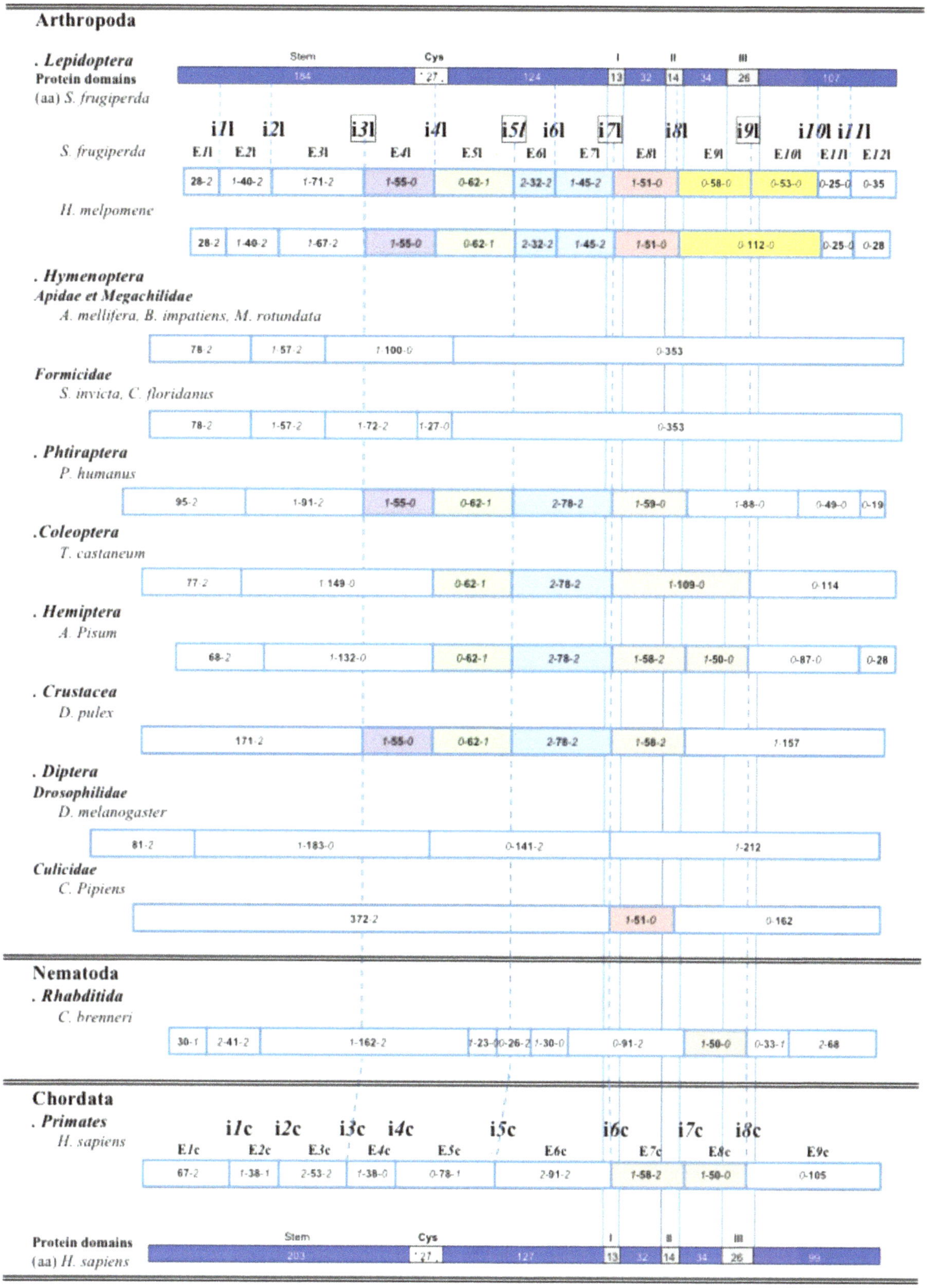

Figure 5. Schematic illustration of the genomic organization of the *fut8* genes identified in animal genomes. Coding exons are represented by rectangles with their relative size in codons (in bold). Italic numbers represent the number of nucleotides belonging to codons that flank the intron insertion site and determine the intron insertion phase. Dashed lines represent conserved intron insertion sites. Conserved exons are in the same color. *S. frugiperda* and *H. sapiens* FUT8 proteins are in blue, specific domains are represented by rectangles with their relative size in aa (Cys-rich domain, in black and motifs I, II and III, in grey).

Table 4. Analysis of intron positions shared with *S. frugiperda fut8* (*inumber*l) in arthropoda and chordata orthologs *fut8* genes.

Protein domain				**Stem**			**Cys-rich**			**Motif I**	**Motif II**	**Motif III**			**Total IS (bp)**	**Nb introns**
Intron insertion sites				**i1**	**i2**	**i3**	**i4**	**i5**	**i6**	**i7**	**i8**	**i9**	**i10**	**i11**		
Arthropoda	**Insecta**	Lepidoptera	*S. frugiperda*	**2/3,459**	**2/344**	**2/1,645**	**0/8,467**	**1/3,290**	**2/733**	**2/1,097**	**0/989**	**0/494**	**0/601**	**0/952**	22,071	11
			B. mori	**2/3,280**	**2/2,841**	**2/5,599**	**0/1,203**	**1/1,197**	**2/355**	**2/483**	**0/1,167**	**0/1,971**	**0/1,188**	**0/795**	20,079	11
			D. plexippus	**2/4,756**	**2/312**	**2/1,263**	**0/2,173**	**1/512**	**2/299**	**2/1,177**	**0/1,393**	**0/468**	**0/590**	**0/71**	13,014	11
			H. melpomene	**2/1,633**	**2/712**	**2/1,821**	**0/7,380**	**1/1111**	**2/204**	**2/146**	**0/795**		**0/129**	**0/248**	14,179	10
		Coleoptera	*T. castaneum*				**0/9,139**	**1/52**		**2/1,524**		**0/8,806**			21,186	5
		Hemiptera	*A. pisum*				**0/67**	**1/91**		**2/118**		**0/62**			752	7
		Phtiraptera	*P. humanus*			**2/342**	**0/82**	**1/215**		**2/295**					3,507	9
		Diptera	*D. melanogaster*				**0/109**			**2/65**					582	3
			C. pipiens							**2/62**	**0/98**				160	2
			*L. longipalpis**							**2/94**	**0/438**				≥3,290	≥3
	Arachnida	Mesostigmata	*M. occidentallis*									**0/210**			1,581	8
		Trombidiformes	*T. urticae*												1,173	6
	Branchiopoda	Cladocera	*D. pulex*			**2/65**	**0/60**	**1/61**		**2/55**					320	5
	Maxilopoda	Siphonostomatoida	*L. salmonis*												0	0
	Chilopoda	Geoplilomorph	*S. maritima*							**2/292**		**0/56**			2,591	6
Chordata	**Acidiacae**	Enterogona	*C. intestinalis A*			**2/836**		**1/906**		**2/820**		**0/447**	**0/5,797**		10,989	10
	Mammalia	Primate	*H. sapiens*			**2/13,118**		**1/52,295**		**2/2,125**		**0/8,711**			130,359	8
		Rodentia	*R. norvegicus*			**2/19,848**		**1/36,598**		**2/1,403**		**0/11,243**			131,971	8
		Artiodactyla	*B. taurus*			**2/14,139**		**1/40,145**		**2/4,795**		**0/11,832**			173,134	8

IS: intronic sequence, Nb: number. The intron phase was defined as follows: phase 0 introns are inserted between two codons; phase 1 introns are inserted after the first nucleotide of the codon, and phase 2 introns are inserted between the second and the third nucleotide of the codon.
(*) The gene is truncated and the 5′ exon is missing.

Table 5. Analysis of intron positions in *fut8* orthologs in hymenoptera (i*number*h).

Protein domain				Stem		Cys-rich			
Intron insertion sites				**i*7*h**	**i*2*h**	**i*3*h**	**i*4*h**	**Total IS (bp)**	**Nb introns**
Insecta	**Hymenoptera**	Apidae	*A. mellifera*	2/49	2/85		0/109	243	3
			B. impatiens	2/101	2/84		0/84	269	3
		Megachilidae	*M. rotundata*	2/101	2/75		0/65	241	3
		Formicidae	*A cephalotes*	*0/92*	2/80	2/96	0/84	352	4
			S. invicta	2/725	2/84	2/130	0/71	1010	4
			C. floridanus	2/116	2/77	2/137	0/75	405	4
			H. saltator	2/145	2/89	2/97	0/271	457	4

The intron phase was defined as described in Table 4. IS: intronic sequence, Nb: number.

between *Drosophila* and *Aedes aegypti* in **Figure 4**) corresponded to differences in gene organization as well.

The total size of the intronic sequences inserted in *fut8* genes varied greatly among insects with more than 10,000 bp in lepidoptera and only 160 bp in diptera (culicidae), with the notable exception of *A. aegypti*, which presented 13,978 bp of intronic sequences. This observation is in agreement with the analysis of the average intron size of the *A. aegypti* genome [48] that shows a 4-fold increase due to intron infiltration by transposable elements compared to other diptera species, such as *A. gambiae* and *D. melanogaster*. Similarly, the intronic sequences identified in *D. plexippus fut8* were about two times smaller than in *B. mori* and *S. frugiperda*, possibly because *D. plexippus* has the smallest genome among lepidoptera [49]. The same diversity was observed in other phyla. For instance, in nematoda, *fut8* has either 7 or 10 introns with relatively short intronic sequences (from 652 bp in *C. brenneri* to 3,676 bp for *Caenorhabditis japonica*). In placozoa and cnidaria, only 3 or 4 introns were found for a total of 682 bp of intronic sequences in *T. adhaerens fut8* and 13,440 bp in *H. magnipapillata fut8*. In vertebrates, the intron number appeared to be constant (8 introns). The main difference, compared to other phyla, was the very large total size of intronic sequences (about 130,000 bp in *Homo sapiens* and *Rattus norvegicus* 173,134 bp in *Bos taurus)* (**Table 4 and Table 6**).

Intron insertion sites in phase 0 were over-represented in gene sequences encoding the most conserved part of the FUT8 proteins: i*4*l and i*4*c in the cysteine-rich domain, and i*8*l (motif II), i*9*l (motif III), i*10*l (SH3) and i*11*l (SH3) in the catalytic domain. Moreover, in human FUT8 structure [9], introns i*4*c and i*8*c/i*9*l are located at the C- and N- terminal domain of alpha helix 3 and alpha helix 11 respectively. Such features are proposed to be characteristic of ancient conserved genes [50].

Many exons were shared by different arthropoda orders, such as phtiraptera, coleoptera, hemiptera, diptera, lepidoptera and even crustacea (**Figure 5**). Remarkably, despite the high conservation of lepidoptera FUT8 sequence with vertebrates and hymenoptera (*S. frugiperda* FUT8 sequence shares 44.24% and 45.93% aa identity with *H. sapiens* and *A. mellifera* respectively), the exon-intron organization of hymenoptera *fut8* genes is order-specific with no shared exons. This is particularly surprising because when the analysis is extended to orthologous *fut8* genes of other phyla, many *Spodoptera* intron insertion sites are still conserved throughout the animal kingdom. Particularly, intron positions i*7*l and i*9*l are common to most of the *fut8* genes analyzed in this work (**Table 4 and Table 6**). As these introns are located in a gene portion encoding motif I and III respectively (highly conserved regions of FUT8), a high selection pressure could be exerted on these parts, such as for instance the presence of regulatory elements that are important for gene expression. The i*8*l insertion site is only found in lepidoptera and some diptera, such as culicidae and psychodoidea (*Lutzomyia longipalpis*). This site is very close to the chordate site i*7*c with only 24 nucleotides between these two sites (**Figure S1**) forming a "near intron pair" as defined by [51]. In contrast to i*8*l, i*7*c is conserved in many arthropods (*A. pisum, P. humanus, D. pulex, Strigamia maritima*), cnidarians (*N. vectensis and H. magnipapillata*) and nematodes (*Brugia malayi and C. brenneri*). Thus, position i*7*c appeared before i*8*l and we hypothesize that the i*8*l site might have resulted from i*7*c sliding. Similarly, the chordate insertion site i*4*c is found in the arthropod *S. maritima* (**Figure S1**). Comparison of the lepidoptera and chordata exon pairs *1*-55-*0*/*0*-62-*1* and *1*-38-*0*/*0*-78-*1*, located between the conserved sites i*3*l and i*5*l, suggests the existence of a common ancestral exon. As hypothesized for i*8*l, position i*4*l could have derived from the sliding of the ancestral

Table 6. Intron positions shared by *S. frugiperda fut8* (i*number*l) and with *fut8* genes identified in chordata (i*number*c).

Protein domain				Stem			Cys-rich				Motif I	Motif II		Motif III				
Intron insertion sites in *Sf9 fut8* gene				i1l	i2l	i3l	i4l		i5l	i6l	i7l	i8l		i9l	i10l	i11l	**Total IS**	**Nb introns**
Specific introns insertion sites in Chordata *fut8* gene								i4c					i7c				**(bp)**	
Chordata	Mammalia	Primate	*H. sapiens*			X		+	X		X		+	X			130,359	8
	Aves	Galliformes	*G. gallus*			X		+	X		X		+	X			36,774	8
			M. gallopavo			X		+	X		X		+	X			37,003	8
		Passeriformes	*T. guttata*			X		+	X		X		+	X			43,603	8
	Reptilia	Squamata	*A. carolinensis*			X		+	X		X		+	X			67,279	8
	Amphibia	Anura	*X. tropicalis*			X		+	X		X		+	X			72,102	8
	Actinopterygii	Cypriniformes	*D. rerio B*			X		+	X		X		+	X			88,524	8
		Tetraodontiformes	*T. rubipres*			X		+	X		X		+	X			23,935	8
		Beloniformes	*O. latipes*			X		+	X		X		+	X			44,674	8
		Perciformes	*O. niloticus*			X		+	X		X		+	X			45,900	8
Echinodermata	Echinoidea	Echinoida	*S. purpuratus*			X		+	X		X		+	X			58,902	8
Hemichordata	Enteropneusta	Enteropneusta	*S. kowalevskii A*			X		+	X		X		+	X			21,376	8
Nematoda	Secernantea	Trichocephalida	*T spiralis*							X	X			X			1,122	7
		Rhabditida	*C. brenneri*										+	X			652	9
		Spirudida	*L. loa*										+				3,347	9
			B. malayi										+				2,417	9
			W. bancrofti										+				2,954	10
		Ascaridida	*A suum*										+				13,589	10
Annelida	Polychaeta		*C. teleta*			X		+			X		+	X			1,183	7
Placozoa	Tricoplacia	Tricoplaciformes	*T. adhaerens**								X		+	X			≥682	≥3
Cnidaria	Hydroida	Anthomedusae	*H. magnipapillata*								X		+	X			13,440	4
	Anthozoa	Actiniaria	*N. vectensis*								X		+	X			2,492	4

IS: Intronic Sequence.
(X): Presence of lepidopteran intron insertion site.
(+): Presence of chordata intron insertion site.
(*) The gene is truncated and the 5′ exon is missing.

position i4c (51 nt between i4l and i4c). However, unlike i7c, i4c is not well conserved outside chordates (**Figure 5**).

To conclude, we propose a model of *fut8* gene evolution in animals (**Figure 6**). In this model, i7c and i9l are considered as ancestral intron insertion sites. Until the arthropoda-chordata split, intron gain seems to have been the most favored event with gain of i7l about 855MYA (Million years ago) and of i3l, i5l, i4c about 783 MYA [52]. After this divergence, intron loss seems to have become rather more common with the consecutive loss in arthropods of i3l (372 MYA), i7c (300 MYA), i5l and i9l (265 MYA). These intron losses could be accompanied by specific intron site gains (for instance, gain of i6l, i10l and i11l in lepidoptera) (**Figure 6**, compare lane 1 and 2) or not (for instance in *D. pulex)*. It has to be noted that the intron site gains and losses identified in arthropods are not observed in chordates. Indeed, insertions of spliceosomal introns are rarely observed during evolution of vertebrates [53]. As *fut8* intron-exon organization is order-specific, these intron losses and gains may be linked to evolutionary innovations, such as appearance of new orders.

Finally, the gain of some intron sites, such as i8l in some diptera (culicidae, *C. pipiens* and psychodidae, *L. longipalpis*), i10l in *C. intestinalis* and i6l in *T. spiralis* (**Figure 6, Figure S1, Table 4 and Table 6**), may be explained by convergent/parallel intron gains, as recently described by [54] for hymenopteran paralogs, suggesting the presence of intron insertion hot spots [55].

Unlike other well studied glycosyltransferases [46,47], we found that this single-copy gene characterized by highly conserved motifs is present from the very first metazoans to vertebrates. This could be explained by FUT8 important function. Evolution of *fut8* gene organization is also very specific and may accommodate different way of regulating its expression for instance in response to the environment of these insects.

Materials and Methods

Cells and viruses

The *Sf9* subclone of the *S. frugiperda Sf21* cell line [56] was maintained at 28°C in TC100 medium (GIBCO) containing 5% heat-inactivated fetal calf serum. Cells were infected with baculoviruses *Ac*MNPV clone 1.2 [57] at a multiplicity of infection (MOI) of 2 plaque-forming units (PFU) per cell. After 1 h incubation at room temperature with the viral suspension, fresh culture medium was added and cells were incubated at 28°C for 5 days. The viral titers were determined by plaque assay [58].

DNA and RNA purification

High molecular weight genomic DNA was extracted from *Sf9* cells (15×10^6 cells) using the Genomic-Tip Kit (QIAGEN) as recommended by the manufacturer. Total cytoplasmic RNA was extracted from 2×10^7 *Sf9* cells using a method previously described [59]. After precipitation, RNA pellets were washed with a cold solution containing 75% ethanol and 25% sodium acetate and resuspended in 50 µl of distilled water. Fifty µg of RNA were then incubated with 15 units of RNase-free DNase I (New England Biolabs) at 37°C for 20 min and then the DNase was inactivated with 5 mM EDTA at 75°C for 10 min. RNA samples were loaded on an RNeasy column (QIAGEN) as recommended by the manufacturer. Purified RNA was immediately stored at −80°C.

Cloning of the *Sf9* α1,6-FucT cDNA

Sf9 fut8 cDNA was cloned by RT-PCR using degenerate primers [60]. The two *ForFut1deg* and *BacFut2deg* degenerate primers were designed based on the conserved aa sequence found in five species in which *fut8* cDNA had been already isolated and characterized: *H. sapiens* (AJ536056), *B. taurus* (BC104555), *D. melanogaster* (AF441264), *Drosophila yakuba* (AJ830719) and *Drosophila pseudoobscura* (AJ830720) (**Table S1**, for primer sequences) (Eurogentech, Belgium). RT was performed as follows: a mixture (final volume of 20 µl) including 1 µg of total RNA from *Sf9* cells, 4 µl of 10 X RT buffer (QIAGEN), 4 µl of 5 mM of each dNTP (QIAGEN), 4 µl of 10 pmoles/ml *BacFut2deg*, 20 units of RNase inhibitor (Roche) and 8 units of Omniscript reverse transcriptase (QIAGEN) was incubated at 37°C for 1 hour. The reverse transcriptase was inactivated at 93°C for 5 min.

Polymerase chain reaction (PCR) amplification was then carried out in a final volume of 20 µl containing 2 µl of 10 X Vent DNA polymerase (New England Biolabs), 2 µl of 10 mM each dNTP (Biolabs), 20 pmoles of each degenerate primer, 1.5 µl of 25 mM $MgSO_4$, 1 unit of Vent Polymerase and 0.5 µl of the RT mixture. Thirty-five cycles of amplification were performed on a Mastercycler apparatus (Eppendorf): denaturation at 94°C for 30 sec, annealing for 1 min at a temperature depending on the primer used, extension at 72°C for 1 min and a final extension of 10 min at 72°C.

PCR products were separated on low melting agarose gels (FMC, Nusieve GTG) and purified with the QIAquick Gel Extraction Kit (QIAGEN). The 5′-ends were adenylated by incubating the DNA with 1 unit of Taq DNA Polymerase (Biolabs) and 5 µl of 10X ThermoPol buffer (20 mM Tris-HCl pH 8.8, 10 mM KCl, 10 mM $(NH_4)_2SO_4$, 2 mM $MgSO_4$, 0.1% Triton X-100) and 200 µM dATP (final volume: 50 µl) at 72°C for 20 min. The PCR fragments were purified with the QIAquick PCR Purification Kit (QIAGEN) before ligation in the pGEM-T Easy vector system (Promega) for sequencing.

The 5′- and 3′- ends of the cDNA were amplified using the 5′/3′-RACE Kit, 2nd generation (Roche), following the manufacturer's protocol, and the exact-match primers *Bac5′RACEFut8* and *For3′RACEFut8* (**Table S1**), designed based on the *fut8* cDNA sequence obtained with the degenerate primers. The AMV reverse transcriptase and total RNA from *Sf9* cells, as before, were used for the RT step. Briefly, 3′-RACE uses the polyA stretch generally present at the 3′-end of each mRNA as hybridization site for the oligo(dT) anchor primer. To specifically amplify the 3′ *fut8* cDNA end, RNA was thus reverse transcribed with the oligo(dT) anchor primer and then the RT product was PCR amplified with *For3′RACEFut8* and the oligo(dT) primers. For the 5′-RACE, the specific *Bac5′RACEFut8* primer was used in the RT step. Then, to specifically amplify the 5′-end of the *fut8* cDNA, a polyA stretch was added at the 5′-end of the cDNA by a terminal transferase and used as a hybridization site for the oligo(dT) anchor primer. The entire 1734 bp ORF was reconstituted in the pGEM-T Easy plasmid, giving the final pGEMFUT8*Sf* construct.

Genomic DNA analysis

Sf9 cell genomic DNA (10 µg) was digested with *Eco*RI or *Hind*III restriction endonucleases. DNA fragments were then separated by electrophoresis on 0.9% agarose gels (SeaKem GTG, Lonza) and transferred to positively charged nylon membranes (Roche). Membranes were pre-hybridized at 68°C in 5X SSC, 0.1% N-laurylsarcosine, 0.02% (w/v) SDS, 1% blocking reagent (Roche) and 100 µg/ml calf thymus DNA for 3 hours. Overnight hybridization was performed at 68°C in the same buffer with a 204 bp *fut8* specific probe (nucleotide 217–420 of the *fut8* cDNA cloned in pGEM-T Easy, **Figure 1**) labeled with digoxigenin using the PCR DIG Probe Synthesis Kit (Roche). After hybridization, filters were washed twice at room temperature with 2X SSC, 0.1% SDS for 5 min and then twice with 0.1X SSC,

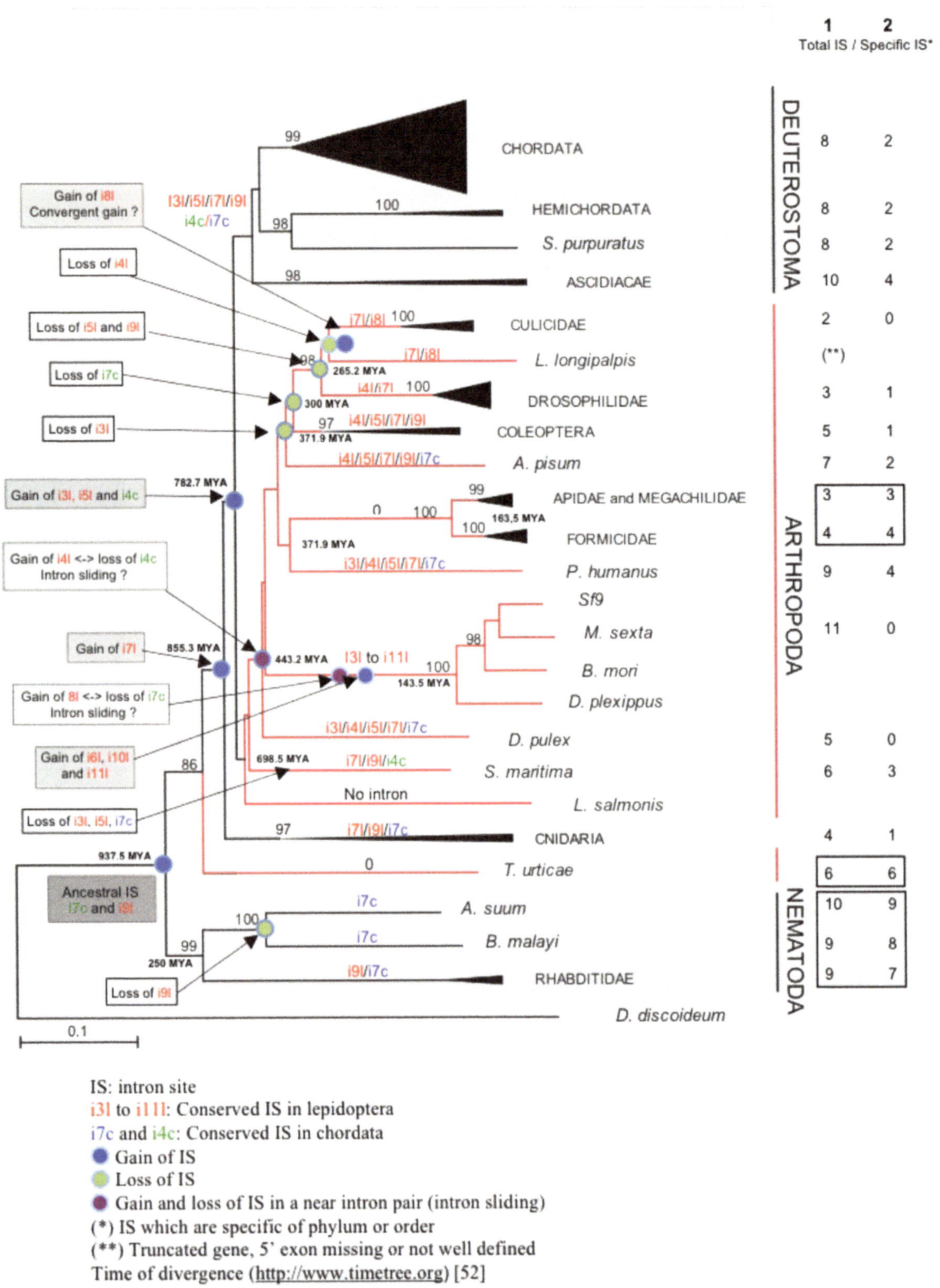

Figure 6. Schematic illustration of the correlation between animal *fut8* gene phylogeny, intron gain/loss and intron position (gene organization). This analysis was carried out by considering only the intron insertion sites in the *fut8* gene sequences encoding the conserved region of FUT8 proteins, between i3l and the stop codon. On the right, column 1 shows the total number of intron insertion sites (IS) identified in the different *fut8* genes, and column 2 shows the number of order- or family-specific intron insertion sites. Intron gains and losses are highlighted (grey and white boxes, respectively) as well as putative intron sliding in "near-intron-pairs" (light grey boxes). Putative ancestral introns are in a dark grey box.

01% SDS at 68°C for 15 minutes. Probe-target hybrids were revealed by incubating the membranes with an alkaline-phosphatase conjugated anti-digoxigenin antibody (Roche) and by a chemoluminescent reaction using CSPD as substrate (Roche). Membranes were exposed to Kodak films for 20 min.

RNA analysis

Total RNA from *Sf9* cells was resolved (1 µg per lane) on 1% agarose gels and transferred to positively charged nylon membranes (Roche). A digoxigenin-labeled riboprobe was synthesized using the DIG Northern Starter Kit (Roche) using as a template the complete FUT8 coding sequence (pos. -1 to pos. 1686 in the *fut8* cDNA, **Figure 1**) cloned in pGEM-T Easy (pGEMFUT8*Sf*). Hybridization and revelation were performed as described for Southern blotting.

PCR analysis of the genomic structure of the *Sf9* α1,6-FucT gene

PCR amplifications were performed using *Sf9* genomic DNA as a template. DNA was isolated as described above using the Genomic-Tip Kit (QIAGEN) and resuspended in water. PCR amplifications were carried out in a final volume of 50 µl containing 10 µl of 5X Phusion DNA polymerase buffer (HF buffer) (Finnzymes, Finland), 1 µl of 10 mM each dNTP (Biolabs), 0.5 µmoles of each primer (**Table S1**; Cloning of introns), 1 unit of Phusion DNA Polymerase and 200 ng of *Sf9* genomic DNA. After denaturation at 98°C for 2 min, a two-step protocol was used with (i) seven cycles of amplification in the following conditions: denaturation at 98°C for 15 sec, annealing and extension for 3 minutes at 72°C and (ii) thirty cycles of amplification with denaturation at 98°C for 15 sec, annealing and extension for 5 min at 68°C. PCR products were isolated on agarose gels (SeaKem GTG, Lonza) and purified with the QIAquick Gel Extraction Kit (QIAGEN). PCR fragments were then adenylated with Taq Polymerase as described above and cloned in pGEM-T Easy (Promega) for sequencing analysis (Operon Eurofins MWG, Germany).

Construction of the recombinant baculovirus expressing soluble and tagged *Sf9* FUT8

To produce soluble recombinant FUT8, the 5′-end of the cDNA encoding the N-terminal domain of FUT8 (from aa 1 to aa 30 at the end of the putative transmembrane domain) was deleted and replaced by the signal peptide sequence of the baculovirus ecdysteroid glycosyltransferase (EGT) [61] followed by a sequence encoding the FLAG epitope (Asp-Tyr-Lys-Asp-Asp-Asp-Asp-Lys) to produce a recombinant protein that can be easily purified and that can be differentiated from the endogenous protein. To this aim, a *Hpa*I-*Bgl*II DNA fragment encoding the FLAG epitope and aa 31 to aa 40 of *Sf9* FUT8 was synthesized using overlapping oligonucleotides and inserted in frame with the EGT signal peptide present in the pUC PS-EGT vector (pUC PS-EGT-FLAG-NterFut8*Sf* construct) [62]. Then, the C-terminal domain of *fut8* cDNA was isolated from the pGEMFut8*Sf* plasmid by *Eco*RI-*BamH*I digestion and inserted in the pUC PS-EGT-FLAG-NterFut8*Sf* construct to obtain the pUC PS-EGT-FLAG-Fut8*Sf* construct. After *BamH*I and *Hind*III digestion, the 1722 bp fragment containing the EGT-FLAG-Fut8 cDNA was isolated by agarose gel electrophoresis (Nusieve GTG, Lonza), purified using the QIAGEN Gel Extraction Kit and ligated in the p119 baculovirus transfer vector digested with *Bgl*II and *Hind*III to produce the p119-EGT-FLAG-Fut8*Sf* construct.

Sf9 cells (4×10^6 cells) were seeded in 25 cm^2 flasks and co-transfected with 500 ng of purified *Ac*SLP10 viral DNA [59] and 5 µg of p119-EGT-FLAG-Fut8*Sf* by lipofection [63] with DOTAP (Roche). The p119 transfer vector is designed for recombination in the P10 locus. *Ac*SLP10 is derived from wild type *Ac*MNPV 1.2 [57] and has only one strong promoter (P10) to drive the expression of a polyhedrin gene [59]. After 5 days of incubation at 28°C, recombinant (i.e., polyhedron-negative) viruses were purified by plaque assay [58]. Viral stocks were generated by propagating viruses in *Sf9* cells (75 cm^2 flasks) and titrated using plaque assays.

Production and purification of soluble FUT8

Sf9 cells seeded in roller bottles at a density of 400,000 cells/mL in serum-free medium were infected at a MOI of 2 PFU per cell. After 3-day incubation at 28°C, supernatants were collected and stored at −80°C before use. Supernatants were then concentrated and diafiltered against TS buffer (50 mM Tris-HCl pH 7.4, 150 mM NaCl) using Centromate cassettes (Pall, 0.1 m^2/30 kDa) before loading on a 5 ml column of anti-FLAG M2 Affinity gel (Sigma) equilibrated with TS buffer. The FLAG-tagged protein was eluted with 30 ml of 100 µg/ml FLAG peptide (in TS buffer). Fractions containing the purified protein were concentrated (AMICON Ultra 50 k, Millipore) and analyzed by PAGE and western blotting.

FUT8 assays

After 2 hours of dialysis against 50 mM sodium acetate buffer (pH 7.5), 5 mg of human apotransferrin (Sigma) were incubated with 120 mU *Arthrobacter urefasciens* neuraminidase (Sigma) and 50 mU *Escherichia coli* β-galactosidase (Sigma) in a final volume of 10 ml at 37°C for 24 hours. After 4 hours dialysis against water, asialo-agalacto apotransferrin was freeze-dried. The absence of galactose and sialic acid residues was verified by gas chromatography-mass spectrometry (GC-MS). To test the enzymatic activity of recombinant *Sf9* FUT8, assays were performed in 50 µl total volume that included 20 µl of purified recombinant protein solution (140 µg/ml), 70 mM cacodylate buffer (pH 7.2) 10 mM L-fucose, 6 mM GDP-[^{14}C]-L-fucose (299 mCi/mMol, Amersham), 10 mM GDP-fucose and 100 µg asialo-agalacto apotransferrin as acceptor, as previously described [34]. After incubation at 30°C for 4 hours, the reaction was stopped with 150 µl of water and samples were precipitated with 1 ml of 5% PTA and processed for scintillation counting. The transfer of [^{14}C]-fucose is expressed in cpm.

In silico α1,6-FucT sequences retrieval and phylogenetic analysis

Only eukaryotic core α1,6-FucT sequences were considered for this study. Homologous *fut8* sequences were searched by querying all genomic and expressed sequence tag (EST) divisions of the National Center for Biotechnology Information (NCBI database, Washington, DC, USA), as described previously for sialyltransferases [22,64] (**Table S2**). DNA and aa sequences were analyzed using DNA Strider [65]. PSORT II (http://expasy.org) was used for the prediction of protein localization sites in cells. Homology searches were performed using the BLAST program [39]. Multiple sequence alignments of the 86 deduced aa sequences homologous to α1,6-fucosyltranferases were generated using ClustalW 2.0.8, Clustal Omega 1.1.0 or MAFFT, which are multiple sequence alignment programs available at http://www.ebi.ac.uk. Gblocks (http://molevol.cmima.csic.es/castresana/Gblocks_server.html) [66] was used to select the 336/785 informative sites. Evolutionary analyses were conducted using the Neighbor Joining (NJ) and Maximum Likelihood (ML) methods implemented in MEGA 5.05 [67]. The branch robustness was tested with 1050 bootstrap replicates. Conservation of synteny and gene order in the metazoan were visualized at the Genomicus web site (version 19.01) (http://www.genomicus.biologie.ens.fr/genomicus-metazoa-19.01/cgi-bin/search.pl) [42].

Supporting Information

Figure S1 Amino acid sequences of α1,6-fucosyltransferases from different phyla were aligned using Clustal Omega 1.1.0 (www.ebi.ac.uk). Letters on grey background indicate the position of intron insertion in the genes. Numbers indicate the intron phase. When the insertion phase is 1 or 2, the aa corresponding to the split codon is in highlighted in grey, when the insertion phase is 0, the two flanking aa are in grey. Putative transmembrane domains determined using http://wolfpsort.org are underlined. Conserved cysteine residues are highlighted in yellow.

Figure S2 The evolutionary history was inferred using the Neighbor-Joining method [68] and 86 amino acid sequences aligned with MAFFT (EBI). All positions containing gaps and missing data were eliminated. The final dataset contained 461 positions selected in 17 blocks by GBlocks [66] (58% of the original 785 positions). The *Dictyostelium discoideum* sequence was used as outgroup. The optimal tree with a branch length sum = 6.93435487 is shown. The percentage of replicated trees (>75%) in which the associated taxa clustered together in the bootstrap test (1050 replicates) is shown next to the branches. The tree is drawn to scale, with branch lengths (next to the branches) in the same units as those of the evolutionary distances used to infer the phylogenetic tree. The evolutionary distances were computed using the p-distance method [69] and refers to the number of amino acid differences per site.

Table S1 Primers used in this study.

Table S2 Nomenclature, name and GenBank accession number of the *fut8* gene sequences used in this study. (*) Nucleotide sequence data available in the Third Party Annotation Section of the DDBJ/EMBL/GenBank databases. (**) Ensembl accession number.

Acknowledgments

We thank the anonymous reviewers for their very interesting comments that helped us improving the manuscript.

Author Contributions

Conceived and designed the experiments: MD-C SJ FM AH-L. Performed the experiments: SJ FM BC MLV PC AO AH-L MD-C. Analyzed the data: MD-C SJ FM AH-L. Contributed reagents/materials/analysis tools: SJ FM BC MLV PC AO. Contributed to the writing of the manuscript: MD-C SJ AH-L.

References

1. Cantarel BL, Coutinho PM, Rancurel C, Bernard T, Lombard V, et al. (2009) The Carbohydrate-Active EnZymes database (CAZy): an expert resource for Glycogenomics. Nucleic Acids Res 37: D233–238.
2. Becker DJ, Lowe JB (2003) Fucose: biosynthesis and biological function in mammals. Glycobiology 13: 41R–53R.
3. Zhao Y, Sato Y, Isaji T. Fukuda T. Matsumoto A, et al. (2008) Branched N-glycans regulate the biological functions of integrins and cadherins. FEBS J 275: 1939–1948.
4. Oriol R, Mollicone R, Cailleau A, Balanzino L, Breton C (1999) Divergent evolution of fucosyltransferase genes from vertebrates, invertebrates, and bacteria. Glycobiology 9: 323–334.
5. Mollicone R, Cailleau A, Oriol R (1995) Molecular genetics of H, Se, Lewis and other fucosyltransferase genes. Transfus Clin Biol 2: 235–242.
6. Liu YC1, Yen HY, Chen CY, Chen CH, Cheng PF, et al. (2011) Sialylation and fucosylation of epidermal growth factor receptor suppress its dimerization and activation in lung cancer cells. Proc Natl Acad Sci U S A 108: 11332–11337.
7. Shields RL, Lai J, Keck R, O'Connell LY, Hong K, et al. (2002) Lack of fucose on human IgG1 N-linked oligosaccharide improves binding to human Fc gamma RIII and antibody-dependent cellular toxicity. J Biol Chem 277: 26733–26740.
8. Wang X, Inoue S, Gu J, Miyoshi E, Noda K, et al. (2005) Dysregulation of TGF-β1 receptor activation leads to abnormal lung development and emphysema-like phenotype in core fucose-deficient mice. Proc Natl Acad Sci USA 102: 15791–15796.
9. Ihara H, Ikeda Y, Toma S, Wang X, Suzuki T, et al. (2007) Crystal structure of mammalian alpha1,6-fucosyltransferase, FUT8. Glycobiology 17: 455–466.
10. Rao ST, Rossmann MG (1973) Comparison of super-secondary structures in proteins. Journal of Molecular Biology. 76: 241–250.
11. Breton C, Oriol R, Imberty A (1998) Conserved structural features in eukaryotic and prokaryotic fucosyltransferases. Glycobiology 8: 87–94.
12. Takahashi T, Ikeda Y, Tateishi A, Yamaguchi Y, Ishikawa M, et al. (2000) A sequence motif involved in the donor substrate binding by alpha1,6-fucosyltransferase: the role of the conserved arginine residues. Glycobiology 10: 503–510.
13. Martinez-Duncker I, Mollicone R, Candelier JJ, Breton C, Oriol R (2003) A new superfamily of protein-O-fucosyltransferases, alpha 2-fucosyltransferases, and alpha 6-fucosyltransferases: phylogeny and identification of conserved peptide motifs. Glycobiology 13: 1C–5C.
14. Okajima T, Xu A, Lei L, Irvine KD (2005) Chaperone activity of protein O-fucosyltransferase 1 promotes notch receptor folding. Science 307: 1599–1603.
15. Javaud C, Dupuy F, Maftah A, Michalski JC, Oriol R, et al. (2000) Ancestral exonic organization of the FUT8, the gene encoding the α6-fucosyltransferase, reveals successive peptide domains which suggest a particular three dimensional core structure for the α6-fucosyltransferase family. Mol Biol Evol 17: 1661–1672.
16. Javaud C, Dupuy F, Maftah A, Julien R, Petit JM (2003) The fucosyltransferasegene family: an amazing summary on the underlying mechanisms of gene evolution. Genetica 118: 157–170.
17. Tomiya N, Narang S, Lee YC, Betenbaugh MJ (2004) Comparing N-glycan processing in mammalian cell lines to native and engineered lepidopteran insect cell lines. Glycoconj J 21: 343–360.
18. Marchal I, Mir AM, Kmiécik D, Verbert A, Cacan R (1999) Use of inhibitors to characterize intermediates in the processing of N-glycans synthesized by insect cells: a metabolic study with Sf9 cell line. Glycobiology 9: 645–654.
19. Juliant S, Lévêque M, Cérutti P, Ozil A, Choblet S, et al. (2013) Engineering the baculovirus genome to produce galactosylated antibodies in lepidopteran cells. Methods Mol Biol 988: 59–77.
20. Lopez M, Coddeville B, Langridge J, Lemoine J, Plancke H, et al. (1997) Microheterogeneity of the oligosaccharides carried by the N-glycosylation sites of the bovine lactotransferrin expressed in Mamestra brassicae using a baculovirus vector. Glycobiology 7: 635–651.
21. Hooker AD, Green NH, Baines AJ, Bull AT, Jenkins N, et al. (1999) Constraints on the transport and glycosylation of recombinant IFN-gamma in Chinese hamster ovary and insect cells. Biotechnol Bioeng 63: 559–672.
22. Petit D, Mir AM, Petit JM, Thisse C, Delannoy P, et al. (2010) Molecular phylogeny and functional genomics of β-galactoside α2,6-sialyltransferases to explain ubiquitous expression of st6gal1 in Amniotes. J Biol Chem 285: 38399–38414.
23. Marchal I, Jarvis DL, Cacan R, Verbert A (2001). Glycoproteins from insect cells: sialylated or not? Biol Chem 382: 151–159.
24. Altmann F, Kornfeld G, Dalik T, Staudacher E, Glössl J (1993) Processing of the asparagine-linked oligosaccharides in insect cells. N-acetylglucosaminyltransferase I and II activities in cultured lepidopteran cells. Glycobiology 3: 619–625.
25. Velardo MA, Bretthauer RK, Boutaud A, Reinhold B, Reinhold VN, et al. (1993) The presence of UDP-N-acetylglucosamine:alpha-3-D-mannoside beta 1,2-N-acetylglucosaminyltransferase I activity in Spodoptera frugiperda cells (IPLB-SF-21AE) and its enhancement as a results of baculovirus infection. J Biol Chem 268: 17902–17907.
26. Geisler C, Jarvis DL (2012) Substrate specificities and intracellular distributions of three N-glycan processing enzymes functioning at a key branch point in the insect N-glycosylation pathway. J Biol Chem 287: 7084–7097.
27. Altmann F, Schwihla H, Staudacher E, Glössl J, März L (1995) Insect cells contain an unusual, membrane bound (beta)-N-acetylglucosaminidase probably involved in the processing of proteins N-glycans. J Biol Chem 270: 17344–17349.

28. Wagner R, Liedtke S, Kretzschmar E, Geyer H, Klenk HD (1996) Elongation of the N-glycans of the fowl-plague virus hemaglutinin expressed in Spodoptera frugiperda (Sf9) cells by coexpression of the human β1,2-N-acetylglucosaminidase I. Glycobiology 6: 165–175.
29. Fabini G, Freilinger A, Altmann F, Wilson IB (2001) Identification of core alpha 1,3-fucosylated glycans and cloning of the requisite fucosyltransferase cDNA from Drosophila melanogaster. Potential basis of the neural anti-horseadish peroxidase epitope. J Biol Chem 276: 28058–28067.
30. Staudacher E, Kubelka V, März L (1992) Distinct *N*-glycan fucosylation potentials of three lepidopteran cell lines. Eur J Biochem 207: 987–993.
31. Kubelka V, Altmann F, Kornfeld G, März L (1994) Structures of the N-linked oligosaccharides of the membrane glycoproteins from three lepidopteran cell lines (Sf-21, IZD-Mb-0503, Bm-N). Arch Biochem Biophys 308: 148–157.
32. Hancock K, Narang S, Pattabhi S, Yushak ML, Khan A, et al. (2007) False positive reactivity of recombinant, diagnostic, glycoproteins produced in High Five insect cells: Effect of glycosylation. J Immunol Methods 330: 130–136.
33. Voss T, Ergülen E, Ahorn H, Kubelka V, Sugiyama K, et al. (1993) Expression of interferon w1 in *Sf9* cells. No evidence for complex-type *N*-linked glycosylation or sialylation. Eur J Biochem. 217: 913–919.
34. Teylaert B, Meurice E, Bobowski M, Harduin-Lepers A, Gaucher C, et al. (2011) Molecular cloning, characterization, genomic organization and promoter analysis of the α1,6-fucosyltransferase gene (Fut8) expressed in the rat hybridoma cell line YB2/0. BMC Biotechnol 11: 1–18.
35. Uozumi N, Yanagidani S, Miyoshi E, Ihara Y, Sakuma C, et al. (1996) Purification and cDNA cloning of porcine brain GDP-L-Fuc: N-acetyl-β-D-glucosaminide α1,6-fucosyltransferase. J Biol Chem 271: 27810–27817.
36. Yanagidani S, Uozumi N, Ihara Y, Miyoshi E, Yamaguchi N, et al. (1997) Purification and cDNA cloning of GDP-L-Fuc: N-acetyl-beta-D-glucosaminide: alpha1,6-fucosyltransferase (alpha1,6-FucT) from human gastric cancer MKN45 cells. J Biochem 121: 626–632.
37. Hayashi H1, Yoneda A, Asada M, Ikekita M, Imamura T (2000) Molecular cloning of mouse alpha-1,6-fucosyltransferase and expression of its mRNA in the developing cerebrum. DNA Seq 11: 91–6.
38. Paschinger K, Staudacher E, Stemmer U, Fabini G, Wilson IB (2005) Fucosyltransferase substrate specificity and the order of fucosylation in invertebrates. Glycobiology 15: 463–474.
39. Altschul SF, Madden TL, Schäffer AA, Zhang J, Zhang Z, et al. (1997) Gapped BLAST and PSI-BLAST: a new generation of protein database search programs. Nucleic Acids Res 25: 3389–3402.
40. Landais I, Ogliastro M, Mita K, Nohata J, Lopez-Ferber M, et al. (2003) Annotation pattern of ESTs from Spodoptera frugiperda Sf9 cells and analysis of the ribosomal protein genes reveal insect-specific features and unexpectedly low codon usage bias. Bioinformatics 19: 2343–2350.
41. Cavener DR Ray SC (1991) Eukaryotic start and stop translation sites. Nucl Acids Res 19: 3185–3192.
42. Louis A, Muffato M, Roest Crollius H (2013) Genomicus: five genome browsers for comparative genomics in eukaryota. Nucleic Acids Research, 41: D700–D705.
43. Senapathy P, Shapiro MB, Harris NL (1990) Splice junctions, branch point sites and exons: sequence statistics, identification and applications to genome project. Methods in enzymology 183: 252–278.
44. Long M, Deutsch M (1999) Association of intron phases with conservation at splice site sequences and evolution of spliceosomal introns. Mol Biol Evol 16: 1528–1534.
45. Yamaguchi Y, Fujii J, Inoue S, Uozumi N, Yanagidani S, et al. (1999) Mapping of the alpha-1,6-fucosyltransferase gene, FUT8, to human chromosome 14q24.3. Cytogenet Cell Genet 84: 58–60.
46. Harduin-Lepers A, Mollicone R, Delannoy P, Oriol R (2005) The animal sialyltransferases and sialyltransferase-related genes: a phylogenetic approach. Glycobiology 15: 805–817.
47. Harduin-Lepers A (2010) Comprehensive Analysis of Sialyltransferases in Vertebrate Genomes Glycobiology Insights 2: 29–61.
48. Nene V, Wortman JR, Lawson D, Haas B, Kodira C, et al. (2007) Genome sequence of Aedes aegypti, a major arbovirus vector. Science 316: 1718–1723.
49. Gregory TR (2007) Animal genome size data. http://www.genomesize.com.
50. de Souza SJ, Long M, Klein RJ, Roy S, Lin S, Gilbert W (1998) Toward a resolution of the introns early/late debate: only phase zero introns are correlated with the structure of ancient proteins. Proc Natl Acad Sci USA 95: 5094–5099.
51. Krauss V, Pecyna M, Kurz K, Sass H (2005) Phylogenetic mapping of intron positions: a case study of translation initiation factor eIF2gamma. Mol Biol Evol 22: 74–84.
52. Hedges SB, Dudley J, Kumar S (2006) TimeTree: a public knowledge-base of divergence times among organisms. Bioinformatics 22: 2971–2972.
53. Kumar A, Bhandari A, Sinha R, Goyal P, Grapputo A (2011) Spliceosomal intron insertions in genome compacted ray-finned fishes as evident from phylogeny of MC receptors, also supported by a few other GPCRs. PLoS One 6: e22046.
54. Klopfstein S, Ronquist F (2013) Convergent intron gains in hymenopteran elongation factor-1α. Mol Phylogenet Evol 67: 266–276.
55. Li W, Tucker AE, Sung W, Thomas WK, Lynch M (2009) Extensive, recent intron gains in Daphnia populations. Science 326: 1260–1262.
56. Vaughn JL, Goodwin RH, Tompkins GJ, McCawley P (1977) The establishment of two cell lines from the insect Spodoptera frugiperda (Lepidoptera: Noctuidae). In Vitro 13: 213–217.
57. Croizier G, Croizier L, Quiot JM, Lereclus D (1988) Recombinaison of Autographa californica and Rachiplusia ou nuclear polyhedrosis viruses in Galleria mellonella L. J Gen Virol 69: 177–185.
58. Summers MD, Smith GE (1987) A manual of methods for baculovirus vectors and insect cell culture procedures. Tex Agr Exp Stat Bull N°1555.
59. Chaabihi H, Ogliastro MH, Martin M, Giraud C, Devauchelle G, et al. (1993) Competition between baculovirus polyhedrin and p10 gene expression during infection of insect cells. J Virol 67: 2664–2671.
60. Saiki RK, Scharf S, Faloona F, Mullis KB, Horn GT, et al. (1985) Enzymatic amplification of beta-globin genomic sequences and restriction site analysis for diagnosis of sickle cell anemia. Science 230: 1350–1354.
61. O'Reilly DR, Miller LK (1989) A baculovirus blocks insect molting by producing ecdysteroid UDP-glucosyl transferase. Science 245: 1110–1112.
62. Marchal I, Cérutti M, Mir AM, Juliant S, Devauchelle G, et al. (2001) Expression of a membrane-bound form of Trypanosoma cruzei trans-sialidase in baculovirus-infected cells: a potential tool for sialylation of glycoproteins produced in the baculovirus-insect cells system. Glycobiology 11: 593–603.
63. Felgner PL, Ringold GM (1989) Cationic liposome-mediated transfection. Nature 337, 387–388.
64. Petit D, Teppa RE, Petit JM, Harduin-Lepers A (2013) A practical approach to reconstruct evolutionary history of animal sialyltransferases and gain insights into the sequence-function relationships of Golgi-glycosyltransferases. Methods Mol Biol 1022: 73–97.
65. Marck C (1988) 'DNA Strider': a 'C' program for the fast analysis of DNA and protein sequences on the Apple Macintosh family of computers. Nucleic Acids Res 16: 1829–1836.
66. Castresana J (2000) Selection of conserved blocks from multiple alignments for their use in phylogenetic analysis. Molecular Biology and Evolution 17: 540–552.
67. Tamura K, Peterson D, Peterson N, Stecher G, Nei M, et al. (2011) MEGA5: molecular evolutionary genetics analysis using maximum likelihood, evolutionary distance, and maximum parsimony methods. Molecular Biology and Evolution 28: 2731–2739.
68. Saitou N, Nei M (1987) The neighbor-joining method: a new method for reconstructing phylogenetic trees. Mol Biol Evol 4: 406–425.
69. Nei M, Kumar S (1992) Molecular Evolution and Phylogenetics. Oxford University Press. 2000.
70. Crooks GE, Hon G, Chandonia JM, Brenner SE (2004) WebLogo: a sequence logo generator. Genome Res 14: 1188–1190.
71. Schneider TD, Stephens RR (1990) Sequence Logos: A New Way to Display Consensus Sequences. Nucl Ac Res 18: 6097–6100.
72. Whelan S, Goldman N (2001) A general empirical model of protein evolution derived from multiple protein families using a maximum-likelihood approach. Molecular Biology and Evolution 18: 691–699.
73. Felsenstein J (1985) Confidence limits on phylogenies: An approach using the bootstrap. Evolution 39: 783–791.

9

Genome-Wide Investigation and Expression Profiling of AP2/ERF Transcription Factor Superfamily in Foxtail Millet (*Setaria italica L.*)

Charu Lata[1,2], Awdhesh Kumar Mishra[3], Mehanathan Muthamilarasan[3], Venkata Suresh Bonthala[3], Yusuf Khan[3], Manoj Prasad[3]*

1 National Research Centre on Plant Biotechnology, New Delhi, India, **2** CSIR-National Botanical Research Institute, Lucknow, Uttar Pradesh, India, **3** National Institute of Plant Genome Research, New Delhi, India

Abstract

The APETALA2/ethylene-responsive element binding factor (AP2/ERF) family is one of the largest transcription factor (TF) families in plants that includes four major sub-families, namely AP2, DREB (dehydration responsive element binding), ERF (ethylene responsive factors) and RAV (Related to ABI3/VP). AP2/ERFs are known to play significant roles in various plant processes including growth and development and biotic and abiotic stress responses. Considering this, a comprehensive genome-wide study was conducted in foxtail millet (*Setaria italica* L.). A total of 171 *AP2/ERF* genes were identified by systematic sequence analysis and were physically mapped onto nine chromosomes. Phylogenetic analysis grouped *AP2/ERF* genes into six classes (I to VI). Duplication analysis revealed that 12 (~7%) *SiAP2/ERF* genes were tandem repeated and 22 (~13%) were segmentally duplicated. Comparative physical mapping between foxtail millet *AP2/ERF* genes and its orthologs of sorghum (18 genes), maize (14 genes), rice (9 genes) and *Brachypodium* (6 genes) showed the evolutionary insights of *AP2/ERF* gene family and also the decrease in orthology with increase in phylogenetic distance. The evolutionary significance in terms of gene-duplication and divergence was analyzed by estimating synonymous and non-synonymous substitution rates. Expression profiling of candidate *AP2/ERF* genes against drought, salt and phytohormones revealed insights into their precise and/or overlapping expression patterns which could be responsible for their functional divergence in foxtail millet. The study showed that the genes *SiAP2/ERF-069, SiAP2/ERF-103* and *SiAP2/ERF-120* may be considered as potential candidate genes for further functional validation as well for utilization in crop improvement programs for stress resistance since these genes were up-regulated under drought and salinity stresses in ABA dependent manner. Altogether the present study provides new insights into evolution, divergence and systematic functional analysis of *AP2/ERF* gene family at genome level in foxtail millet which may be utilized for improving stress adaptation and tolerance in millets, cereals and bioenergy grasses.

Editor: Swarup K. Parida, National Institute of Plant Genome Research (NIPGR), India

Funding: This work was financially supported by core grant of National Institute of Plant Genome Research, New Delhi, India, and Department of Science & Technology, Government of India through INSPIRE Faculty Award [IFA-11LSPA-01]. CL is the recipient of INSPIRE Faculty Award from Department of Science & Technology, while AKM and MM are the recipients of Research Fellowships from Council of Scientific and Industrial Research and University Grants Commission, New Delhi, respectively. The funders had no role in study design, data collection and analysis, decision to publish, or preparation of the manuscript.

* Email: manoj_prasad@nipgr.ac.in

Introduction

Plants frequently confront numerous environmental stresses which ultimately affect their growth and productivity. Therefore, in order to cope with these recurrent challenges, a plant species must acquire stress responsive and adaptive mechanisms that may assist in better survival and yield. Grass species of *Setaria* genus especially *S. italica* (foxtail millet) and *S. viridis* (green foxtail) prove to be excellent examples of stress adaptation and tolerance among graminaceous species [1]. Foxtail millet is a stress tolerant crop having a small genome (~515 Mb; 2n = 2x = 18) with relatively lower repetitive DNA, short life-cycle and inbreeding nature which makes it a perfect model for understanding various biological aspects including architecture, phylogeny and physiology of related Panicoid crops, particularly potential bioenergy grasses which have closely related but relatively composite genomes [2]. Considering its significance as a model system for evolution and biological studies, its genome has recently been sequenced by Beijing Genomics Institute (BGI), China [3] and Joint Genome Institute (JGI), Department of Energy, USA [4] independently [5]. The availability of foxtail millet genome sequence has consequently encouraged plant biology researchers to work towards deciphering its structural and functional genomics that may give new insights for its stress response and adaptation mechanisms and eventually support crop improvement programmes to ensure sustainable food security [6]. However, stress response and adaptation is a complex process as stress may possibly occur at different stages of plant development with different intensities (moderate to severe) and often several stresses may act together, thus increasing the effects manifold. It is hypothesized that plants have evolved an intricate signaling network to survive stress conditions that begins with stress perception, initiation of signal transduction, modulation of stress responsive gene(s) expression and finally its manifestation at cellular and physiological levels. Stress response and adaptation entails differential gene expression which is controlled by specific transcription factors (TFs) that directly regulate majority of downstream multiple stress responsive gene expression in a synchronized manner. Hence TFs are attractive targets for application in plant molecular biology for gene manipulation and crop improvement. Among various TF families, the ethylene responsive TF (ERF) family plays an important role in plant growth and development and also enables them to adapt to changing environmental conditions, and therefore it is important to understand molecular functions of these genes in order to improve plant adaptability and productivity under varied ambiance/environmental changes.

The APETALA2/ethylene-responsive element binding factor (AP2/ERF) superfamily is a large group of TFs which is distinguished by the number of repetitions and the sequence of AP2/ERF DNA-binding domain based on which it is categorized into AP2, ERF and RAV families [7,8]. The AP2/ERF domain was first reported in the *Arabidopsis* homeotic *AP2* gene implicated in floral development [9]. This conserved DNA-binding domain usually consists of 60–70 amino acid residues and is known to interact directly with *cis*-acting elements namely GCC box and/or dehydration responsive element (DRE)/C-repeat element (CRT) present in the promoter regions of downstream target genes [10,11]. The homologous sequences of this domain have been found in homing endonucleases (HNH-endonucleases) of the cyanobacterium *Trichodesmium erythraeum*, the ciliate *Tetrahymena thermophila*, and the viruses Enterobacteria phage RB49 and Bacteriophage Felix 01 and hence it has been postulated that a horizontal transfer of an HNH-AP2 endonuclease from prokaryotes into plants resulted in evolution of the AP2/ERF superfamily [12,13]. Among the threeAP2/ERF families, the members of AP2 family contain two AP2/ERF domains connected by a 25 amino acid linker, whereas the members of ERF subfamily contain a single AP2/ERF domain. The RAV family members contain a single AP2/ERF domain and an additional B3 DNA-binding motif [14]. In addition, the AP2 family is again categorized into AP2 and AINTEGUMENTA (ANT) monophyletic groups in seed plants [15], while the ERF family is further subdivided into ERF and DREB subfamilies [6,7]. The ERF subfamily is characterized by the presence of conserved alanine and aspartic acid at 14th and 19th position respectively in the DNA-binding domain, while the DREB subfamily has valine and glutamine at respective positions [16].

A large number of AP2/ERF TFs have been identified and studied in various plants including *Arabidopsis*, rice, wheat, poplar, barley, castor bean, grape, cucumber, soybean, *Brassica* and *Malus* [8,17–25]. The genome-wide analyses of AP2/ERF TF superfamily have been performed in these crops, both to categorize each family member in an ordered nomenclature system as well as to investigate their expression profiles and chromosomal positions. As already mentioned, two full genome sequences of *Setaria italica* cv. Zhang gu and inbred Yugu 1 are available which have not only provided a useful genomic platform but have also paved pathway for researchers to carry out advanced genetic and genomic studies in this model crop. As AP2/ERF TFs show wide diversity of functions including regulation of several developmental processes such as vegetative and reproductive development, cell proliferation, and responses to various abiotic and biotic stresses and plant hormones [8,10], their superfamily stands as one of the best candidates to examine important traits in foxtail millet. With this aim, a genome wide investigation of foxtail millet AP2/ERF TF superfamily and their expression profiling has been taken up in this study. Hence, this is the first comprehensive report on genome-wide survey, expression profiling and evolutionary analysis of AP2/ERF proteins in foxtail millet (named as *Setaria italica* AP2/ERF; 'SiAP2/ERF').

Materials and Methods

Sequence retrieval and identification of AP2/ERF proteins from *Setaria italica*

The Hidden Markov Model (HMM) profile of the AP2/ERF domain (PF00847) was obtained from Pfam v27.0 database (http://Pfam.sanger.ac.uk/) [26] and searched against the PHYTOZOME database of *Setaria italica* (www.phytozome.net/). All hits with expected (E) values less than 1.0 were retrieved and the non-redundant sequences were examined for the presence of conserved AP2/ERF domain by executing HMMSCAN (http://hmmer.janelia.org/search/hmmscan).

Chromosomal location, gene structure and genomic distribution of *AP2/ERF* genes

The identified AP2/ERF domain-containing proteins were BLASTP searched against *S. italica* genome of PHYTOZOME database with default settings, and physical map was constructed using MapChart [27]. Segmental duplications were calculated based on the method of Plant Genome Duplication Database [28] using MCScan v0.8 [29] and visualized using Circos v0.55 [30]. Tandem duplications were identified manually as described elsewhere [31,32] and marked on the physical map. The exon-intron organizations of the genes were ascertained using Gene Structure Display Server (http://gsds.cbi.pku.edu.cn/) [33].

Phylogenetic analysis, Gene Ontology (GO) annotation, promoter analysis and identification of miRNAs targeting *SiAP2/ERFs*

The amino acid sequences of AP2/ERF proteins were imported into MEGA5 and an unrooted phylogenetic tree based on the Neighbor-joining method was generated after 1000 bootstrap replications [34]. The GO annotation of AP2/ERF protein sequences was performed using Blast2GO [35] and *cis*-regulatory elements were identified using PLACE (http://www.dna.affrc.go.jp/PLACE/) database. Further. the *S. italica* miRNAs reported by Khan et al. [36] were retrieved and searched for their targets in 171 *SiAP2/ERF* transcripts using psRNATarget tool (http://plantgrn.noble.org/psRNATarget/).

Comparative mapping and evolutionary analysis of paralogs and orthologs

The amino acid sequences of SiAP2/ERF proteins that were physically mapped onto foxtail millet genome were BLASTP searched against protein sequences of sorghum, maize, rice and *Brachypodium* (http://gramene.org/; www.phytozome.net). Reciprocal BLAST was also carried out to establish unique relationship between the orthologous genes. Hits with E-value≥ 1e-05 and at least 80% homology were considered significant. The comparative orthologous relationships of *AP2/ERF* genes among foxtail millet, sorghum, maize, rice and *Brachypodium* were finally

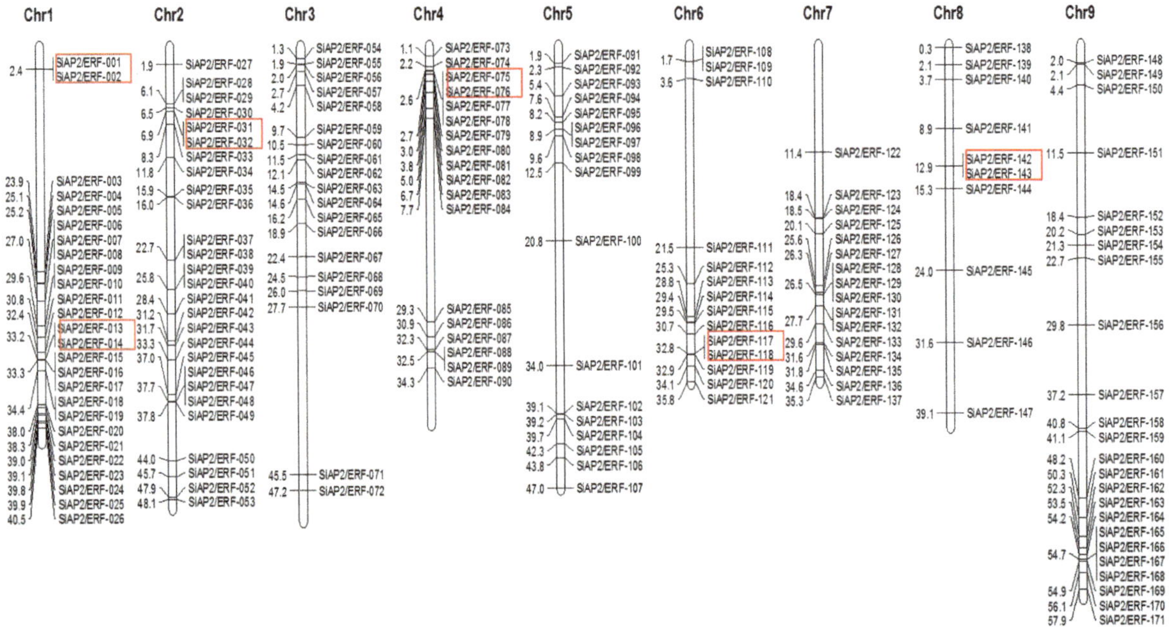

Figure 1. Physical map of 171 *SiAP2/ERF* genes. The bars represent the chromosomes with numbers at the left indicating the physical position (in Mb). The tandemly duplicated gene pairs are indicated within boxes.

illustrated using Circos [30]. For estimating the synonymous (Ks) and non-synonymous (Ka) substitution rates, the corresponding amino-acid as well as cDNA sequences of paralogous and orthologous SiAP2/ERF proteins were analyzed using PAL2NAL (http://www.bork.embl.de/pal2nal/) [37]. Time (million years ago, Mya) of duplication and divergence was calculated using a synonymous mutation rate of 1 substitutions per synonymous site per year as $T = Ks/2\lambda$ ($\lambda = 6.5 \times 10^{-9}$) [38,39].

Tissue-specific expression profiling using RNA-seq data

S. italica Illumina RNA-HiSeq reads from 4 tissues namely spica, stem, leaf and root, retrieved from European Nucleotide Archive [SRX128226 (spica); SRX128225 (stem); SRX128224 (leaf); SRX128223 (root)] [40] and were mapped onto the gene sequences of *Setaria italica* using CLC Genomics Workbench v.4.7.1 (http://www.clcbio.com/genomics). Normalization of the mapped reads was done using RPKM (reads per kilobase per million) method. The heat map for tissue-specific expression profile was generated based on the RPKM values for each gene in all the tissue samples using TIGR MultiExperiment Viewer (MeV4) software package [41,42].

Plant materials, growth conditions and stress treatments

Seeds of drought tolerant foxtail millet cultivar IC-403579 [8] were obtained from National Bureau of Plant Genetic Resources (NBPGR), Hyderabad, India. The seeds were sown in composite soil (peat compost: vermiculite: sand, 2:2:1) in glass house at National Phytotron Facility, Indian Agricultural Research Institute (IARI), New Delhi, India at 28±1°C day/23±1°C night temperature with 70±5% relative humidity and natural sunlight during June–July, 2013. For stress treatments, two week old seedlings were exposed to 20% polyethylene glycol (PEG 6000) (drought), 250 mM NaCl (salt), 100 µM abscisic acid (ABA), 100 µM salicylic acid (SA), 100 µM methyl jasmonate (MeJA) and 100 µM ethephon (Et) for 1 h (early) and 24 h (late) based on our previous studies (Lata et al. 2010; Lata et al. 2011a; Lata et al. 2011b). The plants were supplemented with water and Hoagland solution on alternate days. Unstressed plants were maintained as control. After stress treatments, whole seedlings were carefully harvested and immediately frozen in liquid nitrogen and stored at −80°C until RNA isolation. Three independent experiments were conducted for precision and reproducibility, and for each experiment, ~100 mg seedling samples were collected by random sampling.

RNA extraction and expression analysis using qRT-PCR

Total RNA was isolated from the 14-day old unstressed and stressed (1 h and 24 h) foxtail millet cv. IC-403579 seedlings using TRIzol Reagent (Sigma, USA) following manufacturer's instructions. DNA contamination was removed from the RNA samples using RNase-free DNaseI (1 U µl^{-1}, Fermentas). The quality and purity of the RNA preparations were determined by measuring the OD_{260}/OD_{280} absorption ratio (1.9–2.0), and the integrity of the preparations was determined by electrophoresis in a 1.2% agarose gel containing formaldehyde as described in previous studies [43,44]. RNA concentrations were measured by a spectrophotometer (Nanodrop, USA). About 1 µg of total RNA was used to synthesize first strand cDNA primed with OligodT in a 20 µl reaction mix using 200 U/µl of PrimeScript M-MuLV reverse transcriptase (Takara Bio Inc., USA) following manufacturer's instructions. Quantitative real time (qRT) PCR was performed using SYBR Premix ExTaq II (Tli RNaseH Plus) (Takara Bio Inc., USA) on Mastercycler ep realplex system (Eppendorf) in triplicate. The constitutive gene RNA Polymerase II (RNA POL II; Accession No Si033113m) from foxtail millet was used as endogenous control which gave an amplification product of 146 bp [45]. The qRT-PCR primers of the *SiAP2/ERF* genes were designed from non-conserved regions of the corresponding genes using GenScript real-time PCR (TaqMan) Primer Design tool (www.genscript.com) using default parameters (Table S1). The PCR mixtures and reactions were used as detailed previously [45]. Melting curve analysis (60 to 95°C after 40 cycles) and

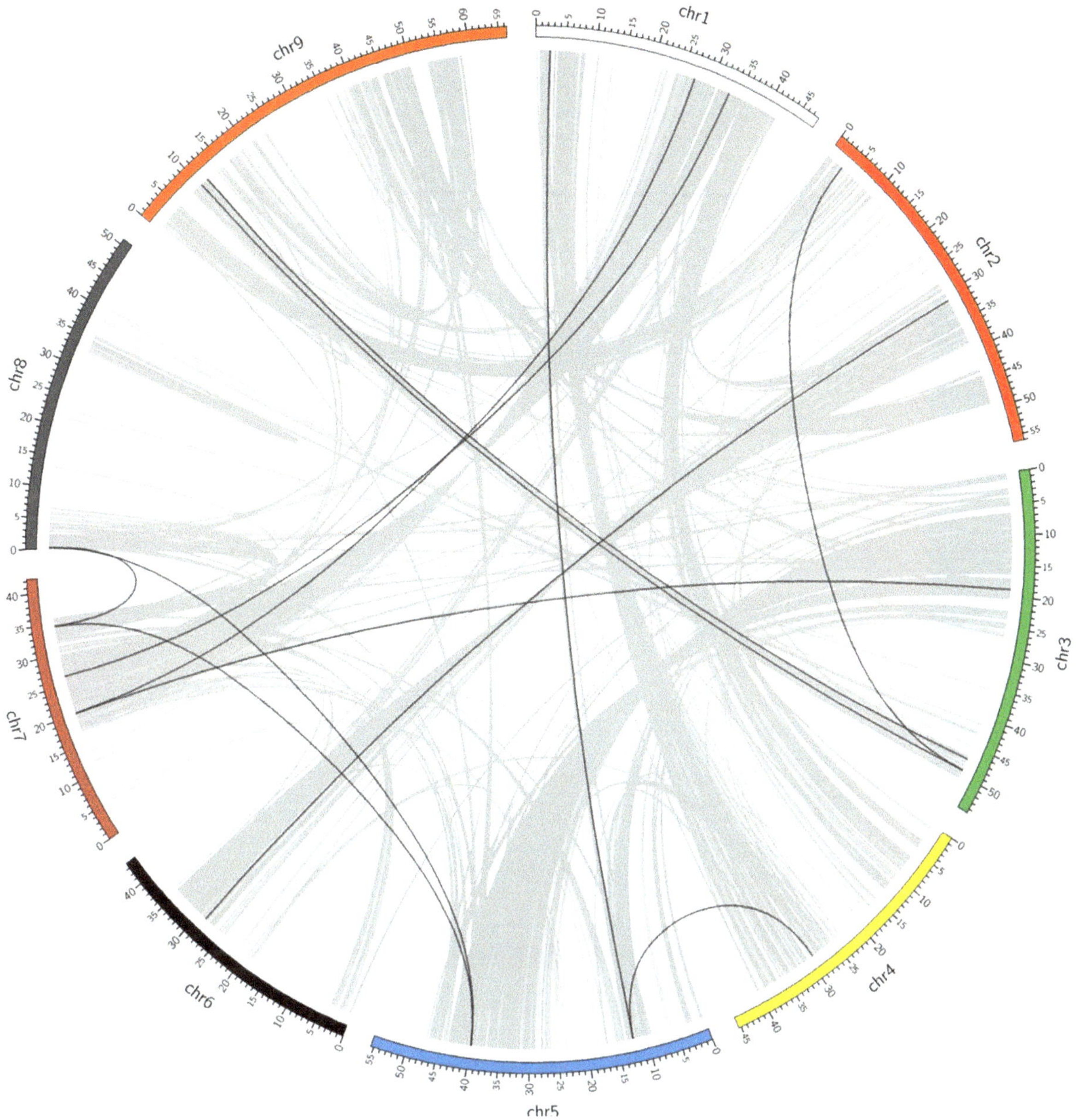

Figure 2. Distribution of segmentally duplicated *SiAP2/ERF* genes on foxtail millet chromosomes. Grey lines indicate collinear blocks in whole foxtail millet genome, and black lines indicate duplicated *SiAP2/ERF* gene pairs.

agarose gel electrophoresis were performed to check the amplification specificity of *AP2/ERF* genes normalized to the internal control RNA POL II and were analyzed using $2^{-\Delta\Delta C_t}$ method [46]. qRT-PCR data analysis was done according to previous studies [45,46]. The PCR cycling conditions were: initial denaturation at 95°C for 2 min, 95°C for 15 s, and 60°C for 1 min for 40 cycles followed by melting curve analysis using default parameters i.e. 95°C for 15 s, 60°C for 15 s, 95°C for 15 s with ramp time of 20 min.

Identification of molecular markers and homology modeling of SiAP2/ERF proteins

The presence of various types of DNA-based markers including simple sequence repeats (SSRs) [47], EST-derived SSRs (eSSRs) [48] and intron length polymorphic (ILPs) markers [49] retrived from FmMDb (http://www.nipgr.res.in/foxtail.html) [50] were searched in the *SiAP2/ERF* genes using in-house perl script. For homology modeling, all the SiAP2/ERF proteins were queried against the Protein Data Bank (PDB) [51] to identify the best template with similar amino acid sequence and known 3D structure. The data was fed in Phyre2 server (Protein Homology/AnalogY Recognition Engine; http://www.sbg.bio.ic.ac.uk/phyre2) for predicting the three-dimensional structure of proteins by homology modeling under 'normal' mode [52]. Active site was predicted using COACH server (http://zhanglab.ccmb.med.umich.edu/COACH/) and highlighted using UCSF Chimera 1.8.

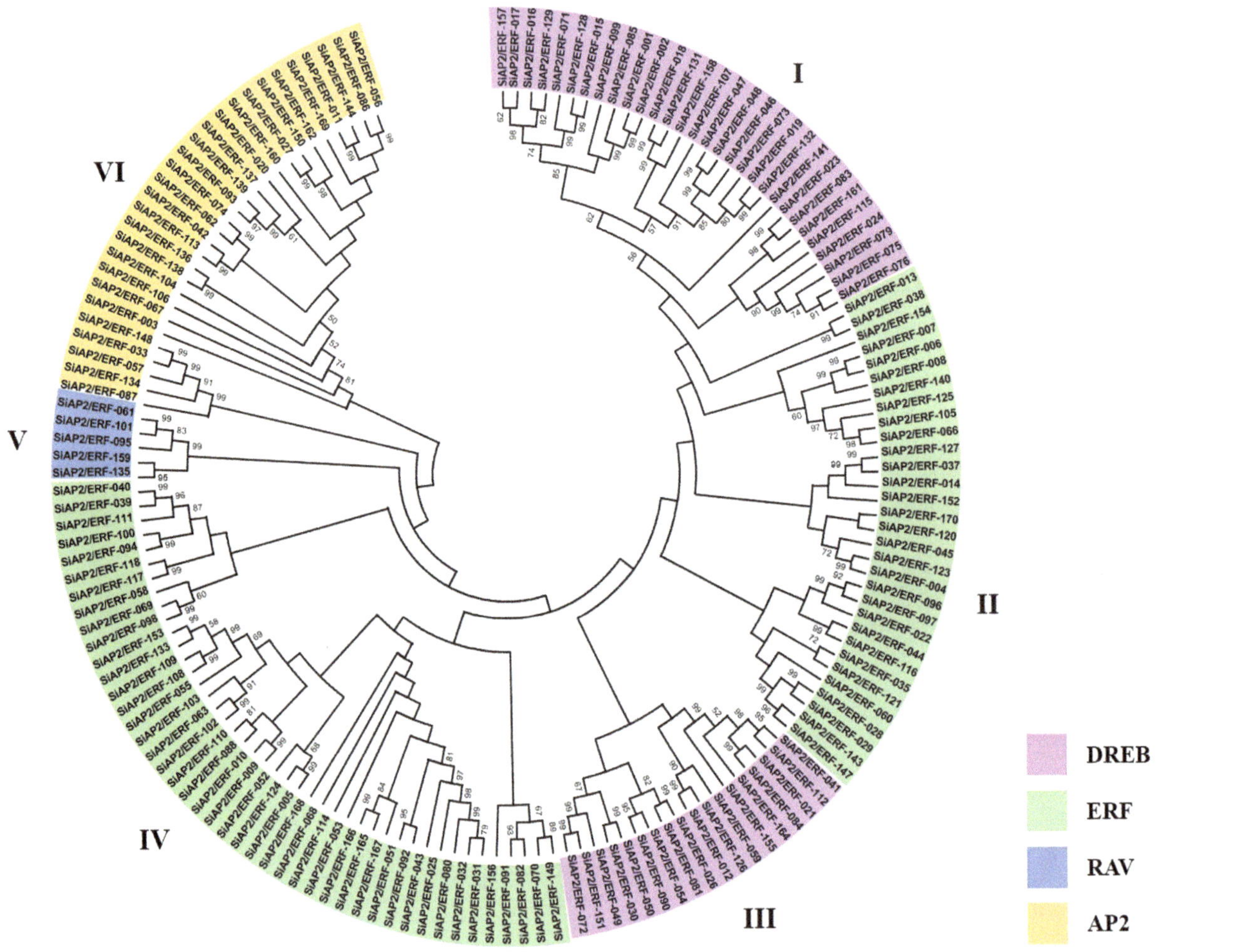

Figure 3. Phylogenetic tree of 171 *SiAP2/ERF* genes. Bootstrap confidence values from 1000 replicates are indicated at each branch. The different classes of SiAP2/ERF proteins are highlighted in different colors.

Results and Discussion

Identification of the AP2/ERF family transcription factors in foxtail millet genome

The HMM BLAST identified a total of 186 AP2/ERF protein sequences from foxtail millet. Fifteen proteins were found to be splice variants of primary transcripts, removal of which led to the identification of a total of 171 putative SiAP2/ERF proteins (Table S2) which represents approximately 0.4407% of all annotated genes (38801 genes total) in the *Setaria* genome [3] which is very similar to those present in poplar (0.4390%) and rice (0.4315%) however approximately 0.10% smaller than that of Arabidopsis (0.5481%) [7,8,18,53,54]. Among splice variants, Si022619m gene was found to encode a maximum of 8 alternate

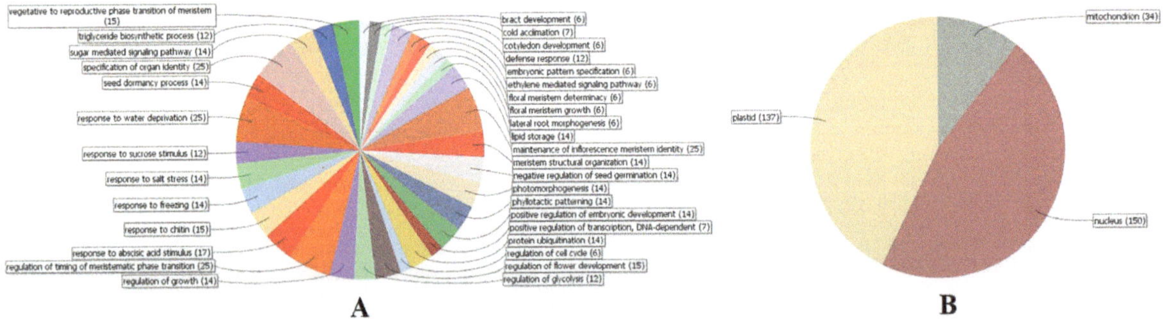

Figure 4. Gene Ontology annotation of SiAP2/ERF proteins. The Blast2GO output defining the (**A**) biological processes and (**B**) cellular localization of 171 SiAP2/ERF proteins.

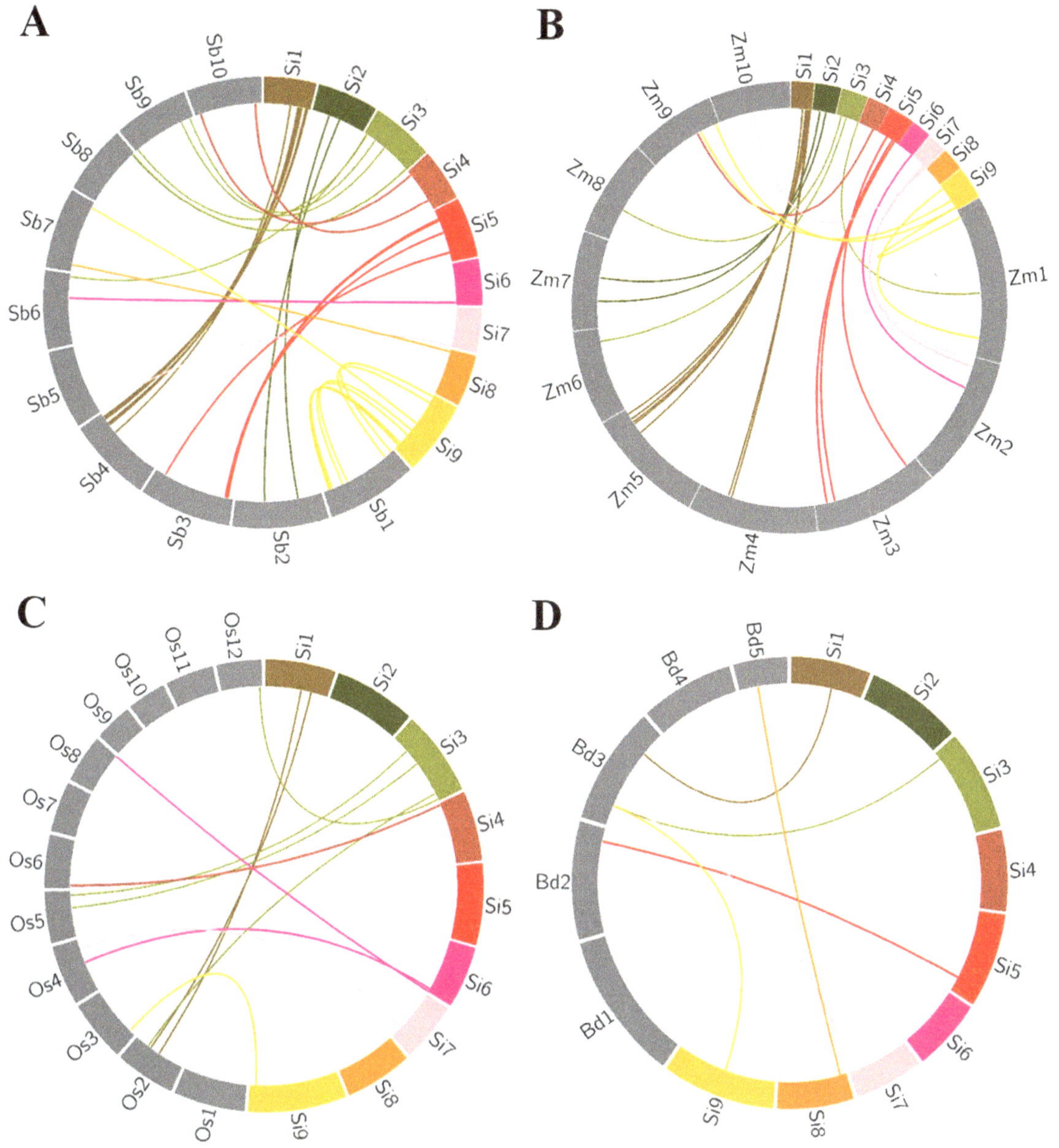

Figure 5. Comparative physical mapping showing the degree of orthologous relationships of *SiAP2/ERF* genes located on nine chromosomes of foxtail millet with (A) sorghum, (B) maize, (C) rice and (D) *Brachypodium*.

transcripts (Si022998m, Si022997m, Si022990m, Si022995m, Si022996m, Si022991m, Si022621m, Si022989m), followed by Si006802m and Si010289m which has 2 splice variants each (Si006941m, Si006905m in Si006802m and Si010292m, Si010301m in Si010289m) (Table S2). Genes Si030514m, Si036615m and Si036647m comprised of one alternate transcript Si030506m, Si036606m and Si036938m, respectively. The number conforms with the number of AP2/ERFs reported in Foxtail millet Transcription Factor Database [55]. Of note, the identification of pseudogenes among these 171 AP2/ERFs require further experimentations. In addition, the respective gene sequences encoding these proteins were retrieved and the presence of AP2/ERF domain was ascertained (Table S3). Due to lack of proper annotation, the existing identities of the genes were highly disordered and therefore for convenience, all 171 genes were assigned consecutive numbers from SiAP2/ERF-001 to SiAP2/ERF-171 in the order of their chromosomal locations. All *SiAP2/ERF* genes varied greatly in the size and sequence of their encoded proteins as well as in their physico-chemical properties (Table S2). Additionally all *SiAP2/ERF* genes were characterized by the presence of one or two highly conserved AP2/ERF DNA-binding domains and a B3 domain in case of RAV proteins (Table S3). The lengths of the identified proteins vary from 88 to 691 amino acids. ExPASy analysis revealed large variation in iso-electric point (pI) values ranging from 4.26 to 11.7 and molecular weight ranging from 9.92 to 72.31 kDa. Interestingly, this wide variation in pI and molecular weight revealed the presence of putative novel variants of SiAP2/ERFs and this is in accordance to previous genome-wide reports on DCL, AGO, RDR, C_2H_2 zinc finger and MYB genes in foxtail millet [56–58]. The characteristic features of SiAP2/ERF protein sequences are summarized in Table S2. Dual targeting is a term used to infer the ability of proteins to localize

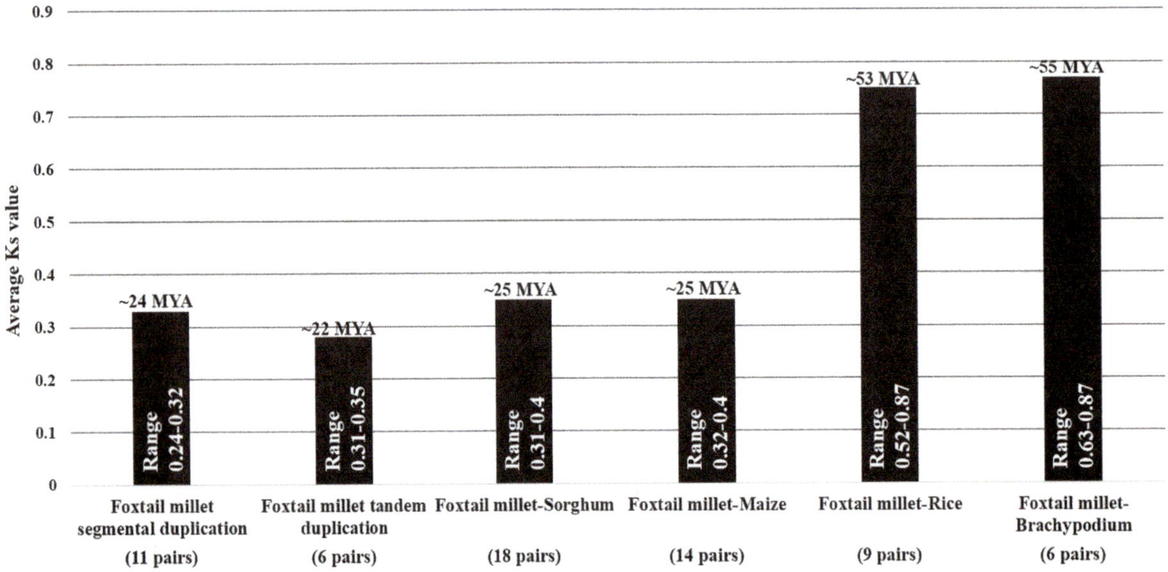

Figure 6. Time of duplication and divergence (MYA) based on synonymous substitution rate (Ks) estimated using paralogous and orthologous *SiAP2/ERF* gene pairs.

into more than one cellular compartment and it can also be viewed as a post translational regulatory mechanism [59]. Localization of 171 AP2/ERF proteins was determined using Blast2GO. Majority of the SiAP2/ERF proteins were predicted to be dual targeted or localized to nucleus, plastid and/or mitochondrion except SiAP2/ERF-025, SiAP2/ERF-032,

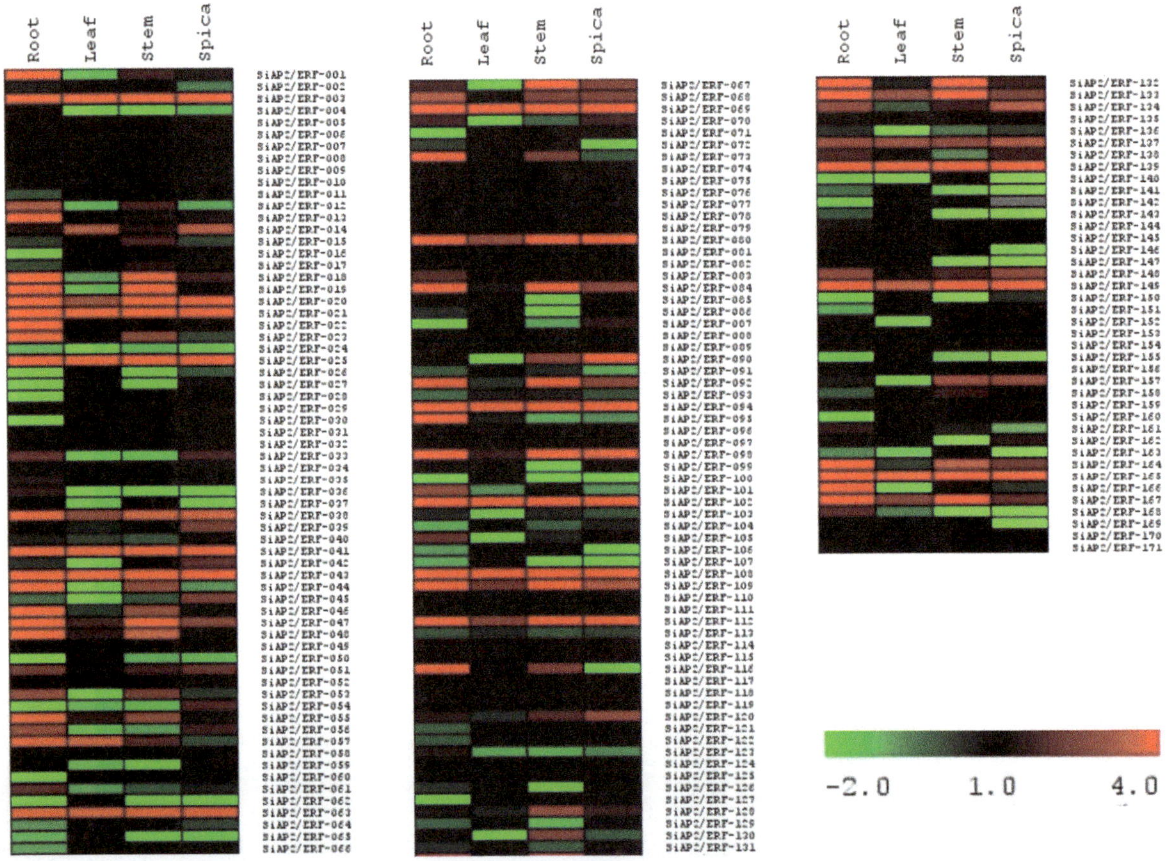

Figure 7. Heat-map showing the expression pattern of *SiAP2/ERF* genes in four tissues namely leaf, root, stem and spica. The color scales for fold-change values are shown at the bottom right.

SiAP2/ERF-035, SiAP2/ERF-040, SiAP2/ERF-051, SiAP2/ERF-055, SiAP2/ERF-063, SiAP2/ERF-065, SiAP2/ERF-091, SiAP2/ERF-100, SiAP2/ERF-108, SiAP2/ERF-121, SiAP2/ERF-122, SiAP2/ERF-153, SiAP2/ERF-165, SiAP2/ERF-166 (nucleus); SiAP2/ERF-024, SiAP2/ERF-059, SiAP2/ERF-066, SiAP2/ERF-078, SiAP2/ERF083 and SiAP2/ERF096 (plastid); and SiAP2/ERF-034, SiAP2/ERF-075, SiAP2/ERF-077, SiAP2/ERF-079, SiAP2/ERF-111 and SiAP2/ERF-159 (intracellular membrane-bound organelle) localized. Further, AP2/ERF superfamily was divided into four major families on the basis of nature and number of DNA-binding domains, namely AP2, ERF, DREB and RAV. The AP2/ERF proteins of *Setaria italica* were also classified into these families. Out of 171 genes, 28 belong to AP2, 90 to ERF, 48 to DREB and 5 to RAV (Table S3) indicating that foxtail millet genome supports large ERF and DREB subfamilies similar to Chinese cabbage genome [60].

Chromosomal distribution and structure of SiAP2/ERF proteins

Physical mapping of *SiAP2/ERF*s on all 9 chromosomes of foxtail millet revealed an uneven distribution of *SiAP2/ERF* genes in the genome (Figure 1). Among all chromosomes, chromosome 2 contained highest number of *AP2/ERF* genes (27; ~16%) followed by chromosome 1 (26; ~15%), while minimum number of genes were assigned on chromosome 8 (10; ~6%). The precise position (in bp) of each *SiAP2/ERF* on foxtail millet chromosomes is detailed in Table S2. Distribution pattern of these genes on individual chromosomes also pointed certain physical regions with a relatively higher accumulation of *AP2/ERF* gene clusters. As for example, *SiAP2/ERF* genes located on chromosomes 1, 6, 7 and 9 appear to congregate at the lower end of the arms as compared to chromosome 3 and 4 where these genes appear to cluster together at the upper end of the arm (Figure 1). It has recently been reported that foxtail millet genome underwent whole-genome duplication similar to other grass species about 70 million years ago [3] and hence occurrence of such large number of *SiAP2/ERF* genes in foxtail millet genome suggests huge amplification of this gene family during the course of evolution. Hence, duplication of these genes was studied and found that 12 (~7%) *SiAP2/ERF* genes were tandem repeated (Figure 1) and 22 (11 pairs; ~13%) were segmentally duplicated (Figure 2). The tandem duplicated genes included six clusters (2 genes each) including two clusters on chromosome 1 and one each on chromosomes 2, 4, 6 and 8. The distance between these genes ranged from 6.2 kb to 32.2 kb. Among the segmentally duplicated gene pairs, three genes namely, SiAP2/ERF-001 (chromosome 1), SiAP2/ERF-013 (chromosome 1) and SiAP2/ERF-072 (chromosome 5) were duplicated twice in the genome forming six paralogs in chromosomes 2, 4, 5, 7 and 9 (Figure 2). Chromosomal localization study of *SiAP2/ERF* genes thus indicates that tandem- and segmental-duplication may be one of the contributing factors in evolution of new genes in foxtail millet genome. Moreover, analysis of *SiAP2/ERF* gene structures indicated highly diverse distribution of intronic regions (from 1 to 10 in numbers) among the exonic sequences suggesting significant evolutionary changes in the foxtail millet genome. Interestingly, 89 (~52%) *SiAP2/ERF* genes were found to be intronless (Figure S1). Similar results were also observed in case of *Arabidopsis* [7] and *Lotus corniculatus* [61]. Further, the shortest *SiAP2/ERF* gene was merely 266 bp (SiAP2/ERF-156), whereas the longest one was identified as SiAP2/ERF-150 with 3.7 kb genomic sequence.

Phylogenetic analysis of SiAP2/ERF proteins

Phylogenetic analysis is essential for understanding the evolutionary history of crop species. Therefore, to understand the evolutionary significance of domain architecture, a phylogenetic tree was constructed with 171 SiAP2/ERF proteins. The phylogenetic analysis clustered all the SiAP2/ERFs into distinct clades (AP2, ERF, DREB and RAV) comprising of 28, 90, 48, and 5 proteins, respectively, according to their domain composition (Figure 3). Interestingly, the DREB formed two clades intervened

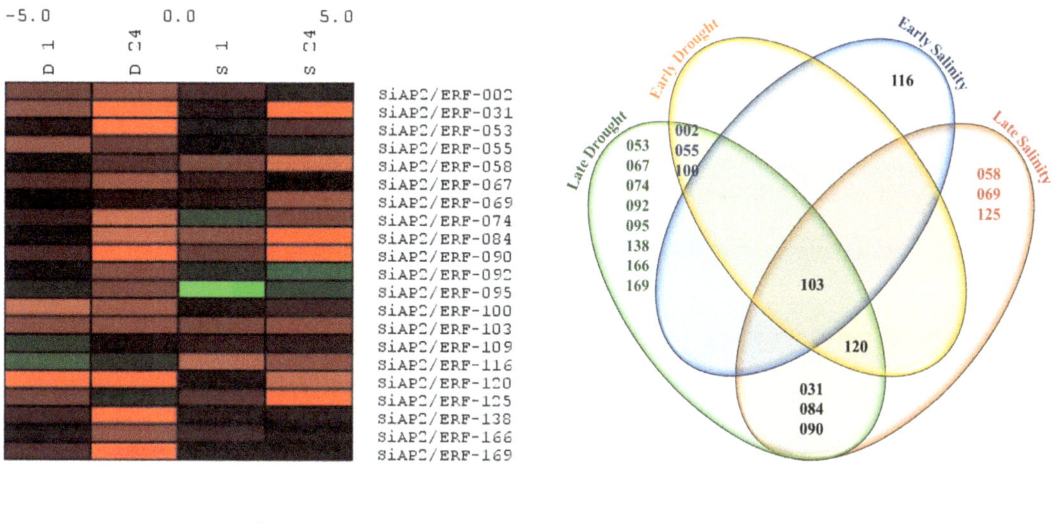

A B

Figure 8. Expression profile of 21 *SiAP2/ERF* genes in response to dehydration and salinity stresses. (**A**) Heat map showing differential gene expression in response to dehydration (D) and salinity (S) stress across two time points (1 h and 24 h). The heat-map has been generated based on the fold-change values in the treated sample when compared with its unstressed control sample. The color scale for fold-change values is shown at the top. (**B**) Venn diagram showing stress-specific higher-expression of *SiAP2/ERF* genes during early and late stresses. The common subset of genes regulated by two stresses is marked by the overlapping circle. The numbers provided in the venn diagram corresponds to the *SiAP2/ERF* ID listed in Table S2.

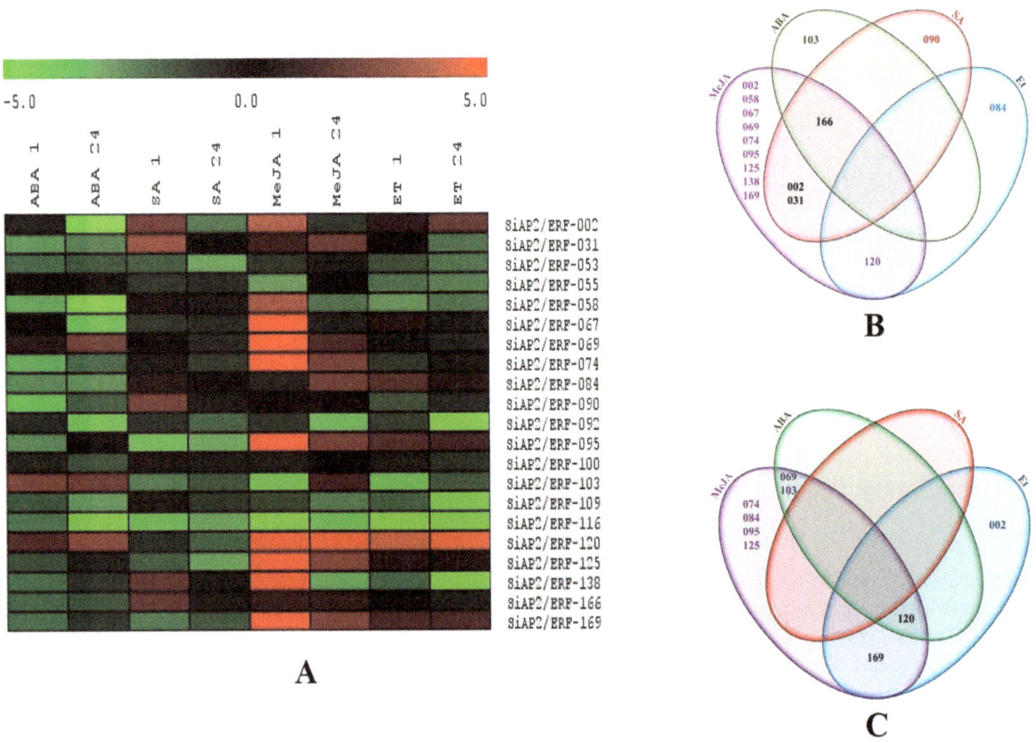

Figure 9. Expression profile of 21 *SiAP2/ERF* genes in response to Abscisic acid (ABA), salicylic acid (SA), methyl jasmonate (MeJA) and Ethephone (Et) treatments. (**A**) Heat map showing differential gene expression in response to ABA, SA, MeJA, Et treatments across two time points (1 h and 24 h). The heat-map has been generated based on the fold-change values in the treated sample when compared with its treated control sample. The color scale for fold-change values is shown at the top. (**B**) Venn diagram showing stress-specific higher-expression of *SiAP2/ERF* genes during early hormonal treatment. (**C**) Venn diagram showing stress-specific higher-expression of *SiAP2/ERF* genes during early hormonal treatment. The common subset of genes regulated by the hormonal treatments is marked by the overlapping circle. The numbers provided in the venn diagram corresponds to the *SiAP2/ERF* ID listed in Table S2.

by ERFs. Although similar observation was not reported in genome-wide studies of AP2/ERF conducted in other plants, in Chinese cabbage it has been found that the DREB clade was intervened by AP2 [60]. Further in-depth *in silico* analysis is requisite for finding the possible reason for such observation. The tree was divided into six groups based on the distribution of AP2, ERF, DREB and RAV. The derivation of statistically reliable pairs of possible homologous proteins sharing similar functions from a common ancestor was confirmed owing to high bootstrap values observed for a good number of the internal branches of the unrooted phylogenetic tree. Close association of SiAP2/ERF sub-families with their corresponding counterparts in other plant systems in terms of expression and/or biological and regulatory functions may be an implication of sequence conservation and also evidence to their similar *in planta* roles. Such phylogeny-based function prediction is obviously a rational systematic approach to facilitate identification of orthologous genes and has near-perfectly been applied for prediction of AP2/ERF proteins in other plant species such as rice, grapes and *Brassica* [20,21,60]. Thus, members of the sub-families of AP2/ERF superfamily in foxtail millet also have similar regulatory roles as those of their orthologs in other crop species.

Gene Ontology annotation

The GO analysis performed using rice protein sequences as reference showed the putative participation of SiAP2/ERF proteins in diverse biological, cellular and molecular processes (Figure 4; Table S4). The analysis of biological processes mediated by SiAP2/ERF depicted that a predominant of SiAP2/ERF proteins were involved in stress responses, such as response to water deprivation, salt stress and freezing. In addition, SiAP2/ERF proteins were also evidenced to participate in regulation of timing of meristematic phase transition, specification of organ identity and maintenance of inflorescence meristem identity. The molecular processes of SiAP2/ERF proteins clearly showed that all the 171 proteins possess sequence-specific DNA binding transcription factor activity (Table S4). Further, cellular component analysis revealed the localization of SiAP2/ERF proteins in nucleus, plastids, mitochondria and other intracellular membrane-bound organelles. These are in concordance with the experimental findings reported earlier [8,10,61,62].

Promoter analysis and miRNA targets of *SiAP2/ERF* genes

Cis-regulatory elements are DNA sequences that are situated upstream of genes in the promoter region and act as TF-binding sites. These are known to play crucial roles in determining tissue-specific as well as stress-responsive gene expression [63]. There have also been reports that multi-stimuli genes are closely correlated with *cis*-regulatory elements in their promoter sequences [64]. Therefore, a comprehensive promoter analysis of all the 171 *SiAP2/ERF* genes was conducted in order to further understand transcriptional regulation and support functional prediction of the respective proteins (Table S5). A total of 300 cis-regulatory elements were found to be present in one or the other *SiAP2/ERF* gene. The cis-regulatory elements CACTFTPPCA1, CAATBOX1, EBOXBNNAPA, MYCCON-

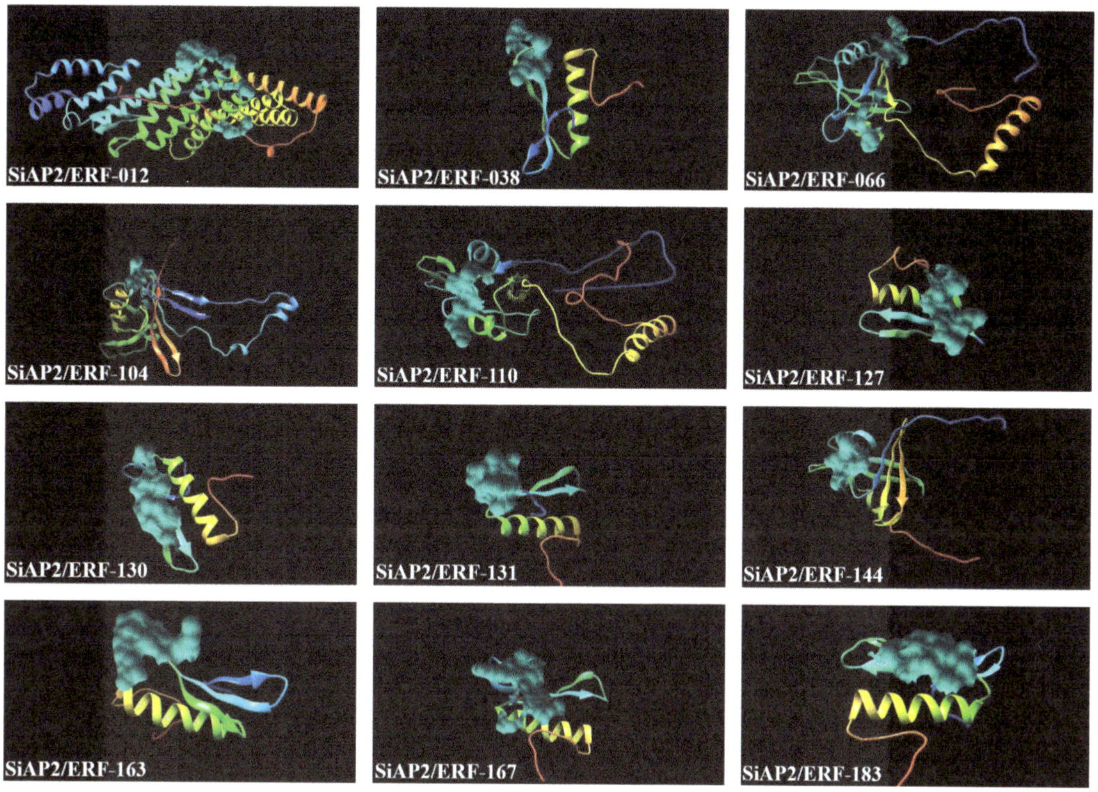

Figure 10. Predicated structures of SiAP2/ERF proteins. The structures of 12 SiAP2/ERF proteins with greater than 90% confidence level were shown along with its potential active site.

SENSUSAT, WRKY71OS, GT1CONSENSUS, ARR1AT, DOFCOREZM, GTGANTG10, RAV1AAT and GATABOX were present in all the 171 genes whereas HSE, VSF1PVGRP18, GMHDLGMVSPB, ABREAZMRAB28, PALBOXLPC, DR5GMGH3, SITEIIAOSPCNA, ABREBNNAPA, ABASEED1, ABAREG2, O2F3BE2S1, OPAQUE2ZM22Z, RGATAOS, CEREGLUBOX3PSLEGA, CPRFPCCHS, LREBOXIIPCCHS1, AGL2ATCONSENSUS, TOPOISOM, NONAMERATH4, PALBOXPPC, WRECSAA01, SORLIP4AT, NONAMERMOTIFTAH3H4, ABREMOTIFIOSRAB16B, ACIIIPVPAL2, ABRE3OSRAB16, B2GMAUX28, AT1BOX, SPHZMC1, D3GMAUX28, INTRONUPPER, WINPSTPIIIK, JASE1ATOPR1, OCETYPEIINTHISTONE and MSACRCYM were unique to *SiAP2/ERF-020, SiAP2/ERF-023, SiAP2/ERF-024, SiAP2/ERF-026, SiAP2/ERF-029, SiAP2/ERF-036, SiAP2/ERF-038, SiAP2/ERF-059, SiAP2/ERF-063, SiAP2/ERF-066, SiAP2/ERF-067, SiAP2/ERF-069, SiAP2/ERF-071, SiAP2/ERF-072, SiAP2/ERF-073, SiAP2/ERF-073, SiAP2/ERF-082, SiAP2/ERF-083, SiAP2/ERF-091, SiAP2/ERF-091, SiAP2/ERF-091, SiAP2/ERF-093, SiAP2/ERF-094, SiAP2/ERF-099, SiAP2/ERF-110, SiAP2/ERF-112, SiAP2/ERF-128, SiAP2/ERF-132, SiAP2/ERF-140, SiAP2/ERF-152, SiAP2/ERF-157, SiAP2/ERF-159, SiAP2/ERF-166, SiAP2/ERF-169 and SiAP2/ERF-170*, respectively (Table S5). In addition, putative microRNAs (miRNAs) targeting the *SiAP2/ERF* genes were also detected using psRNATarget server. The analysis showed five *SiAP2/ERF* genes to be targeted by *Setaria italica* miRNAs (Sit-miRs) (Table S6; Figure S2). The miRNAs identified in this study would be helpful in interpreting the post-transcriptional control of gene regulation during various physiological and stress-induced cellular responses in this otherwise naturally stress tolerant crop.

Orthologous relationships of *SiAP2/ERF* genes between foxtail millet and other grass species

To derive orthologous relationships of SiAP2/ERFs, comparative mapping approach was followed wherein the physically mapped *AP2/ERF* genes of foxtail millet were compared with those in the chromosomes of related grass genomes namely sorghum, maize, rice and *Brachypodium* (Figure 5). Maximum orthology of genes annotated on the foxtail millet chromosomes was exhibited with sorghum (18; ~11%) followed by maize (14; ~8%), rice (9; 5%) and least with *Brachypodium* (6; 4%). The extensive synteny among foxtail millet, sorghum and maize at gene level supports their close evolutionary relationships [3,4]. Intriguingly, most of *SiAP2/ERF* genes showed syntenic bias towards particular chromosomes of sorghum, maize, rice and *Brachypodium* and this suggests that the chromosomal rearrangement events like duplication and inversion predominantly shaped the distribution and organization of *AP2/ERF* genes in these grass genomes. The comparative mapping information thus offers a useful preface for understanding the evolution of *AP2/ERF* genes among grasses including foxtail millet. In addition, this study would be helpful in selecting candidate *SiAP2/ERF* genes and utilize them in genetic improvement of related grass family members. As for example, AP2-like ethylene-responsive transcription factors PLETHORA 1 and 2 are essential for QC specification and stem cell activity in roots of Arabidopsis [65]. It is thus likely that its orthologous foxtail millet gene (SiAP2/ERF-011; Phytozome ID: Si016558m) and rice gene (LOC_Os02g40070.1) may also be involved in similar function.

Duplication and divergence rate of the *SiAP2/ERF* genes

Whole genome duplications such as tandem and segmental duplications usually give rise to multiple copies of genes in a gene family. Such gene duplication events have been reported in various plant TF families including MYB, NAC and AP2/ERF [21,56–58,66–68]. Association of Darwinian positive selection in duplication and divergence of *AP2/ERF* in foxtail millet was explored by estimating the ratios of non-synonymous (Ka) versus synonymous (Ks) substitution rate (Ka/Ks) for 6 tandem and 11 segmentally duplicated gene-pairs as well as between orthologous gene-pairs of *SiAP2/ERF* with those of sorghum (18 pairs), maize (14 pairs), rice (9 pairs) and *Brachypodium* (6 pairs) (Figure 6) (Tables S7–S12). The Ka and Ks are a measure to examine the course of divergence after duplication, and the Ka/Ks ratio is a measure of the selection pressure to which a gene pair is subjected wherein Ka/Ks <1 means purifying selection, Ka/Ks = 1 stands for neutral selection, and Ka/Ks >1 signifies accelerated evolution with positive selection [38]. The Ka/Ks ratio for tandem duplicated gene-pairs in foxtail millet AP2/ERF genes ranged from 0.10 to 0.15 with an average of 0.13, whereas Ka/Ks for segmentally duplicated gene-pairs ranged from 0.03 to 0.14 with an average of 0.09 (Tables S7–S8). The data indicated that the duplicated *SiAP2/ERF* genes were under strong purifying selection pressure and had gone through substitution elimination and enormous selective constraint by natural selection during the course of evolution since their Ka/Ks ratios estimated as <1. Further, the duplication event of these *SiAP2/ERF* tandemly and segmentally duplicated genes may be estimated to have occurred around ~22 and ~24 Mya, respectively (Figure 6). Among the orthologous gene-pairs of *SiAP2/ERF* with those of other grass species, the average Ka/Ks value was maximum between *Brachypodium* and foxtail millet (0.4) and rice and foxtail millet (0.4), and least for sorghum-foxtail millet and maize-foxtail millet gene-pairs (0.2) (Tables S9–S12). The relatively higher rate of synonymous substitution between *Brachypodium*-foxtail millet and rice-foxtail millet *AP2/ERF* genes pointed their earlier divergence around 53–55 Mya from foxtail millet as compared to sorghum and maize *AP2/ERF* genes (Figure 6). Remarkably, the *AP2/ERF* gene-pairs between sorghum and foxtail millet, and maize and foxtail millet (average Ka/Ks = 0.2) appear to have undergone extensive intense purifying selection in comparison to foxtail millet-rice and foxtail millet-*Brachypodium* (average Ka/Ks = 0.4 for both) *AP2/ERF* genes. This is in agreement to their recent time of divergence, around 25 Mya. The estimation of tandem and segmental duplication time (average of 23 Mya) of foxtail millet *AP2/ERF* genes in between the divergence time of foxtail millet-rice (53 Mya), foxtail millet-*Brachypodium* (55 Mya) and foxtail millet-maize and foxtail millet-sorghum (both 25 Mya) orthologous *AP2/ERF* gene-pairs are comparable to evolutionary studies involving the protein-coding genes annotated from the recently released draft genome sequences of foxtail millet [3,4]. Interestingly, though the *SiAP2/ERF* gene-pairs showing segmental (Ka/Ks = 0.09) and tandem duplication (Ka/Ks = 0.13) events are not under similar evolutionary pressure, both set of gene pairs revealed that these events took place almost at similar time (22 Mya for tandem and 24 Mya for segmentally duplicated gene pairs). Therefore, overall, it can be concluded that the segmental and tandem duplication events including the divergence events of *SiAP2/ERF* genes with other grass species have played a major role in evolution of this gene family in foxtail millet. This is also in agreement with earlier genome-wide studies conducted for important gene families in foxtail millet [56–58,67,68].

In silico tissue-specific expression profiling of *SiAP2/ERF* genes

Tissue-specific expression data at a given developmental stage is helpful in identifying genes involved in defining precise nature of individual tissues. Therefore, in order to examine tissue-specific expression profiles of 171 *SiAP2/ERF* genes, a heat map was generated based on the RPKM values for each gene in all tissue samples using RNA-Seq data. A differential expression for all 171 transcripts was observed in 4 tissue samples namely root, leaf, stem and spica (Figure 7). A relative comparison of expression profiles of all 171 SiAP2/ERF showed a relatively higher expression of SiAP2/ERF-020, SiAP2/ERF-021, SiAP2/ERF-025, SiAP2/ERF-041, SiAP2/ERF-043, SiAP2/ERF-063, SiAP2/ERF-069, SiAP2/ERF-094, SiAP2/ERF-108, SiAP2/ERF-139 and SiAP2/ERF-165 in all the four tissues suggesting their importance as potential targets for further functional characterization. In general, majority of the SiAP2/ERFs exhibited root-specific expression (56; ~33%) followed by expression in stem (47; ~27%), then spica (43; ~25%) and least in leaves (26; ~15%). The results indicated that *AP2/ERF* genes in foxtail millet are mostly expressed in roots as confirmed by earlier studies [25,60]. The tissue-specific expression profiling of *SiAP2/ERFs* would further aid the combinatorial involvement of these genes in transcriptional regulation of various tissues, while ubiquitously expressed *SiAP2/ERFs* might control a broad set of genes at transcriptional level. The heat map data also facilitates the overexpression studies of *SiAP2/ERFs* across the tissues to impart stress tolerance to both foxtail millet as well as related grass species.

Expression profiling of *SiAP2/ERFs* during abiotic stresses and phytohormone treatments

Gene expression studies can provide essential indications regarding functions of a gene. In order to analyze the role of *AP2/ERF* genes in foxtail millet, we examined the expression profiles of 21 selected genes representing different sub-families using quantitative real-time (qRT) PCR analysis in response to drought (20% PEG 6000), salt (250 mM NaCl), 100 μM ABA, 100 μM SA, 100 μM MeJA and 100 μM Et during early (1 h) and late (24 h) durations of treatments. The heat map illustration of expression profiles of 21 selected *SiAP2/ERF* genes under drought and salinity is shown in Figure 8. The qRT-PCR analysis demonstrated an overall differential expression patterns to one or more stresses for the genes under study (Figs. 8, 9). The *SiAP2/ERF* genes, in general, were up-regulated by drought and salt treatments except SiAP2/ERF-116 which was down-regulated under drought stress and SiAP2/ERF-092 and SiAP2/ERF-095 which were down-regulated under salt stress at both time points. Only SiAP2/ERF-103 was co-regulated as it was induced by both stresses at all time points. However, 8 *SiAP2/ERF* genes were activated exclusively at late drought stress and 3 at late salinity stress suggesting their role in stress adaptation (Figure 8). The variability in gene expression patterns observed in this study indicated that *SiAP2/ERFs* might play an important role in regulating a complex web of stress responsive pathways for stress adaptation and tolerance towards multiple abiotic stresses.

Phytohormones or plant growth regulators not only play a crucial role in regulation of various plant processes including growth and development but also in signaling and gene expression during environmental stresses both abiotic and biotic. Therefore it was attempted to analyze the expression patterns of the selected 21 *SiAP2/ERF* genes under various hormone treatments. A hierarchical clustering demonstrated overlapping and specific gene expression patterns in response to various phytohormones

(Figure 9). No single gene was exclusively induced in all hormone treatments indicating their treatment-specific roles. However, several genes were exclusively repressed (SiAP2/ERF053, SiAP2/ERF-055, SiAP2/ERF-092, SiAP2/ERF-109 and SiAP2/ERF-116) in all hormone treatments indicating that these genes may be a part of general hormone response. Overall majority of the *SiAP2/ERF* genes were down-regulated in response to ABA except SiAP2/ERF-069, SiAP2/ERF-103 and SiAP2/ERF-120 confirming the previous reports that *AP2/ERF* genes (mostly ERFs and DREBs) are generally regulated in an ABA-independent manner with a few exceptions [11,43]. The regulation of certain *AP2/ERF* genes by SA, MeJA or Et suggests their potential roles in biotic stress responses. Several genes were found to be regulated exclusively by a specific or more than one hormone treatments (Figure 9). As for example, as many as 5 genes were specifically up regulated by MeJA at both early and late time points, while SiAP2/ERF-120 was induced by MeJA and Et at all time points. Phytohormones generally act synergistically or antagonistically to each-other thus influencing signaling response for maintaining cellular homeostasis [69]. Thus *SiAP2/ERF* TFs also act as important mediators of this signaling process. The differential expression patterns of *SiAP2/ERF* genes in this investigation again underlines the intimidating task of understanding the global milieu associated with any stress response. However, as an outcome of this study, we are able to compare their expression profiles during several environmental stress stimuli at early and late time points for precise identification of potential candidate genes for crop improvement programmes. In this regard, *SiAP2/ERF-069, SiAP2/ERF-103* and *SiAP2/ERF-120* may be considered as potential candidate genes for further functional validation as well for utilization in crop improvement programs for stress resistance since these genes were up-regulated under drought and salinity stresses in ABA dependent manner. It can thus be concluded that certain members of *AP2/ERF* gene family in foxtail millet exhibit stimulus-specific and temporal responses and hence expanding current knowledge on molecular basis of stress tolerance and adaptation conferred on plants by them.

Identification of markers in *SiAP2/ERF* genes

Marker-assisted selection (MAS) is a combination of conventional breeding and molecular biology and offers a methodology for accelerating the procedure of crop improvement. The tagging of useful genes, such as those involved in plant hormone synthesis, and those responsible for conferring stress resistance to plants, namely drought and salinity, has been a major target for improving crop growth and productivity [70,71]. With the use of molecular markers, it is now easy to trace important alleles either in segregating or natural populations. Some of the recent studies have shown the importance of *AP2/ERF* TFs, especially DREB TFs, in marker-aided breeding and crop-improvement strategies [16,71]. Considering this, the presence of previously reported DNA-based molecular markers such as SSRs [47], eSSRs [48] and ILPs [49] were searched for their presence in all the 171 *SiAP2/ERF* genes. The analysis identified 54 SSRs and 1 ILP marker in *SiAP2/ERF* genes (Table S13). These markers would be useful in genotyping and MAS for crop improvement.

Homology modeling of *SiAP2/ERF* proteins

Three dimensional protein models of twelve proteins were constructed by sequence similarity searching against the PDB database using BLASTP. These 12 proteins were selected owing to their higher homology to the known protein sequences in the PDB and Phyre2 was used for homology modeling of their predicted structures. The protein structure of all the 12 SiAP2/ERFs were modelled at 90% confidence and the potential active sites were identified (Figure 10). The 3D structure revealed the presence of conserved AP2/ERF domain of about 60–70 amino acids in all the SiAP2/ERF proteins with a typical three-dimensional conformation ordered into a layer of three antiparallel β-sheets followed by a parallel α-helix. Further examination of the AP2/ERF domain showed the presence of two regions namely YRG and RAYD. The YRG region was 20-amino acid long N-terminal stretch rich in basic and hydrophilic residues and was reported to play a crucial role in establishing direct contact with the DNA [72]. Conversely, RAYD region comprises about 40 amino acids and this region was reported to mediate protein-protein interactions through α-helix. Moreover, reports also indicate that RAYD region is involved in DNA binding through interactions of hydrophobic face of the α-helix with the major groove of DNA [72]. The AP2 sub-family members possess two AP2/ERF domains separated by a linker sequence of 25 amino acids which is responsible for positioning of the DNA-binding domains [73]. The molecular modeling thus proved that all the predicted protein structures were highly consistent and this data would offer a preliminary foundation for comprehending the molecular functions of SiAP2/ERF proteins.

Conclusion

The AP2/ERF TFs are important regulators of various plant processes including growth, development and stress responses and thus have been subjected to intensive investigations in various crop plants (Figure S3). However, to the best of our knowledge, no such study has been taken up in otherwise naturally stress tolerant model panicoid C_4 crop *Setaria italica*. The present study identified 171 AP2/ERF TFs in the foxtail millet genome. Isolation and identification of these functional TF genes are expected to aid knowledge towards understanding the molecular genetic basis for foxtail millet stress adaptation and genetic improvement, and may also provide functional gene resources for genetic engineering approaches. To date, only one gene representing this TF superfamily has been characterized from foxtail millet [10]. Hence the present comprehensive study would assist in explicating AP2/ERF family gene function in regulations of stress signaling pathways, and defense responses as well as in providing new opportunities to discover foxtail millet stress tolerance and adaptation mechanisms. The *in silico* structure prediction might provide basic resources to study the molecular regulation of foxtail millet development and stress tolerance. However, extensive *in planta* characterization of putative candidate *SiAP2/ERF* genes is must to further explore its biological roles.

Supporting Information

Figure S1 Gene structures of 171 SiAP2/ERF proteins. Exons and introns are represented by green boxes and black lines, respectively.

Figure S2 Diagrammatic representation of alignment between the miRNA and the SiAP2/ERF targets.

Figure S3 Distribution of AP2/ERFs in sequenced plant genomes.

Table S1 List of primers used in quantitative real time-PCR expression analysis of *SiAP2/ERF* genes.

Table S2 Characteristic features of *SiAP2/ERF* Transcription factor gene family members identified in *Setaria italica*.

Table S3 Summary of functional domains present in the SiAP2/ERF proteins.

Table S4 Blast2GO annotation details of SiAP2/ERF protein sequences.

Table S5 Characteristics of the promoter region of *SiAP2/ERF* genes.

Table S6 List of putative *Setaria italica* miRNAs targeting *SiAP2/ERF* transcripts.

Table S7 The Ka/Ks ratios and estimated divergence time for tandemly duplicated *SiAP2/ERF* genes.

Table S8 The Ka/Ks ratios and estimated divergence time for segmentally duplicated *SiAP2/ERF* genes.

Table S9 The Ka/Ks ratios and estimated divergence time for orthologous SiAP2/ERF proteins between foxtail millet and sorghum.

Table S10 The Ka/Ks ratios and estimated divergence time for orthologous SiAP2/ERF proteins between foxtail millet and maize.

Table S11 The Ka/Ks ratios and estimated divergence time for orthologous SiAP2/ERF proteins between foxtail millet and rice.

Table S12 The Ka/Ks ratios and estimated divergence time for orthologous SiAP2/ERF proteins between foxtail millet and *Brachypodium*.

Table S13 Details of SiAP2/ERF transcription factor-based markers.

Acknowledgments

Timely assistance from Mr. Rohit Khandelwal is appreciated.

Author Contributions

Conceived and designed the experiments: MP CL. Performed the experiments: CL AKM MM VSB YK. Analyzed the data: MP CL. Contributed to the writing of the manuscript: CL MM MP.

References

1. Muthamilarasan M, Prasad M (2014) Advances in *Setaria* genomics for genetic improvement of cereals and bioenergy grasses. Theor Appl Genet DOI:10.1007/s00122-014-2399-3.
2. Lata C, Gupta S, Prasad M (2013) Foxtail millet: a model crop for genetic and genomic studies in bioenergy grasses, Crit Rev Biotechnol 33: 328–343.
3. Zhang G, Liu X, Quan Z, Cheng S, Xu X, et al. (2012) Genome sequence of foxtail millet (*Setaria italica*) provides insights into grass evolution and biofuel potential. Nature Biotechnol 30: 549–554.
4. Bennetzen JL, Schmutz J, Wang H, Percifield R, Hawkins J, et al. (2012) Reference genome sequence of the model plant *Setaria*. Nature Biotechnol 30: 555–561.
5. Lata C, Prasad M (2013) *Setaria* genome sequencing: An overview. J Plant Biochem Biotechnol 22: 257–260.
6. Muthamilarasan M, Theriappan P, Prasad M (2013) Recent advances in crop genomics for ensuring food security. Curr Sci 105: 155–158.
7. Sakuma Y, Liu Q, Dubouzet JG, Abe H, Shinozaki K, et al. (2002) DNA-binding specificity of the ERF/AP2 domain of Arabidopsis DREBs, transcription factors involved in dehydration- and cold-inducible gene expression, Biochem Biophys Res Commun. 290: 998–1009.
8. Nakano T, Suzuki K, Fujimura T, Shinshi H (2006) Genome-wide analysis of the ERF gene family in Arabidopsis and rice. Plant Physiol 140: 411–432.
9. Jofuku KD, Boer BGW, Montagu MV, Okamuro JK (1994) Control of Arabidopsis flower and seed development by the homeotic gene APETALA2. Plant Cell 6: 1211–25.
10. Lata C, Yadav A, Prasad M (2011) Role of plant transcription factors in abiotic stress tolerance. In: Shanker A. and Venkateshwarulu B. (eds) Abiotic Stress Response in Plants, INTECH Open Access Publishers, 269–296.
11. Mizoi J, Shinozaki K, Yamaguchi-Shinozaki K (2012) AP2/ERF family transcription factors in plant abiotic stress responses. Biochimica et Biophysica Acta 1819: 86–96.
12. Magnani E, Sjölander K, Hake S (2004) From endonucleases to transcription factors: evolution of the AP2 DNA binding domain in plants, Plant Cell 16: 2265–2277.
13. Shigyo M, Hasebe M, Ito M (2006) Molecular evolution of the AP2 subfamily, Gene 366: 256–265.
14. Saleh A, Pagés M (2003) Plant AP2/ERF transcription factors. Genetika 35: 37–50.
15. Shigyo M, Ito M (2004) Analysis of gymnosperm two-AP2-domain-containing genes. Development Genes Evoln 214: 105–14.
16. Lata C, Prasad M (2011) Role of DREBs in regulation of abiotic stress responses in plants. J Exp Bot 62: 4731–4748.
17. Riechmann JL, Meyerowitz EM (1998) The AP2/EREBP family of plant transcription factors. Biol Chem, 379: 633–646.
18. Zhuang J, Cai B, Peng RH, Zhu B, Jin XF, et al. (2008) Genome-wide analysis of the AP2/ERF gene family in *Populus trichocarpa*. Biochemical Biophysical Res Comm 371: 468–474.
19. Gil-Humanes J, Piston F, Martin A, Barro F (2009) Comparative genomic analysis and expression of the APETALA2-like genes from barley, wheat, and barley-wheat amphiploids. BMC Plant Biol 9: 66.
20. Dietz KJ, Vogel MO, Viehhauser A (2010) AP2/EREBP transcription factors are part of gene regulatory networks and integrate metabolic, hormonal and environmental signals in stress acclimation and retrograde signalling. Protoplasma 245: 3–14.
21. Licausi F, Giorgi FM, Zenoni S, Osti F, Pezzotti M, et al. (2010) Genomic and transcriptomic analysis of the AP2/ERF superfamily in *Vitis vinifera*. BMC Genomics 11: 719.
22. Zhuang J, Chen J-M, Yao Q-H, Xiong F, Sun C-C, et al. (2011) Discovery and expression profile analysis of AP2/ERF family genes from *Triticum aestivum*. Mol Biol Rep 38: 745–753.
23. Zhuang J, Yao Q-H, Xiong A-S, Zhang J (2011) Isolation, Phylogeny and Expression Patterns of AP2-Like Genes in Apple (Malus × domestica Borkh). Plant Mol Biol Rep 29: 209–216.
24. Zhang G, Chen M, Chen X, Xu Z, Guan S, et al. (2008) Phylogeny, gene structures, and expression patterns of the ERF gene family in soybean (*Glycine max* L.). J Exp Bot 59: 4095–4107.
25. Xu W, Li F, Ling L, Liu A (2013) Genome-wide survey and expression profiles of the AP2/ERF family in castor bean (*Ricinus communis* L.). BMC Genomics 14: 785.
26. Punta M, Coggill PC, Eberhardt RY, Mistry J, Tate J, et al. (2012) The Pfam protein families database. Nucleic Acids Res 40: D290–D301.
27. Voorrips RE (2002) MapChart: software for the graphical presentation of linkage maps and QTLs. J Hered 93: 77–78.
28. Lee TH, Tang H, Wang X, Paterson AH (2012) PGDD: a database of gene and genome duplication in plants. Nucleic Acids Res 41: D1152–D1158.
29. Tang H, Bowers JE, Wang X, Ming R, Alam M, et al. (2008) Synteny and Collinearity in Plant Genomes, Science 320: 486–488.
30. Krzywinski M, Schein J, Birol I, Connors J, Gascoyne R, et al. (2009) Circos: an information aesthetic for comparative genomics. Genome Res 19: 1639–1645.
31. Shiu S-H, Bleecker AB (2003) Expansion of the Receptor-Like Kinase/Pelle Gene Family and Receptor-Like Proteins in Arabidopsis. Plant Physiol 132: 530–543.

32. Du D, Zhang Q, Cheng T, Pan H, Yang W, et al. (2012) Genome-wide identification and analysis of late embryogenesis abundant (LEA) genes in *Prunus mume*. Mol Biol Rep 40: 1937–1946.
33. Guo AY, Zhu QH, Chen X, Luo JC (2007) GSDS: a gene structure display server. Yi Chuan 29: 1023–1026.
34. Tamura K, Peterson D, Peterson N, Stecher G, Nei M, et al. (2011) MEGA5: Molecular evolutionary genetics analysis using maximum likelihood, evolutionary distance, and maximum parsimony methods. Mol Biol Evol 28: 2731–2739.
35. Conesa A, Gotz S (2008) Blast2GO: a comprehensive suite for functional analysis in plant genomics. Int J Plant Genomics 2008: 619832.
36. Khan Y, Yadav A, Suresh BV, Muthamilarasan M, Yadav CB, et al. (2014) Comprehensive genome-wide identification and expression profiling of foxtail millet [*Setaria italica* (L.)] miRNAs in response to abiotic stress and development of miRNA database. Plant Cell Tiss Organ Cult 118: 279–292.
37. Suyama M, Torrents D, Bork P (2006) PAL2NAL: robust conversion of protein sequence alignments into the corresponding codon alignments. Nucleic Acids Res 34: W609–W612.
38. Lynch M, Conery JS (2000) The evolutionary fate and consequences of duplicate genes. Science 290: 1151–1155.
39. Yang Z, Gu S, Wang X, Li W, Tang Z, et al. (2008) Molecular evolution of the cpp-like gene family in plants: insights from comparative genomics of Arabidopsis and rice. J Mol Evol 67: 266–277.
40. Cochrane G, Alako B, Amid C, Bower L, Cerdeño-Tárraga A, et al. (2013) Facing growth in the European Nucleotide Archive. Nucleic Acids Res 41: D30–D35.
41. Saeed AI, Bhagabati NK, Braisted JC, Liang W, Sharov V, et al. (2006) TM4 microarray software suite. Methods Enzymol 411: 134–193.
42. Saeed AI, Sharov V, White J, Li J, Liang W, et al. (2003) TM4: a free, open-source system for microarray data management and analysis. Biotechniques 34: 374–378.
43. Lata C, Sahu PP, Prasad M (2010) Comparative transcriptome analysis of differentially expressed genes in foxtail millet (*Setaria italica* L.) during dehydration stress. Biochem Biophys Res Commun 393: 720–727.
44. Lata C, Bhutty S, Bahadur RP, Majee M, Prasad M (2011) Association of a SNP in a novel DREB2-like gene *SiDREB2* with stress tolerance in foxtail millet [*Setaria italica* (L.)]. J Exp Bot 62: 3387–3401.
45. Kumar K, Muthamilarasan M, Prasad M (2013) Reference genes for quantitative Real-time PCR analysis in the model plant foxtail millet (*Setaria italica* L.) subjected to abiotic stress conditions. Plant Cell Tiss Organ Cult 115: 13–22.
46. Livak KJ, Schmittgen TD (2001) Analysis of relative gene expression data using realtime quantitative PCR and the $2^{-\Delta\Delta Ct}$ method. Methods 25: 402–408.
47. Pandey G, Misra G, Kumari K, Gupta S, Parida SK, et al. (2013) Genome-wide development and use of microsatellite markers for large-scale genotyping applications in foxtail millet [*Setaria italica* (L.)]. DNA Res 20: 197–207.
48. Kumari K, Muthamilarasan M, Misra G, Gupta S, Subramanian A, et al. (2013) Development of eSSR-markers in *Setaria italica* and their applicability in studying genetic diversity, cross-transferability and comparative mapping in millet and non-millet species. PLoS ONE 8: e67742.
49. Muthamilarasan M, Venkata Suresh B, Pandey G, Kumari K, Parida SK, et al. (2014) Development of 5123 intron-length polymorphic markers for large-scale genotyping applications in foxtail millet. DNA Res 21: 41–52.
50. Suresh BV, Muthamilarasan M, Misra G, Prasad M (2013) FmMDb: a versatile database of foxtail millet markers for millets and bioenergy grasses research. PLoS ONE 8: e71418.
51. Berman HM, Westbrook J, Feng Z, Gilliland G, Bhat TN, et al. (2000) The protein data bank. Nucleic Acids Res 28: 235–242.
52. Kelley LA, Sternberg MJE (2009) Protein structure prediction on the Web: a case study using the Phyre server. Nature Protocols 4: 363–371.
53. Zhuang J, Peng R-H, Cheng Z-M, Zhang J, Cai B, et al. (2009) Genome-wide analysis of the putative AP2/ERF family genes in *Vitis vinifera*. Scientia Horticul. 123: 73–81.
54. Rashid M, Guangyuan H, Guangxiao Y, Hussain J, Xu Y (2012) AP2/ERF transcription factor in rice: genome-wide canvas and syntenic relationships between monocots and eudicots. Evol Bioinform Online 8: 321–355.
55. Bonthala VS, Muthamilarasan M, Roy R, Prasad M (2014) FmTFDb: a foxtail millet transcription factors database for expediting functional genomics in millets. Mol Biol Rep 41: 6343–6348.
56. Yadav CB, Muthamilarasan M, Pandey G, Prasad M (2014) Identification, characterization and expression profiling of Dicer-like, Argonaute and RNA-dependent RNA polymerase gene families in foxtail millet. Plant Mol Biol Rep DOI:10.1007/s11105-014-0736-y.
57. Muthamilarasan M, Bonthala VS, Mishra AK, Khandelwal R, Khan Y, et al. (2014) C_2H_2 type of zinc finger transcription factors in foxtail millet define response to abiotic stresses. Funct Integr Genomics 14: 531–543.
58. Muthamilarasan M, Khandelwal R, Yadav CB (2014) Identification and molecular characterization of MYB transcription factor superfamily in C_4 model plant foxtail millet (*Setaria italica* L.). PLOS ONE 9: e109920.
59. Karniely S, Pines O (2005) Single translation-dual destination: mechanisms of dual protein targeting in eukaryotes. Embo Reports 6: 420–425.
60. Song X, Li Y, Hou X (2013) Genome-wide analysis of the AP2/ERF transcription factor superfamily in Chinese cabbage (*Brassica rapa* ssp. pekinensis). BMC Genomics 14: 573.
61. Sun Z-M, Zhou M-L, Xiao X-G, Tang Y-X, Wu Y-M (2014) Genome-wide analysis of AP2/ERF family genes from *Lotus corniculatus* shows LcERF054 enhances salt tolerance. Funct Integr Genomics 14: 453–466.
62. Chen J, Xia X, Yin W (2009) Expression profiling and functional characterization of a DREB2-type gene from *Populus euphratica*. Biochem Biophys Res Commun 378: 483–487.
63. Le DT, Nishiyama R, Watanabe Y, Vankova R, Tanaka M, et al. (2012) Identification and expression analysis of cytokinin metabolic genes in soybean under normal and drought conditions in relation to cytokinin levels. PLoS ONE 7(8): e42411.
64. Fang YJ, You J, Xie KB, Xie WB, Xiong LZ (2008) Systematic sequence analysis and identification of tissue-specific or stress-responsive genes of NAC transcription factor family in rice. Mol Genet Genomics 280: 547–563.
65. Aida M, Beis D, Heidstra R, Willemsen V, Blilou I, et al. (2004) The PLETHORA genes mediate patterning of the Arabidopsis root stem cell niche. Cell 119: 109–120.
66. Cannon SB, Mitra A, Baumgarten A, Young ND, May G (2004) The roles of segmental and tandem gene duplication in the evolution of large gene families in *Arabidopsis thaliana*. BMC Plant Biol 4: 10.
67. Puranik S, Sahu PP, Mandal SN, B VS, Parida SK, et al. (2013) Comprehensive genome-wide survey, genomic constitution and expression profiling of the NAC transcription factor family in foxtail millet (*Setaria italica* L.). PLoS ONE 8: e64594.
68. Mishra AK, Muthamilarasan M, Khan Y, Parida SK, Prasad M (2014) Genome-wide investigation and expression analyses of WD40 protein family in the model plant foxtail millet (*Setaria italica* L.) PLoS ONE 9: e86852.
69. Zeller G, Henz SR, Widmer CK, Sachsenberg T, Ratsch G, et al. (2009) Stress induced changes in the *Arabidopsis thaliana* transcriptome analyzed using whole genome tiling arrays. Plant J 58: 1068–1082.
70. Lopez CG, Banowetz GM, Peterson CJ, Kronstad WE (2003) Dehydrin expression and drought tolerance in seven wheat cultivars. Crop Science 43: 577–582.
71. Lata C, Prasad M (2014) Association of an allele-specific marker with dehydration stress tolerance in foxtail millet suggests SiDREB2 to be an important QTL. J Plant Biochem Biotechnol 23: 119–122.
72. Okamuro JK, Caster B, Villarroel R, van Montagu M, Jofuku KD (1997) The AP2 domain of APETALA2 defines a large new family of DNA binding proteins in Arabidopsis. Proc Natl Acad Sci USA 94: 7076–7081.
73. Wolfe SA, Nekludova L, Pabo CO (2000) DNA recognition by Cys2His2 zinc finger proteins. Annu Rev Biophys Biomol Struct 29: 183–212.

10

Conservation Genetics of the Philippine Tarsier: Cryptic Genetic Variation Restructures Conservation Priorities for an Island Archipelago Primate

Rafe M. Brown[1]*, Jennifer A. Weghorst[2], Karen V. Olson[1], Mariano R. M. Duya[3], Anthony J. Barley[1], Melizar V. Duya[4], Myron Shekelle[5], Irene Neri-Arboleda[6], Jacob A. Esselstyn[7], Nathaniel J. Dominy[8], Perry S. Ong[9], Gillian L. Moritz[10], Adrian Luczon[9], Mae Lowe L. Diesmos[11], Arvin C. Diesmos[12], Cameron D. Siler[13]

1 Biodiversity Institute and Department of Ecology and Evolutionary Biology, University of Kansas, Lawrence, Kansas, United States of America, **2** Biodiversity Institute, University of Kansas, Lawrence, Kansas, United States of America, **3** Institute of Biology, University of the Philippines Diliman, Quezon City, Philippines, **4** Diliman Science Research Foundation Inc., University of the Philippines Diliman, Quezon City, Philippines, **5** Research Institute of EcoScience, Ewha Womans University, Seoul, Republic of Korea, **6** Thomas Glen Erin Animal Hospital, Mississauga, Ontario, Canada, **7** Museum of Natural Science & Department of Biological Sciences, Louisiana State University, Baton Rouge, Louisiana, United States of America, **8** Department of Anthropology, Dartmouth College, Hanover, New Hampshire, United States of America, **9** Institute of Biology, University of the Philippines, Diliman, Quezon City, Philippines, **10** Department of Biological Sciences, Dartmouth College, Hanover, New Hampshire, United States of America, **11** University of Santo Tomas, Espana, Manila, Philippines, **12** Herpetology Section, Zoology Division, Philippine National Museum, Ermita, Manila, Philippines, **13** Sam Noble Oklahoma Museum of Natural History and Department of Biology, University of Oklahoma, Norman, Oklahoma, United States of America

Abstract

Establishment of conservation priorities for primates is a particular concern in the island archipelagos of Southeast Asia, where rates of habitat destruction are among the highest in the world. Conservation programs require knowledge of taxonomic diversity to ensure success. The Philippine tarsier is a flagship species that promotes environmental awareness and a thriving ecotourism economy in the Philippines. However, assessment of its conservation status has been impeded by taxonomic uncertainty, a paucity of field studies, and a lack of vouchered specimens and genetic samples available for study in biodiversity repositories. Consequently, conservation priorities are unclear. In this study we use mitochondrial and nuclear DNA to empirically infer geographic partitioning of genetic variation and to identify evolutionarily distinct lineages for conservation action. The distribution of Philippine tarsier genetic diversity is neither congruent with expectations based on biogeographical patterns documented in other Philippine vertebrates, nor does it agree with the most recent Philippine tarsier taxonomic arrangement. We identify three principal evolutionary lineages that do not correspond to the currently recognized subspecies, highlight the discovery of a novel cryptic and range-restricted subcenter of genetic variation in an unanticipated part of the archipelago, and identify additional geographically structured genetic variation that should be the focus of future studies and conservation action. Conservation of this flagship species necessitates establishment of protected areas and targeted conservation programs within the range of each genetically distinct variant of the Philippine tarsier.

Editor: Adam Stow, Macquarie University, Australia

Funding: Support for fieldwork was provided by the University of Kansas Biodiversity Institute and Department of Ecology and Evolutionary Biology (to RMB, CDS, and JAE), the National Geographic Society (NGS 8446-08 to RMB, JAW, MS and INA), funds from the Primate Action Fund (to MS), Ewha Womans University (Ewha Global Top5 Grant 2013 to MS), the David and Lucile Packard Foundation (Fellowship in Science and Engineering no. 2007-31754, to NJD), and U.S. National Science Foundation (DEB 0743491 to RMB). The funders had no role in study design, data collection and analysis, decision to publish, or preparation of the manuscript.

Competing Interests: The authors have declared that no competing interests exist.

* Email: rafe@ku.edu

Introduction

Biodiversity-rich tropical forests are being degraded worldwide, and the pace of forest destruction is exceptionally rapid in insular Southeast Asia [1]. With only 4–8% of its original forest remaining [2], the Philippines has been designated as both a global conservation biodiversity hotspot [3] and a Megadiverse nation [4]—a distinction shared only with Madagascar. Within this archipelago, the Philippine tarsier, a small endemic nocturnal primate, has been enlisted as a flagship species for an emerging societal conservation movement and an expanding ecotourism industry [1,5].

Traditionally, taxonomy and conservation have been inextricably linked and most conservation strategies have targeted formally named taxonomic units: species or subspecies [6]. Although most conservation efforts have targeted these taxonomic entities, conserving finer-grained genetic diversity across a species' range [7,8] is essential to preserving metapopulation dynamics, preventing inbreeding depression, avoiding population collapses, and ultimately ensuring against extinction [9–11]. Unique evolutionary lineages or genetically defined "Evolutionarily Significant Units" [12] are appropriate targets of conservation programs aimed at preserving genetic diversity among and within species; targeting empirically defined distinct evolutionary lineages has the added benefit of potentially removing the subjectivity sometimes associated with traditional taxonomy [13].

Despite being the focus of a disproportionate number of intensely focused studies, multiple lines of evidence suggest that numerous primate taxa await discovery and formal taxonomic description. For example, between the 1975 and 1999 nocturnal primate species diversity grew worldwide by 2.85-fold increase [14–18], and has since climbed by an additional 1.69-fold increase [19]. Unique among nocturnal primates, tarsiers are found only in insular Southeast Asia (See Appendix S1). Ten species are recognized currently, with several new taxa recently proposed through bioacoustic analysis and molecular data [15]. The Philippine tarsier (Fig. 1) has been the focus of recent attempts to understand morphological variation [16], clarify taxonomy [17], and establish the conservation status of populations [18,19]; these efforts have met with limited success and left unanswered the questions of appropriate targets for conservation action.

In the Philippines, many morphologically indistinguishable yet genetically differentiated "cryptic" species [20–22] have recently been discovered, suggesting that the codistributed Philippine tarsier might also harbor hidden evolutionary lineages (unrecognized species, genetic variants, putative taxa, etc.), which may be unprotected and not incorporated into current conservation planning.

We assessed the conservation genetics of Philippine tarsier to determine (1) how many distinct evolutionary lineages can be identified, (2) whether genetic structure conforms to biogeographical predictions and/or (3) expectations derived from current taxonomy (three subspecies). Due to limited natural history data or consensus from other sources of information (morphology, bioacoustics, ecology), we argue that genetic data should be used to distinguish evolutionarily distinct tarsier population groups as objective, empirically defined conservation priorities.

Materials and Methods

Mitochondrial data collection

Genetic material from *T. syrichta* was nondestructively sampled (ear and tail-tip biopsies) from throughout as much of the species' range as feasible (targeting ranges of all described subspecies; Appendix S1) on the large islands of Mindanao, Samar, Leyte, Bohol, and Dinagat, Philippines (Figs. 1, 2; Appendix S2). Deceased animals were salvaged from illegal roadside animal dealers, local Provincial or City Environmental Natural Resources (PENRO, CENRO) officers or university administrators (confiscations of poached animals by students, animal traders, and bush meat hunters), and from indigenous hunters in forested areas. Salvaged animals were prepared as museum specimens and deposited at the Leyte State University (ViSCA) Natural History Museum, the National Museum of the Philippines, or the University of Kansas Biodiversity Institute (Appendix S2). We sampled 77 individuals of *T. syrichta* from 17 localities for genetic material from most of the large islands within the Mindanao Pleistocene Aggregate Island Complex (PAIC) [2], including representatives of all recognized subspecies. We sequenced the mitochondrial 12S ribosomal RNA (12S), Cytochrome B (CytB), and NADH Dehydrogenase Subunit 2 (ND2) gene fragments and nine nuclear microsatellites.

Genomic DNA was extracted from tissues following Fujita's guanidine thiocyanate method [23]. Molecular data were derived from all available genetic samples and included efforts to determine the sequence of the 12S mitochondrial gene fragment, the ND2, and the CytB gene region. Sequence identity was confirmed via examination of amino acid translations for internal stop codons for Cytb and ND2 sequence alignments, GenBank BLAST searches, and alignment comparisons with annotated, published comparative sequence data available on GenBank. All sequences were deposited in GenBank under accession numbers KM217270–99 (12S), KM217300–23 (CytB), and KM217324–49 ND2 (Appendix S2). Primers used for amplifying mitochondrial gene sequences include: 12S (Tars12s_AJB.F1, 5′–TCA-CAACGTCTTGCTCAACC–3′ and Tars12s_AJB.R2, 5′–TTGAGGAGGGTGACGGGCGG–3′); ND2 (TarND2_AJB.f2, 5′–ACTTTCTAATTCAAGCGACAGCCTCC–3′ and TarN-D2_AJB.r2, 5′–TGGGGGATATGGGTAAAAGTAGGGTGG–3′); CytB (TarsCytB.F1, 5′–CACATCTGCCGAGACGTAAA–3′ and TarsCytB.R1, 5′–TGGGGTGGAGTGTTTAGAGG–3′). We used the following thermal profiles for the mitochondrial fragment amplification: 1 min at 95°C followed by 38 cycles of 94°C for 30 sec, 30 sec at the primer specific annealing temperature (60°C for 12 s; 61°C for ND2; 62°C for CytB), and 72°C for 1 min, and a final extension phase at 72°C for 10 min.

Amplified products were visualized on 1.5% agarose gels and PCR products were purified with 1 µL of a 20% dilution of ExoSAP-IT (US78201, Amersham Biosciences, Piscataway, NJ) on the following thermal profile: 30 min at 37°, followed by 15 min at 80°. Cycle sequencing reactions were run using ABI Prism BigDye Terminator chemistry (Ver. 3.1; Applied Biosystems, Foster City, CA), and purified with Sephadex Medium (NC9406038, Amersham Biosciences, Piscataway, NJ) in Centri-Sep 96 spin plates (CS-961, Princeton Separations, Princeton, NJ). Purified products were analyzed using an ABI Prism 3130xl Genetic Analyzer (Applied Biosystems). Sequence contigs were assembled and edited using Sequencher 4.8 (Gene Codes Corp., Ann Arbor, MI).

Sequencing proved highly problematic for our many degraded tissue samples and approximately half our samples would not amplify for any targeted gene regions. Of the half that could be amplified for >2 target regions, initial sequence alignments were produced in Muscle [24] and manual adjustments were made in MacClade 4.08 [25]. We estimated the phylogeny for each mitochondrial gene fragment independently using likelihood and concatenated mitochondrial sequences following no observation of statistically significant incongruence between datasets. Exploratory analyses of the combined dataset of 36 individuals inferred relationships that did not strongly conflict with topologies inferred from individual gene fragments.

Microsatellite data collection

We sampled a total of 66 individuals from 17 localities of *Tarsius syrichta* across the southern Philippines. Each locality was represented by 1–27 individuals (mean = 3.88). DNA was extracted using QIAGEN DNeasy extraction kits and we used nine microsatellite makers previously developed for *Tarsius syrichta*: T5, T6, T22, T34, T35, T43, T69 [26] and *Tarsius* spp.: T42, T50 [27]. PCR reaction volumes and conditions for the PCR amplifications followed the protocol of Schuelke [28]. Conditions of the PCR amplification were as follows: 94°C for 30 s, 56°C for 45 s, 72°C for 45 s, followed by 8 cycles at 94°C for 30 s, 53°C for 45 s, 72°C for 45 s, and a final extension at 72°C for 10 min. Primers T42, T43 and T45 were fluorescently labeled with 6-FAM by Integrated DNA Technologies (Coralville, Iowa) for PCR. Primer annealing temperatures for these three loci were as follows: T42 at 53°C [27], T43 at 54°C and T45 at 57°C [26].

Philippines
Current Dry Land
120 m isobath
Luzon
Luzon PAIC
15°
10°
Samar
Leyte
Dinagat
Siargao
Bohol
Zamboanga Peninsula
Mindanao
Borneo
Mindanao PAIC
121°
125°
200 km
Bohol
29—Leyte
13—Bohol
8—Bohol
27—Bohol
26—Bohol
32—Leyte
24—Bohol
25—Bohol
4—Bohol
1—Bohol
Samar-Leyte
30—Leyte
31—Leyte
39—Samar
37—Samar
NC012774
AB371090
28—Leyte
Dinagat-Caraga
Surigao del Norte—42
Dinagat—40
Surigao del Norte—53
Surigao del Norte—55
Surigao del Norte—54
Surigao del Norte—51
0.0070
Mindanao
Agusan del Sur—59
Misamis Oriental—61
Agusan del Sur—58
Sarangani—64
Sarangani—63
Agusan del Sur—57
Misamis Oriental—60
Zamboanga
Zamboanga—65
Zamboanga—66

Figure 1. Phylogeographic relationships of *Tarsius syrichta* (see Appendix S1 for taxonomic summary) estimated from a combined, partitioned, RAxML ML analysis of mitochondrial (12S, CytB, ND2) gene fragments. Black circles at nodes correspond to ML bootstraps ≥ 70% and Bayesian PP ≥95%.

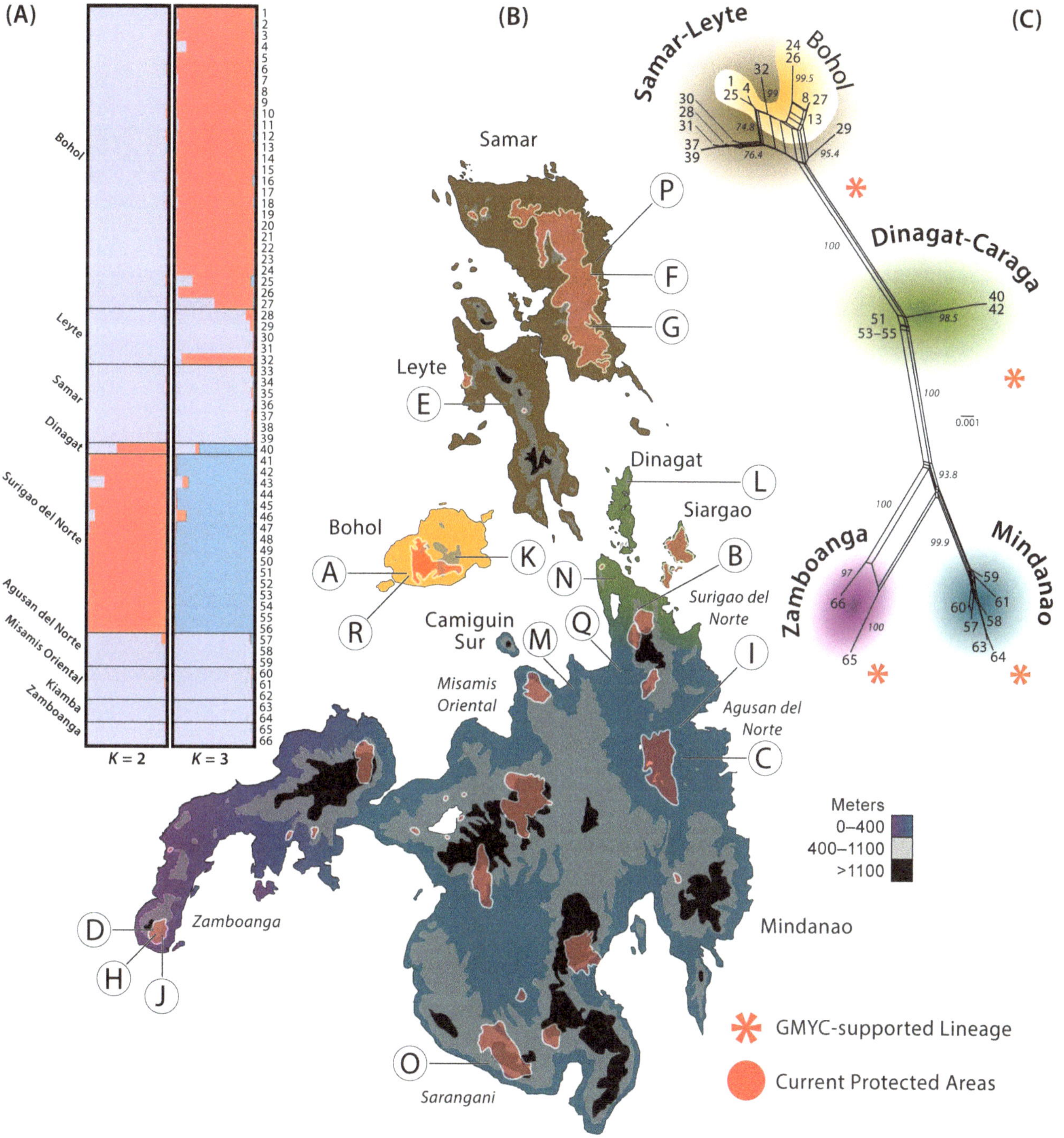

Figure 2. DISTRUCT visualization of STRUCTURE analyses (A) assigning individuals to major population groupings (genetically distinct evolutionary lineages) for Philippine tarsier demes (*K*=2 and 3 populations). Mindanao faunal region (B; see Fig. 1, inset) with sampling (17 sites, 66 individuals) labeled with letters corresponding to full localities listed in Appendix S2, protected areas shaded red. SplitsTree gene network (C; numbers at internodes = ML bootstrap replicates), and results of GMYC analyses (red asterisks denote lineages delineated by the Yule-coalescent), with numbers at tips corresponding to individual samples in Structure plots (A) and cluster shading corresponding to islands on map (B).

Subsequently, 2.1 µl of the PCR product was added to 9 µl of formimide and .15 µl was electrophoresed with the LIZ-500 size standard and analyzed on a 3730 Genetic Analyzer (Applied Biosystems). Fragments were sized with GeneMapper version 4.1 (Applied Biosystems). We replicated data collection for all samples a minimum of two times per locus in order to monitor for allelic dropout and false alleles. In cases where we detected allelic dropout or inconsistent genotypes (n = 2), the sample was replicated once more in order to minimize genotyping errors. In a few instances of ambiguous genotypes, despite these efforts, samples were simply excluded from subsequent analyses (n = 8). Microsatellite data were archived at Dryad (doi:10.5061/dryad.r7468).

Phylogenetic analyses and lineage delimitation: mitochondrial data

Partitioned Bayesian analyses of the combined dataset were conducted in MrBayes v3.1.2 [29]. As the mtDNA genes sampled in this study included incomplete sections for some targeted genes, datasets were partitioned by gene only. The Akaike Information Criterion (AIC) implemented in jModeltest [30] was used to select the model of nucleotide substitution for each subset (Table 1). We ran four independent Metropolis-coupled MCMC analyses, each with four chains and the default heating scheme (temp = 0.2). All analyses were run for 20 million generations, sampling every 1000 generations. To assess stationarity, all sampled log-likelihoods and parameter values from the cold Markov chain were plotted against generation time and compared among independent runs using Tracer v1.4 [31]. These runs demonstrated patterns consistent with stationarity after 4 million generations, hence the first 20% of samples were discarded as burn-in.

Partitioned maximum likelihood (ML) analyses were conducted in RAxMLHPC v7.0 [32] for the combined dataset, using the same partitioning strategy as implemented in the Bayesian analyses. The more complex model (GTR + I + Γ) was used for all subsets, and 100 replicate ML inferences were performed for each run. Each inference was initiated with a random starting tree, employed the rapid hill-climbing algorithm and support was assessed with 1000 bootstraps [33].

To generate an ultrametric phylogeny for our Generalized Mixed Yule-Coalescent (GMYC) analysis, we ran a strict clock divergence time analysis of the mtDNA data, partitioning the data by gene, and fixing the mean substitution rate to 1.0 in BEAST v1.7.5 [34]. We first removed all individuals with identical sequences, however, since zero-length branches can cause the GMYC analysis to over-partition the dataset. BEAST analyses were run for 100 million generations, sampling every 10,000. Convergence was assessed with Tracer as described above. We then used the ultrametric phylogeny with the "single threshold" GMYC model [35] to determine the number of evolutionary lineages in the mtDNA dataset.

Population structure and gene flow: nuclear data

As currently recognized, *Tarsius syrichta* spans the Mindanao PAIC in the southern Philippines. As an initial starting point, our *a priori* hypothesis of population structure, given our understanding that all these islands are part of the same faunal region [2], were that genetic diversity might be partitioned among individual islands. In order to test these expectations, we used the Bayesian clustering method of the program Structure v2.3.3 [27,28,36–39] with allelic data from nine nuclear loci, to estimate the number of populations among *T. syrichta* samples, as well as infer the probabilities of individuals belonging to each of the estimated populations. We considered individuals with q values between 0.10 and 0.90 to be admixed [39–41]. Due to the absence of *a priori* knowledge of inter-locus relationships, analyses were run under the correlated allele frequency model and the admixture ancestry model [36,40]. All other parameters were left with default settings. Taking a conservative approach to evaluating population boundaries for this study, we considered the number of possible populations, represented by K, to range from one (a single geographically non-structured species) to seven (each island sampled in this study, and allowing for additional population genetic complexity that may be present on the large island of Mindanao [2]). We conducted 10 independent runs in Structure for each of the seven values of K (1–7). We ran each analysis for 10 million iterations, with a burn-in of 500,000. The rate of change of K (ΔK) was evaluated in Structure Harvester (Web v0.6.93) [42] following the method of Evanno et al. [43] to determine the number of preferred genetic clusters. The results of structure analyses were visualized using the program *Distruct* v1.1 (Fig. 2) [44].

To assess whether there is evidence of recent migration among populations, we estimated the level of historical gene flow within a Bayesian framework in the program BAYESASS v1.3 [45] for the four evolutionary lineages supported in phylogenetic analyses (Fig. 1). Four independent runs were conducted, each with 20,000,000 iterations, a sampling frequency of 2,000, and a burn-in of 2,000. Trace files were assessed using the program Tracer v1.5 [31] for evidence of convergence prior to result summary.

To visualize population genetic structure, we generated phylogenetic networks for the mitochondrial dataset by employing the Neighbor-Net algorithm [46] in the program SplitsTree v4.10 [47] (Fig. 2). To assess the support for inferred splits in the network, a bootstrap analysis was conducted with 1000 pseudoreplicates.

Results

Genetic variation and gene networks

Of the 77 samples collected, 66 could be characterized for microsatellite variation (Tables 2, 3) but, due to sample degradation, only 36 could be sequenced for mitochondrial DNA. As

Table 1. Models of evolution selected by AIC and applied for partitioned, phylogeographic analyses[1].

Partition	AIC Model	Number of Characters
12S	GTR + Γ	618
Cytb	HKY + Γ	557
ND2	HKY + Γ	538

[1]The model GTR + Γ was used for partitioned RAxMLHPC analyses.

Table 2. Microsatellite variation in 66 individuals of *Tarsius syrichta* from 17 localities (Fig. 2).

Population	N	H_O	H_E	L_P	F_{IS}	P value
Bohol	27	0.58674±0.27641	0.74946±0.16146	9	−0.12254	0.859238
Samar-Leyte	12	0.53519±0.24444	0.80535±0.07257	6	0.10995	0.214076
Dinagat-Caraga	17	0.53184±0.19296	0.63060±0.12853	4	0.0764	0.304008
E. Mindanao	8	0.63294±0.17635	0.83488±0.12227	7	0.48611	0.008798
W. Mindanao	2	0.56250±0.32043	0.75000±0.17817	8	0.42857	0.339198

Abbreviations include: N, number of samples; H_O, observed heterozygosity; H_E, expected heterozygosity; L_P, number of polymorphic loci; F_{IS} inbreeding coefficient.

Table 3. Locus-specific microsatellite variation in 66 individuals of *Tarsius syrichta* from 17 localities (Fig. 2).

		Bohol	Samar-Leyte	Dinagat-Caraga	E. Mindanao	W. Mindanao
Locus T22	H_O	0.62963	0.75	0.64286	0.5	0.5
	H_E	0.76799	0.80435	0.5291	0.83333	0.83333
	A_N	6	9	6	6	3
	HWE	**0.0064**	0.335	1	0.0562	0.3301
Locus T35	H_O	0.69231	0.75	0.6875	0.5	n/a
	H_E	0.7911	0.7971	0.68347	0.53333	n/a
	A_N	8	8	5	5	n/a
	HWE	**0.0211**	0.7219	0.4304	0.279	n/a
Locus T5	H_O	0.84615	0.58333	0.41176	0.375	0.5
	H_E	0.89819	0.84783	0.58824	0.84167	0.5
	A_N	10	9	5	5	2
	HWE	0.2585	**0.0039**	0.1024	**0.0046**	n/a
Locus T6	H_O	0.46154	0.08333	0.35294	0.5	0
	H_E	0.79864	0.82246	0.67023	0.93333	0.66667
	A_N	11	7	7	9	2
	HWE	**0.0**	**0.0**	**0.0134**	**0.0**	0.3346
Locus T34	H_O	0.57692	0.4	0.23529	0.875	0.5
	H_E	0.73454	0.73684	0.36898	0.925	0.83333
	A_N	6	5	5	10	3
	HWE	**0.0161**	**0.0204**	**0.0245**	0.1925	0.3305
Locus T42	H_O	0.85185	0.66667	0.53846	0.85714	0.5
	H_E	0.8833	0.86928	0.77231	0.89011	0.5
	A_N	11	8	5	8	2
	HWE	**0.0**	**0.0064**	0.4807	0.4357	1
Locus T69	H_O	0.18519	0.33333	0.45455	0.75	1
	H_E	0.53319	0.66013	0.58009	0.79167	0.83333
	A_N	5	4	6	5	3
	HWE	**0.0009**	0.0862	**0.048**	0.2565	n/a
Locus T43	H_O	0.88889	0.83333	0.875	0.625	0.5
	H_E	0.89727	0.9058	0.7379	0.90833	0.83333
	A_N	13	11	6	9	3
	HWE	0.562	0.0843	0.1855	**0.0**	0.3384
Locus T45	H_O	0.14815	0.41667	0.58824	0.71429	1
	H_E	0.44095	0.80435	0.7451	0.85714	1
	A_N	6	7	8	6	4
	HWE	**0.0**	**0.0037**	0.0542	0.1938	1

Abbreviations include: H_O, observed heterozygosity; H_E, expected heterozygosity; A_N, number of alleles; *HWE*, deviation from Hardy–Weinberg equilibrium (p-value).

anticipated, the results of network analyses corroborate the major results observed from phylogenetic analyses (Figs. 1, 2); these same mitochondrial clades were identified as distinct in the GMYC lineage delineation procedure.

Phylogenetic and lineage delimitation analyses

GMYC analysis of the mitochondrial phylogeny identified three putative evolutionary lineages; each of these three lineages was supported in Bayesian and ML phylogeographic analyses and phylogenetic networks of mitochondrial sequence data, and were provisionally adopted as the following distinct evolutionary groups (Fig. 1): (1) Bohol, Samar, and Leyte island, (2) Dinagat and Northeast Mindanao (Caraga Region) island, and (3) Mindanao island, composed of (*a*) eastern and (*b*) western subclades; substantial genetic divergence was found among groups (2.1–4.7% in mtDNA; Table 4), with markedly less genetic variation within groups (0.0–1.1%) and minimal gene flow between most groups (Table 5).

Assignment of individuals to population clusters

The *ad hoc* ΔK test of the results of Structure analyses of microsatellite loci preferred a K of two (LnP[K] = −2341.58; Dinagat-Caraga vs. all remaining samples), suggesting an initial estimate of only two distinct populations as defined by our nuclear data alone. All individuals were assigned to one of these two genetic clusters with q values above 0.90. With the exception of populations from Dinagat Island and the northeastern peninsula of Mindanao Island (Surigao del Norte Province; Fig. 2), all sampled populations were inferred to be part of one cohesive genetic group. Evaluating a three-population model in structure reveals a similar pattern of unambiguous individual assignments to populations; however, the model of $K = 3$ which was close to maximally preferred included additional support for the distinctiveness of populations on Bohol (Fig. 2). All Structure analyses provided support for the cryptic Dinagat-Caraga population (Fig. 2).

Discussion

New Philippine tarsier evolutionary lineages: novel conservation priorities

The unambiguous support across all analyses for a genetically distinct Dinagat-Caraga tarsier lineage, mirrored in all analyses of all loci, is novel, and identifies a range-restricted "cryptic" evolutionary entity with unique conservation concerns. Although moderately forested, Dinagat and northeast Mindanao are impoverished economically, lack low-elevation protected areas (Fig. 2), and have become the focus of particularly intensive mining operations—all of which threaten the remaining suitable habitat of this newly documented evolutionary lineage. The Dinagat-Caraga tarsier should therefore be regarded as tantamount to the conservation importance of celebrated Philippine flagship species (e.g., Philippine eagle, tamaraw, golden-spotted monitor lizard, etc.: [2]). Nearby Siargao Island is protected and may harbor the same genetic variant identified here from Dinagat; intensive studies of these populations are urgently needed.

We anticipate future studies will further refine the known distribution of the Dinagat-Caraga tarsier and extend its range to nearby Siargao Island. Additionally it is not clear if the Surigao del Norte (NE Mainland Mindanao, Caraga Region) samples identified here occur naturally on Mindanao or are the result of Dinagat and/or Siargao island animals confiscated from smugglers by regional wildlife enforcement and subsequently released on Mindanao (PSO, *personal communication* with Surigao del Norte Provincial Environmental Natural Resources Office [PENRO] staff).

Geographical distribution of genetic variation

Our empirical findings are more complex than is reflected in current taxonomy [18,48], stand in contrast to past biogeographic paradigms [2,49], uncover a novel conservation target (Dinagat-Caraga lineage; Figs. 1, 2), and identify a conservation research priority of urgent concern (potential differentiation between western and eastern Mindanao). The moderate level of sequence divergence (2.4%; Table 4) detected between eastern and western Mindanao (both supported as distinct lineages in the GMYC analysis, but with marginally significant migration rates detected in analysis of microsatellite data; Table 4) may indicate that separate tarsier conservation programs for these populations are warranted if future studies confirm their distinctiveness. An east–west sampling transect across Mindanao will be necessary to investigate the genetic relationships of these populations which, given the gap in our sampling, may simply be the extremes of natural geographically based genetic structure.

Our results from analyses of mitochondrial data do not differentiate the populations on Bohol from those on Samar-Leyte (currently recognized as separate subspecies: [48]). Considering the suboptimal $K = 3$ results, Structure analyses of microsatellites identified Bohol potentially as distinct. However, given that these analyses potentially are sensitive to unequal sampling, and that we possessed many more samples from Bohol than from Samar-Leyte, we suspect this result may be an artifact; future studies with better sampling from Samar and Leyte islands, and additional microsatellite loci will be necessary to investigate further genetic relations within the Bohol-Samar-Leyte clade. At present, in our efforts to adhere to objective, empirically defined evolutionary lineages, we stop short of defining Bohol as a distinct conservation target, and

Table 4. Uncorrected mitochondrial sequence divergence (%) among Philippine tarsier (*Tarsius syrichta*) evolutionary lineages shown below diagonal.

	Bohol	Samar-Leyte	Dinagat-Caraga	E. Mindanao	W. Mindanao
Bohol	0.0–0.9	0.018	0.015	0.018	0.026
Samar-Leyte	0.4–1.4	0.0–1.1	0.019	**0.065**	**0.056**
Dinagat-Caraga	2.1–3.5	2.1–3.4	0.0–2.3	0.021	0.029
E. Mindanao	3.4–4.6	3.5–4.6	2.7–4.6	0.1–0.8	**0.051**
W. Mindanao	3.8–4.6	3.7–4.6	3.0–4.7	2.0–2.4	0.0–1.0

Percentages on the diagonal represent intraspecific (or within clade) genetic diversity. Mean inferred migration rates (inferred in BAYESASS) between regionally partitioned diversity shown above diagonal. Migration rates above the 0.05 threshold bolded for emphasis.

Table 5. Identities by geographical region and numbers (in parentheses) of Philippine tarsier (*Tarsius syrichta*) evolutionary lineages, putative taxa, or conservation targets inferred (new data and analyses) or predicted (biogeography, taxonomy) from each of the five available sources of information.

Divergent mtDNA lineages & GMYC results (4)	Distinct nuDNA groups (3)	Islands (5)	Taxonomy (3)
Bohol-Samar-Leyte	Dinagat-Caraga	Bohol	Bohol
Dinagat-Caraga	Bohol- Samar-Leyte-east Mindanao-west Mindanao	Samar	Samar-Leyte
Mindanao (a: east Mindanao, b: west Mindanao)		Leyte	Caraga-Mindanao-Zamboanga
		Dinagat	
		Mindanao	

point to the nested nature of the Bohol samples within the Bohol-Samar-Leyte tarsier group (Figs. 1, 2).

Combining mitochondrial and nuclear gene signals to establish targets for conservation action

Under current implementation of international conservation status assessment (IUCN, 2014), prioritization of populations for conservation action follows the recognition of named taxonomic units, i.e., species and subspecies. In the face of differing taxonomic arrangements, biogeographic expectations, and differing geographic patterns of nuclear versus mitochondrial genetic variation, how should the Philippine tarsier be prioritized for implementation of applied conservation measures? Currently, one tarsier sanctuary has been established on Bohol Island and another is under consideration for construction on Leyte. Would these two, and only these two efforts adequately conserve genetic components of Philippine tarsier diversity? We argue that they would not. Would establishment of conservation programs based on current tarsier taxonomy (one effort on Bohol, another on Samar-Leyte, and a third on Mindanao) properly conserve the genetic variation we have documented? Similarly, we argue that such an approach would fail to conserve the genetic variation elucidated here.

Brandon-Jones et al. [17] characterized the three subspecies (one from Bohol, another from Leyte-Samar, and a third from Mindanao) as "dubious" taxa. They wrote: "Groves (2001) recognized no subspecies for *Tarsius syrichta*. Hill (1955) recognized *T. s. syrichta/carbonarius/fraterculus* as a poorly defined subspecies, perhaps synonymous, with *T. s. syrichta*. Museum specimen variation seemed insignificant to Niemitz (1984), but an inadequate basis for judgment, according to Musser, and Dagosto (1987)." Our study sheds some light on the uncertainty of Brandon-Jones et al. [17] and suggests that a detailed study employing other lines of evidence is still needed to revise Philippine tarsier taxonomy. One possible outcome of further study could be that our evolutionary groups are not sufficiently distinct based on new lines of evidence (morphology, bioacoustics, whole genome data) to warrant taxonomic separation, and that Philippine tarsiers should therefore be classified as a single taxon, *T. syrichta* [16,50,51]. Alternatively, further investigation might find two or more of our conservation priority groups warrant taxonomic separation, in which case, the epithets applied by Hill [52] will be available (as species or subspecies). Three names are available for a subset of the evolutionary lineages we identify: (1) Samar-Leyte, (2) Bohol, and (3) eastern Mindanao (but not including western Mindanao [Zamboanga], or the novel Dinagat-Caraga lineage). For the immediate future, we argue that applied conservation efforts based upon the combined intersection of our variable genetic results are superior to those based upon existing taxonomy.

A number of genetic and social system phenomena could conceivably account for the differences in lineage-specific support we have inferred between mitochondrial and nuclear loci. Possibilities include persistence of genetic polymorphism, lineage sorting, nuclear gene flow between distinct mitochondrial lineages, sex-biased dispersal, mitochondrial gene sweeps, or an undocumented Philippine tarsier mating system. Given the present sampling, we are unable to distinguish between these possibilities, which may provide intriguing questions for future research. However, from a practical and applied conservation perspective, we argue that a single, clearly optimal, and objective solution results from the combination of patterns observed here in mitochondrial and nuclear gene loci.

Given our variable inferences, we advocate a combined, cumulative approach towards identification and recognition of tarsier evolutionary lineages for conservation planning. Thus, we argue that in order to maximally preserve genetic variation, it is the *combined* results of our variable sources of information that likely will have the best chance of maximally preserving genetic variation in the Philippine tarsier. As such, we advocate a multi-tiered conservation program involving conservation programs and protected areas in the ranges of each distinctive population lineages identified here, minimally including (1) Bohol-Samar-Leyte, (2) Dinagat-Caraga, and (3) Mindanao. Until a comprehensive taxonomic study using multiple lines of evidence (molecular, morphological and/or bioacoustic data) can be undertaken, we find that the existing evidence provides an inadequate basis for distinguishing between taxonomic alternatives. Instead, we emphasize that the archipelago's tarsier populations are partitioned into at least the three genetic variants, which we have empirically defined here for practical conservation purposes. We caution primatologists from taking taxonomic action until multiple lines of evidence converge on a meaningful solution [20,22], ideally with information from throughout the genome [53] and more robust sampling along multiple transects from throughout the range of this species (i.e., eastern versus western Mindanao). Nevertheless, our results necessitate a refined conservation strategy in order to effectively preserve the geographical distribution of genetic variation in this flagship species. Such an approach will greatly enhance the prospects for continued survival of this endemic primate and, combined with many other recent discoveries in the country, will contribute to the recognition of the archipelago as a globally significant biodiversity conservation priority [2,5,49].

Supporting Information

Appendix S1 Review of Philippine tarsier taxonomy.

Appendix S2 Vouchers, locality data, GPS coordinates, and Genbank numbers for included samples.

Acknowledgments

We thank the Philippine Department of Environment and Natural Resources for facilitating research, sample collection, and export permits (in particular T. M. Lim, C. Custodio, J. deLeon, and A. Tagtag) necessary for this and related research. Sampling protocols were approved by the University of Kansas Animal Care and Use Committee (IACUC 158-01 to RMB) and protocols to capture, sedate, and harvest ear biopsies from wild tarsiers were approved by the Dartmouth Animal Care and Use Committee (IACUC 10-11-12 and 11-06-06AT to NJD). Thanks are due to J. Quilang for assistance with data and comments on the manuscript. We thank N. Antoque, J. Cantil, and V. Yngente for assistance in the field and anonymous reviewers for comments on previous versions of the manuscript.

Author Contributions

Conceived and designed the experiments: RMB JAW CDS JAE MS MLD ACD. Performed the experiments: RMB MRMD LVD INA JAE NJD PSO AL MLD ACD CDS. Analyzed the data: KVO AJB CDS. Contributed reagents/materials/analysis tools: RMB KVO AJB CDS JAE LVD GLM. Wrote the paper: RMB JAW CDS JAE MS GLM. Obtained permission and executed field surveys: RMB MRMD LVD MS INA JAE NJD PSO GLM AL MLD ACD CDS.

References

1. Sodhi NS, Koh LP, Brook BW, Ng PK (2004) Southeast Asian biodiversity: an impending disaster. Trends Ecol Evol 19: 654–660.
2. Brown RM, Diesmos AC (2009) Philippines, Biology. In: Gillespie R, Clague D, editors. Encyclopedia of islands. Berkeley: University of California Press. pp. 723–732.
3. Myers N, Mittermeier RA, Mittermeier CG, da Fonseca GAB, Kent J (2000) Biodiversity hotspots for conservation priorities. Nature 403: 853–858.
4. Mittermeier RA, Gil PR, Mittermeier CG (1997) Megadiversity: Earth's biologically wealthiest nations. Washington, DC: Conservation International, Cemex.
5. Posa MR, Diesmos AC, Sodhi NS, Brooks TM (2008) Hope for threatened tropical biodiversity: lessons from the Philippines. Bioscience 58: 231–240.
6. Mace G (2004) The role of taxonomy in species conservation. Proc Royal Soc London B 359: 711–719.
7. Evans BJ, Supriatna J, Andayani N, Setiad MI, Cannatella DC, et al. (2003) Monkeys and toads define areas of genetic endemism on the island of Sulawesi. Evolution 57: 1436–1443.
8. Melnick DJ, Morales JC, Honeycutt RL (1999) Conservation genetics: applying molecular methods to maximize the conservation of taxonomic and genetic diversity. In: Raven P, Williams T, editors. Nature and human society: The quest for a sustainable world. Washington, DC: National Academies Press. pp. 264–275.
9. Brook BW, Tonkyn DW, O'Grady JJ, Frankham R (2002) Contribution of inbreeding to extinction risk in threatened species. Conserv Ecol 6: 16. Available: http://www.consecol.org/vol6/iss1/art16. Accessed 11 December 2013.
10. Hanski I (1998) Metapopulation dynamics. Nature 396: 41–49.
11. Saccheri I, Hanski I (2006) Natural selection and population dynamics. Trends Ecol Evol 21: 341–347.
12. Ryder OA (1986) Species conservation and systematics—the dilemma of subspecies. Trends Ecol Evol 1: 9–10.
13. Moritz C (1994) Defining 'evolutionary significant units' for conservation. Trends Ecol Evol 9: 373–375.
14. Bearder SK (1999) Physical and social diversity among nocturnal primates: a new view based on long-term research. Primates 40: 267–282.
15. Groves C, Shekelle M (2010) The genera and species of Tarsiidae. Int J Primatol 31: 1071–1082.
16. Groves C (2001) Primate taxonomy. Washington, DC: Smithsonian Institution Press.
17. Brandon-Jones D, Eudey AA, Geissmann T, Groves CP, Melnick DJ, et al. (2004) Asian primate classification. Int J Primatol 25: 97–164.
18. Dagosto M, Gebo DL, Dolino CN (2003) Natural history of the Philippine tarsier. In: Wright P, Simons E, Gursky S, editors. Tarsiers: Past, present, and future. New Brunswick: Rutgers University Press. pp. 237–259.
19. Shekelle M, Arboleda I (2008) *Tarsius syrichta*. In: IUCN 2014. IUCN red list of threatened species. Version 2014.1. Available: www.iucnredlist.org. Accessed 6 June 2014.
20. Barley A, White J, Diesmos AC, Brown RM (2013) The challenge of species delimitation at the extremes: diversification without morphological change in Philippine sun skinks. Evolution 67: 3556–3572. doi:10.1111/evo.12219.
21. Esselstyn JA, Evans BJ, Sedlock JL, Khan FAA, Heaney LR (2012) Single-locus species delimitation: a test of the mixed Yule–coalescent model, with an empirical application to Philippine round-leaf bats. Proc Royal Soc London B 279: 3678–3686. doi:10.1098/rspb.2012.0705.
22. Welton LJ, Siler CD, Oaks JR, Diesmos AC, Brown RM (2013) Multilocus phylogeny and Bayesian estimates of species boundaries reveal hidden evolutionary relationships and cryptic diversity in Southeast Asian monitor lizards. Molec Ecol 22: 3495–3510.
23. Esselstyn JA, Garcia HJD, Saulog MG, Heaney LR (2008) A new species of *Desmalopex* (Pteropodidae) from the Philippines, with a phylogenetic analysis of the Pteropodini. J Mammal 89: 815–825.
24. Edgar RC (2004) MUSCLE: multiple sequence alignment with high accuracy and high throughput. Nuc Acid Res 32: 1792–1797.
25. Maddison DR, Maddison WP (2005) MacClade: analysis of phylogeny and character evolution, Version 4.08. Sunderland: Sinauer Associates.
26. Merker S, Driller C, Neri-Arboleda I, Zischler H (2008) Polymorphic microsatellite markers for Philippine tarsiers (*Tarsius syrichta*). Conserv Genet 9: 1349–1351.
27. Merker S, Driller C, Perwitasari-Farajallah D, Zahner R, Zischler H (2007) Isolation and characterization of 12 microsatellite loci for population studies of Sulawesi tarsiers (*Tarsius* spp.). Mol Ecol Notes 7: 1216–1218.
28. Schuelke M (2000) An economic method for the fluorescent labeling of PCR fragments. Nat Biotechnol 18: 233–234.
29. Ronquist F, Huelsenbeck JP (2003) MRBAYES 3: Bayesian phylogenetic inference under mixed models. Bioinformatics 19: 1572–1574.
30. Posada D (2008) jModeltest: Phylogenetic model averaging. Molec Biol Evol 25: 1253–1256.
31. Rambaut A, Drummond AJ (2007) Tracer v1.4. Available: http://beast.bio.ed.ac.uk/Tracer.
32. Stamatakis A (2006) RAxML-VI-HPC: maximum likelihood-based phylogenetic analyses with thousands of taxa and mixed models. Bioinformatics 22: 2688–2690.
33. Stamatakis A, Blagojevic F, Nikolopoulos D, Antonopoulos C (2007) Exploring new search algorithms and hardware for phylogenetics: RAxML meets the IBM cell. J VLSI Signal Proc 48: 271–286.
34. Drummond AJ, Rambaut A (2007) BEAST: Bayesian evolutionary analysis by sampling trees. BMC Evol Biol 7: 214. Available: http://www.biomedcentral.com/1471-2148/7/214. Accessed 2013 December 13.
35. Fujisawa T, Barraclough TG (2013) Delimiting species using single-locus data and the generalized mixed Yule coalescent approach: a revised method and evaluation on simulated data sets. Syst Biol 62: 707–724.
36. Falush D, Stephens M, Pritchard JK (2003) Inference of population structure using multilocus genotype data: linked loci and correlated allele frequencies. Genetics 164: 1567–1587.
37. Falush D, Stephens M, Pritchard JK (2007) Inference of population structure using multilocus genotype data: dominant markers and null alleles. Molec Ecol Notes 7: 574–578. doi: 10.1111/j.1471-8286.2007.01758.x.
38. Hubisz MJ, Falush D, Stephens M, Pritchard JK (2009) Inferring weak population structure with the assistance of sample group information. Mol Ecol Resources 9: 1322–1332.
39. Pritchard JK, Stephens M, Donnelly P (2000) Inference of population structure using multilocus genotype data. Genetics 155: 945–959.
40. Spinks PQ, Thomson RC, Barley AJ, Newman CE, Shaffer HB (2010) Testing avian, squamate, and mammalian nuclear markers for cross amplification in turtles. Conserv Genet Resour 2: 127–129. doi: 10.1007/s12686-010-9184-7.
41. Vähä J-P, Primmer CR (2006) Efficiency of model-based Bayesian methods for detecting hybrid individuals under different hybridization scenarios and with different numbers of loci. Molec Ecol 15: 63–72. doi: 10.1111/j.1365-294X.2005.02773.x.
42. Earl DA, vonHoldt BM (2012) STRUCTURE HARVESTER: a website and program for visualizing STRUCTURE output and implementing the Evanno method. Conserv Genet Resour 4: 359–361. doi: 10.1007/s12686-011-9548-7.
43. Evanno G, Regnaut S, Goudet J (2005) Detecting the number of clusters of individuals using the software STRUCTURE: a simulation study. Molec Ecol 14: 2611–2620.
44. Rosenberg NA (2004) Distruct: a program for the graphical display of population structure. Molec Ecol Notes 4: 137–138.

45. Rannala B, Mountain JL (1997) Detecting immigration by using multilocus genotypes. P Natl Acad Sci USA 94: 9197–9201.
46. Bryant D, Moulton V (2004) Neighbor-Net: an agglomerative method for the construction of phylogenetic networks. Mol Biol Evol 21: 255–265.
47. Huson DH, Bryant D (2006) Application of phylogenetic networks in evolutionary studies. Mol Biol Evol 23: 245–267.
48. Shekelle M, Gursky-Doyen S, Richardson MC (2013) Family Tarsiidae (Tarsiers). In: Mittermeier RA, Rylands AB, Wilson DE, editors. Handbook of the mammals of the world - Volume 3. Washington DC: Lynx Edicions. pp. 63–64.
49. Brown RM, Siler CD, Oliveros CH, Esselstyn JA, Diesmos AC, et al. (2013) Evolutionary processes of diversification in a model island archipelago. Ann Rev Ecol Evol Syst 44: 411–435. doi: 10.1146/annurev-ecolsys-110411-160323.
50. Groves CP (1998) Systematics of tarsiers and lorises. Primates 39: 13–27.
51. Niemitz C (1984) Taxonomy and distribution of the genus *Tarsius* Storr, 1780. In: Niemitz C, editor. The biology of tarsiers. New York: Gustav Fischer Verlag. pp. 1–16.
52. Hill WCO (1955) Primates: comparative anatomy and taxonomy. II. Haplorhini: Tarsioidea. Edinburgh: Edinburgh University Press.
53. Funk CW, McKay JK, Hohenlohe PA, Allendorf FW (2012) Harnessing genomics for delineating conservation units. Trends Ecol Evol 27: 489–496.

11

Fc Receptors for Immunoglobulins and Their Appearance during Vertebrate Evolution

Srinivas Akula, Sayran Mohammadamin[¤], Lars Hellman*

Department of Cell and Molecular Biology, Uppsala University, The Biomedical Center, Uppsala, Sweden

Abstract

Receptors interacting with the constant domain of immunoglobulins (Igs) have a number of important functions in vertebrates. They facilitate phagocytosis by opsonization, are key components in antibody-dependent cellular cytotoxicity as well as activating cells to release granules. In mammals, four major types of classical Fc receptors (FcRs) for IgG have been identified, one high-affinity receptor for IgE, one for both IgM and IgA, one for IgM and one for IgA. All of these receptors are related in structure and all of them, except the IgA receptor, are found in primates on chromosome 1, indicating that they originate from a common ancestor by successive gene duplications. The number of Ig isotypes has increased gradually during vertebrate evolution and this increase has likely been accompanied by a similar increase in isotype-specific receptors. To test this hypothesis we have performed a detailed bioinformatics analysis of a panel of vertebrate genomes. The first components to appear are the poly-Ig receptors (PIGRs), receptors similar to the classic FcRs in mammals, so called FcRL receptors, and the FcR γ chain. These molecules are not found in cartilagous fish and may first appear within bony fishes, indicating a major step in Fc receptor evolution at the appearance of bony fish. In contrast, the receptor for IgA is only found in placental mammals, indicating a relatively late appearance. The IgM and IgA/M receptors are first observed in the monotremes, exemplified by the platypus, indicating an appearance during early mammalian evolution. Clearly identifiable classical receptors for IgG and IgE are found only in marsupials and placental mammals, but closely related receptors are found in the platypus, indicating a second major step in Fc receptor evolution during early mammalian evolution, involving the appearance of classical IgG and IgE receptors from FcRL molecules and IgM and IgA/M receptors from PIGR.

Editor: Nikolas Nikolaidis, California State University Fullerton, United States of America

Funding: This work was made possible by grants from the Swedish Technical Natural Sciences Research Council (VR-NT). The funders had no role in study design, data collection and analysis, decision to publish, or preparation of the manuscript.

Competing Interests: The authors have declared that no competing interests exist.

* E-mail: Lars.Hellman@icm.uu.se

¤ Current address: Salahaddin University, College of Science, Biology Department, Erbil, Iraqi Kurdistan

Introduction

Immunoglobulins (Igs) are only found in jawed vertebrates and there are strong indications that the complexity of the adaptive immune system has increased gradually during vertebrate evolution. The effector functions of the Igs have separated into different Ig classes thereby increasing the regulatory potential of the immune system. Mammals express up to six different Ig classes: IgM, IgD, IgG, IgE, IgA and IgO, and the total number of isotypes can sometimes exceed 15 (Figure 1) [1]. Of these six Ig classes only IgM and IgD have been found in fish, which tend to have only two to three Ig classes and isotypes. In fish, the list of Ig classes now also includes IgW, IgNAR, IgT and IgZ [2–6]. In general, amphibians have four to five classes of Igs: IgM, IgD, IgA/IgX, IgY and one additional class named IgF or IgP [7,8]. Neither IgA nor IgY is found in fish and IgG and IgE have not been identified in reptiles, amphibians or birds, suggesting they are unique for mammals [9]. Birds have only three Ig classes: IgM, IgA and IgY (Figure 1). However, this low number is most likely a result of a loss of isotypes as massive losses and re-expansion of genes and gene families have occurred in birds [10]. Interestingly, other members of the reptile lineage have instead experienced massive expansions like the Chinese alligator, which has 10 isotypes, although neither IgG nor IgE [4]. All major classes in placental mammals are also present in monotremes, the egg-laying mammals. The platypus has been shown to express six classes and eight different isotypes: IgM, IgG1, IgG2, IgA1, IgA2, IgE, IgD and IgO (Figure 1) [1,11–14]. Marsupials, as represented by the American opossum have IgM, IgG, IgA and IgE, but no IgD gene [15,16]. The gene for IgD has probably been lost in this lineage (Figure 1) [17]. These findings indicate that the number of Ig classes and isotypes has increased during vertebrate evolution from two to three in fish, to sometimes more than fifteen in mammals (Figure 1). The most important steps in this increase have probably been the duplication of IgM forming the early ancestor of IgY. This seems to have occurred at the emergence of the tetrapods, The second step was most likely a duplication of IgM, forming the ancestor of IgA/X. This isotype is also first observed in amphibians. The third major step was the duplication of IgY forming IgG and IgE. However, losses of classes and isotypes have also occurred as exemplified by the loss of IgD in birds and marsupials. Interestingly, the chicken has only one light chain isotype, whereas mammals have two and many fish species have three, indicating that birds have lost two light chain loci. The increase in the number of different Ig classes and isotypes observed in many tetrapods has likely been accompanied by a similar

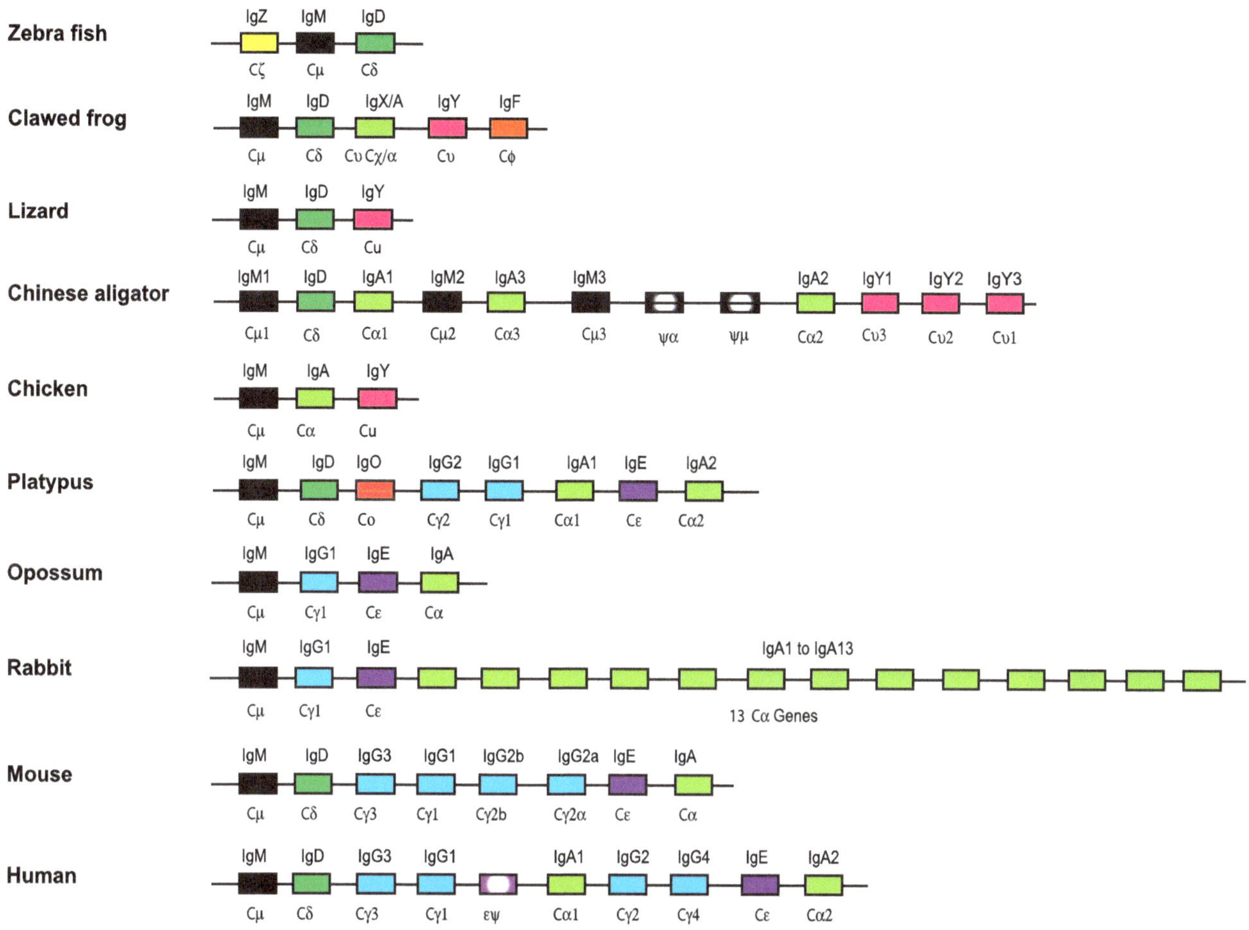

Figure 1. The immunoglobulin heavy chain locus of a panel of selected vertebrates from fish to humans. The locus depicts every single gene as a block and without individual exons. The figure is not to scale and the genes have been color coded; IgM in black, IgD in dark green, IgA and IgX in light green, IgY in magenta, IgG in blue, IgE in purple, IgO in red, IgZ in yellow, IgF in orange and pseudogenes in shaded grey or purple.

increase in isotype-specific receptors. However, how this increase has occurred is still only partly understood.

Humans and mice have a complex set of Fc-specific receptors that match the number of different Ig-isotypes and subclasses (Figures 2 and 3). Four major types of classical Fc receptors for IgG: FcγRI, FcγRII, FcγRIII and FcγRIV, which show varying affinities for the four IgG isotypes, have been identified [18,19]. Additionally, one high-affinity receptor for IgE (FcεRI), one for both IgM and IgA (Fcα/μR), one for IgM (FcμR) and one receptor for IgA (FcαRI) are found in almost all placental mammals studied [20–22]. The α chains of these receptors, which are the Ig-binding subunits, are all related in structure and it's likely they originate from one or a few common early ancestors via successive gene duplications [23]. In humans, all α chains of the IgG and IgE receptors have two extracellular Ig-like domains except for FcγRI, which contains three such domains. The opossum has recently been shown to have all of the classical IgG and IgE receptors except FcγRI, the high-affinity receptor for IgG, indicating that the two-domain receptors were the first to appear during mammalian evolution [24]. All of the IgG and IgE receptors, except one variant of FcγRIII in humans, FcγRIIIB, also have a transmembrane region that anchors the receptor in the membrane [18]. FcγRIIIB is a glycosylphosphatidylinositol (GPI) anchored activating receptor that has no cytoplasmic tail. In some of these receptors the cytoplasmic tail region is mediating the signal to the cell. However, for many of them the signaling function is found in another subunit of the receptor, the γ chain [25]. This non Ig-domain containing signaling subunit is a member of a small family of related molecules including the T-cell receptor (TCR) zeta chain, DAP10 and DAP12 [26–29]. The latter two proteins serve as signaling components of the NK cell receptor and related Ig-domain containing receptors [29]. The biological functions of the FcRs are regulated by immunoreceptor tyrosine-based activation motifs (ITAMs) and immunoreceptor tyrosine-based inhibitory motifs (ITIMs), which activate or inhibit the cellular function, respectively [18,20]. These motifs are located in the cytoplasmic part of the γ chain and in the case of FcγRIIB, in the cytoplasmic part of the α chain. FcRs with ITAMs elicit cell activation, endocytosis and phagocytosis, whereas receptors with ITIMs have an inhibitory effect on cell activation. The FcεRI also has a separate β chain, which most likely has a role in enhancing signaling, given it has one cytoplasmic ITAM motif [26]. The β chains show no homology to the α chains, are located on another chromosome, and are members of the MS4A family, which includes molecules like CD20, HTm4 and at least 18 other members in the human genome [30].

The FcμR has only one Ig domain and is relatively distantly related to the other receptors. Its cytoplasmic part has a well conserved Ser-Tyr rich region instead of an ITIM or ITAM, and

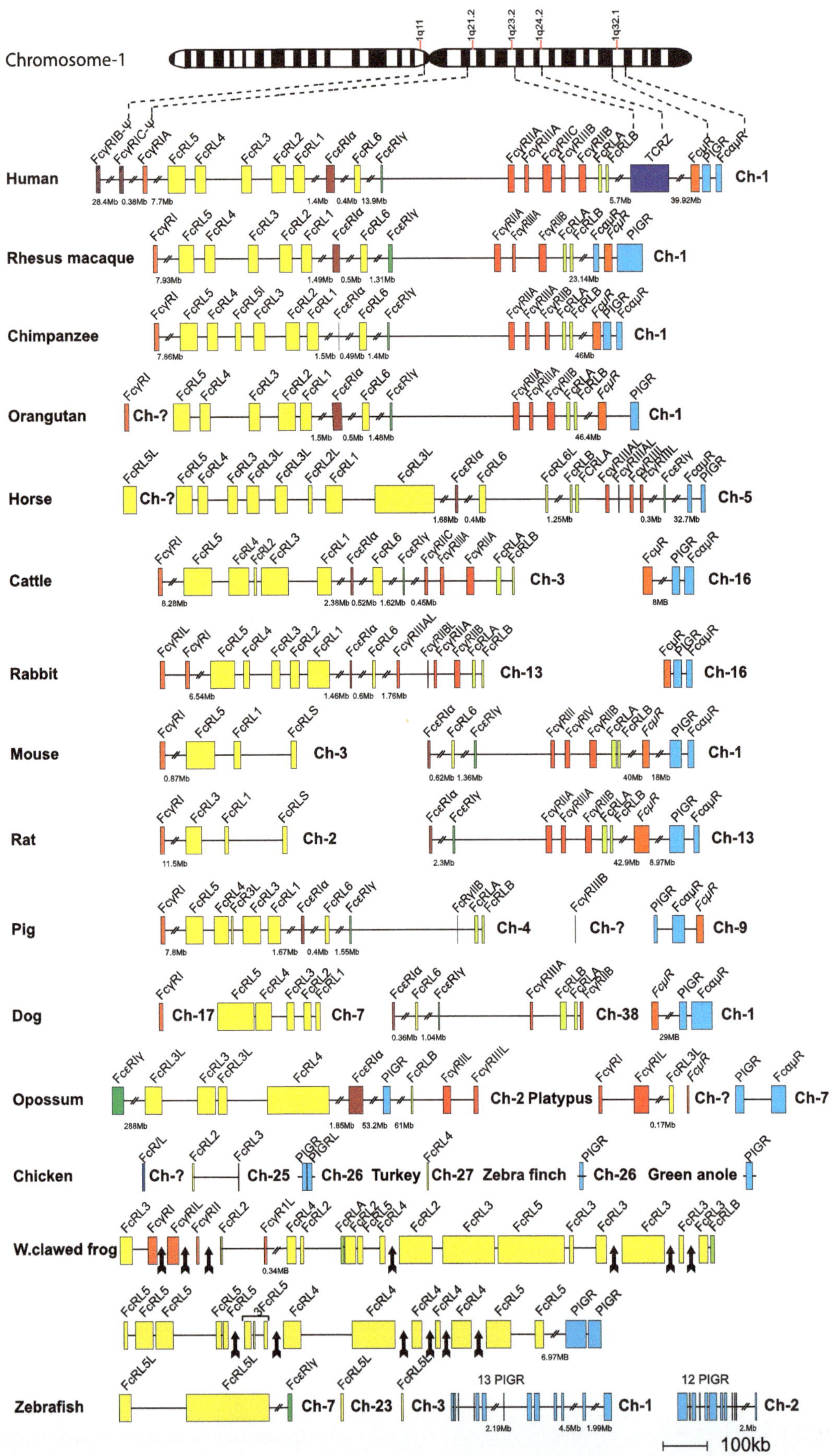
Chromosome-1
1q11
1q21.2
1q23.2
1q24.2
1q32.1
Human
FcγRIB-ψ FcγRIC-ψ FcγRIA FcRL5 FcRL4 FcRL3 FcRL2 FcRL1 FcεRIα FcRL6 FcεRIγ FcγRIIA FcγRIIIA FcγRIIC FcγRIIIB FcγRIIB FCRLA FCRLB TCRZ FcμR PIGR FcαμR
Ch-1
28.4Mb 0.38Mb 7.7Mb 1.4Mb 0.4Mb 13.9Mb 5.7Mb 39.92Mb
Rhesus macaque
FcγRI FcRL5 FcRL4 FcRL3 FcRL2 FcRL1 FcεRIα FcRL6 FcεRIγ FcγRIIA FcγRIIIA FcγRIIB FCRLA FCRLB FcαμR FcμR PIGR
Ch-1
7.93Mb 1.49Mb 0.5Mb 1.31Mb 23.14Mb
Chimpanzee
FcγRI FcRL5 FcRL4 FcRL5l FcRL3 FcRL2 FcRL1 FcεRIα FcRL6 FcεRIγ FcγRIIA FcγRIIIA FcγRIIB FCRLA FCRLB FcμR PIGR FcαμR
Ch-1
7.86Mb 1.5Mb 0.49Mb 1.4Mb 46Mb
Orangutan
FcγRI FcRL5 FcRL4 FcRL3 FcRL2 FcRL1 FcεRIα FcRL6 FcεRIγ FcγRIIA FcγRIIIA FcγRIIB FCRLA FCRLB FcμR PIGR
Ch-? Ch-1
1.5Mb 0.5Mb 1.48Mb 46.4Mb
Horse
FcRL5L FcRL5 FcRL4 FcRL3 FcRL3L FcRL3L FcRL2L FcRL1 FcRL3L FcεRIα FcRL6 FcRL6L FCRLB FCRLA FcγRIIIAL FcγRIIIAL FcγRIIIL FcγRIIIL FcεRIγ FcαμR PIGR
Ch-? Ch-5
1.68Mb 0.4Mb 1.25Mb 0.3Mb 32.7Mb
Cattle
FcγRI FcRL5 FcRL4 FcRL2 FcRL3 FcRL1 FcεRIα FcRL6 FcεRIγ FcγRIIC FcγRIIIA FcγRIIA FCRLA FCRLB FcμR PIGR FcαμR
Ch-3 Ch-16
8.28Mb 2.38Mb 0.52Mb 1.62Mb 0.45Mb 8MB
Rabbit
FcγRIIL FcγRI FcRL5 FcRL4 FcRL3 FcRL2 FcRL1 FcεRIα FcRL6 FcγRIIIAL FcγRIIBL FcγRIIA FcγRIIB FCRLA FCRLB FcμR PIGR FcαμR
Ch-13 Ch-16
6.54Mb 1.46Mb 0.6Mb 1.76Mb
Mouse
FcγRI FcRL5 FcRL1 FcRLS FcεRIα FcRL6 FcεRIγ FcγRIII FcγRIV FcγRIIB FCRLA FCRLB FcμR PIGR FcαμR
Ch-3 Ch-1
0.87Mb 0.62Mb 1.36Mb 40Mb 18Mb
Rat
FcγRI FcRL3 FcRL1 FcRLS FcεRIα FcεRIγ FcγRIIA FcγRIIIA FcγRIIB FCRLA FCRLB FcμR PIGR FcαμR
Ch-2 Ch-13
11.5Mb 2.3Mb 42.9Mb 8.97Mb
Pig
FcγRI FcRL5 FcRL4 FcRL3L FcRL3 FcRL1 FcεRIα FcRL6 FcεRIγ FcγRIIB FCRLA FCRLB FcγRIIIB PIGR FcαμR FcμR
Ch-4 Ch-? Ch-9
7.8Mb 1.67Mb 0.4Mb 1.55Mb
Dog
FcγRI FcRL5 FcRL4 FcRL3 FcRL2 FcRL1 FcεRIα FcRL6 FcεRIγ FcγRIIIA FCRLB FCRLA FcγRIIB FcμR PIGR FcαμR
Ch-17 Ch-7 Ch-38 Ch-1
0.36Mb 1.04Mb 29MB
Opossum
FcεRIγ FcRL3L FcRL3 FcRL3L FcRL4 FcεRIα PIGR FCRLB FcγRIIL FcγRIIIL
Ch-2 Platypus
FcγRI FcγRIIL FcRL3L FcμR PIGR FcαμR
Ch-? Ch-7
288Mb 1.85Mb 53.2Mb 61Mb 0.17Mb
Chicken
FcRIL FcRL2 FcRL3 PIGR PIGRL FcRL4 PIGR PIGR
Ch-? Ch-25 Ch-26 Turkey Ch-27 Zebra finch Ch-26 Green anole
W.clawed frog
FcRL3 FcγRI FcγRIIL FcγRII FcRL2 FcγR1L FcRL4 FcRL2 FcRLA FcRL2 FcRL5 FcRL4 FcRL2 FcRL3 FcRL5 FcRL3 FcRL3 FcRL3 FcRL3 FcRL3 FCRLB
0.34MB
FcRL5 FcRL5 FcRL5 FcRL5 FcRL5 3FcRL5 FcRL4 FcRL4 FcRL4 FcRL4 FcRL4 FcRL5 FcRL5 PIGR PIGR
6.97MB
Zebrafish
FcRL5L FcRL5L FcεRIγ FcRL5L FcRL5L
Ch-7 Ch-23 Ch-3
13 PIGR
Ch-1
2.19Mb 4.5Mb 1.99Mb
12 PIGR
Ch-2
2.Mb
100kb

Figure 2. The major Fc receptor gene loci in a panel of different vertebrates. Each horizontal line corresponds to a chromosome on which different Fc receptor genes are located. Genes are color coded. The Fc receptor-like (FcRL) genes are shown in yellow except the FcRLA and B that are in light green, classical IgG receptors in red (pseudogenes in striped red), IgE receptor in dark green, the IgM receptor in orange, the poly-Ig receptor (PIGR) and the IgA/IgM receptor in blue. The figure contains the genes identified for the following animal species: Zebrafish (*Danio rerio*), Western clawed frog (*Xenopus silunarana* and *Xenopus tropicalis*), Chicken (*Gallus gallus*), Turkey (*Meleagris gallopovo*), Zebra Finch (*Taeniopygia guttata*), Green anole (*Anolis carolinennsis*), Platypus (*Ornithorhynchus anatinus*), Opossum (*Monodelphis domestica*), Rat (*Ratuus norvegius*), Mouse (*Mus musculus*), Horse (*Equus caballus*), Rabbit (*Oryctologus cunicullus*), Pig (*Sus scrofa*), Dog (*Canis lupus familaris*), Cattle (*Bos taurus*), Orangutan (*Pongo abelli*), Chimpanzee (*Pan troglodytes*), Rhesus macaque (*Macaca mulatta*), Human (*Homo sapiens*).

is located in the same chromosomal region as the IgG and IgE receptors [31].

In addition to all the receptors described above where the α chains have Ig-like domains, there are at least two FcRs of a totally unrelated nature, the IgD and the low-affinity IgE receptor, also named CD23. Both of these are lectins and will not be discussed more in this communication.

In addition to the classical FcRs, a new family of related receptors was discovered upon the completion of full genome sequences from a number of mammalian species [32,33]. Eight different such Fc receptor-like (FcRL) genes have been identified in the human genome, FcRL1-FcRL6 as well as FcRLA and FcRLB. Duplications of these genes are denoted FcRLxL, where *x* represents one of the original FcRL1-6. Interestingly, the number of extracellular Ig-like domains differs quite significantly between these receptors; from nine in FcRL5 to three in FcRL1 and 6. In humans, all of them are located in the same chromosomal region as the majority of the classical FcRs. In the same region we find FcRLA and FcRLB. Both of these members lack a transmembrane region and are found in the cytoplasm of the cell [32,33]. The ligands for these potential receptors have been a mystery until very recently, where FcRL4 and 5 were found to bind IgA and IgG, respectively, and FcRL6 was found to interact with MHC class II molecules [34–36]. However, the other FcRL molecules still await identifying ligands and functions. In contrast to the majority of the classical FcRs, the FcRL molecules seem to have an overrepresentation of ITIMs in their cytoplasmic tails, which indicates that they primarily have a dampening function during immune responses. Interestingly, and in accordance with such a hypothesis, genetic studies have indicated involvement of variants of FcRL3 in autoimmune disease [37].

The poly-Ig receptor (PIGR) is one additionally related receptor that is also found in the same chromosomal region as the classical FcRs, and is the transport receptor for IgA and IgM [21,38]. This molecule has two to five extracellular Ig-domains depending on the species [21,39–41].

Other Ig-domain containing receptors like the natural killer Ig-like receptors (KIRs), the leukocyte Ig-like receptors (LILRs) and the leukocyte-associated Ig-like receptors (LAIRs), are also distantly related to the classical FcRs and the FcRL receptors [42,43]. The mouse homologs of the LILRs in humans are named PIRs. The ligands for the KIRs and LILRs of mammals are primarily MHC class I molecules [44,45]. Interestingly, the FcαR shows higher sequence relatedness (35%) to the KIRs and LILRs than to the classical FcRs (only 20% identity) and is also located within the LRC on chromosome 19 in humans together with the KIRs and LILRs [45,46]. Homologs to the genes in the LRC gene complex have been found in early ancestors of vertebrates, the tunicates [47]. A large number of related genes have also been identified in several species of fish including catfishes, indicating an early appearance of these receptors during vertebrate evolution [48]. The LRC genes in chicken are named Ig-like receptors (CHIRs) [49], which all have one or two extracellular Ig-like domains.

Generally, Ig domains have been classified as V, C1, C2 or I domains depending on features such as the spacing of cysteine bridges and the number of beta sheets. V domains are generally found in variable regions of Igs and TCRs as well as in CD2 CD4, CD80 and CD86. C1 domains are found in constant regions of Igs, TCR and in MHC class I and II. C2 domains are found in CD2, CD4, CD80, VCAM and ICAM, and I domains are found in VCAM, ICAM, NCAM, MADCAM and numerous other diverse protein families (EMBL-EBI InterPro). All Ig domains of CHIRs, KIRs, LILRs, the FcRL and the classical FcRs are classed as C2 domains and the Ig domains of the PIGR, IgM, and IgA/IgM receptors are V type domains [49,50].

An interesting possibility is that all of these Ig-domain receptors within the FcR locus and the LRC originate from a common ancestor. These two loci may be the result of the genome duplications that most likely occurred during early vertebrate evolution [47,51]. Due to fact that receptors from both the LRC and the FcR loci can be positive and negative regulators, depending on the presence of ITIMs or ITAMs, they have been named paired receptors [52].

In order to obtain a more detailed picture of the appearance of all these receptors and how their appearance correlates to the increase in Ig-isotype repertoire, we used bioinformatics to screen for related receptors in a panel of vertebrate genomes. The results presented in this communication show that the transport receptor for IgA and IgM, the PIGR, the FcRL family and the signaling component the common γ chain appeared with the bony fish, indicating a major event in the evolution of these receptors at the base of the bony fish. A second major event in FcR evolution apparently occurred during early mammalian evolution with the appearance of the IgM, the IgA/M receptors and the formation of a separate subfamily of FcRL molecules which today includes all of the classical receptors for IgG and IgE.

Materials and Methods

Identifying the gene loci of FcRs and their sequences was conducted using NCBI (National center for Biotechnology Information) and Ensemble databases (http://www.ncbi.nlm.nih.gov/) (Ensemble http://www.ensembl.org/index.html). Human FcR protein sequences were initially used as query sequences using TBLASTN (translation BLAST, Translate Basic Local Alignment Search Tool). Later sequences obtained from the studied species were used to screen for potential additional homologous genes to complete the picture. The gene loci were identified from the resulting data and the sizes of the genes, including the distance between the neighboring genes, were calculated and used to construct maps in scale of the loci. The gene arrangement and receptor domain figures were made in Adobe illustrator. The gene accession numbers used in constructing the figures are listed in Supporting Information S1.

Phylogenic analysis

In order to study the relationship between the various sequences, the entire coding region of the proteins were run in

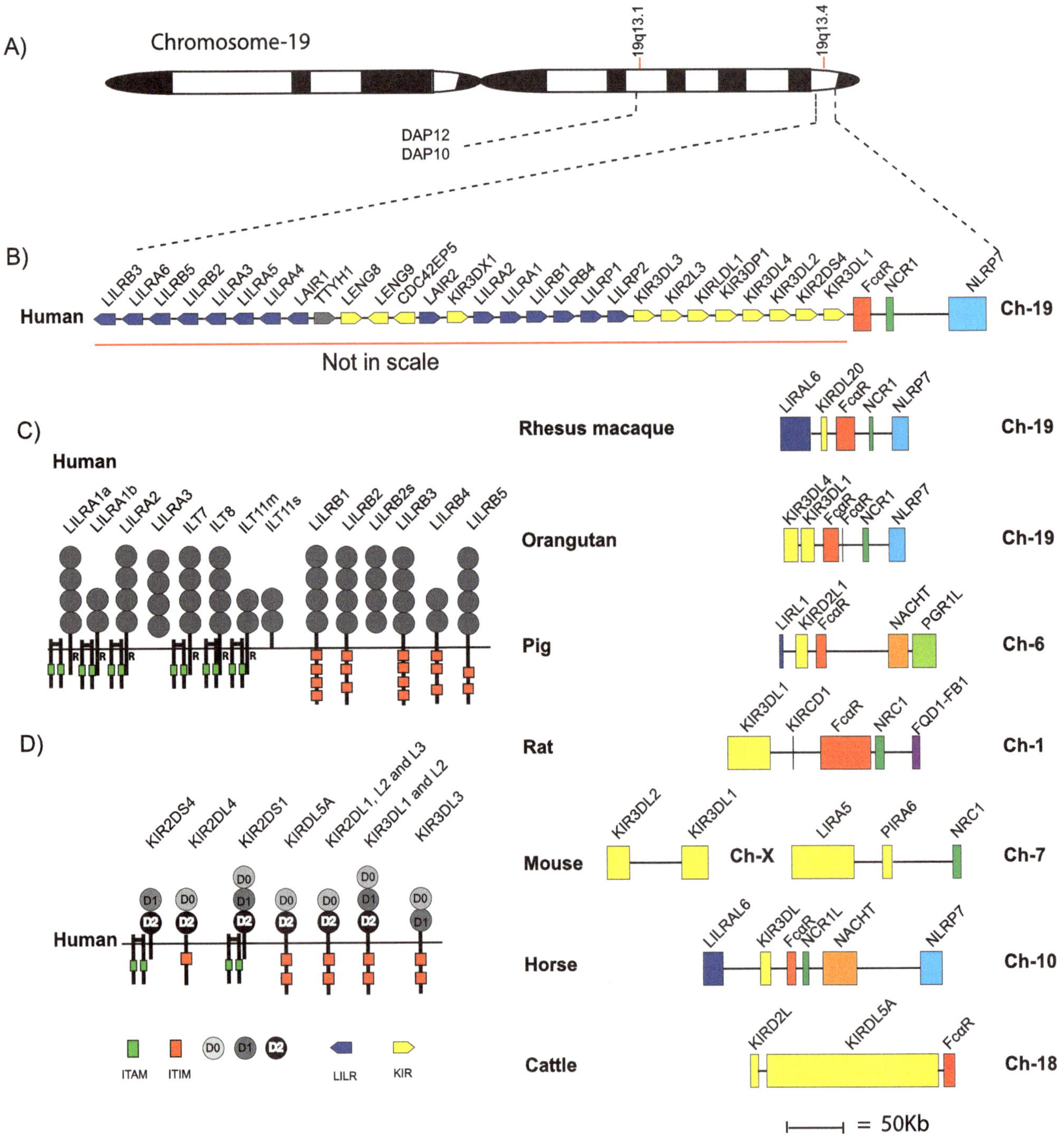

Figure 3. The FcαR locus and neighboring genes. Panel A shows a schematic drawing of chromosome 19, panel B a schematic drawing of the LRC locus, panel C the domain structure of the LILRs of this locus (an 'R' represents arginine in the transmembrane region) and panel D the domain structure of the KIRs. In the detailed map of the region encoding the IgA receptor each horizontal line corresponds to a chromosome on which different Fc receptor genes are located. Genes are color coded. The IgA receptor in red, the KIRs and LILRs in yellow, the neighboring genes that are used as markers for the chromosomal region, NCR1 in dark green, the NACHT in orange, the NLRP7 in light blue, the PGRL1 in light green and the FQD1-FB1 in dark blue. The overall structure of the LRC locus and the protein domain structures have been adopted from Espeli et al 2010 and Brown et al 2004 [45,68].

multiple alignments using MAFFT. For phylogeny analyses we used the PHYLIP, MrBayes, MEGA and CLUSTALW programs [53,54,55].

Alignments. The selected Fc receptors of vertebrates were aligned using version 7 of MAFFT (http://mafft.cbrc.jp/alignment/server/), using the E-INS_I strategy for optimal results for sequences with conserved motifs and carrying multiple domains, with default parameters [56].

Phylogeny. Clustal X: MAFFT alignment files was imported into Clustal X version 2.0 [55] to generate a neighbor joining tree with 100 bootstrap replicates. The tree file was viewed in Fig tree (http://tree.bio.ed.ac.uk/software/figtree/). Phylip: The Phylip package (v3.69)(http://evolution.genetics.washington.edu/phylip.html) was used for constructing Maximum-likelihood and Distance trees [53]. For the distance method PROTDIST, FITCH (JTT matrix Model without using an out-group species) were used. For the bootstrap analyses SEQBOOT, PROTDIST, NEIGHBOR and CONSENSE was used to generate 100 replicate data sets from the Phylip package (v3.69). For Maximum-likelihood trees using PROML and viewed in Fig tree. MrBayes: The phylogenetic analysis was performed using a Bayesian approach as used in MrBayes version 3.2. [54]. Markov Chain Monte Carlo (MCMC) analysis was used to approximate the posterior probabilities of the trees. The phylogenetic tree was drawn in FigTree 1.3.1. MEGA (Molecular Evolutionary Genetics Analysis Version 5.2.2)[57]. Multiple alignment file obtained from MAFFT converted into MEGA format and phylogenetic analysis was performed using Neighbor joining and Maximum-likelihood methods in both methods using boot replications was 500. The distances were calculated by JTT matrix method.

The individual domains of the FcR for a few selected key species were also analyzed for their relatedness using the MEGA Neighbor joining and MEGA Maximum-likelihood methods. The domains were color coded according to a pattern that previously been used by several other labs [24,32,58]. The results are shown in Figure S4.

Results

Human FcR protein sequences were used as query sequences to identify similar sequences in a large panel of vertebrate genomes in the NCBI and Ensemble databases using the TBLASTN (translation BLAST) algorithm. The different gene loci were identified, the size of the genes and the distances between genes were calculated and used to produce maps drawn to scale of these chromosomal regions as presented in figures 2 and 3.

Mammals

All the primates that we have analyzed (human, macaque, chimpanzee and orangutan) have their classical FcRs and the FcRL molecules (except FcαR), on chromosome 1 (Figure 2). However, these genes are not found in a small well-defined region but spread over a large portion of chromosome 1. In contrast, the IgA receptors are located together with some of the NK cell receptors, the KIRs, and the LILRs on chromosome 19 (Figure 3). Interestingly, homologs to the human IgA receptor were only found in placental mammals and not in any of the other vertebrates, indicating a relatively late appearance. The FcαR is located on chromosome 19 in humans, macaques and orangutans, on chromosome 6 in pigs, chromosome 1 in rats, chromosome 10 in horses and on chromosome 18 in cattle (Figure 3). However, no IgA receptor has been identified in the mouse genome. The region, which in other placental animals harbors the FcαR, is broken up into two fragments in mice. These products have ended up on different chromosomes, by a process that most likely resulted in the loss of the FcαR gene (Figure 3).

Within the cluster of genes on chromosome 1 in primates, which contains all the classical receptors and the FcRL molecules (except the specific IgA receptor), a few differences are found between species in gene numbers and organization. In the human genome there is a duplication involving the low-affinity receptors for IgG (FcγRII, and FcγRIII), as well as two additional genes related to the high-affinity receptor for IgG, FcγRI. The new genes for the low-affinity receptors have been named FcγRIIC and FcγRIIIB and the new high-affinity receptor genes, FcγRIB and FcγRIC. These two latter genes are inactivated pseudogenes (Figure 2). The chimpanzee has one extra FcRL5L receptor gene, which is probably the result of a recent gene duplication. In the orangutan, FcγRI is still on an unassigned contig and therefore cannot definitely be assigned to chromosome 1.

In the horse, all the classical FcRs, the FcRL1-6 and FcRLA and B are also found on one chromosome, here on chromosome 5. However, FcγRI has not yet been identified, which may be due to an incomplete genome. In many of the placental mammals this locus has been broken up into several fragments, which have ended up on different chromosomes. This indicates that in many placental mammals, frequent translocations in this region have occurred (Figure 2). In rats and mice, the loci are separated into two fragments. In the rat, FcγRI, FcRL1 and FcRLs are on chromosome 2, whereas the intracellular proteins FcRLA and FcRLB, the PIGR, the FcαμR, the FcμR and the FcεRIα are on chromosome 13 (Figure 2). In the mouse, FcγRI, FcRL5, FcRL1 and FcRLS are on chromosome 3. The remaining receptors are on chromosome 1 (Figure 2). FcRLS is a soluble mosaic protein containing a scavenger receptor cysteine rich type B domain (SRCR) [32].

Interestingly, the rabbit has all the genes of this locus on chromosome 13, except for FcαμR, the FcμR and PIGR, which are located on chromosome 16. A duplication of the high-affinity receptor for IgG, FcγRI has also occurred in the rabbit genome (Figure 2). Cattle show a similar organization as rabbits with FcαμR, the FcμR and PIGR on one chromosome, 16, whereas the other receptors are on chromosome 3. Pigs also show a similar organization with FcαμR, FcμR and PIGR on chromosome 9 and the other receptors on chromosome 4. However, one of the genes FcγRIIIB is still on an unassigned contig (Figure 2). Dogs show the most complicated pattern with these genes separated onto four different chromosomes. The FcRL receptors are on chromosome 7, the intracellular proteins, FcRLA and FcRLB, and the classical FcRs on chromosome 38, FcγRI on chromosome 17 and FcμR, FcαμR and PIGR on chromosome 1 (Figure 2).

In marsupials almost all of the different families identified in humans are present. In the opossum, all the classical receptors, FcγRIIIL, FcγRIIL, FcεRI, the FcRL receptors, FcRL3L, FcRL3, FcRL4 and the intracellular proteins FcRLA, FcRLB, as well as the PIGR, and FcεRIγ are found on chromosome 2 (Figure 2). However, no copy of the three-domain receptor for IgG, the FcγRI was identified as previously also shown by Fayngerts et al. [24]. The opossum genome sequence has gone through peculiar changes and sequences that appeared in the draft genome have sometimes disappeared in the final assembly. This makes a detailed analysis complicated as genes may have been deleted by mistake.

The picture gets more complicated when analyzing the monotremes. Despite a relatively incomplete platypus genome, two genes with homology to the high-affinity IgG receptor, the FcγRI are found (Figure 2). These genes FcγRIα, FcγRIαL, plus two genes with homology to FcRL3L and the PIGR are found on chromosome 7. In the same region there are pieces of two, two-domain receptors, indicating that both two- and three-domain receptors were already present during early mammalian evolution. No copy of the the IgA receptor was found in the platypus genome but a copy of the FcμR and very recently the FcαμR was found indicating a slightly earlier appearance of these latter receptors, possibly during early mammalian evolution.

Reptiles and Birds

Reptiles and birds add a layer of complexity to the picture. There are only a few genomes that are close to completion and in birds, massive losses and re-expansions of genes and gene families have occurred, which complicates the analyses even more [10]. In the chicken genome there are no close homologs to the classical FcRs, and no FcRLA or B. Instead, the chicken has undergone a massive expansion of KIR-related CHIRs found in the LRC region in mammals, where a large number of them are found on micro chromosome 31 [49]. Of the other receptors found on chromosome 1 in primates, the PIGR, one FcRL2L gene as well as one additionally related sequence FcR/L are found (Figure 2). The FcRL2 and FcRL3 are situated on chromosome 25, two copies of the PIGR are located on chromosome 26 and FcR/Lon a yet unknown chromosome in the chicken. In the incomplete turkey genome there is a FcRL4-related gene on chromosome 27. The zebra finch has one copy of the PIGR on chromosome 26. Looking at a reptile, the green anole, there is a contig that harbors the PIGR (Figure 2).

Amphibians

In the Western clawed frog (*Xenopus tropicalis*), there are three gene copies of a receptor, showing similarity to the the α chain of FcγRI (Figure 2). All three copies have three extracellular Ig-like domains and have been named FcγRI and FcγRIL and FcγRIL. However, they are relatively distantly related and have therefore likely appeared via two successive gene duplications early in amphibian evolution (Figure 4). Additionally, one copy of the PIGR, FcRLA and B and a very large number of FcRL genes were found (Figure 3). Very recently a copy of a two-domain receptor appeared in the current version of the *Xenopus* genome. A massive expansion of the FcRL genes has resulted in at least 31 genes, which all show similarity to the human FcRL genes 3, 4 and 5 (Figures 2 and 5). The clusters of FcRL3L, FcRL4L, FcRL5L and the FcγRI receptors are found on different scaffolds. The fact that the genome sequence is incomplete and that the genes are scattered over a large number of scaffolds makes the analysis difficult, where the likelihood that more genes will be found is high.

Bony fish

In the zebrafish only the PIGRs, one FcεRIγ chain gene and four FcRL5L genes are found. In zebrafish, the PIGR and FcεRIγ chain are located on chromosome 2, and the four FcRL5L genes are shared on chromosome 3 and 7, with two on each (Figure 2). Interestingly, a massive duplication of the PIGRs has occurred in the zebrafish. Thirteen very similar PIGR genes are found in a cluster on chromosome 1 and twelve on chromosome 2 (Figure 2). The 25 copies in the zebrafish all have two or three extracellular Ig-like domains and they are no greater than 20% different in amino acid sequence (data not shown), indicating a relatively recent expansion.

Cartilagous fish and lampreys

The genomes of a number of shark species (elephant, ground, requiem, carpet, nurse, grey, cat, dogfish, tiger, horn and angel) and the lamprey were analyzed for genes homologous to the various receptors. However, no homologs of either the PIGRs, the FcεRIγ chain, the FcRL or the classical receptors were found even after extensive screening using both mammalian and fish sequences, indicating that all of them first appear with the bony fish. Interestingly, no direct homologs to the LRC genes with only C2 domains were found in any of the shark genomes or the lamprey also indicating that they appear with the bony fish.

Phylogenetic analysis

All the FcR sequences represented in figures 2 and 3 were analyzed for sequence relatedness using Neighbor-joining, Distance, and Maximum-likelihood methods. In this study we used the entire sequence of the receptors. The MrBayes tree is presented in figure 4, whereas the Neighbor-joining analysis, Maximum likelihood and the Distance trees are presented as supplementary figures S1, S2 and S3. The three V domain-containing receptors (the IgM, the PIGR and the IgM/IgA receptors) are the most distantly related members of this group of proteins, which spontaneously form an out-group. Regarding the classical receptors, all of the IgG and the IgE receptors form separate sub-branches but cluster together in one major branch, including the three-domain receptor FcγRI. The FcRL sequences are more closely related to the IgG and IgE receptors than to the IgM and IgA receptors. In several of the trees some of the platypus sequences cluster with the classical Fc receptors, in particular the FcγRI sequences. This latter finding indicates that the platypus sequences are actually the first step in the formation of a separate group of receptors that now constitute the classical IgG and IgE receptors.

Interestingly, the Western clawed frog sequences form separate branches in the various trees, indicating that the individual members have appeared after the separation from the other vertebrate lineages. Another prominent finding is that all the classical Fc receptors cluster with FcRL sequences, often closest with FcRLA and B or as in the case with the three platypus FcγRI-like sequences, with platypus FcRL3 (Figure 4). All three, three-domain receptors in the Western clawed frog genome that were identified during the screening with the human FcγRI sequence also cluster with frog FcRL3, 4, 5 and the FcRLA and B (Figures 4, S1, S2 and S3).

Domain structure

Individual domains of the majority of the FcRs from a few selected key species were analyzed for their relatedness by a phylogenetic analysis using Neighbor joining and Maximum likelihood methods. The result from the two analyzes were almost identical and therefore one is found as Figure S4. In total, 333 domains were included in the analysis. In Figure 5, the result from this analysis has been translated into a figure showing the number of domains and their relationship by color-coding the individual domains. Domain type 1 is coded in red. The zebrafish FcRL molecules have domains that are related to the domain 1 of mammals and are therefore coded in a lighter shade of red. Domain type 2 is shown in dark blue, domain type 3 in yellow and type 4 and 5 that are closely related and shown in light blue and green, respectively. A number of the Xenopus FcRL domains are closely related to the type 3 domains of mammals and are therefore coded in yellow. The V type domains of the PIGR, the IgM and the IgA/IgM receptors form a separate branch in the phylogenetic tree and are shown in grey in figure 5. The summary figure 5 shows the number of extracellular Ig-like domains and the intracellular region with ITAMs and ITIMs. The question of ITAMS and ITIMs is not easy to resolve as several motifs are non-canonical ITIMs with similarity to the consensus. We have therefore depicted consensus and closely related sequences separately in figure 5. Interestingly, many of the FcRL molecules lacked consensus ITIMs and in Xenopus some FcRL5 molecules had up to 2 consensus ITIMs whereas other members of the same subfamily had no ITIMs, or even closely related sequences. Due to the large number of FcRL genes in Xenopus, a few of them are shown in figure 5 and a number of additional members are depicted in figure S5.

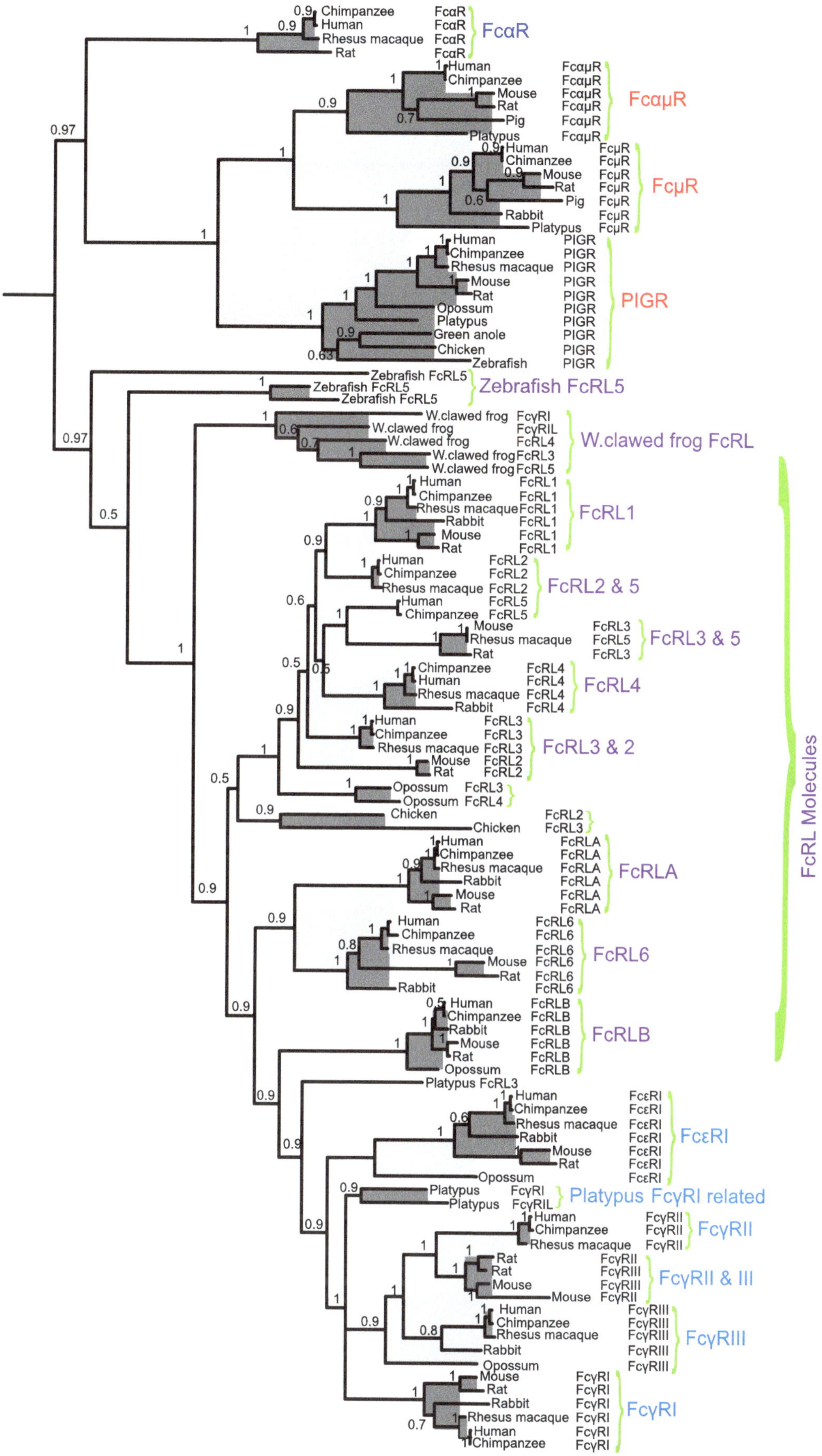
FcαR
FcαμR
FcμR
PIGR
Zebrafish FcRL5
W.clawed frog FcRL
FcRL1
FcRL2 & 5
FcRL3 & 5
FcRL4
FcRL3 & 2
FcRLA
FcRL6
FcRLB
FcεRI
Platypus FcγRI related
FcγRII
FcγRII & III
FcγRIII
FcγRI
FcRL Molecules

Figure 4. A phylogenetic tree of Fc receptor sequences from a panel of vertebrates analyzed with the MrBayes program. The Maximum-likelihood tree was based on the Bayesian methods of phylogenetic interference. Robustness of nodes was tested with the posterior probabilities based on MCMC analysis as implemented in the MrBayes program. Node supported posterior values are given in the phylogenetic tree. Groups of genes that are more closely related, thereby forming a sub-branch, are indicated by light or dark grey shading to make them more easily visible.

In addition, the domain analysis showed that the D1, D2 and D3 domain types are all present in the same order in FcγRI and FcRL3, 4, 5 and A and B. The D1 and D2 domain types are also present in the same orientation in the two domain FcRs in mammals, indicating a close relationship in structure of FcRL and the classical FcRs in mammals (Figure 5).

Discussion

Based on our screening of a large panel of vertebrate genomes, we can now conclude that the first components to appear within the FcR locus are the PIGRs, the FcRL molecules and the signaling subunit, the γ chain as they are found in all bony fish species analyzed. Interestingly, no genes with clear similarities to either of these genes/gene families were found in cartilagous fish and lamprey, indicating that they appeared at the base of the bony fishes. The receptors homologous to the mammalian members of the LRC locus, the KIRs and LILRs seems also to appear with the bony fish as no clear homologs with only C2 domains can be found in sharks or lamprey. This indicates a major evolutionary step in the emergence of a complex adaptive immune system of jawed vertebrates occured at the base of the bony fish. Immunoglobulins and T-cell receptors first appeared with the jawed vertebrates through the introduction of the recombination activating genes (RAGs) and the formation of a split V gene facilitating variability generation by RAG mediated recombinations [59]. The second major step may then have been the appearance of the various Fc receptors and the KIRs and LILRs at the base of the bony fish.

Interestingly, relatively few FcRs and FcRL genes are found in fishes and chicken. Instead there has been an extensive expansion of the LRC locus genes. A very large number of LRC members are found in the channel catfish and also more than 100 copies of the chicken homologs, the CHIRs, are found on the micro chromosome 31 in chickens [48,49]. Several of the CHIRs on micro chromosome 31, including CHIR-AB1, have been identified as a receptor for IgY but none have been found to bind either IgM or IgA [49,60]. An additionally related single gene at the tip of chromosome 20 has four extracellular C2-like Ig domains, which has been shown to be a high-affinity receptor for IgY [61]. This gene has been named ggFcR and requires the common FcR γ chain for signaling [61]. This gene shows the closest homology to the FcRL molecules and is possibly one of a few remnants of this locus in birds. Therefore in chickens, it appears as if the CHIRs, which are KIR-related, have taken over many of the functions performed by the classical receptors in mammals.

Ig-domain containing molecules have experienced a remarkable increase in numbers during vertebrate evolution. They have become part of many vital biological processes with a particular important role in immunity and the central nervous system. One interesting possibility is that whole genome duplications, so called tetraploidizations, have been a key component in their expansion and thereby have also been central for the evolution of both FcRs and NK cell receptors [51]. The question of the role of tetraploidizations for the origin of the FcR, the LRC and related loci has been addressed in detail by Zucchetti et al, where they, by analysis of a primordial locus in tunicates, trace the expansion and diversification of the putative four copies of this locus in mammals including humans [47]. The early locus in the vase tunicate *Ciona intestinalis*, which pre-dates the two genome duplication that most likely occurred during early vertebrate evolution, has been broken into two and is now found on two chromosomes, 4 and 10 [47]. Paralogous regions to this original locus are now found on five or six chromosomes in humans, and not only four, possibly due to secondary translocations. The two loci representing the original locus in tunicates contains a number of Ig-domain containing molecules including the cell adhesion molecules, nectins and NCAM and molecules similar to CD66 (Ceacam), CD166, CD22 and nephrin [47]. Several of these have one or a few ITIMs in their cytoplasmic tails [47]. Interestingly, a number of these Ig-domain containing proteins in *Ciona* contain both V and C type domains, or V and I type domains, indicating that an early locus contained Ig domains of several types. The *Ciona* Ig domain containing members of this ancestral locus have between one to six Ig domains [47].

A finding that may favor a common origin of the FcR locus and the LRC is the presence of genes for the signaling components, FcεRIγ, DAP10, DAP12 and the TCR zeta chain in both chromosomal regions (Figures 2 and 3). These signaling components are structurally unrelated to the Ig-domain containing genes, which makes them good markers for such a process as they cannot have appeared as a result of a gene duplication from related genes in the chromosomal region. The FcεRIγ and TCR zeta genes are located on chromosome 1 and DAP10 and 12 are located on the same chromosome as the LRC locus in primates (Figures 2 and 3). A second finding that also supports genome duplications as a contributing factor is the close relationship in domain structure between members of both of these loci [50]. The majority of the genes in both loci contain C2 type domains as represented by both the classical FcRs for IgG and IgE, the FcRL molecules and the KIRs and LILRs [50].

Although all the domains of the KIRs, LILRs, the classical FcRs and FcRL molecules are of the C2 type, some genes in the FcR locus also contain V type domains. Interestingly, all the Ig domains of the PIGR are V type domains [21]. The single Ig domain of the FcαμR is closely related to the first domain (D1) of the PIGR [21] and the IgM receptor also contains a single Ig domain that belongs to the V type of Ig domains. The related genes in the tunicate *Ciona intestinalis* has V, C and I type domains indicating that both types of domains were present in such a locus during early chordate evolution.

The PIGR, which is of key importance for the transport of IgA and IgM over epithelial layers in mammals, has been found in all vertebrates from bony fish to humans. This receptor is probably also needed for the interaction of IgM with various immune cells and also for the transport of Igs across epithelial layers in fish. Interestingly, the PIGR in fish has only two or three Ig-like extracellular domains compared to five in mammals [38–41]. The second and third domains of the mammalian PIGR seems to be an addition later in vertebrate evolution, where these regions appear to be of importance for efficient transport of dimeric IgA [62]. One step in this process has been an internal duplication involving domains two and three, leading to an exon containing two domains, which is in contrast to all other Ig domains within various PIGR genes, which are encoded by a separate exon. This

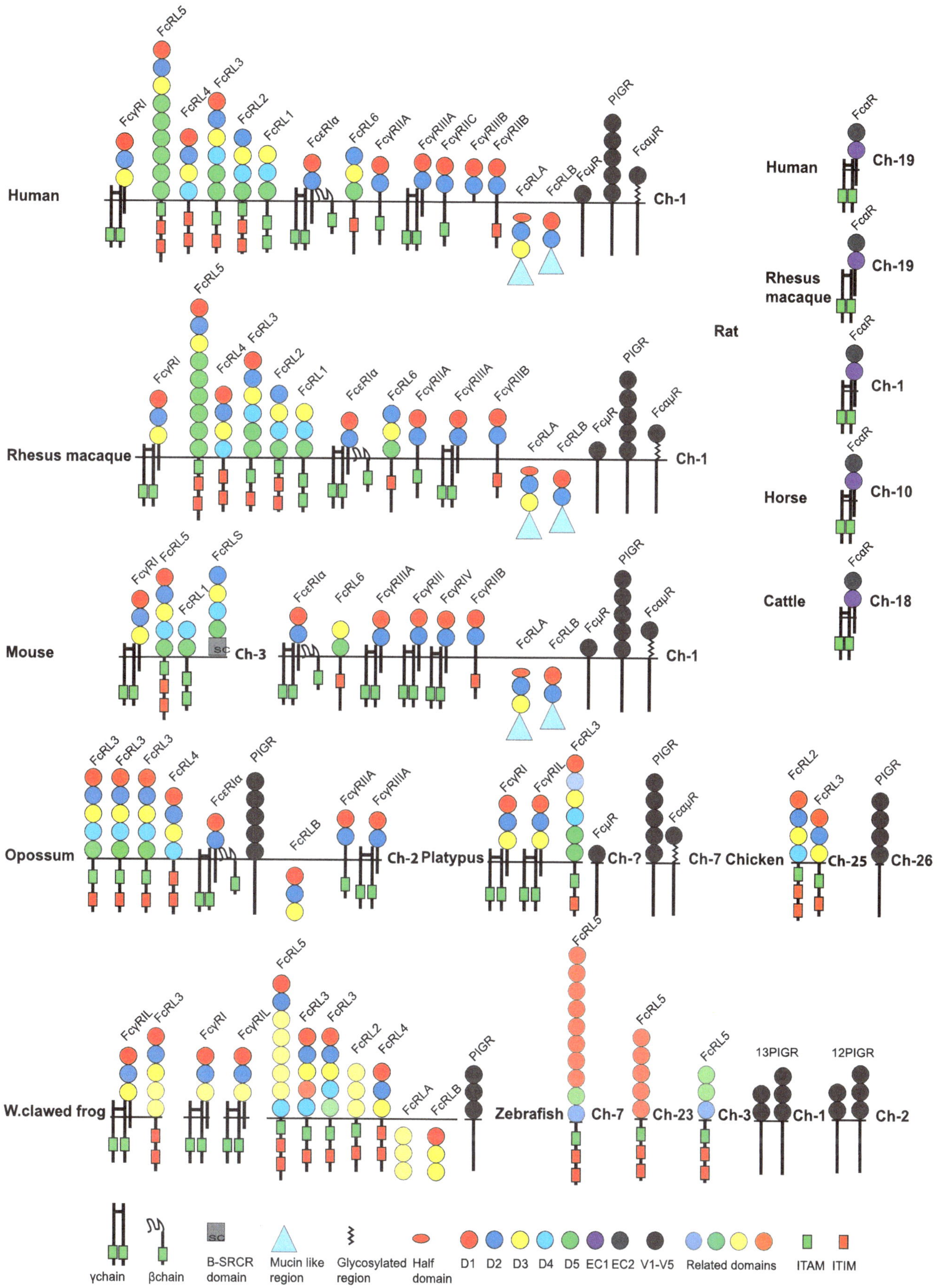
Human
FcγRI
FcRL5
FcRL4
FcRL3
FcRL2
FcRL1
FcεRIα
FcRL6
FcγRIIA
FcγRIIIA
FcγRIIC
FcγRIIIB
FcγRIIB
FcRLA
FcRLB
FcμR
PIGR
FcαμR
Ch-1
Rhesus macaque
Ch-1
Mouse
FcRLS
SC
Ch-3
FcγRIII
FcγRIV
Opossum
Ch-2
Platypus
FcγRIL
Ch-?
Ch-7
Chicken
Ch-25
Ch-26
W.clawed frog
Zebrafish
Ch-7
Ch-23
Ch-3
13PIGR
12PIGR
Ch-1
Ch-2
Human
FcαR
Ch-19
Rhesus macaque
Ch-19
Rat
Ch-1
Horse
Ch-10
Cattle
Ch-18
γchain
βchain
B-SRCR domain
Mucin like region
Glycosylated region
Half domain
D1 D2 D3 D4 D5 EC1 EC2 V1-V5
Related domains
ITAM ITIM

Figure 5. A summary of domain structures and signaling motifs of the various vertebrate Fc receptors. The Ig-like domains are depicted as filled circles with color-coding according to the similarities in sequence based on phylogenetic analyses [24,58]. The domain type D1, D2, D3, D4 and D5 show a relatively conserved pattern in most tetrapods and have therefore been color-coded in red dark blue, yellow, light blue and green. A phylogenetic analysis of all the individual domains presented in figure 5 and a few additional receptors are presented in supplementary figure S4. The color-coding in figure 5 is based on this supplementary figure. The extracellular regions, the transmembrane regions and cytoplasmic tails are not to scale in order to show the positions of potential signaling motifs like ITAMs (green boxes) and ITIMs (red boxes), which regulate the biological function the Fc receptors. Non-consensus ITIMs are also indicated as boxes with half red half white. Some of the intracellular proteins contain C-terminal mucin-like regions that are depicted as blue triangles.

duplication did not appear or was later reverted in the chicken genome [63]. The massive expansion which has occurred among the PIGR genes in zebrafish is also striking. This fish species now has at least 25 closely related copies of this gene with members having either two or three Ig domains with the genes found on two different chromosomes (Figure 3). The functional significance of this massive increase in PIGRs is not known. Noteworthy is the absence of both the PIGR and the FcRL receptors in cartilagous fish. This indicates that both of these receptors appeared with the bony fish.

Evidence from the mammalian PIGR indicates that the J chain structure is important for efficient transport over epithelial layers of IgA and IgM [2,21]. Interestingly, the J chain has been found in cartilagous fish and in lungfish but seems to be absent in most bony fish [2,64]. However, despite an apparent lack of the J chain in the PIGR of carp, IgM transport at numerous epithelial sites in this fish species seems to be active [39]. IgM is present essentially in the same form in all vertebrates, and is probably the ancestor of all other isotypes. The fact that the PIGR is one of the first genes within the FcR locus to appear during vertebrate evolution may therefore seem logical.

The PIGR seems to be one of the first receptors for IgM to appear during vertebrate evolution. However, what is the origin of the other IgM receptors Fcα/μR and FcμR? Curiously, no evidence, even after extensive screening, has been obtained for the presence of Fcα/μR in fish, amphibians, birds and reptiles. The first evidence for the dual IgM/A receptor and the IgM receptor seems to be with the monotremes as one copy of each is found in the platypus genome, which indicates that these receptors appeared during early mammalian evolution. The Fcα/μR is the receptor that is most closely related to the PIGR, particularly to the ligand binding D1 domain of the PIGR, which is of the V type Ig domains, clearly indicating a common origin [21,65]. The IgA/M receptor has only one Ig domain and a stalk structure that is extensively glycosylated, which makes it unique among these receptors [21,65]. The IgM receptor also has only one Ig domain and this is of the V type.

The question then arises about the origin of these V and C2 type domain receptors. Are they products of gene duplications from an early two or three domain receptor containing both V and C2 type domains, or possibly primordial genes having only one type of domain? The domain structure of the PIGR and the FcRL molecules differs quite significantly, which may indicate a different evolutionary origin, favoring the second alternative. Conversely, the presence of receptors in tunicates, having V, C and I type domains, may instead favor the first alternative [47].

The question of the origin of the IgA receptor is also interesting as this receptor is only found in placental mammals and located in the LRC locus together with KIRs and LILRs. The FcαRI binds its ligand at the interphase between the CH2 and CH3 of human IgA, which also differs from the binding region of the classical FcRs of mammals, which binds IgE in the upper part of the CH3 domain and IgG in the upper part of the CH2 domain [22]. The chicken IgY receptor, CHIR-AB1, also binds IgY in a similar position as the IgA receptor [66]. Both FcαRI and CHIR-AB1 share an additional similar feature having a positively charged amino acid in the transmembrane region, which is involved in the interaction with the signaling component the FcεRIγ or DAP12 [49]. Therefore, the most likely explanation is that it has its origin from a regional duplication involving KIRs or LILRs, which like the IgY receptors in chickens, has gained Ig specificity.

As previously described, whole genome duplications may have had a major impact on the diversification of these receptors. However, regional duplications have most likely been equally important. Two examples are found among the IgG receptors in humans, where two new low-affinity and two new high-affinity receptors, FcγRIIC, FcγRIIIB, FcγRIB and FcγRIC, have been generated by what appears as very recent regional duplications (Figure 2). Regional duplications have also most likely been the key player in the amplification of PIGRs in zebrafish and the massive expansion of FcRL3, 4 and 5 related genes in *Xenopus* (Figure 2).

The presence of FcRL molecules, with no members of the classical FcRs in fish, indicates that FcRL molecules were among the first identifiable members of this locus. The fact that they are found in the same chromosomal region as the IgE and IgG receptors in mammals, as well as that the classical FcRs are first found in marsupials, strengthens the possibility that the FcRL molecules are the ancestors of the mammalian IgG and IgE receptors. The identification of IgA and IgG binding of FcRL4 and 5, respectively, also supports this idea [34,35]. In addition, the phylogenetic analysis clearly favors this scenario as the mammalian IgG and IgE receptors are found in a separate branch, close to the FcRL cluster (Figure 4). Finally, there are similarities in domain structure between the IgG and IgE receptors and FcRL3, 4, 5 and A and B, indicating that the three first domains of FcγRI and the two domains of FcγRII, III and IV came from a FcRL3, 4, 5, A or B gene (Figure 5). Interestingly, an IgM receptor has been identified in the channel catfish (*Ictalurus punctatus*), which is most likely a soluble receptor as it lacks both a transmembrane region and a classical GPI anchor [67]. This molecule has three Ig domains and, in the phylogenetic analysis, is most closely related to the FcRL molecules. This latter finding further supports the notion that the classical FcRs in mammals originate from FcRL molecules [67]. Related molecules to this IgM receptor have also been found in other fish species, where non-membrane anchored soluble molecule forms also exist [67]. An important subsequent question arises about its function, as most classical receptors are cell bound, whereas these fish IgM receptors seem to be entirely soluble molecules. Their functions are a mystery as these soluble receptors do not seem to be able to signal to cells.

The initial analysis of FcRs in a non-placental mammal, the opossum, showed the presence of most of the classical receptors except the IgA, the A/M, the IgM and the high-affinity IgG receptor [24]. The presence of the two-domain receptors and the lack of the high-affinity receptor for IgG indicated that the two-domain receptors were the first IgG and IgE specific receptors to appear during mammalian evolution. However, the identification of three-domain receptors in both a monotreme, the platypus, and in an amphibian, the Western clawed frog, contested this idea. Interestingly, in the Maximum-likelihood and the Distance trees the platypus FcγRI and FcγRIL genes clustered with the classical receptors and in particular the FcγRI, indicating that these receptors in the platypus may be early representatives of the classical receptors. The isotype specificity of these two- and three-domain receptors is not known, nor whether they even bind Igs at all. They may actually be the missing link to the evolution of the complex set of receptors seen in placental mammals and the appearance of these receptors from early FcRL family members. This is an obviously very interesting area of investigation. Also of interest is that one of these three domain receptors has 2 ITIMs whereas the other has no motifs that can be classified as ITIMs or ITAMs. This receptor may therefore connect with the γ chain for signaling similar to many of the classical FcRs in placental mammals.

Interestingly, the FcRL molecules in Xenopus did, in the various trees form a separate branch in the phylogenetic tree indicating that the individual members have appeared after the separation from the other vertebrate lineages. An alternative explanation for this clustering is that there has been extensive homogenization by gene conversion within these lineages.

In summary, the analysis of the origin of the FcRs, using a large panel of vertebrate genomes, provides strong support for the notion that the complexity in FcRs has increased in parallel with the complexity of the Ig isotypes. This began with the appearance of PIGRs, the FcRL molecules and the signaling subunit, the γ chain. These three receptors/receptor families seem to have appeared with the bony fish. The end result of this expansion is a complex set of receptors for almost all the different isotypes including three to four for IgG, one for IgE, one for IgA, one for IgM and a shared receptor for IgM and IgA. The appearance of the various Ig receptors also seems to have taken different routes in different vertebrate lineages. In mammals, most of the FcRs have developed from the FcRL molecules of the FcR locus, whereas in birds, multiple IgY receptors have developed within the LRC locus. Although distantly related genes to the LRC locus in mammals are found in tunicates, clearly identifiable homologs to the LCR locus genes of mammals with only C2 domains are also first seen in bony fish, indicating a major evolutionary event concerning the evolution of FcRs and LRC genes at the base of the bony fish. The first solid evidence for a direct homolog to the classical FcRs is found among the two-domain receptor homologs in marsupials. However, the three domain receptors in the platypus also clustered with the classical receptors in the Maximum-likelihood and Distance trees indicating that they may be early ancestors of the subgroup of FcRL genes that formed the classical receptors. The close structural relationship between FcRL molecules and in particular FcRL3, 4, 5, A and B involving domain types D1, D2 and D3 also provides strong support for FcRL molecules being the direct ancestors of the various classical IgG and IgE receptors in marsupials and placental mammals. A second major step in the evolution of FcRs may therefore have occurred in early mammals when one or a few FcRL genes formed the ancestor of the classical FcRs and also sequences from the PIGR duplicated and became an early ancestor of the IgA/M and IgM receptors. The analysis of the possible isotype specificity of the various related receptors in platypus may shed important light on the origin of the classical IgG and IgE receptors in placental mammals. This may also give clues to the origin of the role of IgE in triggering mast cell granule release through its interaction with IgE specific Fc receptors (work in progress).

Supporting Information

Supporting Information S1 Accession numbers for all genes used in the alignment figures.

Figure S1 Clustal X NJ tree. The rooted neighbor-joining tree of vertebrate Fc receptors was analyzed in Clustal X with 100 replicates.

Figure S2 Phylip ML tree. The rooted Maximum-likelihood tree of vertebrate Fc receptors was calculated using likelihood parameters with 100 replicates using in Phylip programs. Node bootstrap values are indicated.

Figure S3 Phylip distance tree. The rooted distance matrix tree of vertebrate Fc receptors was calculated using likelihood parameters with 100 replicates using in Phylip programs. Node bootstrap values are indicated in the phylogenetic tree.

Figure S4 MEGA Domain Tree. The individual domains of all receptors shown in figure 5 were analyzed individually for their sequence relatedness. The individual domain types were color coded with the same colors as in figures 5. The rooted maximum likelihood tree was constructed by using MEGA5.2 soft wear. Bootstrap analysis was performed with 500 replicates. Bootstrap values are indicated in the phylogenetic tree.

Figure S5 Domain structure of a number of additional FcRL molecules from the frog (Xenopus) not presented in figure 5. A marked difference in the presence of ITIMs is observed among the various FcRL members in frog. Some have up to 2 consensus ITAMs whereas many have not got a single canonical site or distantly related site.

Acknowledgments

We would like to thank Michael Thorpe for critically reading and for linguistic revision of the manuscript.

Author Contributions

Conceived and designed the experiments: SA SM LH. Performed the experiments: SA SM. Analyzed the data: SA SM LH. Wrote the paper: SA LH.

References

1. Zhao Y, Cui H, Whittington CM, Wei Z, Zhang X, et al. (2009) Ornithorhynchus anatinus (platypus) links the evolution of immunoglobulin genes in eutherian mammals and nonmammalian tetrapods. J Immunol 183: 3285–3293.
2. Zhang YA, Salinas I, Li J, Parra D, Bjork S, et al. (2010) IgT, a primitive immunoglobulin class specialized in mucosal immunity. Nat Immunol 11: 827–835.
3. Dooley H, Flajnik MF (2006) Antibody repertoire development in cartilaginous fish. Dev Comp Immunol 30: 43–56.
4. Zhang C, Du Pasquier L, Hsu E (2013) Shark IgW C Region Diversification through RNA Processing and Isotype Switching. J Immunol 191: 3410–3418.
5. Hansen JD, Landis ED, Phillips RB (2005) Discovery of a unique Ig heavy-chain isotype (IgT) in rainbow trout: Implications for a distinctive B cell developmental pathway in teleost fish. Proc Natl Acad Sci U S A 102: 6919–6924.
6. Danilova N, Bussmann J, Jekosch K, Steiner LA (2005) The immunoglobulin heavy-chain locus in zebrafish: identification and expression of a previously unknown isotype, immunoglobulin Z. Nat Immunol 6: 295–302.

7. Zhao Y, Pan-Hammarstrom Q, Yu S, Wertz N, Zhang X, et al. (2006) Identification of IgF, a hinge-region-containing Ig class, and IgD in Xenopus tropicalis. Proc Natl Acad Sci U S A 103: 12087–12092.
8. Schaerlinger B, Bascove M, Frippiat JP (2008) A new isotype of immunoglobulin heavy chain in the urodele amphibian Pleurodeles waltl predominantly expressed in larvae. Mol Immunol 45: 776–786.
9. Warr GW, Magor KE, Higgins DA (1995) IgY: clues to the origins of modern antibodies. Immunol Today 16: 392–398.
10. International Chicken Genome Sequencing C (2004) Sequence and comparative analysis of the chicken genome provide unique perspectives on vertebrate evolution. Nature 432: 695–716.
11. Vernersson M, Aveskogh M, Munday B, Hellman L (2002) Evidence for an early appearance of modern post-switch immunoglobulin isotypes in mammalian evolution (II); cloning of IgE, IgG1 and IgG2 from a monotreme, the duck-billed platypus, Ornithorhynchus anatinus. Eur J Immunol 32: 2145–2155.
12. Belov K, Hellman L (2003) Immunoglobulin genetics of Ornithorhynchus anatinus (platypus) and Tachyglossus aculeatus (short-beaked echidna). Comp Biochem Physiol A Mol Integr Physiol 136: 811–819.
13. Vernersson M, Aveskogh M, Hellman L (2004) Cloning of IgE from the echidna (Tachyglossus aculeatus) and a comparative analysis of epsilon chains from all three extant mammalian lineages. Dev Comp Immunol 28: 61–75.
14. Vernersson M, Belov K, Aveskogh M, Hellman L (2010) Cloning and structural analysis of two highly divergent IgA isotypes, IgA1 and IgA2 from the duck billed platypus, Ornithorhynchus anatinus. Mol Immunol 47: 785–791.
15. Aveskogh M, Pilstrom L, Hellman L (1999) Cloning and structural analysis of IgM (mu chain) and the heavy chain V region repertoire in the marsupial Monodelphis domestica. Dev Comp Immunol 23: 597–606.
16. Aveskogh M, Hellman L (1998) Evidence for an early appearance of modern post-switch isotypes in mammalian evolution; cloning of IgE, IgG and IgA from the marsupial Monodelphis domestica. Eur J Immunol 28: 2738–2750.
17. Wang X, Olp JJ, Miller RD (2009) On the genomics of immunoglobulins in the gray, short-tailed opossum Monodelphis domestica. Immunogenetics 61: 581–596.
18. Nimmerjahn F, Ravetch JV (2008) Fcgamma receptors as regulators of immune responses. Nat Rev Immunol 8: 34–47.
19. van der Poel CE, Spaapen RM, van de Winkel JG, Leusen JH (2011) Functional characteristics of the high affinity IgG receptor, FcgammaRI. J Immunol 186: 2699–2704.
20. Rivera J, Fierro NA, Olivera A, Suzuki R (2008) New insights on mast cell activation via the high affinity receptor for IgE. Adv Immunol 98: 85–120.
21. Klimovich VB (2011) IgM and its receptors: structural and functional aspects. Biochemistry (Mosc) 76: 534–549.
22. Bakema JE, van Egmond M (2011) The human immunoglobulin A Fc receptor FcalphaRI: a multifaceted regulator of mucosal immunity. Mucosal Immunol 4: 612–624.
23. Davis RS, Ehrhardt GR, Leu CM, Hirano M, Cooper MD (2005) An extended family of Fc receptor relatives. Eur J Immunol 35: 674–680.
24. Fayngerts SA, Najakshin AM, Taranin AV (2007) Species-specific evolution of the FcR family in endothermic vertebrates. Immunogenetics 59: 493–506.
25. Yamashita T, Suzuki R, Backlund PS, Yamashita Y, Yergey AL, et al. (2008) Differential dephosphorylation of the FcRgamma immunoreceptor tyrosine-based activation motif tyrosines with dissimilar potential for activating Syk. J Biol Chem 283: 28584–28594.
26. Blank U, Ra C, Miller L, White K, Metzger H, et al. (1989) Complete structure and expression in transfected cells of high affinity IgE receptor. Nature 337: 187–189.
27. Rodewald HR, Arulanandam AR, Koyasu S, Reinherz EL (1991) The high affinity Fc epsilon receptor gamma subunit (Fc epsilon RI gamma) facilitates T cell receptor expression and antigen/major histocompatibility complex-driven signaling in the absence of CD3 zeta and CD3 eta. J Biol Chem 266: 15974–15978.
28. Weissman AM, Hou D, Orloff DG, Modi WS, Seuanez H, et al. (1988) Molecular cloning and chromosomal localization of the human T-cell receptor zeta chain: distinction from the molecular CD3 complex. Proc Natl Acad Sci U S A 85: 9709-9713.
29. Lanier LL (2009) DAP10- and DAP12-associated receptors in innate immunity. Immunol Rev 227: 150–160.
30. Liang Y, Tedder TF (2001) Identification of a CD20-, FcepsilonRIbeta-, and HTm4-related gene family: sixteen new MS4A family members expressed in human and mouse. Genomics 72: 119–127.
31. Kubagawa H, Oka S, Kubagawa Y, Torii I, Takayama E, et al. (2009) Identity of the elusive IgM Fc receptor (FcmuR) in humans. J Exp Med 206: 2779–2793.
32. Davis RS (2007) Fc receptor-like molecules. Annu Rev Immunol 25: 525–560.
33. Ehrhardt GR, Cooper MD (2011) Immunoregulatory roles for fc receptor-like molecules. Curr Top Microbiol Immunol 350: 89–104.
34. Wilson TJ, Fuchs A, Colonna M (2012) Cutting edge: human FcRL4 and FcRL5 are receptors for IgA and IgG. J Immunol 188: 4741–4745.
35. Franco A, Damdinsuren B, Ise T, Dement-Brown J, Li H, et al. (2013) Human Fc receptor-like 5 binds intact IgG via mechanisms distinct from those of Fc receptors. J Immunol 190: 5739–5746.
36. Schreeder DM, Cannon JP, Wu J, Li R, Shakhmatov MA, et al. (2010) Cutting edge: FcR-like 6 is an MHC class II receptor. J Immunol 185: 23–27.
37. Tsubata T (2012) Role of inhibitory BCR co-receptors in immunity. Infect Disord Drug Targets 12: 181–190.
38. Asano M, Komiyama K (2011) Polymeric immunoglobulin receptor. J Oral Sci 53: 147–156.
39. Rombout JH, van der Tuin SJ, Yang G, Schopman N, Mroczek A, et al. (2008) Expression of the polymeric Immunoglobulin Receptor (pIgR) in mucosal tissues of common carp (Cyprinus carpio L.). Fish Shellfish Immunol 24: 620–628.
40. Feng LN, Lu DQ, Bei JX, Chen JL, Liu Y, et al. (2009) Molecular cloning and functional analysis of polymeric immunoglobulin receptor gene in orange-spotted grouper (Epinephelus coioides). Comp Biochem Physiol B Biochem Mol Biol 154: 282–289.
41. Hamuro K, Suetake H, Saha NR, Kikuchi K, Suzuki Y (2007) A teleost polymeric Ig receptor exhibiting two Ig-like domains transports tetrameric IgM into the skin. J Immunol 178: 5682–5689.
42. Parham P, Abi-Rached L, Matevosyan L, Moesta AK, Norman PJ, et al. (2010) Primate-specific regulation of natural killer cells. J Med Primatol 39: 194–212.
43. Barrow AD, Trowsdale J (2008) The extended human leukocyte receptor complex: diverse ways of modulating immune responses. Immunol Rev 224: 98–123.
44. Rajalingam R (2011) Human diversity of killer cell immunoglobulin-like receptors and disease. Korean J Hematol 46: 216–228.
45. Espeli M, Niederer HA, Traherne JA, Trowsdale J, Smith KG (2010) Genetic variation, Fcgamma receptors, KIRs and infection: the evolution of autoimmunity. Curr Opin Immunol 22: 715–722.
46. Ding Y, Xu G, Yang M, Yao M, Gao GF, et al. (2003) Crystal structure of the ectodomain of human FcalphaRI. J Biol Chem 278: 27966–27970.
47. Zucchetti I, De Santis R, Grusea S, Pontarotti P, Du Pasquier L (2009) Origin and evolution of the vertebrate leukocyte receptors: the lesson from tunicates. Immunogenetics 61: 463–481.
48. Stafford JL, Bengten E, Du Pasquier L, Miller NW, Wilson M (2007) Channel catfish leukocyte immune-type receptors contain a putative MHC class I binding site. Immunogenetics 59: 77–91.
49. Viertlboeck BC, Gobel TW (2011) The chicken leukocyte receptor cluster. Vet Immunol Immunopathol 144: 1–10.
50. Nikolaidis N, Klein J, Nei M (2005) Origin and evolution of the Ig-like domains present in mammalian leukocyte receptors: insights from chicken, frog, and fish homologues. Immunogenetics 57: 151–157.
51. Sundstrom G, Larsson TA, Brenner S, Venkatesh B, Larhammar D (2008) Evolution of the neuropeptide Y family: new genes by chromosome duplications in early vertebrates and in teleost fishes. Gen Comp Endocrinol 155: 705–716.
52. Kuroki K, Furukawa A, Maenaka K (2012) Molecular recognition of paired receptors in the immune system. Front Microbiol 3: 429.
53. Felsenstein J (1981) Evolutionary trees from DNA sequences: a maximum likelihood approach. J Mol Evol 17: 368–376.
54. Ronquist F, Teslenko M, van der Mark P, Ayres DL, Darling A, et al. (2012) MrBayes 3.2: efficient Bayesian phylogenetic inference and model choice across a large model space. Syst Biol 61: 539–542.
55. Larkin MA, Blackshields G, Brown NP, Chenna R, McGettigan PA, et al. (2007) Clustal W and Clustal X version 2.0. Bioinformatics 23: 2947–2948.
56. Katoh K, Standley DM (2013) MAFFT multiple sequence alignment software version 7: improvements in performance and usability. Mol Biol Evol 30: 772–780.
57. Tamura K, Peterson D, Peterson N, Stecher G, Nei M, et al. (2011) MEGA5: molecular evolutionary genetics analysis using maximum likelihood, evolutionary distance, and maximum parsimony methods. Mol Biol Evol 28: 2731–2739.
58. Guselnikov SV, Ramanayake T, Erilova AY, Mechetina LV, Najakshin AM, et al. (2008) The Xenopus FcR family demonstrates continually high diversification of paired receptors in vertebrate evolution. BMC Evol Biol 8: 148.
59. Fugmann SD (2010) The origins of the Rag genes-from transposition to V(D)J recombination. Semin Immunol 22: 10–16.
60. Viertlboeck BC, Schweinsberg S, Schmitt R, Herberg FW, Gobel TW (2009) The chicken leukocyte receptor complex encodes a family of different affinity FcY receptors. J Immunol 182: 6985–6992.
61. Viertlboeck BC, Schmitt R, Hanczaruk MA, Crooijmans RP, Groenen MA, et al. (2009) A novel activating chicken IgY FcR is related to leukocyte receptor complex (LRC) genes but is located on a chromosomal region distinct from the LRC and FcR gene clusters. J Immunol 182: 1533–1540.
62. Norderhaug IN, Johansen FE, Krajci P, Brandtzaeg P (1999) Domain deletions in the human polymeric Ig receptor disclose differences between its dimeric IgA and pentameric IgM interaction. Eur J Immunol 29: 3401–3409.
63. Wieland WH, Orzaez D, Lammers A, Parmentier HK, Verstegen MW, et al. (2004) A functional polymeric immunoglobulin receptor in chicken (Gallus gallus) indicates ancient role of secretory IgA in mucosal immunity. Biochem J 380: 669–676.
64. Tacchi L, Larragoite E, Salinas I (2013) Discovery of J Chain in African Lungfish (Protopterus dolloi, Sarcopterygii) Using High Throughput Transcriptome Sequencing: Implications in Mucosal Immunity. PLoS One 8: e70650.
65. Shimizu Y, Honda S, Yotsumoto K, Tahara-Hanaoka S, Eyre HJ, et al. (2001) Fc(alpha)/mu receptor is a single gene-family member closely related to polymeric immunoglobulin receptor encoded on Chromosome 1. Immunogenetics 53: 709–711.
66. Purzel J, Schmitt R, Viertlboeck BC, Gobel TW (2009) Chicken IgY binds its receptor at the CH3/CH4 interface similarly as the human IgA: Fc alpha RI interaction. J Immunol 183: 4554–4559.

67. Stafford JL, Wilson M, Nayak D, Quiniou SM, Clem LW, et al. (2006) Identification and characterization of a FcR homolog in an ectothermic vertebrate, the channel catfish (Ictalurus punctatus). J Immunol 177: 2505–2517.

68. Brown D, Trowsdale J, Allen R (2004) The LILR family: modulators of innate and adaptive immune pathways in health and disease. Tissue Antigens 64: 215–225.

12

Key Processes for *Cheirolophus* (Asteraceae) Diversification on Oceanic Islands Inferred from AFLP Data

Daniel Vitales[1]*, Alfredo García-Fernández[2,3], Jaume Pellicer[4], Joan Vallès[1], Arnoldo Santos-Guerra[5], Robyn S. Cowan[4], Michael F. Fay[4], Oriane Hidalgo[1,4], Teresa Garnatje[2]

1 Laboratori de Botànica – Unitat associada CSIC, Facultat de Farmàcia, Universitat de Barcelona, Barcelona, Catalonia, Spain, **2** Institut Botànic de Barcelona (IBB-CSIC-ICUB), Barcelona, Catalonia, Spain, **3** Área de Biodiversidad y Conservación, Universidad Rey Juan Carlos, Móstoles, Madrid, Spain, **4** Jodrell Laboratory, Royal Botanic Gardens, Kew, Richmond, Surrey, United Kingdom, **5** Unidad de Botánica-ICIA, Puerto de la Cruz, Canary Islands, Spain

Abstract

The radiation of the genus *Cheirolophus* (Asteraceae) in Macaronesia constitutes a spectacular case of rapid diversification on oceanic islands. Twenty species – nine of them included in the IUCN Red List of Threatened Species – have been described to date inhabiting the Madeiran and Canarian archipelagos. A previous phylogenetic study revealed that the diversification of *Cheirolophus* in Macaronesia started less than 2 Ma. As a result of such an explosive speciation process, limited phylogenetic resolution was reported, mainly due to the low variability of the employed molecular markers. In the present study, we used highly polymorphic AFLP markers to i) evaluate species' boundaries, ii) infer their evolutionary relationships and iii) investigate the patterns of genetic diversity in relation to the potential processes likely involved in the radiation of *Cheirolophus*. One hundred and seventy-two individuals representing all Macaronesian *Cheirolophus* species were analysed using 249 AFLP loci. Our results suggest that geographic isolation played an important role in this radiation process. This was likely driven by the combination of poor gene flow capacity and a good ability for sporadic long-distance colonisations. In addition, we also found some traces of introgression and incipient ecological adaptation, which could have further enhanced the extraordinary diversification of *Cheirolophus* in Macaronesia. Last, we hypothesize that current threat categories assigned to Macaronesian *Cheirolophus* species do not reflect their respective evolutionary relevance, so future evaluations of their conservation status should take into account the results presented here.

Editor: João Pinto, Instituto de Higiene e Medicina Tropical, Portugal

Funding: This work was subsidized by the Spanish Ministry of Science (projects CGL2010-22234-C02-01 and 02/BOS) http://www.idi.mineco.gob.es/ and the Generalitat de Catalunya (Ajuts a grups de recerca consolidats, 2009/SGR/00439) http://www.gencat.cat/. DV benefited from a FPU grant from the Spanish Ministry of Education http://www.mecd.gob.es. JP benefited from a Beatriu de Pinós postdoctoral fellowship with the support of the Secretary for Universities and Research of the Ministry of Economy and Knowledge (Government of Catalonia) http://www.gencat.cat/ and the co-fund of Marie Curie Actions (European Union 7th R&D Framework Programme) http://ec.europa.eu/. AG-F benefited from a postdoctoral contract (Project CGL2010-22234-C02-01). The funders had no role in study design, data collection and analysis, decision to publish, or preparation of the manuscript.

Competing Interests: The authors have declared that no competing interests exist.

* Email: dvitales@ub.edu

Introduction

In the last two decades, the Macaronesian archipelagos (Canary Islands, Cape Verde, Azores, Madeira and Savages) have attracted much interest from researchers studying plant diversification and radiation processes [1]. These volcanic islands provide a wide variety of ecological conditions, geological ages and geographical isolation scales [2–4], which promote the existence of a mosaic of habitats that represent an excellent natural laboratory in which to study selection forces and evolutionary processes. The Macaronesian archipelagos have been recognized as a hotspot of plant diversity [5], and have rapidly become a popular model system for scientists to test many speciation hypotheses, both empirically and theoretically. The wide diversity of habitats found in Macaronesia – spanning from xerophytic coastal cliffs to subalpine belts – has served to demonstrate the role of adaptive radiation in a range of plant groups (e.g. *Argyranthemum* Webb, [6]; *Sonchus* L. alliance, [7]; *Aeonium* Webb & Berthel., [8]; *Echium* L., [9]; *Tolpis* Adans., [10]). Indeed, niche pre-emption through adaptive radiation is the most prevailing hypothesis to explain the high degree of endemism and monophyly within Macaronesian lineages [11,12]. Furthermore, during the geological history of Macaronesian archipelagos, several islands have emerged, disappeared and/or changed their relative geographical position (see [4]), promoting complex isolation-connection and colonisation processes between islands and between islands and the continent. This complexity in volcanic archipelagos has given rise to numerous study cases examining the relative importance of vicariance versus dispersal in shaping insular biotas [13–15]; the role of islands as regions from which taxa might colonise continents and other archipelagos [16,17] and the different stages of colonisation and radiation processes in relation to the ontogeny phases of oceanic islands [18,19].

Occasionally, island radiations occur over a short period of time, resulting in ecologically and morphologically distinct taxa, but leading to poor molecular differentiation [9,20–23]. These cases, in which explosive species radiation takes place are among the most interesting and least understood evolutionary events, maybe due to the difficulty in carrying out species level analyses [24–26]. Rapid island radiations have been generally associated to some features – such as small population size, release from previous ecological constraints, and adaptation to new niches – recurrently observed in a wide variety of species and island archipelagos [27]. However, since traditional phylogenetic studies usually provide little resolution in delimiting taxonomical boundaries and untangling the relationships among these rapidly evolving species, the precise role of morphological, life history, and physiological traits and their genetic basis in explosive plant radiations remain essentially unsolved [28].

The Macaronesian *Cheirolophus* Cass. (Asteraceae) complex comprises 20 species (out of the 30 constituting the whole genus), with *Cheirolophus massonianus* (Lowe) A.Hansen & Sunding occurring in the Madeiran archipelago, and the remaining species distributed across the western Canary Islands (see Fig. 1). Although they are usually associated with humid basalt cliffs, a few taxa have adapted to inhabit very diverse ecological zones of the archipelago [29,30]. Most of the island *Cheirolophus* are subshrubs, shrubs or even arborescent shrubs, showing a clear increase in woodiness relative to their continental congeners and a general shift towards inflorescences with white to purple flowers arranged in a candelabrum-like pattern. Some species are relatively widespread throughout a single island (e.g. *C. webbianus* (Sch.Bip.) Holub from Tenerife), whereas others are found on two different islands (e.g. *C. teydis* (C.Sm.) G.López from Tenerife and La Palma and *C. massonianus* from Madeira and Porto Santo). However, most species are narrow endemics occurring in a small number of restricted localities with only a few hundred or less individuals. Consequently, nine Macaronesian species have been included in the IUCN Red List of Threatened Species [31] as vulnerable, endangered or critically endangered. In addition, 17 of the species or subspecies endemic to the Canary Islands are included in the 2010 Red List of Spanish Vascular Flora [32]. In those cases where many related taxa are under threat and conservation strategies must be prioritized, taxonomic knowledge becomes of special relevance and efforts should be made to establish accurately the boundaries of the species concept. To achieve this goal, the recent "unified species concept" advocates the use of diverse lines of evidence (e.g. monophyly at one or multiple DNA loci, morphological diagnosis, ecological distinctiveness, etc.) so that a higher degree of corroboration in taxonomic delimitation is attained [33]. In recent times, conservation genetics has become an essential approach for evaluating the status and the level of threat at the species and population levels, identifying the genetic basis of processes which may potentially lead into an extinction vortex and informing authorities about management priorities for endangered species and/or populations [34].

Several investigations based on morphological, cytogenetic, isozymes and DNA sequencing data have been previously used to infer evolutionary relationships and taxonomy for the 29 currently recognized species of *Cheirolophus* [35–40]. These studies revealed that the genus arose in the Mediterranean region with subsequent dispersal towards Macaronesia, but they failed to reconstruct the relationships among insular congeners. A recent phylogenetic study featuring relaxed-clock dating analyses and diversification tests (involving sequences of various nuclear and plastid regions and sampling several populations per species, [41]) has evidenced a recent origin of Macaronesian *Cheirolophus* radiation. The age of this group was estimated to be 1.74 Ma (95%HPD 0.82–2.93), implying a diversification rate of 0.34–2.84 species Myr^{-1}. Indeed, this diversification rate is comparable to those exhibited by the Hawaiian *Bidens* L. (0.3–2.3 species Myr^{-1}) or Macaronesian *Echium* (0.4–1.5 species Myr^{-1}), considered as the fastest plant radiations on volcanic islands documented to date [22]. Bayesian phylogeographic and ancestral range analyses were also applied in the same study, highlighting the major role of allopatric speciation to explain current diversity in Macaronesian *Cheirolophus* [41]. Geographical isolation among populations and long distance dispersal (both intra and inter-island colonisations) were proposed as the main forces driving the rapid diversification of the group, but introgression and emerging ecological adaptation were also suggested as additional factors reinforcing the speciation process. Despite such efforts, the employed nrDNA and cpDNA markers provided very low variability within the Macaronesian species, thus hampering accurate inferences about the role played by those evolutionary mechanisms potentially involved in the radiation of *Cheirolophus*.

Amplified fragment length polymorphism (AFLP) analyses have been found to provide insights into interspecific relationships in diverse animal and plant groups when other methodologies have failed in this attempt (e.g. [42–49]). Indeed, several authors have suggested that this technique can be especially useful in reconstructing phylogenetic relationships among species that have diverged or radiated recently [50–53]. Although some considerations must be taken into account for its use, AFLP are currently widely employed in molecular ecology and evolution research [54–59]. In fact, this DNA fingerprinting approach has proven to be particularly suitable for evaluating the genetic structure of plant species in different oceanic archipelagos, and hence elucidating the potential evolutionary forces – such as gene flow or genetic drift – that influence the distribution of genetic diversity among individuals, populations, and species [60–62].

AFLP fingerprinting – complemented with morphological and other molecular data – was employed here with the main objective of unravelling the explosive radiation that *Cheirolophus* underwent in the Canarian and Madeiran archipelagos. We applied phylogenetic and population genetic approaches to (i) evaluate the taxonomic identity of the Macaronesian species and to (ii) disentangle the evolutionary relationships between them. Additionally, we studied the patterns of genetic diversity within and between populations to (iii) infer the role of the evolutionary processes potentially involved with this explosive radiation (i.e. geographic isolation, ecological adaptation and introgression).

Materials and Methods

Sampling strategy

One hundred and seventy-two individuals from 29 populations of 20 Macaronesian *Cheirolophus* species were sampled, covering the whole taxonomic diversity recognized for the Madeiran and Canarian archipelagos (see Fig. 1 and Table S1 for further geographical details of sampling sites). One to 12 individuals per population were included in the study depending on the material availability and the uneven success of DNA extraction and AFLP procedures. We decided to incorporate those populations containing extremely scarce sampling (one or two individuals) because of their distinctiveness: (1) it was the only material available for the taxon (*C. dariasii* (Svent.) Bramwell; *C. cf. webbianus*); (2) the inclusion of that particular population was essential to understand the distribution of the species (*C. teydis* from La Palma; *C. massonianus* from Porto Santo); or (3) the additional subpopulation completes the sampling of extremely local species (*C. metlesicsii* Parada). Leaf material for DNA extraction was collected from plants in the field, dried in silica-gel

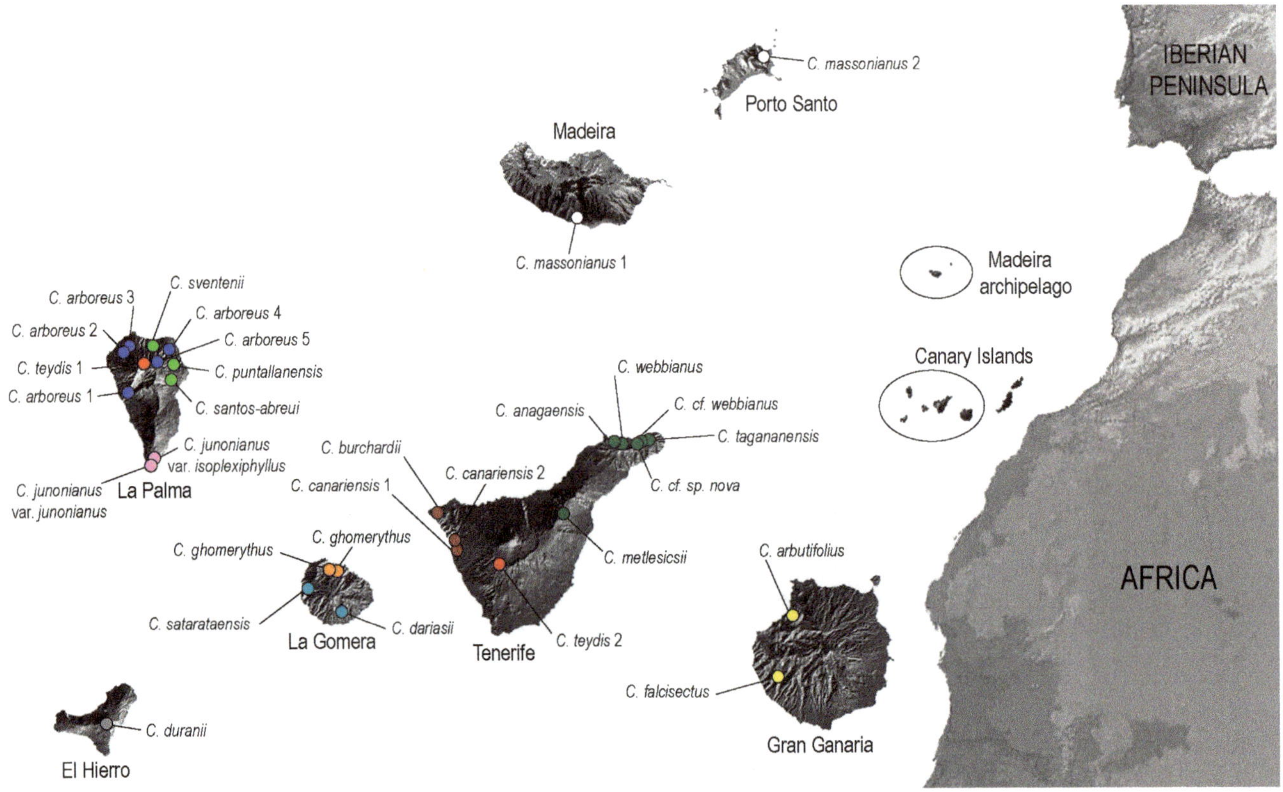

Figure 1. Geographic location of the 29 sampled populations of Macaronesian *Cheirolophus* species. Colour coding circles correspond to genetic structure derived from Bayesian mixture analysis of AFLP markers implemented in BAPS.

and stored at room temperature. Insular Cabildos of Tenerife, Gran Canaria, La Gomera, La Palma and El Hierro provided all the necessary permits to collect samples from protected natural areas in the respective islands. The Canarian Council of Education, University and Sustainability issued the authorization to collect samples of the protected species listed on Table S1. Samples used of the endangered *C. massonianus* were provided by the Botanical Garden of Madeira.

DNA extraction, AFLP protocol and nrDNA sequencing

Genomic DNA was extracted from fragments of silica-gel dried leaf tissue following the protocol of Doyle and Doyle [63] with slight modifications. DNA samples were cleaned using QIAquick columns (Qiagen, Valencia, CA, USA) and their quality and DNA concentration were determined using NanoDrop ND-1000 spectrophotometry (ThermoScientific, Wilmington, DE, USA). The AFLP technique was carried out following the protocol described in Vos et al. [64] in accordance with the modified AFLP Plant Mapping Protocol (Applied Biosystems Inc. Foster City, CA, USA) using *Eco*RI and *Mse*I with 500 ng of isolated genomic DNA per sample. Eighteen primer pair combinations were tested on six individuals from three different populations to screen those producing the most informative and readable profiles. Three pairs of primers were selected giving the most polymorphic and scorable polymorphic pattern: EcoRI-CTT/MseI-AC; EcoRI-CTC/MseI-AA; and EcoRI-CAG/MseI-AT. The success of each step was tested by running the PCR products on a 1.5% agarose gel. Fragments were run on an ABI Prism 3100 Genetic Analyzer (Applied Biosystems Inc.) with 10 μL of High Dye (deionized formamide) and 0.2 μL of GeneScan 500 ROX Size Standard per sample. Amplified fragments were genotyped as present/absent using GeneMarker AFLP/Genotyping software (version 1.9; SoftGenetics, LLC., State College, PA, USA).

For an initial scoring, all alleles within a range of 50 to 490 bp were considered. Afterwards, visual correction was carried out to eliminate erroneous peaks (low intensity or no reproducibility). AFLP error rates were calculated following [65]. Twenty-five random samples per primer combination were replicated to ensure reproducibility, repeating all parts of the AFLP protocol (extraction, digestion, pre-selective and selective PCR). All alleles with an error rate >5% were eliminated. In order to test the occurrence of size homoplasy, we also calculated the correlation between AFLP fragment sizes and frequencies using AFLP-SURV 1.0 [66].

We also employed some nuclear ribosomal DNA sequences to examine the potential role of genetic introgression in the radiation of Macaronesian *Cheirolophus*. Only those populations (see Table S2) showing clear evidences of interspecific gene flow according to our AFLP data – together with some of their putative parental species – were studied. Sequencing procedures, GenBank accessions and any other information about the material and methods employed can be found in [41]

AFLP genetic diversity and population differences

The use of AFLP data (dominant markers) for estimating allelic frequencies implies the consideration of an outcrossing mating system and near random mating. This means that those populations would be under Hardy-Weinberg equilibrium [67]. *Cheirolophus* has a predominantly outcrossing mating system and is pollinated by generalist insects, so one expects near random mating in the studied populations.

To estimate genetic diversity in each population showing more than one sampled individual, the following parameters were calculated: a) private alleles; b) rare alleles, present in <10% of the samples; and c) unbiased heterozygosity (Hj), calculated using TFPGA v.1.3 [68]. Further measures of genetic diversity were estimated through: (i) the frequency-down-weighted marker values (DW) index of [69] using AFLPDAT [70]; (ii) the band richness (Br), which is the number of phenotypes expected at each locus, and can be interpreted as an analogue of the allelic richness, ranging from 1 to 2 [71]; and (iii) the percentage of polymorphic loci (PLP) with a significance of 1% (P = 0.99). Br and PLP indices were calculated according to the rarefaction method of Hurlbert [72], and conditioned to the smallest population size (N = 3) allowed by the software AFLPDIV v.1.0 (http://www.pierroton.inra.fr/genetics/labo/Software/Aflpdiv/). As a consequence, Br and PLP were not calculated for populations having less than three individuals. To assess whether genetic diversity indexes (i.e. Hj, DW, Br and PLP) differed among islands, diversity range values were compared in a non-parametric Kruskal-Wallis test using package RCMDR [73] implemented in R software [74].

Pairwise F_{ST} values were estimated for each pair of populations included in this study (Table S1) using AFLP SURV v.1.0 [75]. Significance was evaluated through 10000 permutations.

Phylogenetic analyses

To address questions of species delimitation and evolutionary relationships, AFLP data generated from the three selected primer pairs were combined into one large matrix and analysed together. Phylogenetic trees were reconstructed for this combined dataset using neighbor-joining (NJ) and Bayesian inference methods. NJ trees were built using Nei-Li distances [76] and 1000 replicates with NTSYS PC v.2.1 software [77]. The support for specific nodes for the NJ tree was calculated using bootstrap [78] with 10000 replicates. Bayesian inference was performed in MrBayes v.3.1.2 [79] using a F81-like model for restriction sites [80,81] and running four independent chains each of length 10 million sampling every 100 trees. Convergence was assessed using Tracer v.1.4.1 [82] and a burn-in of 1 million trees was discarded. The remaining trees were used to construct a 50% majority rule consensus tree.

Additionally, non-tree-building approaches have been recommended to avoid conflicting phylogenetic signals in those cases in which the analyses of the data do not exhibit a bifurcating tree-like behaviour [83]. Specifically, network methods have been proposed to resolve the uncertainty of these processes [83,84]. We used the Neighbor-Net method [85] carried out with SplitsTree v.4.10 [83] to construct a distance-based network for the AFLP dataset using the Jaccard coefficient [86], which is restricted to shared band presence rather than shared absence.

Genetic structure: spatial patterns, Bayesian clustering and PCoA

We performed Mantel tests to evaluate the spatial effect in the genetic differences between populations using two similarity matrices. Genetic distance matrices – with and without *C. massonianus* from Madeira – were constructed with F_{ST} values between populations, and geographical matrices were calculated by the spatial distance (X and Y coordinates) between populations using ArcGIS v.9.1 (ESRI, Redlands, CA, USA). Mantel tests were performed with the vegan package [87] implemented in R software [74] using 100000 permutations and considering a *p*-value limit of 0.05. Because most populations were from the Canary Islands and only two of them are from Madeiran archipelago, distant from the remainder (>500 km), the Mantel test analysis was repeated without these populations.

Bayesian clustering analysis were carried out using STRUCTURE v.2.3 [88]. We applied the admixture ancestry model and the correlated allele frequencies. Ten independent simulations were run for each possible number of genetic groups (*K*) from $K = 1$ to 29, using a burn-in period of 10^5 generations and run lengths of 5×10^5. To estimate the number of genetic groups (*K*) we selected the *K* value that maximizes the probability of the data L(*K*). We also considered the criterion proposed by Evanno et al. [89] to estimate the best value of *K* for our data set, based on the rate of change in the probability between successive *K* values, ΔK. Bayesian analyses of the genetic structure were also conducted with BAPS (Bayesian Analysis of Population Structure, Spatial Clustering of Groups, [90]), which uses stochastic optimization instead of Markov chain Monte Carlo to find the optimal partition. We performed a mixture analysis of individuals with the geographic origin of the samples used as an informative prior ('spatial clustering of individuals') or without this prior ('clustering of individuals'). BAPS simulations were run with the maximal number of groups (*K*) set from 1 to 29. Each run was replicated 10 times, and the results were averaged according to the resultant likelihood scores. The output of the mixture analyses were used as input for population admixture analysis [90], with the default settings in order to detect admixture between clusters.

Some authors have warned about a degree of over-splitting in the genetic structure analysed with AFLP markers (e.g. [54]) and have suggested the use of simpler statistical analyses to compare the pattern found by Bayesian methods. Similarities among individuals were also studied via Principal Coordinate Analysis (PCoA, [91]) using the Jaccard distance, in order to detect other possible relations that could not be visualized with assignment methods or phylogenetic analyses. This procedure was carried out with R software [74] using the vegan package [87].

Finally, we conducted analysis of molecular variance (AMOVA; [92]) using ARLEQUIN v.3.5 [93] to estimate genetic differentiation following an alternative and widely used non-Bayesian approach that does not assume Hardy-Weinberg equilibrium or independence of markers. A first AMOVA analysis was implemented without taking regional structure into account. We also carried out four independent AMOVAs grouping the populations of *Cheirolophus* according to: i) their current taxonomic affiliation; ii) their island origin; iii) the most plausible model proposed by STRUCTURE; and, iv) the clustering proposed by BAPS. Pairwise genetic distance between individuals using the square Euclidean distance was used for the AMOVA analyses, considering three levels: within populations, between populations within regions and between regions. Only two levels (within and between populations) were considered when no additional grouping structure was applied.

Morphological revision and analyses

Diagnostic morphological characters classically used in taxonomical treatments of Macaronesian *Cheirolophus* were also analysed in this study. Morphological traits were measured from herbarium vouchers, cultivated individuals and/or gathered from the literature depending on the species. In the present study, morphological data was only considered for a given species when at least three different individuals where available for measurements. We obtained accurate data from 16 species for six morphological variables widely employed to distinguish among Canarian *Cheirolophus* taxa (Table S3) (including leaf length, leaf width, size of capitula, plant size, floral colour and leaf shape). Qualitative traits (i.e. floral colour and leaf shape) were coded numerically (+1 for whitish and −1 for rose-colored flowers; +1 for entire, −1 for divided and 0 for intermediate/both leaf shapes). All the variables were standardized by subtracting the character mean from each species measure and then dividing by the character standard deviation. The resulting matrix was analysed by Principal Components Analysis (PCA) using RCMDR [73] implemented in R software [74].

In order to correlate genetic AFLP data with morphological traits, we carried out Mantel tests and a generalized analysis of molecular variance (GAMOVA; [94]). Mantel tests were performed with R software [74] and the package vegan [87] computing 100000 permutations and considering a *p*-value limit of 0.05. Complementary to Mantel test, GAMOVA analyses provides a regression-based method of the analysis of molecular variance (AMOVA; [92]) that could be interpreted as a multiple regression. GAMOVA approach is an especially suited tool to identify, associate and evaluate the relationships between the differences in a phenotype or environmental variable of interest between some populations and the genetic variations among the same populations. The significance between genetic differentiation (i.e. F_{ST}) and six morphological traits (see above) was tested by running a regression matrix with a permutation test with 10000 repetitions.

Results

AFLP genotyping and filtering

Initially, 371 alleles were obtained from automatic genotyping. After manual correction, error rates calculation, elimination of small and troublesome alleles and low intensity peaks, a final matrix with 249 (67.1%) alleles were considered for subsequent analyses. The final data set obtained showed an error rate of 3.2%, which is below the maximum error rate percentage accepted for good AFLP reproducibility (5%) [95].

Population genetics: diversity and differentiation

Within-population genetic diversity measures are shown in Table S1. Private fragments were scarce across the studied populations; only three were detected, one each in *C. santos-abreui* A.Santos, *C. satarataensis* (Svent.) Holub and *C. tagananensis* (Svent.) Holub populations. The frequency of rare fragments was, conversely, much higher (10.6%), but most of these were shared by two or more populations. Hj diversity values were similar in all populations, ranging between Hj = 0.0459 (*C. duranii* (Burchard) Holub), and Hj = 0.1539 (*C. arboreus* (Webb & Berthel.) Holub from Los Tilos) in populations with more than three sampled individuals, with a mean value of Hj = 0.07892. The frequency-down-weighted marker values (DW) index showed overall high values, but considerable differences were observed among populations (ranging from 8.65 in *C. junonianus* (Svent.) Holub var. *isoplexiphyllus* to 19.75 in *C. webbianus*). The percentage of polymorphic loci (PLP %) ranged from 10.1% in *C. duranii* to 77.0% in *C. arboreus* (Los Tilos), and band richness (Br) ranged from 1.08 in *C. junonianus* var. *junonianus* to 1.46 in *C. arboreus* (Los Tilos). The DW was the only diversity index showing significant differences among islands (Kruskal-Wallis chi-squared = 15.02, df = 5, *p*-value<0.05); La Palma populations showed the lowest DW values. Pair-wise F_{ST} comparisons were significant (Table S4; mean value F_{ST}±SD = 0.259±0.185 with all populations, mean value F_{ST}±SD = 0.302±0.051 with populations with more than three individuals sampled). The biggest difference was found between *C. puntallanensis* A.Santos and *C. arbutifolius* (Svent.) G.Kunkel (F_{ST} = 0.577). The pair-wise comparisons between populations of the same species were much smaller than those between species (e.g. mean F_{ST} = 0.055 between *C. ghomerythus* (Svent.) Holub populations).

Phylogenetic analyses of Macaronesian *Cheirolophus* based on AFLP data

Trees constructed with the combined AFLP data using NJ and Bayesian estimation had generally similar topologies. Tree-building analyses of AFLP data resulted in similar fully-resolved assignment of individuals into species classically identified by diagnostic morphological and ecological characters (e.g. Bayesian 50% majority rule tree in Fig. 2; other trees not shown). However, species relationships were poorly resolved and only a few groups were partially identified. All trees included a large, strongly supported clade grouping most species from La Palma plus *C. duranii* from El Hierro (posterior probability, PP = 0.99; Fig. 2). Other supported smaller lineages grouped species from the Taganana peninsula (*C. tagananensis*, *C. anagaensis* A.Santos, *C. cf. webbianus* and *C. cf. sp. nova*; PP = 0.99) or species from the Teno mountains (*C. canariensis*, *C. burchardii* Susanna; PP = 1.00), in both cases from Tenerife. The phylogenetic analyses also confirmed the close relationship between *C. satarataensis* and *C. dariasii* (PP = 1.00), until recently considered as subspecies of the same species. The two populations of *C. teydis* from Tenerife and La Palma grouped together but were not embedded in a resolved clade. Similarly, *C. arbutifolius* and *C. falcisectus* Svent. ex Montelongo & Moraleda (both from Gran Canaria) appeared closely related (PP = 0.84) but their phylogenetic position in the genus was not fully resolved.

The Neighbor-Net (NN) analysis, although indicating considerable reticulation in the data, resolved most of the species groups according to their current taxonomic boundaries (Fig. 3). In addition, this NJ network approach clustered most of the species in agreement with the Bayesian reconstruction depicted in Fig. 2. The species from La Palma formed a cluster closely related to the species from El Hierro (*C. duranii*). Those inhabiting eastern Tenerife (*C. webbianus*, *C. tagananensis*, *C. anagensis*, *C. metlesicsii* and *C. cf. sp. nova*) and western Tenerife (*C. canariensis* and *C. burchardii*) were also segregated into regional clades already resolved in the Bayesian and NJ phylogenetic analyses. Neighbor-Net analysis also revealed some phylogenetic conflicts not identified in the tree building analyses. For example, the species from El Hierro (*C. duranii*) showed clear reticulation, with an intermediate position between La Palma cluster and a lineage grouping two species from La Gomera (*C. satarataensis* and *C. dariasii*). Finally, NN analyses suggested a degree of reticulated connection between *C. ghomerythus* (from La Gomera) and *C. massonianus* (from Madeira), but this potential relationship was not supported by any Bayesian approach.

Bayesian clustering and spatial analyses of the genetic structure

Using the matrix of inter-population F_{ST} distances, and the matrix of geographical distances in kilometres, the Mantel test indicated a significant correlation between genetic and geographical distances (r = 0.258; *p*-value<0.05). Similar results were obtained if the Madeiran populations were excluded from the analysis (Mantel test r = 0.326, *p*-value<0.05).

The Bayesian analysis of population genetic structure conducted with STRUCTURE found the highest L(*K*) and Δ*K* values for *K* = 2. This grouping separated La Palma species (plus *C. duranii* from El Hierro) – cluster A – from the rest of the species from Tenerife, La Gomera, Gran Canaria and Madeira (including the *C. teydis* population from La Palma) – cluster B –, showing high

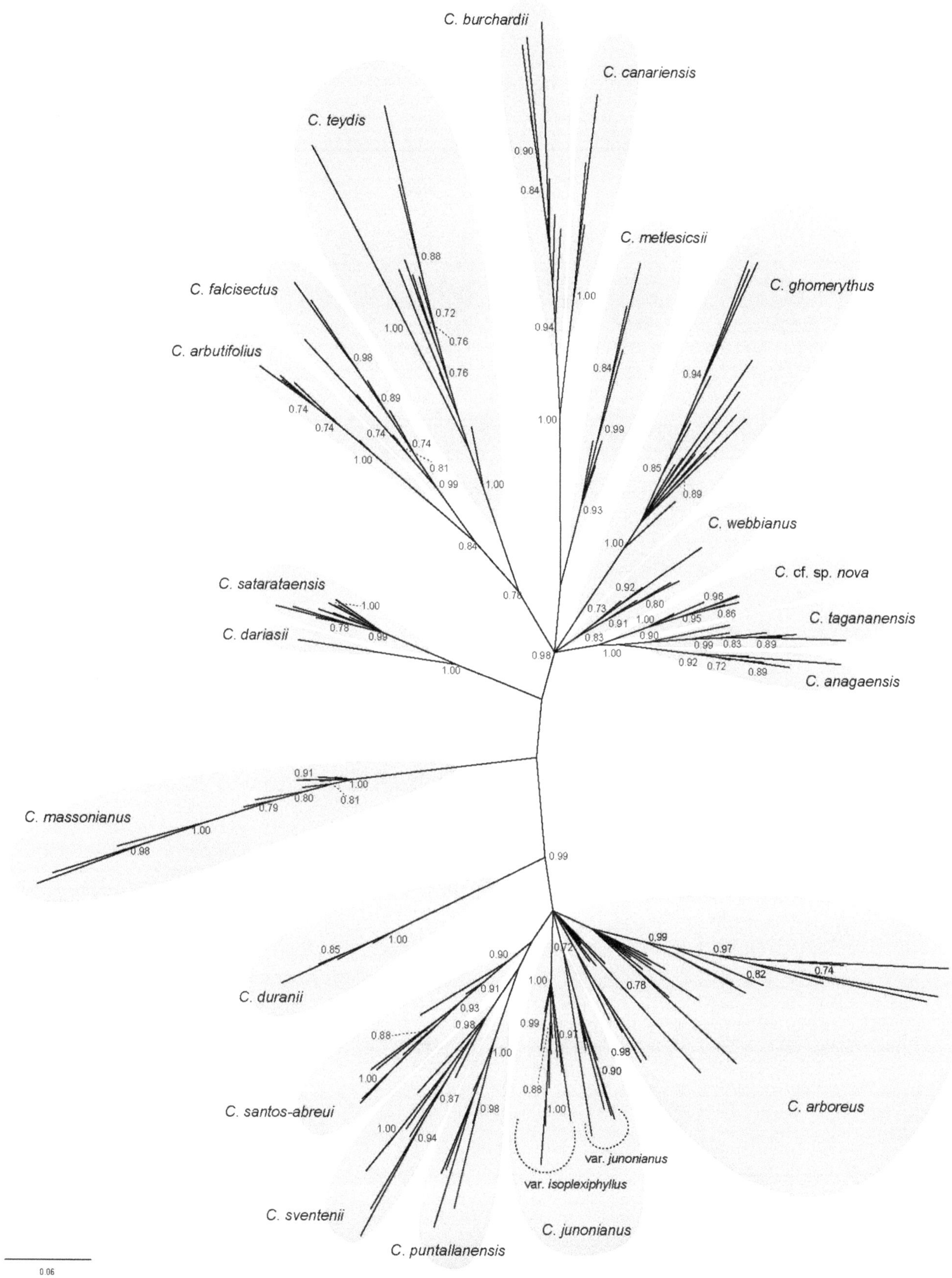

Figure 2. Phylogenetic analysis of AFLP data. Unrooted 50% majority rule tree from Bayesian analysis of the combined AFLP dataset for all Macaronesian species of *Cheirolophus*. Posterior probability values ≥ 70 are shown near each branch.

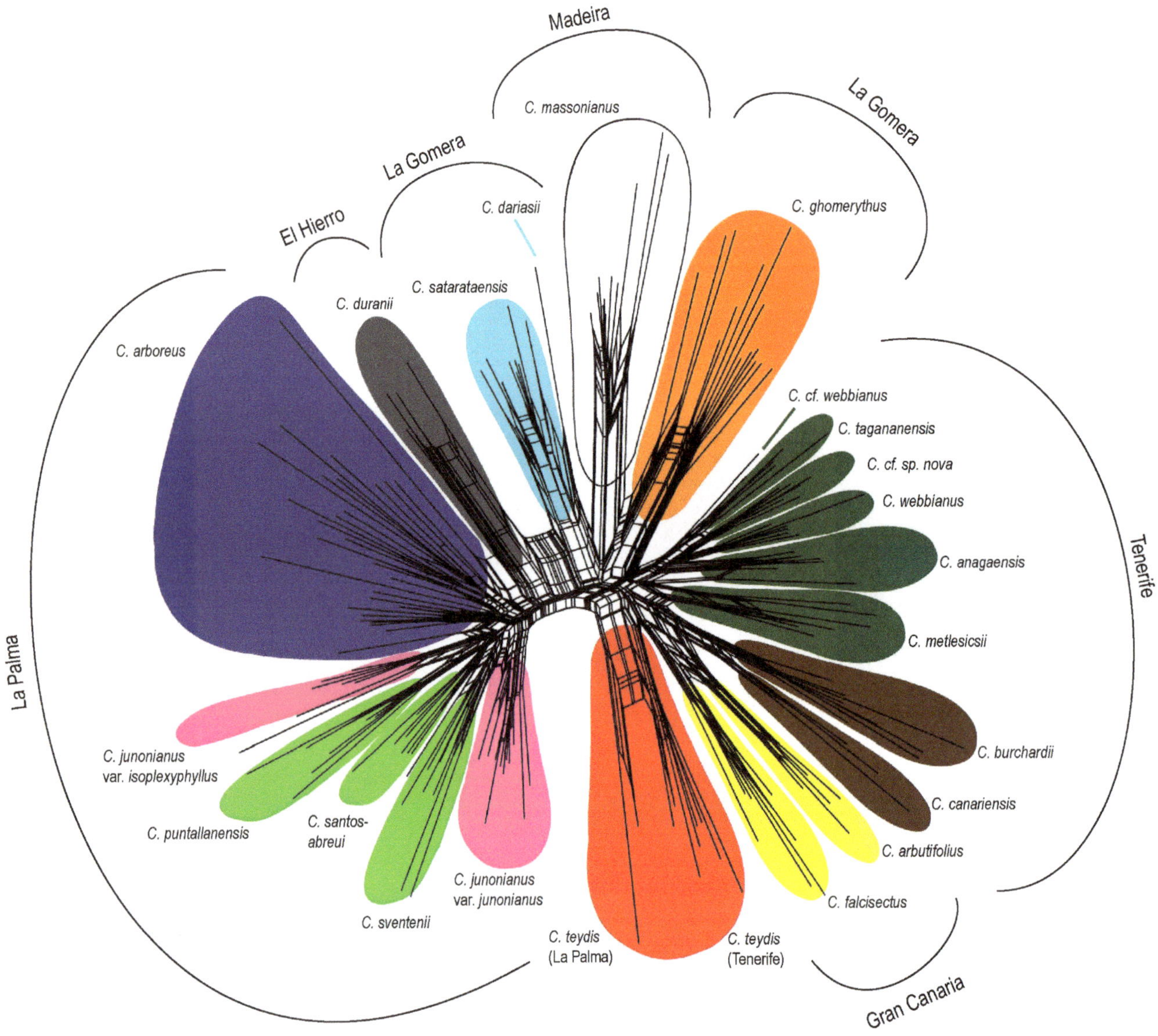

Figure 3. Neighbor-Net of AFLP data obtained from Macaronesian *Cheirolophus*. Colour coding profiles delimitate the different species indicating the genetic clusters assigned by BAPS.

percentages of individual memberships (97% and 98%, respectively) for these predefined groups (Fig. 4). According to the STRUCTURE analyses, the population of *C. teydis* from La Palma, some individuals of *C. arboreus*, *C. duranii* from El Hierro and one specimen of *C. metlesicsii* were the only ones presenting considerable (>10%) levels of admixture among the two genetic groups detected (between 14.9% and 34.3%). BAPS results showed a more fragmented distribution, with $K = 11$ as the most plausible number of clusters (P = 0.995; Figs. 1 and 3). The mixture analyses with or without spatially informative priors resulted in congruent assignment of individuals and no individual reassignment between the populations was observed. The genetic structure revealed by BAPS proved to be highly congruent with the phylogenetic analyses carried out (Fig. 3), showing at the same time a clear geographical pattern. Except for the populations of *C. teydis* and *C. massonianus*, the occurrence of genetic clusters determined by BAPS appeared to be limited to single islands and, in most cases, even restricted to particular regions within the islands (Fig. 1). No admixed individuals were detected according to this Bayesian clustering approach.

The results of non-hierarchical AMOVA indicated a highly significant level of genetic structure among populations (56.79%; df = 29; p-value<0.01, Table 1). The hierarchical AMOVA analyses also showed significant genetic differentiation explained at all levels and for all grouping schemes tested (all p-values<0.01, Table 1). The model that explained a larger fraction of variation among groups (38.63%) was the one considering the current taxonomic circumscription ($K = 20$). Using geographical origin (i.e. native island source, $K = 7$) as the grouping criterion revealed slightly lower, but also significant, genetic variance (23.59%) among sites. According to the structure proposed by the Bayesian clustering analyses, genetic differentiation among groups explained 18.73% of variation ($K = 2$; STRUCTURE) and 33.29% of variation ($K = 11$; BAPS).

The PCoA using the first three principal coordinates explained 32.8% of the total variation in the data and confirmed several

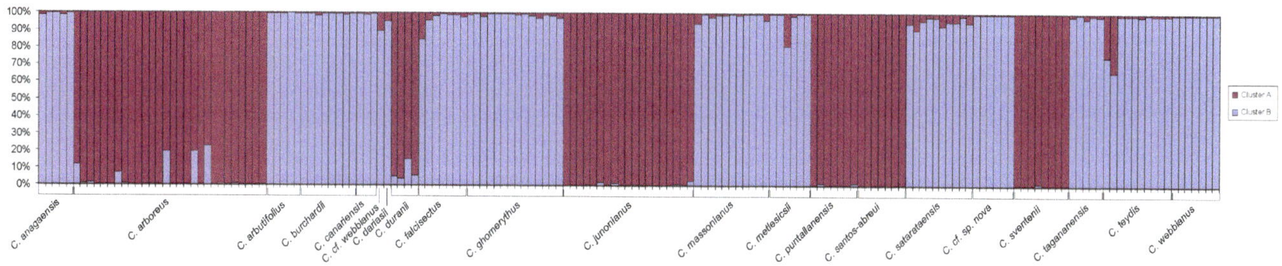

Figure 4. Results of STRUCTURE analyses of the entire AFLP dataset with $K=2$. Bayesian estimation of genetic structure within Macaronesian *Cheirolophus* according to the best model proposed by STRUCTURE ($K=2$).

relationships detected in the phylogenetic and cluster analyses (Fig. S1). The first coordinate (accounting for 18.2% of the total variation) distinguished two main groups of species: the western Canarian taxa occurring on La Palma and El Hierro clearly segregated from another main cluster formed by species inhabiting Tenerife, Gran Canaria, La Gomera and Madeira. The second coordinate (representing 7.9% of variation) segregated *C. ghomerythus* populations in a different cluster, confirming the distinctiveness of this species from the other taxa on La Gomera. The addition of this second axis also separated *C. burchardii*, *C. canariensis* and *C. arbutifolius* accessions in one discrete cluster and *C. massonianus* individuals in another. Finally, plotting a third coordinate (accounting for 6.7% of the total variation) with the first axis, led to the distinction of four discrete clusters of species from: i) Tenerife and Gran Canaria; ii) La Gomera; iii) La Palma and El Hierro; and iv) Madeira.

Morphological analyses and nrDNA sequencing data

The two first principal components represented 60% of the total morphological variation in the data set (Table S3). The first component (PC1) accounted for 31.4% of the total variance, showing the highest coefficients for flower colour and leaf shape traits (Table S3). PC2 represented another 28.4% of the total variance, having the highest coefficients for capitulum and leaf size and plant height traits. No correlation was found between genetic and morphological distances according to Mantel test ($r = 0.00083$, p-value>0.05), nor did any of the GAMOVA analysis show significant results for the morphological variables considered (all p-values>0.05).

Table 1. Analyses of molecular variance (AMOVA) of Macaronesian *Cheirolophus* species based on AFLP markers.

Source of variation	d.f.	Sum of squares	Variance components	Fixation indices	Percentage of variation	P
1. No population structure						
Among populations	28	2758.35	14.84	0.57	56.79	<0.0001
Within populations	143	1614.26	11.29		43.21	<0.0001
2. Species						
Among groups	19	2439.19	10.22	0.39	38.63	<0.0001
Among populations within groups	9	319.16	4.95	0.30	18.70	<0.0001
Within populations	143	1614.26	11.29	0.57	42.68	<0.0001
3. Islands						
Among groups	6	1236.50	6.55	0.24	23.59	<0.0001
Among populations within groups	22	1521.85	9.92	0.47	35.75	<0.0001
Within populations	143	1614.26	11.29	0.59	40.67	<0.0001
4. STRUCTURE						
Among groups	1	557.24	5.38	0.19	18.73	<0.0001
Among population within groups	27	2201.11	12.07	0.52	42.00	<0.0001
Within populations	143	161426	11.29	0.61	39.27	<0.0001
5. BAPS						
Among groups	10	1951.35	9.91	0.33	33.29	<0.0001
Among population within groups	18	807.00	6.58	0.37	24.55	<0.0001
Within populations	143	1614.26	11.29	0.58	42.15	<0.0001

The four possible scenarios considered were: no population structure (1), species delimitation (2), islands groups (3) STRUCTURE clustering (4) and BAPS clustering (5).

The analysis of nrDNA sequences from [41] evidenced a considerable number of heteromorphic sites (double peaks) within potentially introgressed species. Table S2 shows nine positions (four observed in ITS regions and five in ETS region) presenting multiple peaks on the electropherograms in seven Macaronesian species.

Discussion

Species delimitation among Macaronesian *Cheirolophus*

The performance of diverse methods to analyse AFLP data sets has been found to be particularly important in evolutionary studies of plant radiations [45,48,62,96]. Most of the analyses conducted in this study to evaluate the genetic structure of *Cheirolophus* in Macaronesia revealed different but complementary patterns, supporting the complex evolutionary history previously suggested for the genus in the archipelago [41]. Phylogenetic analyses based on AFLP were particularly useful in unravelling taxonomic boundaries among Macaronesian congeners. Both tree-building and network methods provided full support to the current species circumscription mainly based on morphological characteristics. Thus, our results corroborate the distinctiveness of these extraordinarily recent diverged species and support the utility of classical diagnostic characters employed in the taxonomical delimitation of Macaronesian *Cheirolophus*.

Geographic differentiation and long-distance dispersal

AFLP analyses were not able to thoroughly disentangle the phylogeny of Macaronesian *Cheirolophus*, but assignment methods provided further evidence on the evolutionary relationship between different lineages. The model proposed by STRUCTURE ($K=2$) identified a major geographical division, with the western islands (La Palma and El Hierro) clustering independently from the Central-Eastern ones (La Gomera, Tenerife, Gran Canaria), that grouped together with Madeira (Fig. 4). This genetic barrier has been also found in PCoA analyses (Fig. S1) and is supported by the phylogenetic signal (Figs. 2 and 3). Close evolutionary relationships among species from La Palma and El Hierro have already been reported in some studies concerning diverse animal and plant taxa from Macaronesia [97–99]. However, the majority of phylogeographic studies on Canarian flora and fauna have inferred that endemics from El Hierro appear to be more closely related to species from La Gomera than to those from La Palma [100]. Indeed, plastid DNA phylogeographic analysis has recently revealed a clear evolutionary connexion among El Hierro and La Gomera haplotypes [41], which provides evidence for a series of incongruent patterns among AFLP and cpDNA data (see discussion on introgression section below). A model with more clusters was proposed by BAPS ($K=11$), suggesting as well significant influence of allopatric speciation in the evolutionary history of Macaronesian *Cheirolophus*. The species from Tenerife are grouped in three clusters corresponding to well differentiated regions of the island. One cluster comprised the taxa inhabiting mainly the eastern part of Tenerife (i.e. *C. webbianus*, *C. tagananensis*, *C. anagaensis*, *C. metlesicsii* and *C. cf. sp. nov.*). Another cluster groups *C. canariensis* and *C. burchardii* from the Teno mountains, at the western end of the island. This eastern/western pattern of geographic distribution in Tenerife has been recovered in other studies of several plant groups (see [101], for a review). In some cases, this segregation has been explained by disjunct evolution of lineages in two palaeo-islands of Tenerife [102,103], currently corresponding to Anaga and Teno massifs at different ends of the island. According to previous phylogeographic and dating analyses [41], this geographic splitting of *Cheirolophus* lineages from eastern and western Tenerife could be related to connection-isolation cycles caused by volcanic activity during the last 2 My [104]. A third cluster included the populations of *C. teydis* from Tenerife (Las Cañadas) and La Palma, all of them living around the subalpine zone (1800–2200 m) on both islands. The representatives from Gran Canaria were placed together by BAPS in a separate cluster. This result may support the recent hypothesis of a single colonisation event for this island and the subsequent diversification process giving rise to *C. arbutifolius* and *C. falcisectus* [41]. Our Neighbor-Net and PCoA analyses pointed to an evolutionary closeness among Gran Canaria and Tenerife lineages, but we were not able to reconstruct their accurate phylogeographic relationship. Nonetheless, in agreement to the BAPS results, the species from La Gomera are grouped in two different clusters showing a clear geographic distribution pattern: one including *C. ghomerythus* from the northern coast of the island and the other one embedding *C. satarataensis* and its former subspecies *C. dariasii*, from the south and south-west of La Gomera. Certain evolutionary closeness among the two lineages from La Gomera was detected in the PCoA analyses, but their monophyly could not be confirmed. BAPS method also differentiated four geographically-related clusters of species in La Palma: *C. junonianus* from the south of the island; *C. arboreus* from the north and the west; *C. santos-abreui*, *C. puntallanensis* and *C. sventenii* (A.Santos) G.Kunkel from the north and north-eastern part of La Palma; and *C. teydis* from the summits of this island (and from Las Cañadas in Tenerife). Finally, *C. duranii* from El Hierro and *C. massonianus* from Madeira grouped in two separate clusters, suggesting as well that geographic isolation has been involved in the radiation of the genus.

According to these results, sporadic long-distance dispersal (LDD) events have to be considered to explain the numerous intra and inter-island colonisations resulting in the current distribution of genetic and taxonomic diversity in Macaronesian *Cheirolophus*. Unfortunately, straightforward evidences of LDD are very difficult to obtain [105], and we have not been able to provide direct proofs of these events occurring on our group of study. However, several supporting evidences (discussed below) suggest that Macaronesian *Cheirolophus* might have the potential to undergo successful LDD events. Birds and lizards have been found to be involved in seed long- dispersal of multiple species in the Canary Islands [106,107] and other oceanic archipelagos [108,109]. In the Galapagos and Azores, finches (granivorous birds from family Fringillidae) have been reported as legitimate seed dispersers of dry-fruited plants – including Asteraceae – implicated in LDD events between islands [109–111]. The goldfinch (*Carduelis carduelis* L.) has been observed in the Canary Islands feeding on *Cheirolophus* seeds [112], which suggests that bird-mediated dispersal could also be an important LDD mechanism in Macaronesian species. The role of birds in LDD events in *Cheirolophus* was also pointed by Garnatje et al. [113], who inferred that seabirds could have mediated the re-colonisation event from the Balearic Islands towards the continent in *C. intybaceus* (Lam.) Dostál. Finally, viable seeds from another Centaureinae species have been recovered from lizard guts discarded by predatory birds in the Canarian archipelago [114], supporting this kind of secondary seed dispersal as a likely mechanism involved in the LDD of Macaronesian *Cheirolophus*. All these data suggest that *Cheirolophus* achenes may be able to be transported through long distances but, given the lack of strong direct evidences in our case study and the inherent stochasticity of LDD process [105], speculation about potential role of these or other animals as LDD vectors must be made with extreme caution.

Furthermore, the capacity for successful long distance colonisation in *Cheirolophus* could also be enhanced by certain intrinsic biological features showed by this group of plants. The genus presents a pseudo-self-compatible mating system [115], that may be able to originate a sexually reproducing population from a single propagule, carrying at the same time more genetic variation than a seed from an autogamous population [116]. In summary, the results of our AFLP analysis are consistent with a role of allopatric divergence in the radiation of Macaronesian *Cheirolophus*; a hypothesis apparently supported by additional bibliographic evidences suggesting a certain degree of ability for successful long-distance colonisation events in the genus.

Geographic-genetic correlation and limited gene flow in Macaronesian *Cheirolophus*

Genetic isolation has been proposed as a major factor determining plant speciation on oceanic islands [117]. The significance found in the spatial explicit analysis (Mantel test) suggests that Macaronesian *Cheirolophus* are influenced by geographic-genetic correlation across species. In particular, our results indicate that the more closely evolutionary-related species are as well geographically closer to each other, which is in good agreement with a gene flow scenario dominated by short distance events [118]. Regular seed dispersal in *Cheirolophus* has been reported to be limited to very short distances (see [32] for some species description). This genus shows relatively heavy and pappus-lacking cypselas that are unable to be transported by the wind, falling by gravity beside the mother plant during dissemination [119]. Another important factor enhancing geographic-genetic correlation across species might be topographic isolation [118]. Most *Cheirolophus* taxa inhabit deep ravines, coastal cliffs or steep slopes [30,32], additionally preventing regular seed dispersal through long distances. The large number of Macaronesian *Cheirolophus* species inhabiting a few restricted and isolated locations (e.g. microendemisms such as *C. anagaensis*, *C. burchardii*, *C. dariasii*, *C. falcisectus*, *C. junonianus*, *C. metlesicsii*, *C. puntallanensis*, *C. santos-abreui* or *C. tagananensis*) could be related to the additive effect of poorly connected habitats with the limited seed dispersal capacity of *Cheirolophus*.

Our analysis across the whole distribution range of Macaronesian *Cheirolophus* revealed overall low levels of genetic diversity. This result is in accordance with the general expectation that endemic species, and particularly island endemics [120], exhibit lower levels of genetic diversity than widespread species [121,122]. Generally, microendemics occupying restricted and isolated populations (e.g. *C. anagaensis*, *C. duranii*, *C. junonianus*, *C. massonianus*, *C. cf. sp. nova*, or *C. tagananensis*) show lower genetic diversity levels than species with numerous, widely distributed populations (e.g. *C. arboreus*, *C. ghomerythus*, *C. sventenii*, or *C. teydis*). However, this expected pattern is contradicted by certain populations of widely distributed species (e.g. *C. arbutifolius*, *C. sataratuensis*) also showing lower genetic diversity values. Considering the limited number of sampled populations per species and the low number of individuals available, these indices could be underestimating genetic diversity, particularly in these widespread species. We found only three populations showing private fragments, probably due to the low sampling size within some populations and the strict choice of polymorphic bands.

Some of the genetic diversity indices studied in the Macaronesian *Cheirolophus* can also be compared with the values reported by Garnatje et al. [113] in the Mediterranean complex of *C. intybaceus*, which includes some taxa distributed in the eastern Iberian Peninsula and the Balearic Islands. Diversification in the *C. intybaceus* complex presumably started in the same period as the Macaronesian radiation [41], but in the Mediterranean group it resulted in only four doubtful species. Heterozygosity levels detected in populations of Macaronesian species are significantly lower than in Mediterranean populations of the *C. intybaceus* complex. Lower genetic variation has been associated with species showing limited geographical distribution, smaller populations and exhibiting little gene flow [123–125]. Our results could be reflecting that Macaronesian *Cheirolophus* – distributed in small populations across islands with steeply dissected topography – have been comparatively more isolated than Mediterranean ones, thus contributing to their progressive genetic divergence. As it has been proposed for other plant groups by Ellis et al. [118] and Knope et al. [22], the combination of certain ability for sporadic long-distance colonisation – see the section above – and poor gene flow capacity – due to both intrinsic and geographic characteristics – could have played an important role enhancing the explosive diversification of *Cheirolophus* in Macaronesia.

Ecological adaptation

Geographic isolation may have been important for enhancing diversification but ecological adaptation is as well a common mechanism contributing – either in allopatry [126] or in sympatry [127] – to island plant speciation. According to Whittaker and Fernández-Palacios [1], the diversification of *Cheirolophus* in the Macaronesian archipelagos can be considered an example of non-adaptive radiation occurred on oceanic islands. Indeed, most species of this group exploit very similar niches in different islands, showing minor morphological differences. However, Macaronesian *Cheirolophus* present as well a few cases of taxa that have apparently adapted their morphology to the significantly diverse ecological conditions found on these archipelagos. *Cheirolophus teydis* is the only species of the genus that occupies the subalpine habitat, showing morphological adaptations to tougher ecological conditions (i.e. rosette-like disposed leaves with reduced laminas; waxy leaf cover; high number of smaller flowers; annual flowering shoots). The species from Gran Canaria – *C. arbutifolius* and *C. falcisectus* – are the result of a diversification process originated after a single colonisation of the island (see above). These species have diverged into different niches allopatrically within the same island – *C. falcisectus* inhabits more xeric habitats and shows clear leaf reduction while *C. arbutifolius* occupies more humid locations and present an arborescent habit and a larger leaf surface – thus suggesting an additional example of ecological differentiation. A similar case of ecological adaptation can be found in La Palma, where *C. junonianus* – from the south of the island and inhabiting significantly more arid localities than the rest of species from the northern part of this island – shows parallel morphological adaptation in size and leaf shape to drought conditions. Equivalent eco-morphological responses have already been reported in other Macaronesian plant taxa that have undergone an adaptive radiation process (e.g. *Argyranthemum* [2]). Interestingly, our morphology-genetics correlation analyses indicate that morphological similarity across the different Macaronesian *Cheirolophus* species is not affected by genetic similarity. Certainly, phylogenetic and clustering analyses are not congruent with a relationship between the main lineages and ecology: some species occupying different niches share the same or close genetic groups/lineages (e.g. *C. falcisectus* and *C. arbutifolius*; or *C. junonianus* and *C. arboreus*), whilst species showing similar habitats and morphological features frequently belong to very different genetic clusters/lineages (e.g. *C. duranii* and *C. tagananesis*; or *C. ghomerythus* and *C. burchardii*). These patterns suggest that the few cases of ecological adaptation (cited above) do not seem to be the result of a single eco-morphologic shift occurring at initial stages of the Macaronesian *Cheirolophus* diversification, but they may correspond to more recent, multiple and independent phenotype-environment differentiation processes.

Clearly, there are not enough data here to discard a vital role of selection in the radiation of *Cheirolophus*. We only measured a few morphological traits usually employed to delimitate taxonomically the Macaronesian species (Table S3), but other potentially important morphological and physiological features could have been missed. In addition, there is no accurate evaluation of the niches occupied by the different species, so there could also be fine-scale ecological variables differentiating habitats formerly considered as equivalent. Therefore, more precise inferences about the relative importance of ecological adaptation in this radiation process would require additional intraspecific sampling, supplementary morphological, physiological and ecological measurements as well as more appropriate tests (see [126,128,129]). Further studies applying these methodologies will improve our understanding of the role played by selection versus neutral differentiation in islands diversifications.

Evidence of interspecific gene flow

Introgression is another mechanism formerly proposed to have played a role in the evolutionary history of Macaronesian *Cheirolophus* [41]. For instance, the species from Madeira, *C. massonianus*, was proposed to have come into contact with a continental congener, resulting in a chloroplast capture event. The genetic imprint of this hybridization event was not detected in the nrDNA regions sequenced in that former study, grouping *C. massonianus* within the Canarian clade and suggesting that introgression – via plastid transfer (see [130]) – did not affect the nuclear genome. From our AFLP data, pair-wise F_{ST} comparisons between Madeiran and Canarian populations showed similar genetic differentiation values to comparisons within Canarian populations (Table S4), apparently supporting the overall closeness among nuclear genomes of species from both archipelagos. Unfortunately, we did not include in this AFLP study any continental species putatively involved in *C. massonianus* hybridization, so our analyses do not allow assessment of whether introgression was limited to the chloroplast genome or also affected the nuclear genome.

In contrast, traces of genetic admixture were detected from AFLP data by STRUCTURE analysis in a population of *C. teydis* from La Palma (Fig. 4). Our results suggest that *C. teydis* originated in Tenerife and colonised subsequently La Palma, where genetic contact with other species from this island may occur. Plastid DNA analyses [41] support this hypothesis, pointing to a process of plastid capture to explain haplotypic differentiation among *C. teydis* populations from Tenerife and La Palma. In the same way, few *C. arboreus* individuals from populations in North-West La Palma show some genetic introgression among both genetic clusters defined by STRUCTURE. It has been reported that some specimens of *C. arboreus* from this part of the island exhibit morphological traits significantly distinct from the type [32]. Indeed, one of these *C. arboreus* populations from NW La Palma (Bco. Briestas) shows a different cpDNA haplotype from the rest of populations [41]. Another potential case of introgression could be affecting the species from El Hierro, *C. duranii*. This species is grouped together with the rest of taxa from La Palma according to our phylogenetic, PCoA and clustering analyses (Figs. 2, 4 and S1). In contrast, plastid DNA phylogenetic analysis [41], revealed a clear evolutionary closeness among *C. duranii* and the species from La Gomera. The Neighbor-Net analysis of our AFLP data mainly supports the evolutionary relationship among *C. duranii* from El Hierro and the rest of species from La Palma, but it also shows a faint reticulation signal between *C. duranii* and *C. satarataensis* from La Gomera (Fig. 3). In this case, certain genetic admixture – albeit very weak – can also be perceived from the STRUCTURE analysis, suggesting as well that this species from El Hierro may present genetic traces from both La Palma and La Gomera *Cheirolophus* species. Finally, the occurrence of introgression events during the radiation process could be also supported by the polymorphic sites observed in nrDNA sequences of the species here mentioned (see Table S2).

These results suggest that introgression might have played certain role in the evolutionary history of *Cheirolophus* in Macaronesia. However, similar patterns can be generated by ancestral polymorphisms and incomplete lineage sorting [131]. The STRUCTURE analysis does provide indication of admixed genotypes in some species, but at $K = 2$ the parent species for introgression are impossible to determine, being perhaps a vestige from some ancestral polymorphisms. According to the phylogeographic analysis performed by Vitales et al. [41], La Palma and El Hierro were colonised by *Cheirolophus* from Tenerife Island, so the genetic cluster A – mainly found in La Palma and El Hierro – proposed by STRUCTURE should be derived from the putative older cluster B, predominant in Tenerife, La Gomera and Gran Canaria (see Fig. 4). The admixture signal observed in *C. duranii* or in NW *C. arboreus* populations fits well with this alternative scenario considering certain retention of ancestral polymorphisms. In contrast, the pattern observed in *C. teydis* from La Palma – showing considerable admixture signal from the derived cluster B – seems more difficult to explain by the only action of ancient polymorphism retention. Heteromorphic positions found in nrDNA are consistent with some genetic reticulation but they can also be attributed to retention of polymorphisms and incomplete concerted evolution [132], especially considering the rapidity of this radiation process. In summary, there are some evidence for genetic introgression during the diversification of Macaronesian *Cheirolophus*, but in most cases there are alternative possible explanations (e.g. ancestral polymorphisms), thus limiting our conclusions about the relative importance of reticulation in the evolutionary history of the group.

Conservation recommendations

The genus *Cheirolophus* is illustrating one of the largest plant radiations in the Canary Islands [133]. Having nine of the 20 extant insular species included in the IUCN Red List [31], *Cheirolophus* shows the highest proportion of endangered taxa for any species-rich lineage in this archipelago. Thus, attending to the major conservational interest of the group, we consider capital to analyse the results of this study from a conservation point of view. Even though Macaronesian *Cheirolophus* are the result of an exceptionally recent diversification, phylogenetic analyses based on AFLP confirmed the evolutionary identity of currently described endemics. Generally, the genetic diversity indexes calculated for the different populations and species of the group were found to be relatively heterogeneous, reflecting the complexity of this radiation process (see above). However, the frequency-down-weighted fragment values per population (DW) provided interesting and useful information about conservation biology of the numerous endangered representatives of the genus. This rarity index (DW), has been employed to assess the genetic distinctiveness of populations and species [134], and can also be used as an indicator of uniqueness and evolutionary relevance for conservation. Previous works suggest that species with high levels of unique genetic information are more likely to be threatened (e.g. [135,136]). However, according to our results, current threat categories assigned to Macaronesian *Cheirolophus* species do not reflect their uneven evolutionary differentiation. *Cheirolophus* taxa considered in higher extinction risk categories (CR and EN) in the Spanish Red List of Endangered Flora [32] and IUCN Red List of Threatened Species [31] have relatively low DW values (e.g. *C. arboreus*, *C. duranii*, *C. metlesicsii*, *C. santos-abreui*). In contrast, those species showing higher DW values - therefore considered more genetically distinct – are assigned to lower extinction risk categories (VU) or even considered unthreatened (e.g. *C. canariensis*, *C. satarataensis*, *C. webbianus*). As resources for conservation are limited, their optimal allocation is essential [137]. Prioritization may be especially important when dealing with rich groups of closely related endemics inhabiting biodiversity hotspots [138–140], as is the case of Macaronesian *Cheirolophus*. We suggest that future evaluations of the endangered status of Macaronesian *Cheirolophus* should take into account the evolutionary distinctiveness results presented here. Predictably, the prioritization in resources allocation among *Cheirolophus* species would change if their genetic differentiation level is considered during conservation assessment process.

Supporting Information

Figure S1 Principal coordinates (PCoA) plot of AFLP data for the Macaronesian *Cheirolophus* populations included in this study. Different symbols correspond to different populations as shown in the legend in the right side. Some populations groups that are well-differentiated and/or mentioned in the text are circled and named.

Table S1 Sampling information and genetic diversity indexes assessed. Taxa, locality, code, number of sampled individuals (N), and genetic diversity indexes assessed by AFLP in 29 populations of *Cheirolophus* from Macaronesia. Genetic indices: number of private fragments (f_u); heterozygosity (Hj); percentage of polymorphic loci for a standardised sample size of three (*PLP* 1%); band richness for a standardised sample size of three (*Br*); and frequency-down-weighted marker values index (DW).

Table S2 Polymorphic positions in nrDNA sequencing of some Canarian *Cheirolophus* taxa. Positions in ITS and ETS sequences of some Canarian *Cheirolophus* specimens where more than one base is represented in a single amplification product, seen as subequal multiple peaks on the electropherograms. Data of sequences from Vitales et al. (2014).

Table S3 Data from diagnostic morphological characters of Canarian *Cheirolophus* species. Character loadings in first two principal components for the analysis of *Cheirolophus* morphological data (high loadings are highlighted in boldface type).

Table S4 Pair-wise F_{ST} distances based on AFLP data calculated among the Macaronesian *Cheirolophus* populations included in this study.

Acknowledgments

Laia Barres, Roser Vilatersana and the Botanical Garden of Madeira are acknowledged for their contribution to supplying *C. massonianus* material. We also thank Miquel Veny for keeping the collections of living plants.

Author Contributions

Conceived and designed the experiments: TG JV JP DV. Performed the experiments: RSC MFF DV. Analyzed the data: AG-F OH DV. Contributed reagents/materials/analysis tools: AS-G MFF RSC OH. Wrote the paper: AG-F TG JV JP DV. Reviewed and improved the manuscript: AS-G OH MFF RSC.

References

1. Whittaker RJ, Fernández-Palacios JM (2007) Island biogeography: Ecology, evolution and conservation. Oxford Uni. Oxford.
2. Humphries CJ (1979) Endemism and evolution in Macaronesia. In: Bramwell D, editor. Plants and islands. London. pp.171–200.
3. Carracedo JC (1994) The Canary Islands: an example of structural control on the growth of large oceanic-islands volcanoes. J Volcanol Geotherm Res 60: 225–241.
4. Fernández-Palacios JM, De Nascimento L, Otto R, Delgado JD, García-del-Rey E, et al. (2011) A reconstruction of Palaeo-Macaronesia, with particular reference to the long-term biogeography of the Atlantic island laurel forests. J Biogeogr 38: 226–246.
5. Caujapé-Castells J (2011) Jesters, red queens, boomerangs and surfers: a molecular outlook on the diversity of the Canarian endemic flora. In: Bramwell D, Caujapé-Castells J, editors. The biology of islands floras. Cambridge: Cambridge University Press. pp.284–234.
6. Francisco-Ortega J, Jansen RK, Santos-Guerra A (1996) Chloroplast DNA evidence of colonization, adaptive radiation, and hybridization in the evolution of the Macaronesian flora. Proc Natl Acad Sci U S A 93: 4085–4090.
7. Kim SC, Crawford DJ, Francisco-Ortega J, Santos-Guerra A (1996) A common origin for woody *Sonchus* and five related genera in the Macaronesian islands: molecular evidence for extensive radiation. Proc Natl Acad Sci U S A 93: 7743–7748.
8. Jorgensen TH, Olesen JM (2001) Adaptive radiation of island plants: evidence from *Aeonium* (Crassulaceae) of the Canary Islands. Perspect Plant Ecol Evol Syst 4: 29–42.
9. García-Maroto F, Mañas-Fernández A, Garrido-Cárdenas JA, Alonso DL, Guil-Guerrero JL, et al. (2009) Delta6-desaturase sequence evidence for explosive Pliocene radiations within the adaptive radiation of Macaronesian *Echium* (Boraginaceae). Mol Phylogenet Evol 52: 563–574.
10. Gruenstaeudl M, Santos-Guerra A, Jansen RK (2012) Phylogenetic analyses of *Tolpis* Adans. (Asteraceae) reveal patterns of adaptative radiation, multiple colonization and interspecific hybridization. Cladistics 1: 1–19.
11. Silvertown J, Francisco-Ortega J, Carine M (2005) The monophyly of island radiations: an evaluation of niche pre-emption and some alternative explanations. J Ecol 93: 653–657.
12. Kim S-C, McGowen MR, Lubinsky P, Barber JC, Mort ME, et al. (2008) Timing and tempo of early and successive adaptive radiations in Macaronesia. PLoS One 3. doi:10.1371.
13. Grehan J (2001) Biogeography and evolution of the Galapagos: integration of the biological and geological evidence. Biol J Linn Soc 74: 267–287.
14. Heads M (2010) The endemic plant families and the palms of New Caledonia: a biogeographical analysis. J Biogeogr 37: 1239–1250.
15. Renner SS, Strijk JS, Strasberg D, Thébaud C (2010) Biogeography of the Monimiaceae (Laurales): a role for East Gondwana and long-distance dispersal, but not West Gondwana. J Biogeogr 37: 1227–1238.
16. Carine MA, Russell SJ, Santos-Guerra A, Francisco-Ortega J (2004) Relationships of the Macaronesian and Mediterranean Floras: molecular evidence for multiple colonizations into Macaronesia and back-colonization of the continent in *Convolvulus* (Convolvulaceae). Am J Bot 91: 1070–1085.
17. Bellemain E, Ricklefs RE (2008) Are islands the end of the colonization road? Trends Ecol Evol 23: 461–468.
18. Stuessy TF, Jakubowsky G, Gomez RS, Pfosser M, Schluter PM, et al. (2006) Anagenetic evolution in island plants. J Biogeogr 33: 1259–1265.
19. Whittaker RJ, Triantis KA, Ladle RJ (2008) A general dynamic theory of oceanic island biogeography. J Biogeogr 35: 977–994.
20. Richardson JE, Weitz FM, Fay MF, Cronk QCB, Linder HP, et al. (2001) Phylogenetic analyses of *Phylica* L. (Rhamnaceae) with an emphasis on island species: evidence from plastid trnL-F and nuclear internal transcribed spacer (ribosomal) DNA sequences. Taxon: 405–427.
21. Mort ME, Crawford DJ, Archibald JK, O'Leary TR, Santos-Guerra A (2010) Plant DNA barcoding: A test using Macaronesian taxa of *Tolpis* (Asteraceae). Nat Hist 59: 581–587.
22. Knope ML, Morden CW, Funk VA, Fukami T (2012) Area and the rapid radiation of Hawaiian *Bidens* (Asteraceae). J Biogeogr 39: 1206–1216.

23. Albertson RC, Markert JA, Danley PD, Kocher TD (1999) Phylogeny of a rapidly evolving clade: the cichlid fishes of Lake Malawi, East Africa. Proc Natl Acad Sci U S A 96: 5107–5110.
24. Linder HP (2008) Plant species radiations: where, when, why? Philos Trans R Soc Lond B Biol Sci 363: 3097–3105.
25. Glor RE (2010) Phylogenetic insights on adaptive radiation. Annu Rev Ecol Evol Syst 41: 251–270.
26. Mzighani SI, Nikaido M, Takeda M, Seehausen O, Budeba YL, et al. (2010) Genetic variation and demographic history of the *Haplochromis laparogramma* group of Lake Victoria–An analysis based on SINEs and mitochondrial DNA. Gene 450: 39–47.
27. Bromham L, Woolfit M (2004) Explosive radiations and the reliability of molecular clocks: island endemic radiations as a test case. Syst Biol 53: 758–766.
28. Kim S-C (2012) Mapping unexplored genomes II: genetic architecture of species differences in the woody *Sonchus* alliance (Asteraceae) in the Macaronesian Islands. J Plant Res 125: 125–136.
29. Bramwell D (1976) The endemic flora of the Canary Islands; distribution, relationships and phytogeography. In: Kunkel G, editor. Biogeography and ecology in the Canary Islands. The Hague: Junk Publishers. pp.207–240.
30. Bramwell D, Bramwell Z (2001) Wild flowers of the Canary Islands. 2nd ed. Madrid: Editorial Rueda.
31. IUCN (2012) IUCN Red List of Threatened Species Version 2012.2. Available: http://www.iucnredlist.org/. Accessed 1 February 2012.
32. Bañares A, Blanca G, Güemes J, Moreno JC, Ortiz S (2010) Atlas y libro Rojo de la Flora Vascular Amenazada de España. Adenda 2010. Madrid: Dirección General de Medio Natural y Política Forestal y Sociedad Española de Biología de la Conservación de Plantas.
33. De Queiroz K (2007) Species concepts and species delimitation. Syst Biol 56: 879–886.
34. DeSalle R, Amato G (2004) The expansion of conservation genetics. Nat Rev Genet 5: 702–712.
35. Dostál J (1976) *Cheirolophus*. In: Tutin TG, Heywod VH, Burges NA, Moore DM, Valentine DH, et al. editors. Flora Europaea. Cambridge: Cambridge University Press. pp.249–250.
36. Bremer K (1994) Asteraceae. Cladistics and classification. Portland: Timber Press.
37. Garnatje T, Susanna A, Messeguer R (1998) Isozyme studies in the genus *Cheirolophus* (Asteraceae: Cardueae-Centaureinae) in the Iberian Peninsula, North Africa and the Canary Islands. Plant Syst Evol 213: 57–70.
38. Susanna A, Garnatje T, Garcia-Jacas N (1999) Molecular phylogeny of *Cheirolophus* (Asteraceae:Cardueae-Centaureinae) based on ITS sequences of nuclear ribosomal DNA. Plant Syst Evol 214: 147–160.
39. Garnatje T, Garcia S, Canela MÁ (2007) Genome size variation from a phylogenetic perspective in the genus *Cheirolophus* Cass. (Asteraceae): biogeographic implications. Plant Syst Evol 264: 117–134.
40. Garnatje T, Hidalgo O, Vitales D, Pellicer J, Vallès J, et al. (2012) Swarm of terminal 35S in *Cheirolophus* (Asteraceae, Centaureinae). Genome 55: 529–535.
41. Vitales D, Garnatje T, Pellicer J, Vallès J, Santos-Guerra A, et al. (2014) The explosive radiation of *Cheirolophus* (Asteraceae, Cardueae) in Macaronesia. BMC Evol Biol 14: 118.
42. Koopman WJM, Zevenbergen MJ, Berg RG Van den (2001) Species relationships in *Lactuca* s.l. (Lactuceae, Asteraceae) inferred from AFLP fingerprints. Am J Bot 88: 1881–1887.
43. Després L, Gielly L, Redoutet B, Taberlet P (2003) Using AFLP to resolve phylogenetic relationships in a morphologically diversified plant species complex when nuclear and chloroplast sequences fail to reveal variability. Mol Phylogenet Evol 27: 185–196.
44. Richardson JE, Fay MF, Cronk QCB, Chase MW (2003) Species delimitation and the origin of populations in island representatives of *Phylica* (Rhamnaceae). Evolution 57: 816–827.
45. Guo Y-P, Saukel J, Mittermayr R, Ehrendorfer F (2005) AFLP analyses demonstrate genetic divergence, hybridization, and multiple polyploidization in the evolution of *Achillea* (Asteraceae-Anthemideae). New Phytol 166: 273–289.
46. Mendelson TC, Shaw KL (2005) Rapid speciation in an arthropod. Nature 433: 375–376.
47. Díaz-Pérez A, Sequeira M, Santos-Guerra A, Catalán P (2008) Multiple colonizations, in situ speciation, and volcanism-associated stepping-stone dispersals shaped the phylogeography of the Macaronesian red fescues (*Festuca* L., Gramineae). Syst Biol 57: 732–749.
48. Meudt HM, Lockhart PJ, Bryant D (2009) Species delimitation and phylogeny of a New Zealand plant species radiation. BMC Evol Biol 9: 111.
49. Joyce DA, Lunt DH, Genner MJ, Turner GF, Bills R, et al. (2011) Repeated colonization and hybridization in Lake Malawi cichlids. Curr Biol 21: R108–9.
50. Bussell JD, Waycott M, Chappill JA (2005) Arbitrarily amplified DNA markers as characters for phylogenetic inference. Perspect Plant Ecol Evol Syst 7: 3–26.
51. Koopman WJM (2005) Phylogenetic signal in AFLP data sets. Syst Biol 54: 197–217.
52. Althoff DM, Gitzendanner MA, Segraves KA (2007) The utility of amplified fragment length polymorphisms in phylogenetics: a comparison of homology within and between genomes. Syst Biol 56: 477–484.
53. García-Pereira MJ, Caballero A, Quesada H (2010) Evaluating the relationship between evolutionary divergence and phylogenetic accuracy in AFLP data sets. Mol Biol Evol 27: 988–1000.
54. Herrmann D, Poncet N, Manel S, Rioux D, Gielly L, et al. (2010) Selection criteria for scoring amplified fragment length polymorphisms (AFLPs) positively affect the reliability of population genetic parameter estimates. Genome 310: 302–310.
55. Crawford LA, Koscinski D, Keyghobadi N (2012) A call for more transparent reporting of error rates: the quality of AFLP data in ecological and evolutionary research. Mol Ecol 21: 5911–5917.
56. Gaudeul M, Rouhan G, Gardner MF, Hollingsworth PM (2012) AFLP markers provide insights into the evolutionary relationships and diversification of New Caledonian *Araucaria* species (Araucariaceae). Am J Bot 99: 68–81.
57. Fiz-Palacios O, Vargas P, Vila R, Papadopulos AST, Aldasoro JJ (2010) The uneven phylogeny and biogeography of *Erodium* (Geraniaceae): radiations in the Mediterranean and recent recurrent intercontinental colonization. Ann Bot: 871–884.
58. Papadopulos AST, Price Z, Devaux C, Hipperson H, Smadja CM, et al. (2013) A comparative analysis of the mechanisms underlying speciation on Lord Howe Island. J Evol Biol 26: 733–745.
59. Ley AC, Hardy OJ (2013) Improving AFLP analysis of large-scale patterns of genetic variation – a case study with the Central African lianas *Haumania* spp (Marantaceae) showing interspecific gene flow. Mol Ecol 22: 1984–1997.
60. Bensch S, Akesson M (2005) Ten years of AFLP in ecology and evolution: why so few animals? Mol Ecol 14: 2899–2914.
61. Meudt HM, Clarke AC (2007) Almost forgotten or latest practice? AFLP applications, analyses and advances. Trends Plant Sci 12: 106–117.
62. Bacon CD, Allan GJ, Zimmer EA, Wagner WL (2011) Genome scans reveal high levels of gene flow in Hawaiian *Pittosporum*. Taxon 60: 733–741.
63. Doyle J, Doyle JL (1987) Genomic plant DNA preparation from fresh tissue-CTAB method. Phytochem Bull 19.
64. Vos P, Hogers R, Bleeker M, Reijans M, Lee TVD, et al. (1995) AFLP: a new technique for DNA fingerprinting. Nucleic Acids Res 23: 4407–4414.
65. Bonin A, Bellemain E, Bronken Eidesen P, Pompanon F, Brochmann C, et al. (2004) How to track and assess genotyping errors in population genetics studies. Mol Ecol 13: 3261–3273.
66. Vekemans X, Beauwens T, Lemaire M, Roldán-Ruiz I (2002) Data from amplified fragment length polymorphism (AFLP) markers show indication of size homoplasy and of a relationship between degree of homoplasy and fragment size. Mol Ecol 11: 139–151.
67. Lynch M, Clarke AC (1994) Analysis of population genetic structure with RAPD markers. Mol Ecol 3: 91–99.
68. Miller MP (1997) Tools for population genetic analyses (TFPGA). Available: http://www.ccg.unam.mx/~vinuesa/tlem09/docs/TFPGADOC.PDF. Accessed 29 September 2012.
69. Schönswetter P, Tribsch A (2005) Vicariance and dispersal in the alpine perennial *Bupleurum stellatum* L. (Apiaceae). Taxon 54: 725–732.
70. Ehrich D (2006) AFLPDAT: a collection of r functions for convenient handling of AFLP data. Mol Ecol Notes 6: 603–604.
71. Coart E, Glabeke S Van, Petit RJ, Bockstaele E Van, Roldán-Ruiz I (2005) Range wide versus local patterns of genetic diversity in hornbeam (*Carpinus betulus* L.). Conserv Genet 6: 259–273.
72. Petit RJ, Mousadik A El (1998) Identifying populations for conservation on the basis of genetic markers. Conserv Biol 12: 844–855.
73. Fox J (2005) The R Commander: A basic statistics graphical user interface to R. Journal Stat Softw 14: 1–42.
74. R Development Core Team (2012) R: A language and environment for statistical computing. Viena: R Foundation for Statistical Computing. Available: http://www.r-project.org/. Accessed 23 April 2012.
75. Weir BS, Cockerham CC (1984) Estimating F-statistics for the analysis of population structure. Evolution 38: 1358–1370.
76. Nei M, Li WH (1979) Mathematical model for studying genetic variation in terms of restriction endonucleases. Proc Natl Acad Sci 76: 5269–5273.
77. Rohlf FJ (1998) NTSYS-pc, Numerical Taxonomy and Multivariate Analysis, Exeter Software, Version 2.0. Available: http://www.exetersoftware.com/cat/ntsyspc/ntsyspc.html. Accessed 5 February 2012.
78. Felsenstein J (1985) Confidence limits on phylogenies: an approach using the bootstrap. Evolution 39: 783–791.
79. Ronquist F, Huelsenbeck JP (2003) MrBayes 3: Bayesian phylogenetic inference under mixed models. Bioinformatics 19: 1572–1574.
80. Luo R, Hipp AL, Larget B (2007) A bayesian model of AFLP marker evolution and phylogenetic inference. Stat Appl Genet Mol Biol 6: 11.
81. Felsenstein J (1981) Evolutionary trees from DNA sequences: a maximum likelihood approach. J Mol Evol 17: 368–376.
82. Rambaut A, Drummond AJ (2007) Tracer. Available: http//beast.bio.ed.ac.uk/Tracer. Accessed 12 September 2011.
83. Huson DH, Bryant D (2006) Application of phylogenetic networks in evolutionary studies. Mol Biol Evol 23: 254–267.
84. Vriesendorp B, Bakker FT (2005) Reconstructing patterns of reticulate evolution in angiosperms: what can we do? Taxon 54: 593–604.
85. Bryant D, Moulton V (2004) Neighbor-net: an agglomerative method for the construction of phylogenetic networks. Mol Biol Evol 21: 255–265.
86. Jaccard P (1901) Étude comparative de la distribution florale dans une portion des Alpes et du Jura. Bull Société Vaudoise Sci Nat 37: 547–579.

87. Oksanen J, Kindt R, Legendre P, O'Hara B, Simpson GL, et al. (2008) The vegan package version 1.15-0. Available: http://cran.r-project.org. Accessed 14 April 2013.
88. Hubisz MJ, Falush D, Stephens M, Pritchard JK (2009) Inferring weak population structure with the assistance of sample group information. Mol Ecol Resour 9: 1322–1332.
89. Evanno G, Regnaut S, Goudet J (2005) Detecting the number of clusters of individuals using the software STRUCTURE: a simulation study. Mol Ecol 14: 2611–2620.
90. Corander J, Marttinen P (2006) Bayesian identification of admixture events using multilocus molecular markers. Mol Ecol 15: 2833–2843.
91. Gower JC (1966) Some distance properties of latent root and vector methods used in multivariate analysis. Biometrika 53: 325–338.
92. Excoffier L, Smouse PE, Quattro JM (1992) Analysis of molecular variance inferred from metric distances among DNA haplotypes: Application to human mitochondrial DNA restriction data. Genetics 131: 479–491.
93. Excoffier L, Laval G, Schneider S (2005) Arlequin (version 3.0): an integrated software package for population genetics data analysis. Evol Bioinform Online 1: 47–50.
94. Nievergelt CM, Libiger O, Schork NJ (2007) Generalized analysis of molecular variance. PLoS Genet 3: e51.
95. Pompanon F, Bonin A, Bellemain E, Taberlet P (2005) Genotyping errors: causes, consequences and solutions. Nat Rev Genet 6: 847–859.
96. López-Sepúlveda P, Tremetsberger K, Ortiz MÁ, Baeza CM, Peñailillo P, et al. (2013) Radiation of the *Hypochaeris apargioides* complex (Asteraceae: Cichorieae) of southern South America. Taxon 62: 550–564.
97. Juan C, Oromi P, Hewitt GM (1995) Mitocondrial DNA phylogeny and sequential colonization of Canary Islands by darking beetles of the genus *Pimelia* (Tenebrionidae). Proc R Soc London B 261: 173–180.
98. Francisco-Ortega J, Barber JC, Santos-Guerra A, Febles-Hernández R, Jansen RK (2001) Origin and evolution of the endemic genera of *Gonosperminae* (Asteraceae: Anthemideae) from the Canary Islands: evidence from nucleotide sequences of the internal transcribed spacers of the nuclear ribosomal DNA. Am J Bot 88: 161–169.
99. Jorgensen TH (2002) The importance of phylogeny and ecology in microgeographical variation in the morphology of four Canarian species of *Aeonium* (Crassulaceae). Biol J Linn Soc 76: 521–533.
100. Sanmartín I, van der Mark P, Ronquist F (2008) Inferring dispersal: a Bayesian approach to phylogeny-based island biogeography, with special reference to the Canary Islands. J Biogeogr 35: 428–449.
101. Trusty JL, Olmstead RG, Santos-Guerra A, Sá-Fontinha S, Francisco-Ortega J (2005) Molecular phylogenetics of the Macaronesian-endemic genus *Bystropogon* (Lamiaceae): palaeo-islands, ecological shifts and interisland colonizations. Mol Ecol 14: 1177–1189.
102. Guillou H, Carracedo JC, Paris R, Pérez Torrado FJ (2004) Implications for the early shield-stage evolution of Tenerife from K/Ar ages and magnetic stratigraphy. Earth Planet Sci Lett 222: 599–614.
103. Juan I, Emerson B, Orom I, Hewitt G (2000) Colonization and diversification: towards a phylogeographic synthesis for the Canary Islands. Trends Ecol Evol 15: 104–109.
104. Ancochea E, Huertas MJ, Cantagrel JM, Coello J, Fuster JM, et al. (1999) Evolution of the Cañadas edifice and its implications for the origin of the Cañadas Caldera (Tenerife, Canary Islands). J Volcanol Geotherm Res 88: 177–199.
105. Nathan R (2006) Long-distance dispersal of plants. Science 313: 786–788.
106. Valido A, Nogales M (1994) Frugivory and seed dispersal by the lizard *Gallotia galloti* (Lacertidae) in a xeric habitat of the Canary Islands. Oikos 70: 403–411.
107. Nogales M, Delgado JD, Medina FM (1998) Shrikes, lizards and *Lycium intricatum* (Solanaceae) fruits: a case of indirect seed dispersal on an oceanic island (Alegranza, Canary Islands). J Ecol 86: 866–871.
108. Nogales M, Heleno R, Traveset A, Vargas P (2012) Evidence for overlooked mechanisms of long-distance seed dispersal to and between oceanic islands. New Phytol 194: 313–317.
109. Heleno RH, Olesen JM, Nogales M, Vargas P, Traveset A (2012) Seed dispersal networks in the Galápagos and the consequences of alien plant invasions. Proc Biol Sci 280: 20122112.
110. Guerrero AM, Tye A (2009) Darwin's Finches as seed predators and dispersers. Wilson J Ornithol 121: 752–764.
111. Heleno R, Blake S, Jaramillo P, Traveset A, Vargas P, et al. (2011) Frugivory and seed dispersal in the Galápagos: what is the state of the art? Integr Zool 6: 110–129.
112. Montelongo Parada V (1982) *Cheirolophus metlesicsii* sp. nov. una nueva especie de Asteraceae de Tenerife. Bot Macaronesica 10: 67–78.
113. Garnatje T, Pérez-Collazos E, Pellicer J, Catalán P (2013) Balearic insular isolation and large continental spread framed the phylogeography of the western Mediterranean *Cheirolophus intybaceus* s.l. (Asteraceae). Plant Biol: 1–10.
114. Padilla DP, González-Castro A, Nogales M (2012) Significance and extent of secondary seed dispersal by predatory birds on oceanic islands: the case of the Canary archipelago. J Ecol 100: 416–427.
115. Crawford DJ, Lowrey TK, Anderson GJ, Bernardello G, Santos-Guerra A, et al. (2009) Genetic diversity in Asteraceae endemic to oceanic islands: Baker's Law and polyploidy. In: Funk VA, Susanna A, Stuessy TF, Bayer RJ, editors. Systematics, Evolution, and Biogeography of Compositae. Vienna: International Association for Plant Taxonomy. pp.139–151.
116. Crawford DJ, Anderson GJ, Bernardello G (2011) The reproductive biology of island plants. In: Bramwell D, Caujapé-Castells J, editors. The biology of islands floras. Cambridge: Cambridge University Press. pp.11–36.
117. Ellis AG, Weis AE, Gaut BS (2006) Evolutionary radiation of the "stone plants" in the genus *Argyroderma* (Aizoaceae): unravelling the effects of landscape, habitat, and flowering time. Evolution 60: 39–55.
118. Kisel Y, Barraclough TG (2010) Speciation has a spatial scale that depends on levels of gene flow. Am Nat 175: 316–334.
119. Sventenius ER (1960) Las centaureas de la sección *Cheirolophus* en las islas macaronésicas. Anu Estud Atl 6: 219–236.
120. Frankham R (1997) Do island populations have less genetic variation than mainland populations? Heredity 78: 311–327.
121. Hamrick JL, Godt MJW (1989) Allozyme diversity in plant species. Plant population genetics, breeding and genetic resources. Sunderland: Sinauer Associates Inc. pp. 43–64.
122. Hamrick JL, Godt MJW (1996) Effects of life history traits on genetic diversity in plants species. Philos Trans R Soc Biol 351: 1291–1298.
123. Hamrick JL, Linhart YB, Mitton JB (1979) Relationships between life history characteristics and electrophoretically detectable genetic variation in plants. Annu Rev Ecol Syst 10: 173–200.
124. Lande R (1980) Genetic variation and phenotypic evolution during allopatric speciation. Am Nat 116: 463–479.
125. Warwick SI, Gottlieb LD (1985) Genetic Divergence and Geographic Speciation in *Layia* (Compositae). Evolution 39: 1236–1241.
126. Schluter D (2000) The ecology of adaptive radiation. Oxford: Oxford University Press.
127. Papadopulos AST, Baker WJ, Crayn D, Butlin RK, Kynast RG, et al. (2011) Speciation with gene flow on Lord Howe Island. Proc Natl Acad Sci U S A 108: 13188–13193.
128. Bouck A, Vision T (2007) The molecular ecologist's guide to expressed sequence tags. Mol Ecol 16: 907–924.
129. Noor MAF, Feder JL (2006) Speciation genetics: evolving approaches. Nat Rev Genet 7: 851–861.
130. Tsitrone A, Kirkpatrick M, Levin DA (2003) A model for chloroplast capture. Evolution 57: 1776–1782.
131. Holder MT, Anderson JA, Holloway AK (2001) Difficulties in detecting hybridization. 50: 978–982.
132. Bailey CD, Carr TG, Harris SA, Hughes CE (2003) Characterization of angiosperm nrDNA polymorphism, paralogy, and pseudogenes. Mol Phylogenet Evol 29: 435–455.
133. Fernández-Palacios JM (2008) The Canaries: an important biogeographical meeting place. J Biogeogr 35: 379–387.
134. Paun O, Schönswetter P, Winkler M, Intrabiodiv Consortium (Institution), Tribsch A (2008) Historical divergence vs. contemporary gene flow: evolutionary history of the calcicole *Ranunculus alpestris* group (Ranunculaceae) in the European Alps and the Carpathians. Mol Ecol 17: 4263–4275.
135. Purvis A, Gittleman JL, Cowlishaw G, Mace GM (2000) Predicting extinction risk in declining species. Proc Biol Sci 7: 1947–1952.
136. Redding DW, Mooers AØ (2006) Incorporating evolutionary measures into conservation prioritization. Conserv Biol 20: 1670–1678.
137. Arponen A (2012) Prioritizing species for conservation planning. Biodivers Conserv 21: 875–893.
138. Caujapé-Castells J, Tye A, Crawford DJ, Santos-Guerra A, Sakai A, et al. (2010) Conservation of oceanic island floras: Present and future global challenges. Perspect Plant Ecol Evol Syst 12: 107–129.
139. Forest F, Grenyer R, Rouget M, Davies TJ, Cowling RM, et al. (2007) Preserving the evolutionary potential of floras in biodiversity hotspots. Nature 445: 757–760.
140. Cadotte MW, Davies TJ (2010) Rarest of the rare: advances in combining evolutionary distinctiveness and scarcity to inform conservation at biogeographical scales. Divers Distrib 16: 376–385.

13

Dealing with Discordant Genetic Signal Caused by Hybridisation, Incomplete Lineage Sorting and Paucity of Primary Nucleotide Homologies: A Case Study of Closely Related Members of the Genus *Picris* Subsection *Hieracioides* (Compositae)

Marek Slovák[1]*, Jaromír Kučera[1], Eliška Záveská[2], Peter Vd'ačný[3]

1 Institute of Botany, Slovak Academy of Sciences, Bratislava, Slovakia, **2** Department of Botany, Charles University, Praha, Czech Republic, **3** Department of Zoology, Comenius University, Bratislava, Slovakia

Abstract

We investigated genetic variation and evolutionary history of closely related taxa of *Picris* subsect. *Hieracioides* with major focus on the widely distributed *P. hieracioides* and its closely related congeners, *P. hispidissima*, *P. japonica*, *P. olympica*, and *P. nuristanica*. Accessions from 140 sample sites of the investigated *Picris* taxa were analyzed on the infra- and the inter-specific level using nuclear (ITS1-5.8S-ITS2 region) and chloroplast (*rpl32-trnL*$^{(UAG)}$ region) DNA sequences. Genetic patterns of *P. hieracioides*, *P. hispidissima*, and *P. olympica* were shown to be incongruent and, in several cases, both plastid and nuclear alleles transcended borders of the taxa and genetic lineages. The widespread *P. hieracioides* was genetically highly variable and non-monophyletic across both markers, with allele groups having particular geographic distributions. Generally, all gene trees and networks displayed only a limited and statistically rather unsupported resolution among ingroup taxa causing their phylogenetic relationships to remain rather unresolved. More light on these intricate evolutionary relationships was cast by the Bayesian coalescent-based analysis, although some relationships were still left unresolved. A combination of suite of phylogenetic analyses revealed the ingroup taxa to represent a complex of genetically closely related and morphologically similar entities that have undergone a highly dynamic and recent evolution. This has been especially affected by the extensive and recurrent gene flow among and within the studied taxa and/or by the maintenance of ancestral variation. Paucity of phylogenetically informative signal further hampers the reconstruction of relationships on the infra- as well as on the inter-specific level. In the present study, we have demonstrated that a combination of various phylogenetic analyses of datasets with extremely complex and incongruent phylogenetic signal may shed more light on the interrelationships and evolutionary history of analysed species groups.

Editor: Simon Joly, Montreal Botanical Garden, Canada

Funding: Funding for this study was provided by The Research and Development Support Agency, Bratislava, Slovakia (grants no. APVV-0239-09 and APVV-0320-10 to Karol Marhold). The funder had no role in study design, data collection and analysis, decision to publish, or preparation of the manuscript.

Competing Interests: The authors have declared that no competing interests exist.

* Email: marek.slovak@savba.sk

Introduction

Identifying patterns of genetic variation within and among closely allied congeners is crucial in understanding their evolutionary relationships and processes involved in their diversification. Exploration of genetic patterns across the entire geographic ranges of each of the constituent species not only enables clarification of modes of their diversification and historical changes in spatial distribution, but may also shed light on their potential reciprocal interactions (e.g. hybridisation and introgression).

In many instances, however, multi-gene phylogenies of closely related and recently diversified taxa are incongruent, often with alleles transcending species or genetic groups, and finally cause the genetic patterns to exhibit a diffusive or mosaic-like structure [1,2,3,4,5]. A suite of various factors and processes might lead to discordant topologies of genetic trees and complex genetic patterns. The most commonly detected evolutionary events affecting genetic variation are hybridization and introgression, gene duplication, incomplete lineage sorting and/or recombination [6,7,8,9]. Eventually, the low level of variation in the involved genetic markers and the different molecular nature of employed coding and non-coding regions in nuclear, mitochondrial and chloroplast DNA may be a source of gene trees incongruence. Subjective taxonomic delimitation of studied taxa not corresponding with real evolutionary patterns must be also taken into account [10,11]. It may prove difficult to correctly diagnose the relative influence of those factors on the studied species, because they all result in very similar complex genetic patterns and discordant phylogenetic inferences [9,12,13].

Studies that have carefully and explicitly analyzed multiple reasons of genetic incongruence are not common [14,15,16,17]. On the contrary, the overwhelming majority of studies were focused on only a subset of possible reasons of incongruence, however, using a variety of different approaches [1,2,3,5]. Reconstruction of species relationships, i.e. species trees based on multiple gene trees, is recently performed using coalescent models that can handle incongruence caused by incomplete lineage sorting [18]. However, these models are applicable only to datasets where no gene flow (hybridization) among species is to be expected [19,20]. Indeed, it is fairly difficult to distinguish between effects caused by hybridisation and incomplete lineage sorting [9]. One way how to overcome these difficulties has been proposed by Joly et al. [9] who developed an approach enabling to test the influence of hybridization in the presence of incomplete lineage sorting, as implemented in the software JML [20].

Members of *Picris* subsect. *Hieracioides* Vassiliev (Compositae) [21] represent an interesting group for investigation of genetic patterns, diversification and reciprocal interactions within and among closely related species that differ in ecological requirements and geographic distribution [22,23,24]. According to the current taxonomic concepts [21,23,24], *Picris* subsect. *Hieracioides* comprises seven morphologically very similar taxa from Europe, Asia and Africa: *P. abyssinica* Sch. Bip., *P. hieracioides* L. including subspecies *P. hieracioides* subsp. *umbellata* (Schrank) Ces. and *P. hieracioides* L. subsp. *hieracioides* (referred as *P. h.* subsp. *umbellata* and *P. h.* subsp. *hieracioides* in the text), *P. hispidissima* (Bartl.) W. D. J. Koch, *P. japonica* Thunb., *P. nuristanica* Bornm., and *P. olympica* Boiss [21,24,25]. Information on overall relationships within the genus *Picris* subsect. *Hieracioides* is still limited. Only three taxa from the subsection *Hieracioides*, namely *P. abyssinica*, *P. hieracioides* and *P. nuristanica*, were marginally analysed in a study focused on the genus *Leontodon*. The two latter *Picris* species were shown to be closely related, while *P. abyssinca* was genetically much more distant and clustered together with other North African *Picris* taxa [26]. All members of the subsection *Hieracioides* are reported to be perennials with homocarpic achenes (i.e. with same morphological type of achenes within capitulum) and two-hooked anchor hairs on the stem and peduncles [21]. Here, we focus on species from Europe and Asia Minor, with main attention paid besides of the widespread and highly variable *P. hieracioides* also to *P. hispidissima*, *P. olympica* and marginally also to *P. japonica* and *P. nuristanica*.

Although our previous investigations on *P. hieracioides*, *P. hispidissima*, *P. japonica*, and *P. nuristanica* revealed that these taxa are very similar morphologically and karyologically as well as they are closely related genetically (AFLP), their evolutionary relationships and mutual interactions were left unresolved [22–24,27,28]. Additionally, our previous studies [23,24] did detect morphologically intermediate plants in mixed populations of *P. hispidissima* and *P. hieracioides* and also found populations that represented genetic admixtures of different *P. hieracioides* subspecies and/or genetic groups, mostly accompanied by transitional morphology. In all cases, the observed intermediate genetic and/or morphological states were hypothetically attributed to hybridisation although this heterogeneity may be also a result of retention of ancestral variation. No studies collecting the relevant data on *P. olympica* have been done prior to the present study. Likewise, overall relationships within *Picris* subsect. *Hieracioides* have not been investigated yet. Thus the specific goals of the present study are:

(1) Clarifying the phylogenetic relationships among *P. hieracioides, P. hispidissima, P. japonica, P. olympica, and P. nuristanica.*

(2) Exploring the overall genetic pattern of *P. hieracioides* in Europe and Asia Minor, the distribution of its infraspecific genetic groups, and comparing its genetic variation with that of *P. hispidissima* and *P. olympica*.

(3) Shed more light on potential interactions within the studied group and on the processes acting during evolutionary history.

Material and Methods

Studied species

Picris hieracioides is a biennial to perennial, morphologically highly variable species that includes two infraspecific taxa, *P. h.* subsp. *umbellata* and *P. h.* subsp. *hieracioides*. The subspecies of *P. hieracioides* are morphologically distinguishable by a combination of several morphological traits on both vegetative and reproductive organs. This was definitely proved also by cultivation experiments under environmentally homogeneous conditions [24]. Beside this, they also differ in longevity, ecological requirements, and geographic distribution [22,24]. Additionally, each subspecies harbours two highly allopatric genetic (AFLP) and morphologically cryptic groups [24]. *Picris hispidissima* is a biennial species (M. Slovák, unpublished data) that differs morphologically from *P. hieracioides* in having a pectinate-ciliate indumentum of involucral bracts and inflation of the peduncle [23]. The last studied species, *P. olympica*, is a perennial characterized by a low caespitose growth with quite short ascending scapose-like stems bearing only a few capitula. It is considered to be morphologically closely related to *P. hieracioides* [21,29]. The investigated species differ conspicuously in their ecological requirements and geographic ranges, with *P. hieracioides* being the most widespread (distributed across the major part of Europe and Asia Minor) and having wide ecological amplitude [21,24]. The nominate *P. h.* subs. *hieracioides* is mainly biennial, growing on open, dry, xerothermous biotopes with no bedrock preference and thriving in lowlands or lower mountains throughout all of Europe. In contrast, *P. h.* subsp. *umbellata* is a short-lived perennial herb inhabiting humid tall herb communities from montane forests to sub-alpine communities across European calcareous mountain ranges. Both *P. hispidissima* and *P. olympica* have restricted distributions. The former occurs in dry, calcareous rocky slopes and crevices in the Adriatic coastal mountains from Croatia to Montenegro, while the latter is restricted to a few high mountain ranges in western Anatolia where it inhabits open, craggy grasslands with alkaline bedrocks at alpine levels [21,29]. All species, but especially *P. hieracioides*, have spread from their natural biotopes into man-made habitats and migrated along anthropogenic corridors ([22,24]; M. Slovák unpublished data).

Sampling strategy and plant material

In order to perform the reconstruction of phylogenetic relationships among the studied species and to reveal their genetic patterns, we aimed to include as many populations as possible from the entire geographic ranges of *P. hieracioides*, *P. hispidissima*, and *P. olympica* (Table S1). This, however, was at the expense of sample size at the population level [30,31,32]. We also co-analyzed a limited number of samples of the closely related Asian taxa *P. japonica* (2 populations) and *P. nuristanica* (1 population). We did not include samples of *P. abyssinica* since it was proved that it is genetically highly divergent from *P. hieracioides* and *P. nuristanica* but is more closely related to other African taxa [26]. Each population of the investigated species had more than several hundreds of individuals, and the taxa under study are neither endangered nor protected and no specific permits were required to collect plant samples at the study sites. We sequenced one to three accessions per population for the nuclear ITS and the plastid *rpl32-trnL*$^{(UAG)}$ regions (Table S1). To cover as high proportion of genetic variation of analysed taxa as possible, but simultaneously with respect to sampling limitation at population level, we decided to analyse more than one individual for populations where non-homogenized ITS sequences indicated existence of more than a single allele. The widespread *P. hieracioides* including populations morphologically corresponding to both subspecies, *P. h.* subsp. *hieracioides* and *P. h.* subsp. *umbellata*, was collected from 122 sample sites covering both subspecies' entire geographic distributions, and henceforth these sites are referred to as populations (Table S1). Twelve populations of *P. hispidissima* from its entire distribution area were sampled [23]. Three of the five currently recognized populations of *P. olympica* were also collected for the purposes of the present study.

Intermediate morphological variation was identified in a few populations of *P. hieracioides* which were mostly located in zones where the two subspecies or their lineages meet [24]. Since the extent of intermediate morphological variation across these populations was wide, it was impossible to unequivocally sort some of them into discrete categories. Therefore, in order to avoid formation of artificial entities, which might lead to subsequent biasness of results and interpretations, we assigned particular populations a priori not to accepted subspecies, but we considered all of them as *P. hieracioides* instead.

Selection of appropriate outgroup taxa was based on previously published studies [21,26] and our preliminary phylogenetic analyses. Thus a priori selected outgroup comprised members of all sections of the genus *Picris* (*P. galilaea* (Boiss.) Eig., *P. capuligera* (Durieu) Walp., *P. pauciflora* Willd., *P. rhagodioloides* (L.) Desf., *P. scaberrima* Ten., *P. sinuata* (Lam.) Lack, and *P. strigosa* M. Bieb.) and two members of the closely related genus *Helminthotheca* (*H. aculeata* (Vahl) Lack and *H. echioides* (L.) Holub). Voucher specimens were deposited in the herbarium of the Institute of Botany at the Slovak Academy of Science (SAV).

DNA sequence markers, DNA extraction and PCR amplification

We utilised sequences of the nuclear ribosomal internal transcribed spacer (ITS1-5.8S-ITS2 region) and the non-coding cpDNA intergenic spacer *rpl32-trnL*$^{(UAG)}$ [33]. Several single-copy nuclear genes (A25, A28 and B12) [34] and another non-coding chloroplast regions (3'*trn*V$^{(UAC)}$-*ndh*C, *ndhF-rpl32* and *trn*H$^{(GUG)}$-*psb*A) [33,35] were tested during preliminary screening. However, these were proved to be unsuitable for the present study due to low levels of variation and/or problems encountered during their PCR amplification.

Total genomic DNA was extracted from silica gel-dried leaf tissue by the DNeasy plant mini kit (Qiagen, Hilden, Germany), following the manufacturer's protocol. The ITS region and the plastid *rpl32-trnL*$^{(UAG)}$ spacer (referred as cpDNA henceforth in the text) were amplified using the standard PCR reaction, employing the following universal primer pairs: P1A, P4 and internal ones, ITS2 and ITS3, if necessary [36,37] and *trnL-retF* and *rpl32-retR* [39].

Amplifications were carried out in a Mastercycler ep Gradient S thermal cycler (Eppendorf, Hamburg, Germany), using PCR reaction volumes of 25 µL consisting of 0.75 U of *Pfu* polymerase (Fermentas, St. Leon-Rot, Germany), 0.2 mM of each dNTP, 0.2 µM of each primer, 1 µL of DNA template, and reaction buffer containing 2 mM of $MgSO_4$ (Fermentas, St. Leon-Rot, Germany). For ITS amplification, we used the following thermo-cycler program: preheating at 94°C for 3 min, then running it for 35 cycles at 94°C for 30 sec, then at 50°C for 30 sec, and 72°C for 60 sec, with the final extension at 72°C for 10 min, and cooling at 4°C. The cpDNA region was amplified using the "*rpl16*" program, as described by [38]. PCR products were purified using the NucleoSpin Extract II kit (Macherey-Nagel, Düren, Germany). Cycle sequencing reactions were carried out using the same primers, and sequencing was performed on an ABI PRISM 3130*xl* sequencer at BITCET Consortium, Comenius University in Bratislava. The sequencing of a few accessions repeatedly failed for the ITS region and these were therefore excluded from further analyses (Table S1).

Sequence alignments and phylogenetic analysis

Sequences were edited and aligned manually using the BioEdit program version 7.0.4.1 [39]. In case of ITS sequences, the electroferograms were carefully inspected for intra-individual polymorphic sites (IPS) having more than one signal (cf [40]). These were labelled with NC-IUPAC ambiguity codes. Polymorphic positions within ITS sequences, in which both bases were detected also separately in different accessions elsewhere in the alignment, were considered additive polymorphic sites (APS) [40].

We believe that the observed intra-individual variation did not arise from PCR errors, because numerous accessions of *P. hieracioides* and also other taxa possessed homogenized ITS sequences. The geographic pattern of homogeneous and heterogeneous ITS sequences was obviously not random, indicating a genuine variation rather than erroneous signals caused by inaccurate amplification. Although the intra-individual ITS variation might be revealed by cloning procedure, this approach, on the other hand, may result in amplification of numerous ITS sequence types which do not represent relevant variability [41]. Incomplete concerted evolution and/or recombination, processes operating on the multiple-copy regions like ITS, increase the number of unwanted sequence types making the ITS phylogenetic analysis even more complex. Furthermore, with our current sampling involving hundreds of accessions exhibiting intra-individual ITS polymorphism, the cloning procedure would be out of the reasonable solution. As long as most of the sequences were possible to obtain by direct sequencing and polymorphic sites could be designated by IUPAC ambiguity codes, we decided to separately analyse two ITS datasets, one including and one excluding the individuals with APS (see below).

Altogether we generated three alignments: the first one was designated as ITS_1 and involved 117 sequences without APS (Alignment S1. ITS_1). The second one was designated as ITS_2 and comprised 216 sequences coming from all accessions analysed for the ITS region and including all sequences with APS (Alignment S2. ITS_2). The second alignment was constructed after considering that the intra-individual polymorphic sites, especially those with APS, may significantly influence the hierarchical structure of the phylogenetic trees.

Finally, the third cpDNA alignment comprised all plastid sequences, in particular 219 accessions (Alignment S3. cpDNA). In order to detect possible incongruence between particular markers, the incongruence length difference (ILD) test [42], as implemented in the "partition homogeneity" test of PAUP* (1000 replicates, Mulpars off, outgroups included), was employed. Although topologies of the ITS and cpDNA trees were shown to be incongruent, statistical support especially at their basal nodes was comparatively poor, causing their main branching patterns to be unsupported. Therefore, we decided to take the advantage of synergistic effect of combining datasets and generated also concatenated alignments. The concatenated alignment,

ITS_cpDNA, contained 117 ITS sequences without APS and corresponding cpDNA sequences (Alignment S4. ITS_cpDNA).

Indels were not coded separately but treated as missing data. The aligned datasets were analysed independently using the following phylogenetic approaches:

(1) Maximum parsimony (MP) phylogenetic analysis was performed with the heuristic search option in PAUP* version 4.0b10 [43]. The following settings were utilised: accelerated character transformation (ACCTRAN), gaps treated as missing data, single-site polymorphisms determined uncertainties, tree construction with stepwise addition, 1000 bootstrap replicates with random taxon addition, tree bisection-reconnection (TBR) branch swapping, and retention of multiple trees found during branch swapping (MULTREES option in effect). The identical sequences were merged in McClade version 4.0 PPC [44] to reduce computation time. Clade support was calculated via bootstrap analyses using 10000 re-samplings done with the fast heuristic search in PAUP*. Bootstrap support was categorized according to the following criteria: strong (>85%), moderate (70%–85%), weak (50%–69%), or poor (<50%).

(2) Bayesian inference (BI) was run in MrBayes version 3.1.2, using the Markov Chain Monte Carlo algorithm (MCMC) [45]. Bayesian analyses were performed on the CIPRES Portal version 1.15 [46]. Prior to Bayesian analyses, the most appropriate nucleotide substitution models were chosen, using the Akaike Information Criterion (AIC) in jMODELTEST version 0.0.1 [47,48]. Evolutionary models were calculated for each part of the datasets separately. Specifically, the ITS datasets contained partitions corresponding to the ITS1 and ITS2 spacers and the 5.8S rRNA gene, while concatenated datasets included besides the three ITS region partitions also a fourth partition represented by the cpDNA region. The following models or model combinations were found to be the most appropriate for the datasets studied: (1) ITS_1 and ITS_2 datasets – the SYM + G model for ITS1 and ITS2 sequences and the K80 model for 5.8S rRNA gene sequences; (2) cpDNA dataset – the TVM + I + G model; (3) concatenated ITS_cpDNA – the SYM + G model for ITS1 and ITS2, the K80 model for the 5.8S rRNA gene sequences, and the TVM + I + G model for the cpDNA partition.

All BI analyses were run with four independent Metropolis-coupled MCMC chains (three heated and one cold chain) for ten to twenty five million generations and sampled every 1000th generation. The first 25% of sampled trees were regarded as 'burn-in' trees and were discarded prior to reconstruction of a 50% majority-rule consensus tree. Stationarity was confirmed by checking convergence diagnostic parameters. Specifically, the average standard deviation of split frequencies was lower than 0.01 in all cases; the plots of generations versus log probability of the observed data showed no obvious trends; and the Potential Scale Reduction Factor (PSRF) approached 1. Finally, topologies and node posterior probability values were compared among the runs. The topologies were stabilized among all datasets with only minute differences in branching pattern of terminal clades. Nodes with posterior probability (PP) values of 0.90 and above were regarded as significant and those with PP values below 0.90 regarded as non-significant.

(3) Net-like approaches were used to identify and display potential contradictory signals in the datasets. All alignments including ITS_2 containing intra-individual polymorphic sites were analysed using the neighbour-net analysis of [49] in SplitsTree version 4.10 with uncorrected P-distance and default settings. To visualize the relationships among cpDNA haplotypes and to detect possible ancestral polymorphism, the cpDNA dataset was subjected to haplotype network analysis based on the parsimony method of [50] using TCS version 1.21 [51] limited to 30 steps of parsimonious connection in creating the network.

(4) In order to precisely specify the amount of phylogenetically informative signal in the datasets, we analysed spectrum of supporting nucleotide positions. There are three groups of supporting positions recognized by Wägelle and Rödding [52]: (1) symmetrical or binary positions have two different character states in functional outgroup and ingroup and thus support both group of a split equally; (2) asymmetrical positions support only one group which possesses the same nucleotide at particular position, while the other group harbours different and more than one character state at this position; (3) noisy positions include same character states present in all sequences of the functional ingroup but also at least in one sequence of the functional out-group and thus represent convergences or chance similarities between in-group and outgroup, or alternatively ingroup autapomorphies.

(5) We also apply a Bayesian coalescent-based approach to estimate a species tree employing *BEAST as implemented in the program BEAST version 1.7.4 [18]. Two input files, one with 117 homogeneous ITS sequences (Alignment S1. ITS_1) and the second one containing corresponding cpDNA sequences, were used for the BEAST analysis. Populations assigned to particular taxa were used as OTU's. The input file for *BEAST was created in BEAUti version 1.7.4, with the following settings: two data partitions (corresponding to the two loci), the best-fit evolutionary model for each partition as determined by jMODELTEST, uncorrelated lognormal clock, a Yule process model for the species tree prior, and other parameters as default. Four independent MCMC analyses were run each for 120 million generations, sampling every 1000th generation. Another MCMC analysis was run with settings suitable for a subsequent JML analysis (see below), i.e. with piecewise constant population size model and 40 million generations sampling every 1000th generation. The computer program Tracer version 1.5 [53] was used to check convergence of all parameters to the stationary distribution in each run and TreeAnnotater version 1.7.4 was employed to set the burn-in (discarding the first 30000 trees) and to calculate the maximum clade credibility tree.

(6) We performed statistical tree topology tests on ML gene trees inferred from the ITS_1, cpDNA and ITS_cpDNA alignments to find out whether discrepancies between topologies shown in particular gene trees and coalescent species tree are statistically significant. To this end, looking at the topology of the coalescent species tree, we enforced the following constrains on gene trees: basal position of *P. olympica* within the subsection *Hieracioides*; sister relationship of *P. nuristanica* and *P. japonica*; and monophyly of *P. hispidissima* and *P. hieracioides* (Table 1). Constrained trees were built under the evolutionary substitution model as specified for each alignment above, using the maximum likelihood (ML) criterion and heuristic search with TBR swapping algorithm and 10 random sequence addition replicates. The site-wise likelihoods for the best unconstrained ML tree and all constrained trees were calculated in the computer program raxmlGUI version 1.3 [54] and consequently were compared

using the approximately unbiased, weighted Shimodaira-Hasegawa, and weighted Kishino-Hasegawa tests as implemented in the computer package CONSEL version 0.1j [55–57]. A p-value of <0.05 was chosen for rejection of the null hypothesis that the log likelihoods of the constrained and best unconstrained trees are not significantly different.

(7) To test whether hybridization influenced species relationships and could be the source of gene tree incongruence, we employed the program JML [20]. This software uses a posterior distribution of species trees, population sizes and branch lengths to simulate replicate sequence datasets under the coalescent with no migration. The minimum pairwise sequence distance between sequences of two species is evaluated on the simulated datasets and compared to the one estimated from the original data (i.e. from the ITS or cpDNA dataset). This procedure, the posterior predictive test, is a good predictor of hybridization events that disturb the bifurcating species tree model. Two separate JML analyses were run to simulate sequence replicates in the ITS and cpDNA datasets. For these analyses, 40,000 species trees resulting from the JML-specified *BEAST analysis were used. Settings for particular simulations involved: (1) relative mutation rate as inferred from the log file generated during the *BEAST analyses (set to 1.016 and 0.385 for the ITS and the cpDNA simulation, respectively); (2) heredity scalar (2 and 1 for the ITS and cpDNA simulation, respectively); and (3) appropriate model of sequence evolution for both markers. In each analysis, 9,000 trees were removed as burn-in and every 10th tree was used for simulations. Based on the original sequence data files, minimum pairwise sequence distances between all pairs of species and exact probabilities of observing these distances in simulations under assumption of no migration were calculated. All pairwise sequence distances with p-value <0.05 were recorded as potential cases of hybridization.

Results

A total of 202 sequences of the ITS region and 205 of the cpDNA spacer of taxa a priori considered as ingroup were obtained. Results of phylogenetic analyses, however, revealed that beside of the 202 ingroup ITS sequences, another three accessions that were a priori considered as outgroup taxa (*P. scaberrima* and *P. strigosa*), appeared to be a part of ingroup. In contrast, all accessions a priori considered as outgroup taxa were shown to be part of real outgroup in the plastid region analysis (see below). Thus, sequences of 11 accessions (7 species) are considered outgroup for the ITS region while 14 accessions (9 species) for the cpDNA spacer. Information on the DNA datasets and details on maximum parsimony analyses are summarized in Table 2.

The ITS region

In addition to homogeneous sequences, we identified numerous accessions containing intra-individual single nucleotide polymorphic sites in the ITS dataset (Table S2). Out of 653 nucleotide positions, 148 were detected to be polymorphic in at least one individual of the subsection *Hieracioides*, with up to 16 intra-individual polymorphic sites per sequence. Forty two polymorphic sites showed an additive pattern, ranging between 1 and 16 APS per individual sequence (Table S2). The heterogeneous sequences varied conspicuously in terms of number of APS with respect to potentially parental ITS variants, ranging from those with fully additive patterns to those comprising almost completely homogenized contigs.

Results of BI, MP and split decomposition analyses based on the ITS_1 dataset (117 sequences without APS) did not show the analysed *Picris* taxa from the subsection *Hieracioides* to form a monophyletic group (Figures 1 and 2A). Specifically, accessions of the presumed outgroup species, *P. scaberrima* and *P. strigosa*, were clustered within the ingroup taxa (Figures 1 and 2A).

The Bayesian 50% majority-rule consensus tree and the strict consensus MP tree based on the ITS_1 dataset (117 sequences without APS) displayed essentially identical topologies with only small differences in terminal positions and/or in clade supports

Table 1. Log likelihoods and p-values of AU, WSH, and WKH tests for tree comparisons considering different topological scenarios.

Topology	Alignment	Log likelihood (−ln L)	Δ (−ln L)[a]	AU[b]	WSH[c]	WKH[d]
Best maximum likelihood tree (unconstrained)	ITS_1	2408.7451	–	0.843	0.948	0.794
	cpDNA	1973.5581	–	0.777	0.903	0.650
	ITS_cpDNA	4977.4663	–	0.798	0.922	0.682
Basal position of *P. olympica* within the subsection *Hieracioides*	ITS_1	2414.1421	5.40	0.212	0.319	0.201
	cpDNA	1973.5582	0.00	0.514	0.726	0.350
	ITS_cpDNA	4979.1512	1.68	0.385	0.582	0.293
Sister relationship of *P. nuristanica* and *P. japonica*	ITS_1	2408.7453	0.00	0.250	0.709	0.206
	cpDNA	1973.5582	0.00	0.377	0.706	0.308
	ITS_cpDNA	4977.4664	0.00	0.511	0.840	0.318
Monophyly of *P. hispidissima* and *P. hieracioides*	ITS_1	2428.4956	19.75	0.013	0.077	0.057
	cpDNA	1981.1009	7.54	0.033	0.139	0.139
	ITS_cpDNA	4992.7916	15.33	0.141	0.205	0.114

Significant differences ($p<0.05$) between the best unconstrained and constrained topologies are indicated in bold.
[a]Δ (−ln L) - differences in log likelihoods between the best gene tree and alternative gene trees.
[b]AU - approximately unbiased test.
[c]WSH - weighted Shimodaira-Hasegawa test.
[d]WKH - weighted Kishino-Hasegawa test.

Table 2. Information for the DNA datasets and details on maximum parsimony analyses.

Alignment	ITS_1[a]	cpDNA[b]	concat1[c]
No. of sequences/No. of sequences in merged dataset	117	219/112	117
Characters in the alignment	653	960	1613
Length variation (including outgroup)	640–648	702–908	1352–1546
No. of variable sites	167	101	263
No. of parsimony informative characters	138	60	207
No. of most parsimonious trees	1515	1512	4130
Tree length in parsimony	252	127	430
Consistency index	0.774	0.898	0.702
Retention index	0.935	0.955	0.895

[a]ITS1 – analysis based on the ITS_1 Alignment S1. ITS_1.
[b]cpDNA - analysis based on the Alignment S3. cpDNA.
[c]concat - analysis based on the concatenated Alignment S4. ITS_cpDNA.

(Figure 1). All ingroup taxa including both aforementioned presumed outgroup species formed a large strongly supported clade (BS = 90%, PP = 1.00), with shallow hierarchical structure comprising numerous sub-clades with various levels of support (Figure 1). All non-*hieracioides* species, *P. hispidissima*, *P. japonica*, *P. olympica*, *P. nuristanica*, *P. scaberrima*, and *P. strigosa*, formed their own strongly supported sub-clades, with *P. olympica* and *P. scaberrima* being the most divergent (Figure 1). *Picris hieracioides* is not monophyletic and forms four ITS groups with allopatric or parapatric spatial distributions (Figures 1, 2A and 3): ITS_1 composed of two sub-clades: ITS_1A – mostly lowland populations morphologically corresponding to the *P. h.* subsp. *hieracioides* morphotype from the Apennine Peninsula, the Balkan Peninsula, western Turkey, rarely from Central and northwestern Europe and a single population morphologically assignable to the *P. h.* subsp. *umbellata* morphotype from northwestern Europe (blue in Figures 1, 2A and 3); ITS_1B – mostly lowland populations of the *P. h.* subsp. *hieracioides* morphotype from Central, northwestern Europe and Balkan Peninsula and a single population of the *P. h.* subsp. *umbellata* morphotype from the Western Carpathians (pale blue in Figures 1, 2A and 3); ITS_2 – mountain populations of the *P. h.* subsp. *umbellata* morphotype from Alps and the Western Carpathians (red in Figures 1, 2A and 3); ITS_3 – mostly mountain populations corresponding to the *P. h.* subsp. *umbellata* morphotype from the Sierra Nevada, Pyrenees, Jura Mts., Alps, Germany and Sweden (orange in Figures 1, 2A and 3); and ITS_4 – an unresolved group of populations assignable to both of *P. hieracioides* morphotypes from the Apennine and the Iberian Peninsulas, Belgium and Poland (brown in Figures 1, 2A and 3). We also endeavoured to analyse the dataset ITS_2 (all 216 ITS sequences including those with APS) using MP and BI. Both approaches, however, resulted in a completely unresolved clade with all accessions from the subsection *Hieraciodes* placed in a basal polytomy (data not shown).

The split decomposition diagram based on the ITS_1 dataset (117 ITS sequences) showed four main splits having various levels of bootstrap support (Figure 2A): (1) *P. olympica* separated from the rest by a set of long parallel and strongly supported edges; (2) lowland populations of *P. hieracioides*, morphologically corresponding to *P. h.* subsp. *hieracioides*, are clearly separated from all ingroup and outgroup taxa (ITS 1A and 1B). The exception represented few populations of the *P. h.* subsp. *hieracioides* morphotype that clustered together with populations attributable to the *P. h.* subsp. *umbellata* morphotype in an unresolved ITS4 group; (3) a large heterogeneous group associating several smaller well supported splits corresponding to *P. hispidissima*, *P. japonica*, *P. nuristanica*, and two lineages composed of mountainous populations of the *P. h.* subsp. *umbellata* morphotype (ITS_2 and ITS_3); and (4) a group of both morphotypes of *P. hieracioides* (ITS_4) and *P. strigosa*, a presumably outgroup species. *Picris scaberrima* formed a long edge linked with outgroup taxa by short parallel splits. Importantly, the second neighbour-net analysis based on the ITS_2 dataset (206 ITS sequences) revealed essentially the same pattern as the previous one (figure not shown). This apparently indicates that heterogeneous sequences did not significantly affect the overall split pattern. Individuals with un-homogenized ITS sequences and numerous APS appeared mostly in the central position of the network, while those possessing more homogenized sequences, with lower number of APS, were preferentially incorporated in terminal positions (figure not shown).

Analyses of spectrum of supporting positions revealed that majority of phylogenetically informative positions are shared by outgroup taxa (Figure 2B). The entire functional ingroup, involving all accessions of the subsection *Hieraciodes* plus *P. strigosa* and *P. scaberrima*, are supported only by 7 nucleotide positions. Within the functional ingroup, the highest support obtained *P. olympica* (9 positions), *P. japonica* (4 positions), the *P. hieracioides* clade with lowland populations corresponding to the *P. h. hieracioides* morphotype (3 positions) and *P. nuristanica* (2 positions). All other above mentioned splits/subclades were supported at maximum by a single position.

The cpDNA spacer

MP and BI analyses yielded trees with different topologies to a certain extent. However, there was no strongly supported incongruence between them. In order to maintain consistency in the presentation of our results we discuss only the topology based on the BI phylogenetic tree.

In contrast to the ITS trees, *P. scaberrima* and *P. strigosa* appeared clustered together with other outgroup taxa. All accessions of the *Hieracioides* group appeared in a largely unresolved clade with five essentially supported cpDNA subclades and two unsupported groups (cpDNA_A–G) placed in a basal polytomy (Figure 4). The topology and variation of non-*hieracioides* taxa in the cpDNA spacer tree were in some aspects

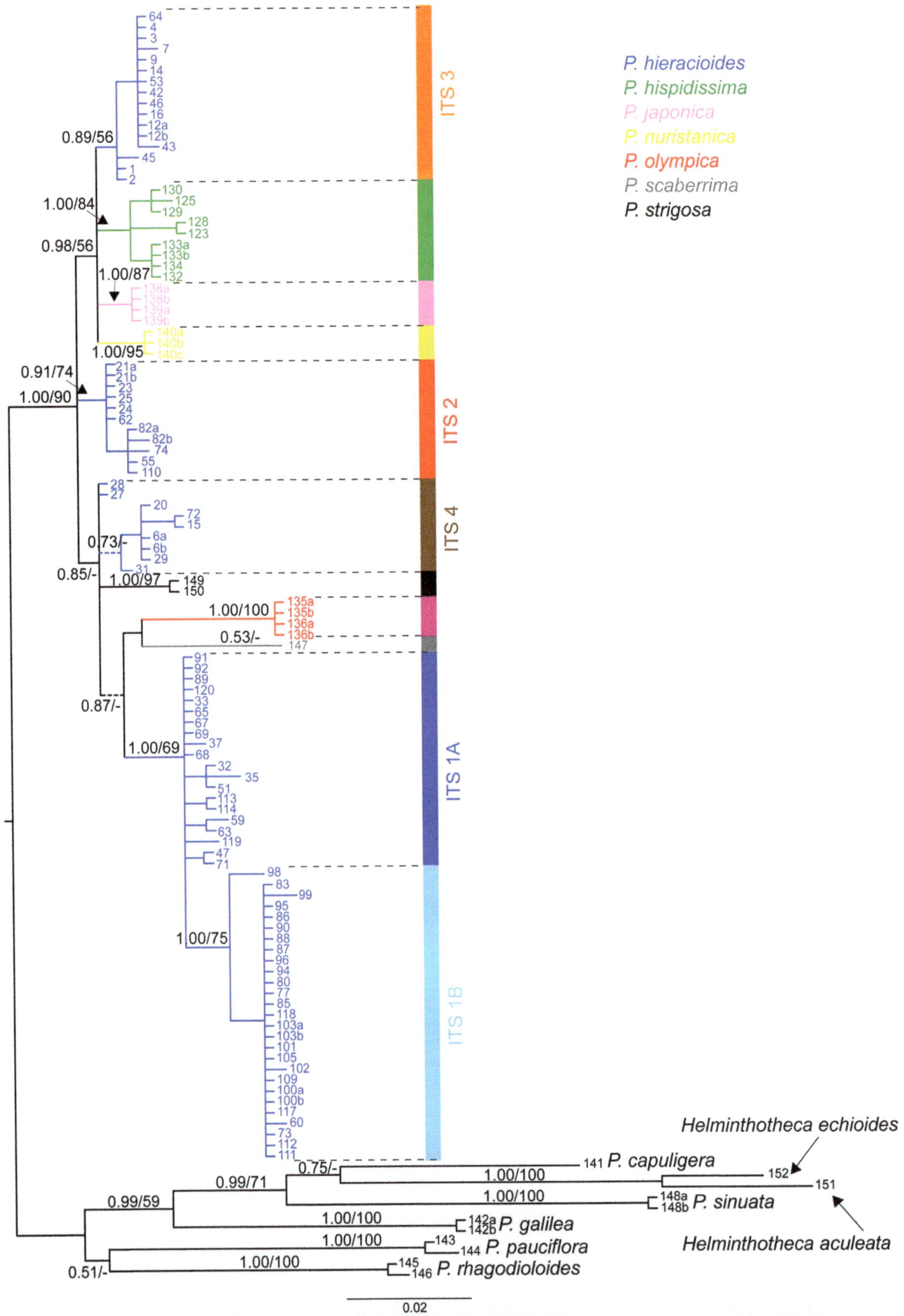
P. hieracioides
P. hispidissima
P. japonica
P. nuristanica
P. olympica
P. scaberrima
P. strigosa
ITS 3
ITS 2
ITS 4
ITS 1A
ITS 1B
0.89/56
1.00/84
0.98/56
1.00/87
1.00/95
0.91/74
1.00/90
0.73/-
0.85/-
1.00/97
1.00/100
0.53/-
0.87/-
1.00/69
1.00/75
0.75/-
0.99/71
0.99/59
0.51/-
Helminthotheca echioides
Helminthotheca aculeata
P. capuligera
P. sinuata
P. galilea
P. pauciflora
P. rhagodioloides
0.02

Figure 1. Majority-rule consensus tree of Bayesian inference based on the reduced dataset_ITS_1. The numbers above branches refer to posterior probability values of Bayesian inference/the bootstrap support as inferred for the maximum parsimony analyses (values ≥50% are shown). Support values for terminal branches are not shown. Dashed lines represent branches collapsed in strict consensus tree of the maximum parsimony analyses. Colour line below the tree indicates affiliation of grouping to the ITS genetic lineages as mentioned in the text. Each accession label includes the population code following Table S1.

congruent and in other discordant with respect to the ITS tree (see Figure 1). Although not forming distinct subclades, *Picris japonica* and *P. nuristanica* possessed their own exclusive haplotypes, similarly as they did in the ITS data (see Figure 5). Contrastingly, *P. olympica* and *P. hispidissima* were heterogeneous and shared haplotypes with *P. hieracioides* (Figures 4, 5 and 6). Repeatedly, *P. hieracioides* was heterogeneous and was found in seven genetic cpDNA geographically correlated groups (Figures 4, 5 and 6): unresolved group cpDNA_A – comprising populations of *P. hieracioides* morphologically corresponding to both subspecies and originating from the entire distribution area, *P. hispidissima* from Croatia and Montenegro, and *P. olympica* (cyan in Figures 4, 5 and 6); cpDNA_B – mountain populations of the *P. h.* subsp. *umbellata* morphotype from the Apennines, Alps, and Carpathians, while populations from lowland of northwestern Europe corresponding morphologically rather to *P. h.* subsp. *hieracioides* and finally *P. hispidissima* from Croatia (violet in Figures 4, 5 and 6); cpDNA_C – *P. hispidissima* and lowland populations of the *P. h.* subsp. *hieracioides* morphotype from Croatia and Montenegro (olive in Figures 4, 5 and 6); unresolved group cpDNA_D – including mountain and lowland populations of *P. hieracioides* from Sierra Nevada, the Carpathians and north-western Europe morphologically corresponding to its both subspecies, *P. japonica* and *P. nuristanica* (pink in Figures 4, 5 and 6); cpDNA_E – *P. hispidissima* from Montenegro (grey in Figures 4, 5 and 6); cpDNA_F – populations of both morphotypes of *P. hieracioides* from the Iberian mountain ranges, the western Alps, the Jura Mts, and northwestern Europe (green in Figures 4, 5 and 6); cpDNA_G – lowland populations of the *P. h.* subsp. *hieracioides* morphotype from the Apennine and the Balkan Peninsula, mountain populations of the *P. h.* subsp. *umbellata* morphotype from the Alps and *P. hispidissima* from Croatia (light brown in Figures 4, 5 and 6). The highest density of rare haplotypes was observed on the Balkan Peninsula (nine haplotypes), the Iberian Peninsula mountains (seven haplotypes), and the Apennine Peninsula (five haplotypes) (Figure 6).

The parsimony-based haplotype network revealed 34 very closely related ingroup haplotypes, mostly connected by one or two steps (Figure 5) and the split decomposition diagram displayed a stellar-like structure with numerous haplotypes in an unresolved basal position (data not shown). Similarly as in the ITS dataset, the majority of supporting positions was harboured by outgroup species (data not shown). Both *hieracioides* clades as well as all their sub-clades were supported extremely weakly, in particular, by two positions at the maximum (data not shown).

Concatenated phylogeny and coalescence based species tree

Although the ILD test revealed significant ($p<0.001$) incongruence between both gene trees, we performed concatenated analyses because our split spectrum analyses show that the observed discordances are not strongly supported. Thus, incongruence between both genetic regions could rather reflect paucity of phylogenetically informative signal (Figures 1, 2AB and 4). Concatenated phylogeny also did not unambiguously resolve evolutionary relationships among ingroup taxa, but still depicted several variably supported (sub)clades. Position of two presumed outgroup taxa, *P. scaberrima* and *P. strigosa*, varied among phylogenetic analyses and remained uncertain (see Figure 7AB and MP/BI phylogenies not shown). Non-*hieracioides* ingroup species, namely *P. japonica*, *P. olympica* and *P. nuristanica*, each clustered in strongly supported sub-clades, repeatedly with *P. olympica* being the most divergent (Figures 7AB). Importantly, although weakly supported, the majority of accessions assigned to morphotypes of both subspecies of *P. hieracioides* as well as to *P. hispidissima* tended to form taxon specific groupings (Figure 7A).

The coalescence-based species tree (Figure 8) showed different topology than the concatenated gene tree and brought more light into the phylogenetic relationships of the analysed taxa: the presumed outgroup taxa, *P. scaberrima* and *P. strigosa*, appeared together in a sister position to the subsection *Hieracioides*; *P. olympica* appeared in a basal position to all taxa from the subsection *Hieracioides*; the Asian *P. japonica* and *P. nuristanica* were clustered together in a sister position to the *P. hispidissima* and *P. hieracioides* clade. Relationships among *P. hispidissima* and *P. hieracioides* were, however, unsupported.

Statistical tree topology tests

Although all tested evolutionary scenarios found in the coalescent species tree (see above and Material and Methods) were not observed in the present gene trees, they cannot be rejected at the significance level of 0.05 by any of the three statistical tree topology tests performed (Table 1). The only exception is the monophyly of *P. hispidissima* and *P. hieracioides*, which can be excluded for the ITS_1 and cpDNA alignment at the significance level of 0.05 only by the AU test but not by the WSH and WKH tests. Thus, the phylogenetic relationships within the subsection *Hieracioides* as unravelled by the Bayesian coalescent-based analysis cannot be very likely excluded also for the gene trees.

Test of hybridization and/or introgression

Only a single case of potential hybridization within the ITS dataset was detected, in between a pair of outgroup taxa, *Helminthotheca aculeata* and *H. echioides* ($p<0.001$). By contrast, 31 cases with $p<0.05$ were found in the cpDNA dataset involving accessions of the corresponding pairs of the following taxa: *P. olympica* vs. *P. hieracioides*, *P. olympica* vs. *P. rhagodioloides*; *P. sinuata* vs. *P.capuligera*; and *H. aculeata* vs. *H. echioides* (Table 3).

Discussion

Intricate phylogenetic relationships within the subsect. *Hieracioides*

Tree-building and network analyses of both the nuclear and plastid markers of the studied *Picris* taxa revealed phylogenies with rather low resolutions and topological discrepancies (see Results), although utilization of these markers has been proposed as one of the most informative approaches even at lower taxonomic levels [35,58]. More light on the evolutionary relationships within the subsection *Hieracioides* was cast by the Bayesian coalescence analysis which generates species instead of gene trees. Although relationships in the present species tree were

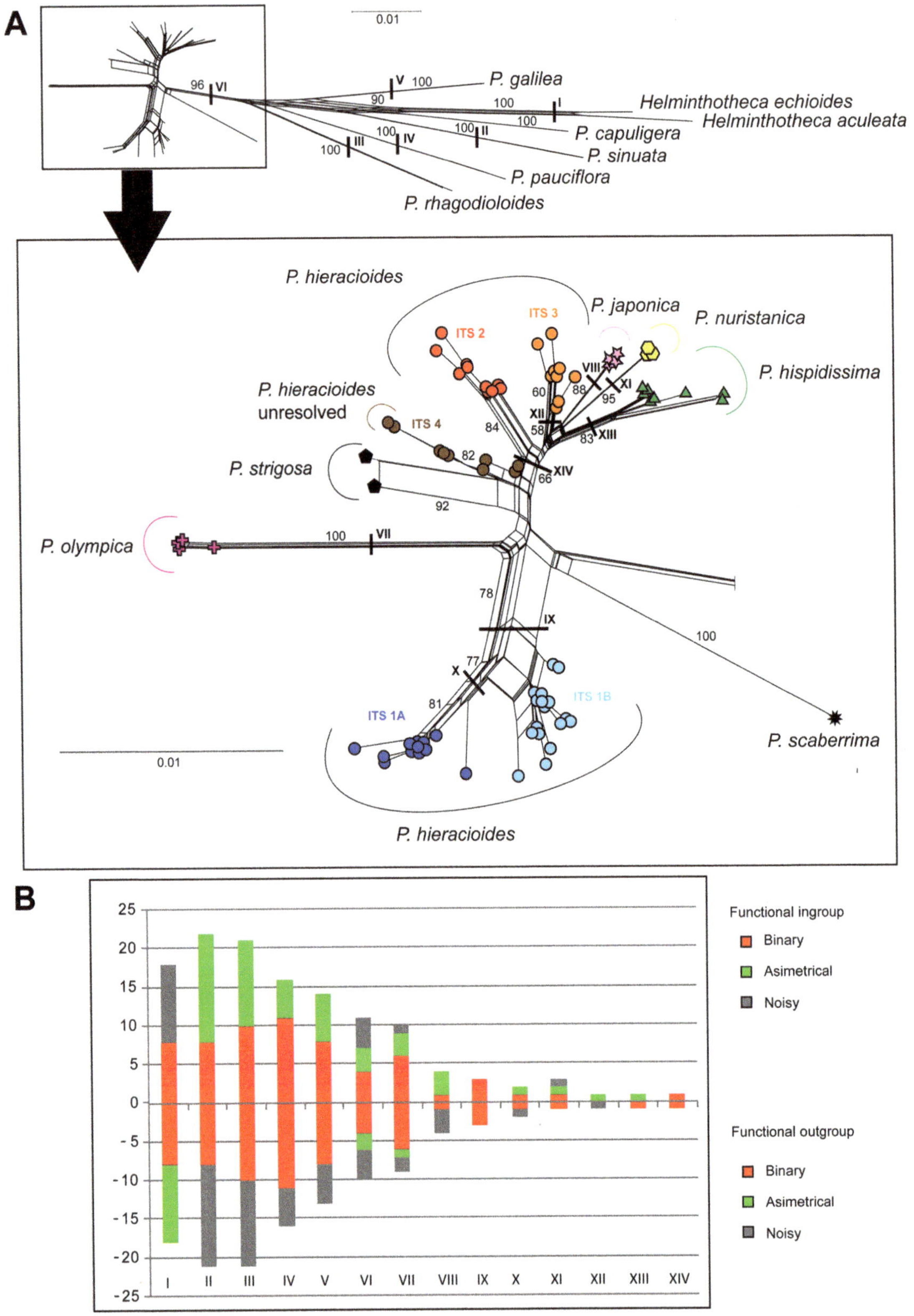

Figure 2. Neighbour-Net diagram (A) and split support spectrum (B) for the ITS_1 dataset. The Neighbour-Net diagram is based on uncorrected *P*-distances. Colour symbols indicate affiliation of accessions to the ITS genetic lineages as mentioned in the text while the shape of symbols indicate affiliation of accessions to taxa. Bootstrap supports of selected important splits are indicated above edges. Column height in the split spectrum represents the number of clade-supporting positions, i.e., putative primary homologies. Column parts above the y-axis represent the ingroup partition while those below the axis correspond to the outgroup partition.

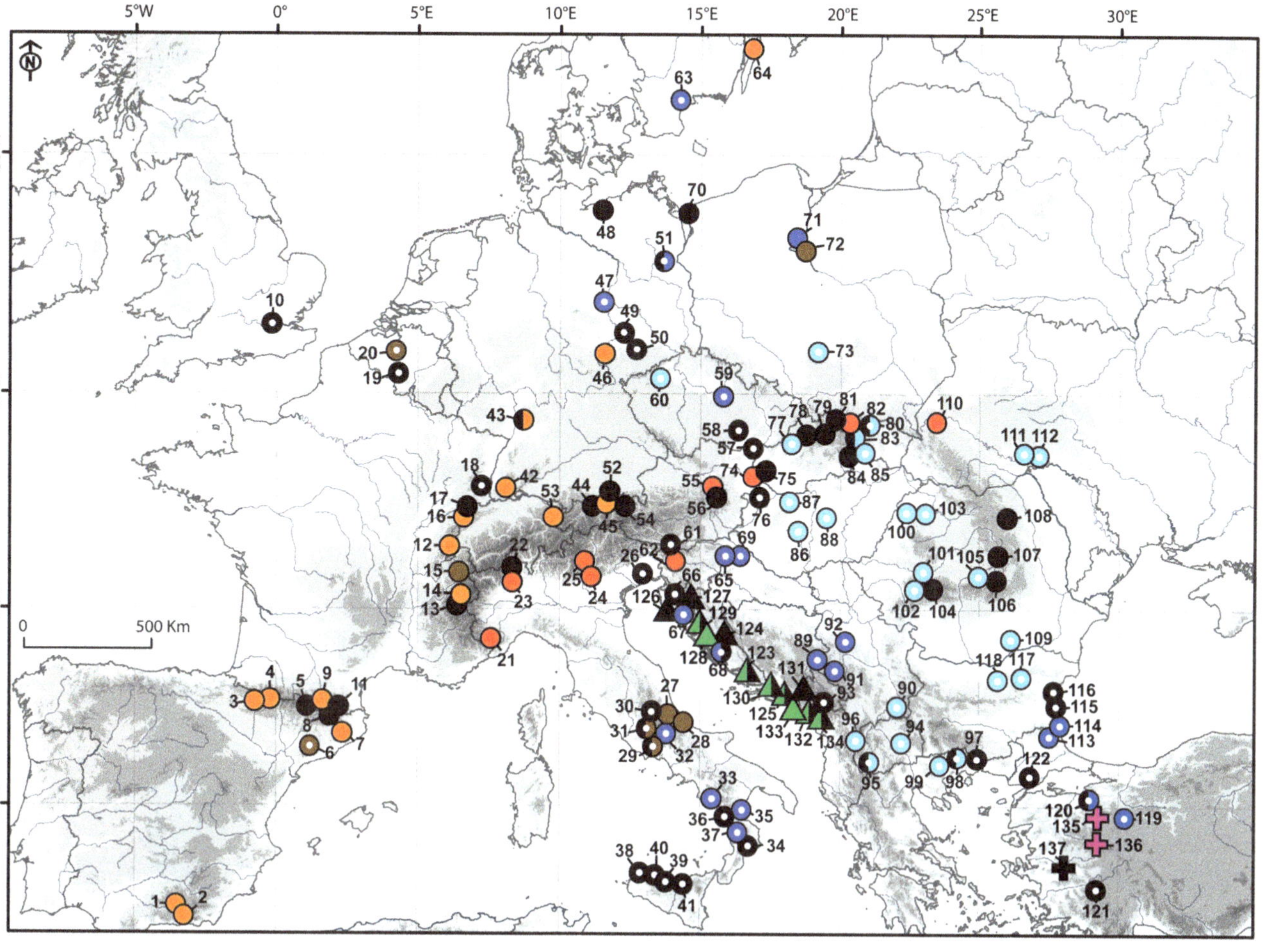

Figure 3. Map displaying geographic distribution of genetic variation in the analysed *Picris* taxa, as inferred from the nuclear ITS sequence data. Coloured symbols indicate affiliation of accessions to the ITS genetic lineages as mentioned in the text. Black symbols indicate populations including individuals with unhomogenised sequences containing additive polymorphic sites. Symbol shapes indicate taxa as follows: *P. hieracioides* – circle, *P. hispidissima* – triangle, and *P. olympica* – square. The circle symbols with empty centres refer to populations morphologically corresponding to *P. hieracioides* subsp. *hieracioides,* while those with solid symbols refer to populations morphologically corresponding to *P. hieracioides* subsp. *umbellata*. The populations of *P. nuristanica*, *P. japonica* and outgroup taxa are not shown. Numbers refer to population codes as denoted in Table S1.

not strongly statistically supported and were not detectable in all gene trees, they could not be rejected by statistical tree topology tests performed on the concatenated dataset (see Results). Thus, in spite of the complex patterns, we might assume the following: (1) the analysed members of the subsection *Hieracioides* have a monophyletic origin, with *P. scaberrima* and *P. strigosa* being their sister taxa; (2) *P. olympica* is basal within the subsection *Hieracioides*, i.e. it is sister to all other members of this subsection; (3) all Asian species are sister to the European ones; and (4) *P. japonica* + *P. nuristanica* are sister to *P. hieracioides* + *P. hispidissima*. Interrelationships among populations of *P. hieracioides* and *P. hispidissima* are left unresolved, however.

Which factor(s) might be responsible for low resolution, varying supports and discordances between gene tree topologies? In the present study, we have shown by the split spectrum analysis that one of the primary and crucial factors responsible for this pattern is a paucity of parsimony informative and node supporting nucleotide positions. Deficiency of phylogenetically informative signal in the datasets analysed was unambiguously evidenced by the low number of primary nucleotide homologies (Figures 2B and 7B). This was indicated also by the presence of short internal branches in all phylogenetic trees (Figures 1 and 4) and the high number of rare plastid haplotypes in derived positions connected mostly with ancestral haplotypes by only one or two mutational steps (Figure 5). Low level of phylogenetically informative positions in the used markers, which are among the most informative [35,58], strongly indicates a recent diversification of the studied taxa and genetic groups [1,59,60,61,62]. It would be desirable to confirm this by molecular dating. However, in case of insufficient amount of phylogenetic information, such an analysis is assumed to be vague and statistically unsupported.

Furthermore, incomplete lineage sorting represents another evolutionary force that might be responsible for the discordances and complex genetic patterns in the six *Picris* taxa [1,3,4,5]. The Bayesian coalescent-based approach considers the incomplete lineage sorting as a main source of uncertainties in inferred species trees [18]. Thus, the unresolved relationships among genetic lineages of *P. hieracioides* and *P. hispidissima* as well as the low nodal support could signalise either influence of this phenomenon or violation of prior assumption by hybridisation or other processes. Likewise, the disjunctive occurrence of the ancestral plastid haplotypes in populations of *P. olympica* and the spatially

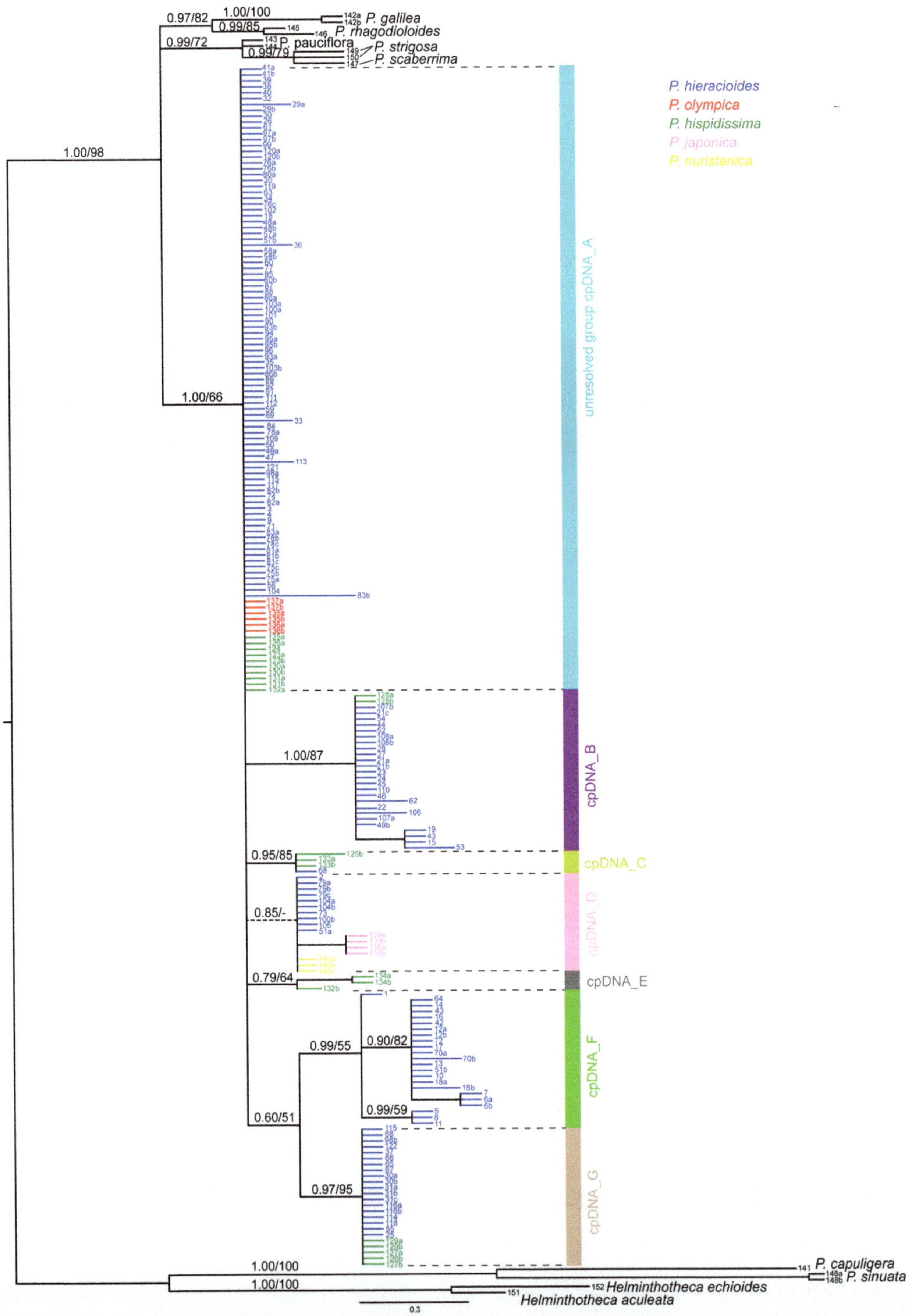
P. galilea
P. rhagodioloides
P. pauciflora
P. strigosa
P. scaberrima
1.00/100
0.97/82
0.99/85
0.99/72
0.99/79
1.00/98
1.00/66
P. hieracioides
P. olympica
P. hispidissima
P. japonica
P. nuristanica
unresolved group cpDNA_A
1.00/87
cpDNA_B
0.95/85
cpDNA_C
0.85/-
cpDNA_D
0.79/64
cpDNA_E
0.99/55
0.90/82
cpDNA_F
0.99/59
0.60/51
0.97/95
cpDNA_G
1.00/100
P. capuligera
P. sinuata
1.00/100
Helminthotheca echioides
Helminthotheca aculeata
0.3

Figure 4. Bayesian majority-rule consensus tree based on the cpDNA intergenic spacer dataset. The numbers above branches refer to posterior probability values of Bayesian inference/the bootstrap support as inferred for the maximum parsimony analyses (values ≥50% are shown). Support values for terminal branches are not shown. Dashed lines represent branches collapsed in strict consensus tree of the maximum parsimony analyses. Colour line below the tree indicates affiliation of groupings to the cpDNA genetic lineages as mentioned in the text. Each accession label includes the population code following Table S1.

distant *P. hieracioides* might be the consequence of incomplete lineage sorting. By contrast, the mosaic pattern detected in the ITS dataset may repeatedly suggest influence of hybridization.

Hybridisation might substantially influence phylogenetic reconstructions at even lower hierarchical levels. Decreased phylogenetic tree resolution, incongruent patterns in different gene trees, and occurrence of alleles transcending taxa or genetic groups are commonly observed in plant phylogenies in which extensive gene exchange between taxa has been documented [2,3,17,41,60, 61,62,63,64,65,66]. Presence of extensive and recurrent gene exchange among investigated *Picris* taxa might be indicated by the following facts. (1) Considerable intra-individual and intra-population ITS sequence heterogeneity and allele transcending clades and taxon borders in *P. hieracioides*, *P. hispidissima*, and *P. olympica*. On the other hand, maintenance of two or more ITS variants within a single genome might be attributed to other molecular-genetic processes [11,67]. (2) Occurrences of heterogeneous ITS sequences predominantly in the contact zones of the studied species or genetic groups, while locations of homogeneous sequences with little or no genetic admixture mostly along the sampled area's margins inhabited mainly by single lineages or species (see Results, Figure 3). (3) Strict self-incompatibility in *P. hieracioides* and *P. hispidissima*, which has been recently documented ([27]; M. Slovák unpublished data). (4) Last but not least, successful crossing between populations of morphotypes and genetic lineages of *P. hieracioides* as well as with *P. hispidissima* obtained during our preliminary field experiments (M. Slovák, unpublished data).

Posterior predictive checking revealed hybridization/introgression at least in five pairs of the studied species. Four pairs were recorded exclusively by simulations on the cpDNA dataset and one pair was detected by simulations on both the plastid and the nuclear dataset. Since the method of posterior predictive checking is powerful unless the hybridization event has occurred very rapidly after the speciation event [9], it might be hypothesised that hybridizations among the *Picris* taxa are rather recent or have occurred later after speciation events. This hypothesis corresponds well mostly in the case of comparatively distantly related species, *P. olympica* with *P. hieracioides*, *P. olympica* with *P. rhagodioloides*, and *P. capuligera* with *P. sinuata*. Such hybridisation

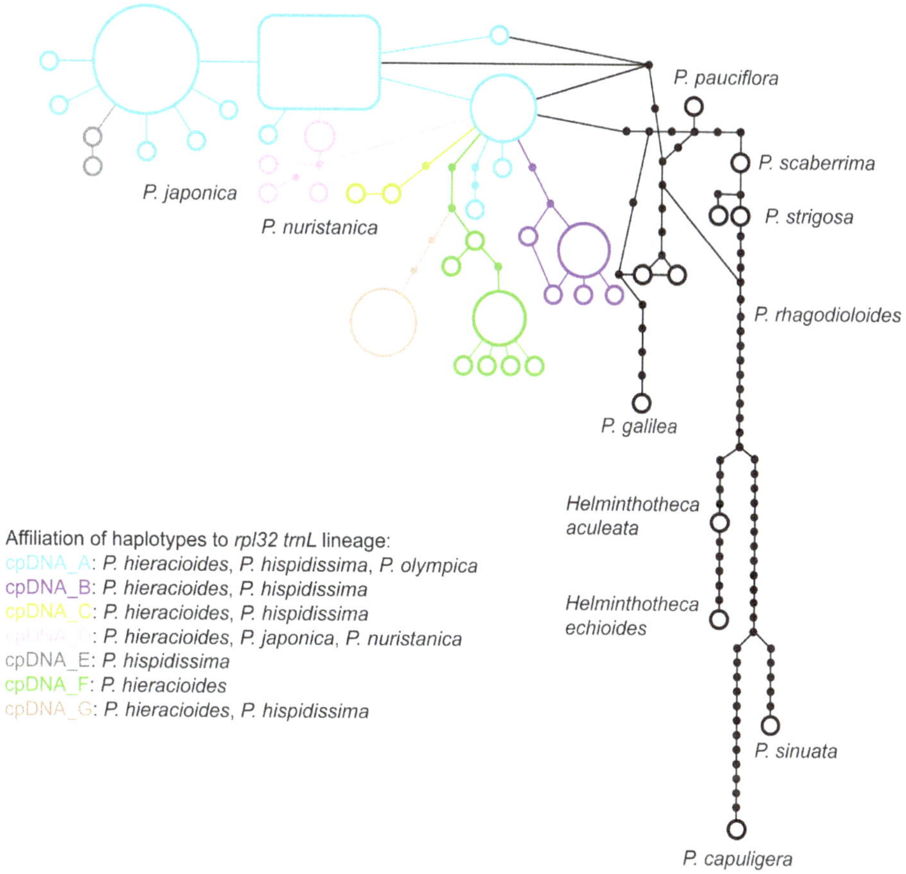

Figure 5. Maximum parsimony network of the cpDNA haplotypes of *Picris* populations. The symbol sizes are proportional to the haplotype frequencies, the lines represent mutational steps, and black dots are unsampled haplotypes. Colour symbols indicate affiliation of accessions to the cpDNA genetic lineages as mentioned in the text.

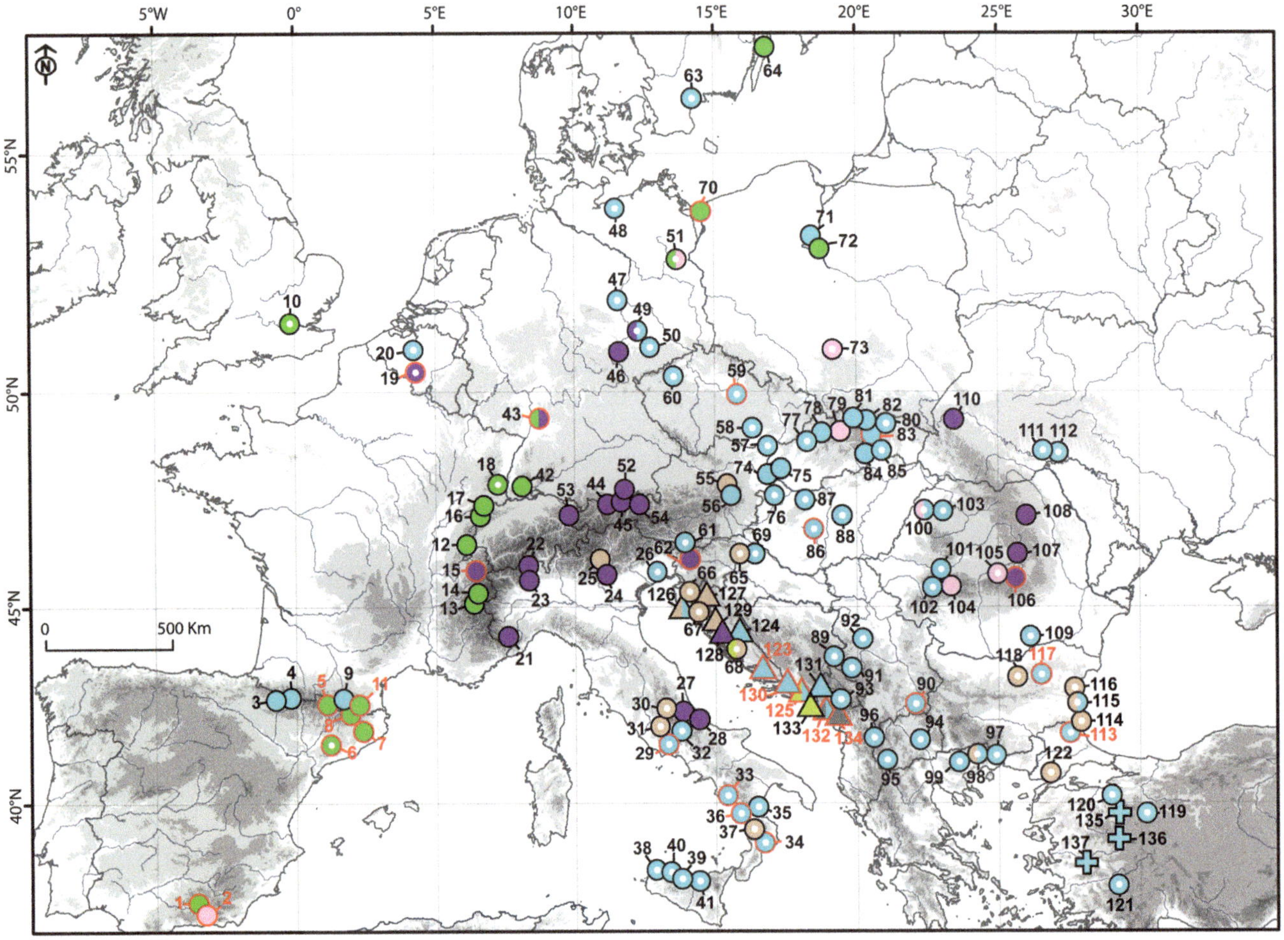

Figure 6. Map displaying geographic distribution of genetic variation in the analysed *Picris* taxa, as inferred from the cpDNA intergenic spacer dataset. Coloured symbols indicate affiliation of accessions to the cpDNA genetic lineages as mentioned in the text. Bi-coloured symbols denote populations possessing two different haplotypes. Symbol shapes indicate taxa as follows: *P. hieracioides* – circle, *P. hispidissima* – triangle, and *P. olympica* – square. The circle symbols with empty centres refer to populations morphologically corresponding to *P. hieracioides* subsp. *hieracioides*, while those with solid symbols refer to populations morphologically corresponding to *P. hieracioides* subsp. *umbellata*. Population of *P. nuristanica* and *P. japonica* and outgroup taxa are not shown. Numbers refer to population codes as denoted in Table S1. The symbols and their numbers highlighted in red indicate populations harbouring rare haplotypes.

events might take part during the Pleistocene glacial/interglacial range shifts, especially, at sites where lowland taxa could have had recurrent contact with mountainous ones [2,68]. Posterior predictive checking, however, did not support hybridisation scenarios among the pairs of closest relative taxa where it would be expected: among subspecies and genetic lineages of *P. hieracioides* as well as between *P. hieracioides* and *P. hispidissima*. One of the most plausible explanations might be that the hybridization events among those taxa happened just after their diversification. Especially in case of the two subspecies of *P. hieracioides*, the divergence of genetic lineages might be shallow causing hybridization to be undetectable [9].

The overall failure to detect hybridization events by simulations on the ITS dataset might be attributed to the presence of recombination and/or concerted evolution within the ITS locus. As stated in [9], some patterns resulting from concerted evolution can potentially bias the estimates of population sizes toward lower values. This causes the loci to coalesce faster and therefore hybridization might be harder to detect. Only one hybridization event was found by both genetic markers, viz., in *H. aculeata* and *H. echioides*. The sample size might be the limitation factor in the posterior predictive checking approach. Since only a single individual per each aforementioned species (outgroup) was sampled, their population sizes cannot be properly estimated for species tree reconstruction. Therefore we assume this result might be a methodological artefact.

Finally, it remains arguable whether some of the studied taxa, but especially *P. hispidissima*, fit into the reticulate evolution scenario or, alternatively, its large genetic heterogeneity arose from genetic erosion. The latter hypothesis might be explained by recurrent introgression with adjacent populations of *P. hieracioides* after their diversification. We are more inclined to the latter scenario because all taxa are morphologically rather well delimited, inhabit ecologically specific biotopes, and trans-taxon alleles are concentrated predominantly in areas of sympatry or parapatry.

Although methods to potentially distinguish between hybridisation and incomplete lineage sorting have been developed, there is still a high risk of confusing these two phenomena or they cannot be discerned in some cases at all [9,12,13]. In the investigated closely related *Picris* taxa from the subsection *Heiracioides*, it is highly tricky or even impossible to discern between these two

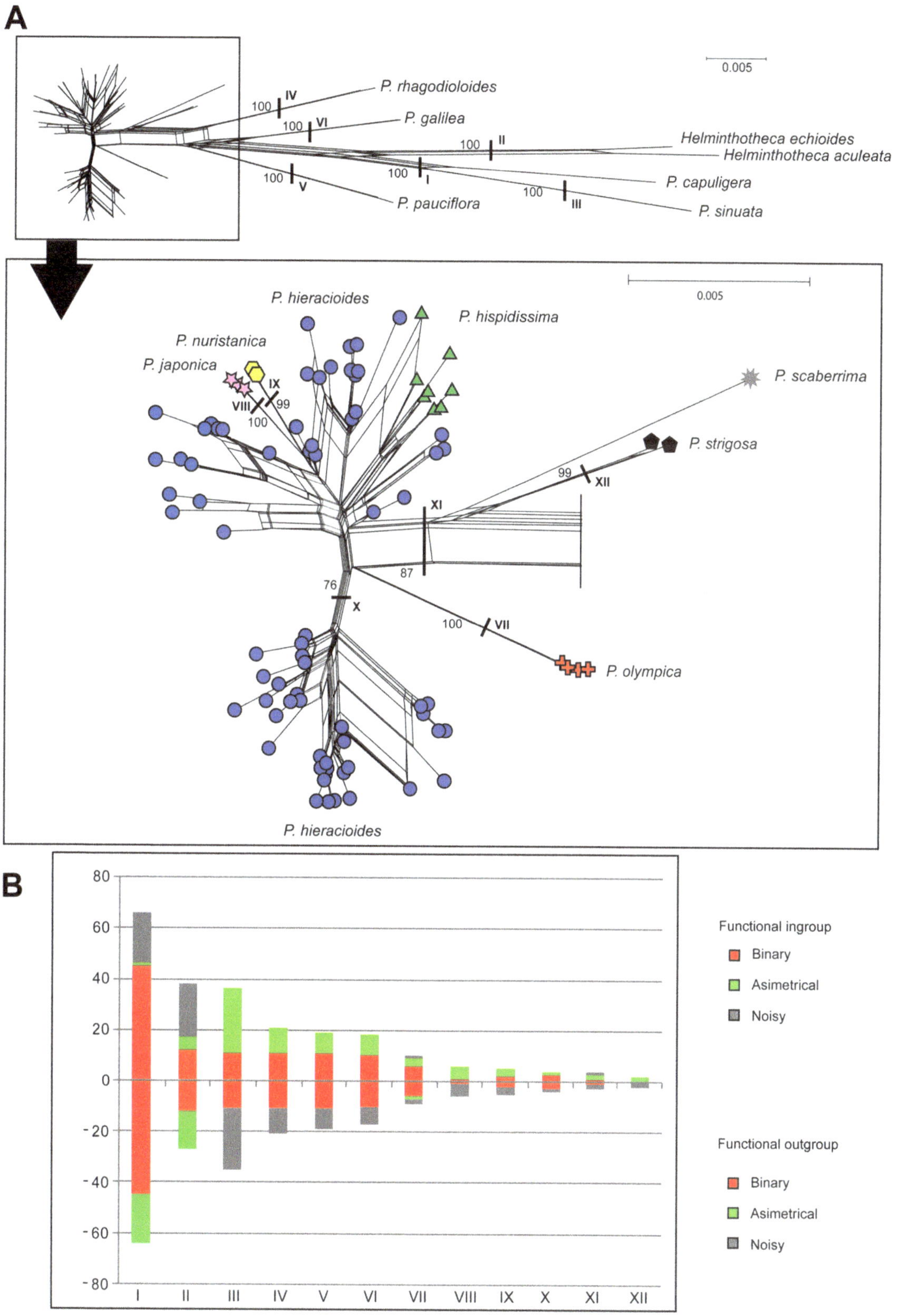

Figure 7. Neighbour-Net diagram (A) and split support spectrum (B) for the concatenated ITS_cpDNA dataset. Coloured symbols indicate affiliation of accessions to taxa. Bootstrap supports of selected important splits are indicated above edges. Column height in the split spectrum represents the number of clade-supporting positions, i.e., putative primary homologies. Column parts above the y-axis represent the ingroup partition while those below the axis correspond to the outgroup partition.

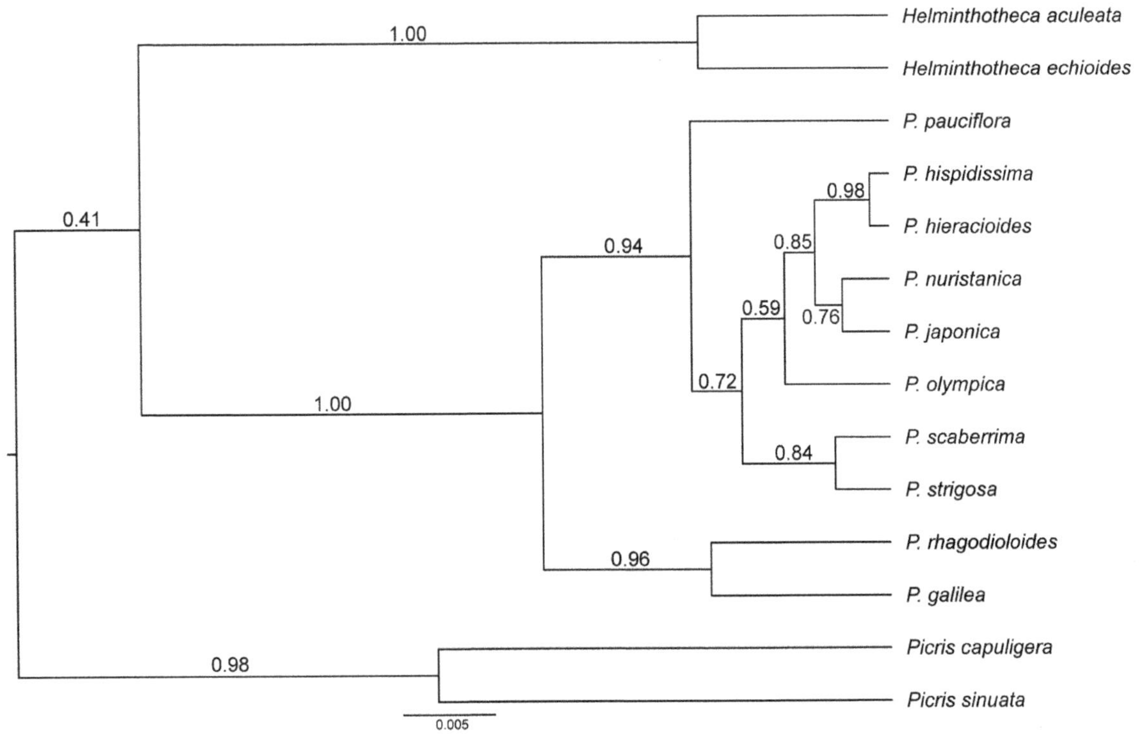

Figure 8. The maximum clade credibility species tree obtained from Bayesian inference in *Beast based on the ITS and cpDNA loci. The posterior probability values are indicated above the branches.

processes. Reasons for this ambiguity might come from weak genetic divergence of analysed taxa, and importantly from presence of gene transfer via introgression which, moreover, most probably occurred repeatedly during their evolutionary history.

Comparison of genetic patterns with results inferred from AFLP and morphology

Spatial distribution of genetic variation in populations of *P. hieracioides* and *P. hispidissima* varied considerably across phylogenetic trees, with both concordant and discordant patterns being detected (Figures 1 and 4). The ITS data were highly consistent with taxonomic/morphological delimitation of the studied taxa. The genetic ITS groups of all studied species as well as within *P. hieracioides* exhibited high levels of concordance with the results of the AFLP analyses [24]. The congruence of genetic patterns resolved by ITS and AFLP, especially, with regard to the phylogenetic signal, has been already demonstrated for various plant groups [69]. The spatial distribution of genetic variation in populations and genetic lineages of *P. hieracioides* and in *P. hispidissima* was, however, clearly blurred by populations harbouring individuals with non-homogenized ITS sequences. Such a diffuse pattern of genetic variation disrupting geographic structure is not unexpected in the nuclear biparentally inherited genetic markers. In fact, this reflects both seed and pollen-mediated gene dispersal [70], especially, in widespread taxa with unrestricted gene flow [71–73]. Numerous heterogeneous ITS sequences did not, however, display APS for some positions variable in the potentially parental haplotypes (Table S2). This pattern can be attributed to in vivo recombination between ITS ribotypes, or to the eventual combination of minor, undetected alleles with more widespread ones [11,67,74]. Detection of such minor ITS ribotypes within taxa/populations might be prevented by preferential amplification of the major ITS copies or by quantitative masking of the rare ribotypes [74]. Another highly plausible explanation on the origin of incomplete homogenization of APS involves concerted evolution [2,40,58,63,64,67]. Its presence in our ITS dataset might be one of the crucial reasons responsible for the failure to detect mutual hybridisation among populations and genetic lineages of *P. hieracioides* and *P. hispidissima* (see discussion above).

In contrast, genetic variation inferred from the plastid cpDNA spacer was, however, only partly congruent with that captured by ITS, AFLPs, and morphology. Further, numerous discrepancies in the number and composition of the genetic cpDNA groups were noted (see 'Results'). Thus, the plastid pattern exhibited an apparently mosaic-like spatial structure and two or more species/genetic groups shared same plastid haplotypes. This was, especially, true for the sympatric/parapatric populations (Figure 6). Complex and incongruent genetic plastid patterns might, beside of the aforementioned hybridisation and incomplete lineage sorting, arise from different attributes of the employed genetic markers such as their evolutionary rate, inheritance mode (bi-parental vs. uni-parental), or number of copies (multiple vs. single copies), which reflects different time horizons of evolution [11]. The nuclear ITS marker evolves faster than plastid ones and therefore enable detection of more recent diversification events [11,75].

Evolution of the studied *Picris* taxa and taxonomic implications

Our results suggested that evolution of the studied *Picris* species was shaped by the interplay of several factors and evolutionary processes. Their diversification was most likely rapid, and may have occurred in an allopatric manner. Since all of these taxa, but especially, *P. hispidissima* and *P. olympica*, have unique habitat and ecology preferences (see Material and Methods), we presume

Table 3. List of the distances among couples of taxa with p-values <0.05, as inferred by posterior predictive distributions.

Gene	Individual 1	Individual 2	Obs. Distance	p-value
cpDNA	*P. olympica1*	*P. h.* subsp. *hieracioides22*	0	0.0288288
	P. olympica 1	*P. h.* subsp. *hieracioides34*	0	0.0288288
	P. olympica2	*P. h.* subsp. *hieracioides22*	0	0.0288288
	P. olympica2	*P. h.* subsp. *hieracioides34*	0	0.0288288
	P. olympica3	*P. h.* subsp. *hieracioides22*	0	0.0288288
	P. olympica3	*P. h.* subsp. *hieracioides34*	0	0.0288288
	P. olympica4	*P. h.* subsp. *hieracioides22*	0	0.0288288
	P. olympica4	*P. h.* subsp. *hieracioides34*	0	0.0288288
	P. olympica1	*P. h.* subsp.*umbellata2*	0	0.0288288
	P. olympica1	*P. h.* subsp.*umbellata4*	0	0.0288288
	P. olympica1	*P. h.* subsp.*umbellata5*	0	0.0288288
	P. olympica1	*P. h.* subsp.*umbellata7*	0	0.0288288
	P. olympica2	*P. h.* subsp.*umbellata2*	0	0.0288288
	P. olympica2	*P. h.* subsp.*umbellata4*	0	0.0288288
	P. olympica2	*P. h.* subsp.*umbellata5*	0	0.0288288
	P. olympica2	*P. h.* subsp.*umbellata7*	0	0.0288288
	P. olympica3	*P. h.* subsp.*umbellata2*	0	0.0288288
	P. olympica3	*P. h.* subsp.*umbellata4*	0	0.0288288
	P. olympica3	*P. h.* subsp.*umbellata5*	0	0.0288288
	P. olympica3	*P. h.* subsp.*umbellata7*	0	0.0288288
	P. olympica4	*P. h.* subsp.*umbellata2*	0	0.0288288
	P. olympica4	*P. h.* subsp.*umbellata4*	0	0.0288288
	P. olympica4	*P. h.* subsp.*umbellata5*	0	0.0288288
	P. olympica4	*P. h.* subsp.*umbellata7*	0	0.0288288
	P. rhagodioloides1	*P. olympica1*	0.00625	0.0405405
	P. rhagodioloides1	*P. olympica2*	0.00625	0.0405405
	P. rhagodioloides1	*P. olympica3*	0.00625	0.0405405
	P. rhagodioloides1	*P. olympica4*	0.00625	0.0405405
	P. sinuata1	*P. capuligera1*	0.0197917	0.0373874
	P. sinuata2	*P. capuligera1*	0.0197917	0.0373874
	H. aculeata1	*H. echioides1*	0.00625	0.000900901
ITS	*H. aculeata1*	*H. echioides1*	0.0306279	0.00135135

that ecological factors also played an important role in their speciation.

The centre of the diversity of the studied species complex, as indicated by the distributions of *P. olympica* and *P. hispidissima* as well as by the accumulation of rare haplotypes, lies in the southern European peninsulas and partly also the Alps and the Carpathians (Figure 6). These regions have been shown to be the most important glacial refugia for the majority of European flora and fauna [76,77]. This might suggest that the evolutionary history of the investigated *Picris* taxa was connected with glacial/interglacial cycles in the Quaternary. This is in concordance with conclusions concerning the origin of the AFLP lineages, as detected within *P. hieracioides* [24].

Genetic data presented herein together with recent distribution [21] indicate that *P. olympica* very likely evolved in alpine levels of high mountain ranges of Anatolia in Asia Minor. Here it very likely survived isolated for a long period, although remote secondary contacts and interaction with other closely related taxa cannot be excluded. Likewise, autecological characteristics and distribution of *P. hispidissima* suggest that evolutionary history of this taxon has undoubtedly been confined with coastal, calcareous mountain ranges of the western part of the Balkan Peninsula. This region has also been repeatedly proven as important evolutionary hotspot and glacial refugium for other plant groups [32,68].

Furthermore, genetic lineages detected within *P. hieracioides* most likely diversified in allopatry or parapatry in southern European peninsulas and adjacent mountainous regions of the Alps and the Carpathians. The mountain-dwelling populations morphologically corresponding to the *P. h.* subsp. *umbellata* most likely evolved in European mountains or their proximity (the Alps, central Apennines, the Carpathians, Iberian mountain ranges and the Jura Mts.) [24]. If it had existed there already during Pleistocene climatic oscillations, it could potentially survive in the peripheral glacial areas together with other mesophylous herbs [78,79]. As proved by the highest cpDNA haplotype richness, diversification centre of populations morphologically assignable to the lowland *P. h.* subsp. *hieracioides* was located more southwards, especially, in lowland and hilly regions of the Apennine Peninsula and plausibly also in the Balkan Peninsula. Subsequently, they colonised other habitats of their present day occurrence in central and north-western Europe as well as in Asia Minor from these southern refugia.

An important taxonomic question is whether the analysed taxa should be recognised as distinct species/subspecies, or whether they rather represent informal phylogeographic lineages within a single heterogeneous taxon. The latter possibility has been favoured in case of several other complex plant species groups [30,32,80,81]. *Picris hieracioides* has been here evidenced to be highly heterogeneous across both markers. Actually, both plastid and ITS nuclear alleles have transcended borders of both subspecies of *P. hieracioides*, as recently delimited in [24]. Moreover, several populations of *P. hieracioides* displayed transitional morphological variation. In addition, from ecological point of view, *P. hieracioides* is characterised by an exceptionally large range of habitat preferences [22,24]. All these facts might indicate that both currently delimited subspecies of *P. hieracioides* are artificial biological units, representing rather ecotypes evolved recurrently in different evolutionary lineages under similar environmental influences than real taxa.

Ecotype hypothesis, however, strongly contradicts results of our previous study focused not only on investigation of wild populations of *P. hieracioides*, but also on cultivation experiments realised under environmentally homogeneous conditions [22,23,24]. These experiments included more than 300 individuals from 32 populations that belonged to both subspecies of *P. hieracioides*, originating from almost the entire studied area [24]. Subsequent comparative morphological investigations revealed that both subspecies retained morphologically stable under homogeneous environmental conditions. Importantly, these analyses showed that environmental conditions have had only a minute impact on the morphological characters identified as taxonomically diagnostic. Likewise, *P. hispidissima* retained its morphological distinctiveness under environmentally homogenous conditions (M. Slovák, unpublished data). Thus, taxonomically important morphological features of these entities are not a consequence of environmental plasticity, but are genetically inherited instead. All three taxa are well defined morphologically, ecologically and partially also genetically (AFLP's) at least within 'pure' populations untouched by their recent mutual secondary contacts.

Both, subspecies of *P. hieracioides* as well as *P. hispidissima* are recently diversified and thus genetically weakly separated taxa. Additionally, their genetic variation has been significantly affected by their reciprocal multiple secondary contacts, taking place after anthropogenically mediated range expansions. Due to the lack of habitat and breeding isolation barriers, hybridisation and introgression among these taxa occur frequently in contact zones. This process leads to the genetic and morphological homogenisation of affected populations, erasing their distinctness in contact zones. Similar cases of taxon fusion caused by hybridisation preceded by migration of populations enhanced by human activities have been reported for instance in members of the genus *Knautia* [80] and *Cerasus* [82]. For the latter case the term "anthropohybridisation" was coined. Establishment of sound taxonomic concept in such difficult cases is very problematic and usually it is a matter of subjective choice. Therefore, at the present state of knowledge, we prefer to leave both subspecies of *P. hieracioides* at currently accepted taxonomic ranks [24]. In contrast, *P. hispidissima* should be included under highly variable *P. hieracioides* as separated subspecies: *P. h.* subsp. *hispidissima* (Bartl.) Slovák and Kučera, comb. nova, hoc loco (basionym: *Crepis hispidissima* Bartl. in Bartling & Wendland, Beitr. Bot. 2: 125, 1825).

Although this study does not provide comprehensive and firm taxonomic conclusions, it represents an important background for future investigations on this intricate and dynamically evolving species complex. Last but not least, the present investigation documents rapid process of taxon erosion, happening due to the anthropogenic disturbance of natural biotopes.

Supporting Information

Table S1 Locality details of studied taxa, number of individuals used for the particular molecular analyses, and GenBank accession numbers.

Table S2 Summary of nucleotide site variation for the nuclear ITS region in the studied taxa.

Alignment S1 ITS_1: comprised 117 ITS sequences without APS. The accession labels in alignments follow the population and taxon codes used in Table S1.

Alignment S2 ITS_2: comprised all 206 ITS sequences. The accession labels in alignments follow the population and taxon codes used in Table S1.

Alignment S3 cpDNA: comprised all 219 rpl32-*trn*L^{UAG} sequences. The accession labels in alignments follow the population and taxon codes used in Table S1.

Alignment S4 ITS_cpDNA: comprised 117 ITS sequences without APS and corresponding plastid ones. The accession labels in alignments follow the population and taxon codes used in Table S1.

Acknowledgments

We would like to thank the following colleagues for their help in sample collection: Ulla-Grilt Andersson (Sweden), Fabio Conti (Italy), Giannantonio Domina (Italy), Rolland Douzet (France), Sandrine Godefroid (Belgium), Iva Hodálová (Slovakia), Nicodemo Passalacqua (Italy), Peter Repa (Slovakia), Åke Svensson (Sweden), Luis Villar (Spain) and Torbjörn Tyler (Sweden). The authors also wish to thank The Royal Botanic Gardens, Kew (United Kingdom) and The National Botanic Garden of Belgium (Belgium) who provided *P. hieracioides* and *P. sinnuata* seeds (see details in Table S1). Our special thanks go to Judita Zozomová-Lihová (Slovakia) for her valuable advice and laboratory assistance during this study, and finally to Karol Marhold and all anonymous reviewers for their valuable comments on previous versions of this manuscript.

Author Contributions

Conceived and designed the experiments: JK MS. Performed the experiments: JK MS. Analyzed the data: MS PV EZ. Contributed reagents/materials/analysis tools: JK MS. Wrote the paper: MS PV EZ.

References

1. Comes HP, Abbott RJ (2001) Molecular phylogeny, reticulation, and lineage sorting in Mediterranean *Senecio* sect. *Senecio* (Asteraceae). Evolution 55: 1943–1962.
2. Nieto Feliner G, Gutiérrez Larena B, Fuertes Aguilar J (2004) Fine-scale geographical structure, intra-individual polymorphism and recombination in

nuclear ribosomal internal transcribed spacers in *Armeria* (Plumbaginaceae). Ann Bot (Oxford) 93: 189–200.

3. Willyard A, Cronn R, Liston A (2009) Reticulate evolution and incomplete lineage sorting among the ponderosa pines. Molec Phylogen Evol 52: 498–511.
4. Schmidt-Lebuhn AN, de Vos JM, Keller B, Conti E (2012) Phylogenetic analysis of *Primula* section *Primula* reveals rampant non-monophyly among morphologically distinct species. Molec Phylogen Evol 65: 23–34.
5. Yu WB, Huang PH, Li DZ, Wang H (2013) Incongruence between nuclear and chloroplast DNA phylogenies in *Pedicularis* section *Cyathophora* (Orobanchaceae). PLoS ONE, doi: 10.1371/journal.pone.0074828
6. Rieseberg LH (1997) Hybrid origins of plant species. Ann Rev Ecol Syst 28: 359–389.
7. Wendel JF, Doyle J (1998) Phylogenetic incongruence: Window into genome history and molecular evolution. In: Soltis D, Soltis P, Doyle J, editors.Molecular systematics of plants II: DNA sequencing. Kluwer, Boston, USA. pp. 265–296.
8. Linder CR, Rieseberg LH (2004) Reconstructing patterns of reticulate evolution in plants. Amer J Bot 91: 1700–1708.
9. Joly S, McLenachan PA, Lockhart PJ (2009) A statistical approach for distinguishing hybridisation and incomplete lineage sorting. Amer Nat 174: 54–70.
10. Funk DJ, Omland KE (2003) Species-level paraphyly and polyphyly: Frequency, causes, and consequences, with insights from animal mitochondrial DNA. Annual Rev Ecol Evol Syst 34: 397–423.
11. Small RL, Cronn RC, Wendel JF (2004) Use of nuclear genes for phylogeny reconstruction in plants. Austral Syst Bot 17: 145–170.
12. Holland BR, Benthin S, Lockhart PJ, Moulton V, Huber KT (2008) Using supernetworks to distinguish hybridisation from lineage-sorting. BMC Evol Biol 8: 202.
13. Yu Y, Than C, Degnan JH, Nakhleh L (2011) Coalescent histories on phylogenetic networks and detection of hybridisation despite incomplete lineage sorting. Syst Biol 60: 138–149.
14. Wiens JJ, Hollingsworth BD (2000) War of the iguanas: Conflicting molecular and morphological phylogenies and long-branch attraction in iguanid lizards. Syst Biol 49: 143–159.
15. Duvall MR, Ervin AB. (2004) 18S gene trees are positively misleading for monocot/dicot phylogenetics. Molec Phylogen Evol 30: 97–106.
16. van der Niet T, Linder HP (2008) Dealing with incongruence in the quest for the species tree: a case study from the orchid genus *Satyrium*. Molec Phyl Evol 47: 154–174.
17. Frajman B, Eggens F, Oxelman B (2009) Hybrid origins and homoploid reticulate evolution within *Heliosperma* (Sileneae, Caryophyllaceae) - A multigene phylogenetic approach with relative dating. Syst Biol 58: 328–345.
18. Heled J, Drummond AJ (2010) Bayesian inference of species trees from multilocus data. Molec Biol Evol 27: 570–580.
19. Kubatko L, Carstens B, Knowles L (2009) STEM: species tree estimation using maximum likelihood for gene trees under coalescence. Bioinformatics 25: 971–973.
20. Joly S. (2012) JML: testing hybridization from species trees. Molec Ecol Res 12: 179–184.
21. Lack HW (1974)Die Gattung *Picris* L. sensu lato im ostmediterran–westasiatischen Raum. PhD thesis, University of Vienna, Austria. 184 p.
22. Slovák M, Marhold K (2007) Morphological evaluation of *Picris hieracioides* L. (Compositae) in Slovakia. Phyton (Horn) 47: 73–102.
23. Slovák M, Urfus T, Vít P, Marhold K (2009) Balkan endemic *Picris hispidissima* (Compositae): Morphology, DNA content and relationship to polymorphic *P. hieracioides*. Pl Syst Evol 278: 178–201.
24. Slovák M, Kučera J, Marhold K, Zozomová-Lihová J (2012) The morphological and genetic variation in the polymorphic species *Picris hieracioides* (Compositae, Lactuceae) in Europe strongly contrasts with traditional taxonomical concepts. Syst Bot 21: 258–278.
25. Lack HW (1979) The genus *Picris* (Asteraceae, Lactuceae) in Tropical Africa. Pl Syst Evol 131: 35–52.
26. Samuel R, Gutermann W, Stuessy TF, Ruas CF, Lack HW, et al. (2006) Molecular phylogenetics reveals *Leontodon* (Asteraceae, Lactuceae) to be diphyletic. Amer J Bot 93: 1193–1205.
27. Slovák M, Šingliarová B, Mráz P (2008) Chromosome numbers and mode of reproduction in *Picris hieracioides* s.l. (Asteraceae) with notes on some other *Picris* taxa. Nordic J Bot 28: 238–244.
28. Slovák M, Vít P, Urfus T, Suda J (2009) Complex pattern of genome size variation in a polymorphic member of the Asteraceae. J Biogeogr 36: 372–384.
29. Lack HW (1975) *Picris*. In: Davis PH, editor.Flora of Turkey and the East Aegean Islands 5.Edinburgh, UK: University Press. pp. 678–684.
30. Frajman B, Oxelman B (2007) Reticulate phylogenetics and phytogeographical structure of *Heliosperma* (Sileneae, Caryophyllaceae) inferred from chloroplast and nuclear DNA sequences. Molec Phylogen Evol 43: 140–155.
31. Prentice HC, Malm JU, Hathaway L (2008) Chloroplast DNA variation in the European herb *Silene dioica* (red campion): postglacial migration and interspecific introgression. Pl Syst Evol 272: 23–37.
32. Bardy KE, Albach DC, Schneeweiss GM, Fischer MA, Schönswetter P (2010) Disentangling phylogeography, polyploid evolution and taxonomy of a woodland herb (*Veronica chamaedrys* group, Plantaginaceae s.l.) in southeastern Europe. Molec Phylogen Evol 57: 771–786.
33. Shaw J, Lickey EB, Schilling EE, Small RL (2007) Comparison of whole chloroplast genome sequences to choose noncoding regions for phylogenetic studies in angiosperms: The tortoise and the hare III. Amer J Bot 94: 275–288.
34. Chapman MA, Chang JC, Weisman D, Kesseli RV, Burke JM (2007) Universal markers for comparative mapping and phylogenetic analysis in the Asteraceae (Compositae). Theor Appl Genet 115: 747–755.
35. Shaw J, Lickey EB, Beck JT, Farmer SB, Liu W, et al. (2005) The tortoise and the hare II: Relative utility of 21 noncoding chloroplast DNA sequences for phylogenetic analysis. Amer J Bot 92: 142–166.
36. Francisco-Ortega J, Fuertes Aguilar J, Gómez Campo C, Santos-Guerra A, Jansen RK (1999) Internal transcribed spacer sequence phylogeny of *Crambe* L. (Brassicaceae): Molecular data reveal two old world disjunctions. Molec Phylogen Evol 11: 361–380.
37. White TJ, Bruns T, Lee S, Taylor JW (1990) Amplification and direct sequencing of fungal ribosomal RNA genes for phylogenetics. In: Innis MA, Gelfand DH, Sninsky JJ, White TJ, editors.PCR Protocols: A Guide to Methods and Applications.New York, USA: Academic Press. pp. 315–322.
38. Timme R, Kuehl EJ, Boore JK, Jansen RK (2007) A comparative analysis of the *Lactuca* and *Helianthus* (Asteraceae) plastid genomes: identification of divergent regions and categorization of shared repeats. Amer J Bot 94: 302–313.
39. Hall TA (1999) BioEdit: a user-friendly biological sequence alignment editor and analysis program for Windows 95/98/NT. Nucleic Acids Symp Ser (Oxford) 41: 95–98.
40. Fuertes Aguilar J, Nieto Feliner G (2003) Additive polymorphisms and reticulation in an ITS phylogeny of thrifts (*Armeria*, Plumbaginaceae). Molec Phylogen Evol 28: 430–447.
41. Záveská E, Fér T, Šída O, Krak K, Marhold K, et al. (2012) Phylogeny of *Curcuma* (Zingiberaceae) based on plastid and nuclear sequences: Proposal of the new subgenus *Ecomata*. Taxon 61: 747–763.
42. Farris JS, Källersjö M, Kluge AG, Bult C (1994) Testing significance of congruence. Cladistics 10: 315–319.
43. Swofford DL (2001) PAUP*. Phylogenetic Analysis Using Parsimony (*and Other Methods), v. 4.0 beta 10. Sunderland: Sinauer Associates.
44. Maddison DR, Maddison WP (2000)MacClade4: analysis of phylogeny and character evolution, version 4.0. Sinauer, Sunderland, Massachusetts, USA. 398 p.
45. Ronquist F, Huelsenbeck JP (2003) MrBayes 3: Bayesian phylogenetic inference under mixed models. Bioinformatics 19: 1572–1574.
46. Miller MA, Pfeiffer W, Schwartz T (2010) "Creating the CIPRES Science Gateway for inference of large phylogenetic trees" in Proceedings of the Gateway Computing Environments Workshop (GCE), 14 Nov. 2010, New Orleans, LA : 1–8 pp.
47. Guindon S, Gascuel O (2003) A simple, fast, and accurate algorithm to estimate large phylogenies by maximum likelihood. Syst Biol 52: 696–704.
48. Posada D (2008) ModelTest: Phylogenetic Model Averaging. Molec Biol Evol 25: 1253–1256.
49. Huson DH, Bryant D (2006) Application of phylogenetic networks in evolutionary studies. Molec Biol Evol 23: 254–267.
50. Templeton AR, Crandall KA, Sing CF (1992) A cladistic analysis of the phenotypic associations with haplotypes inferred from restriction endonuclease mapping and DNA sequence data. III. Cladogram estimation. Genetics 132: 619–633.
51. Clement M, Posada D, Crandall K (2000) TCS: a computer program to estimate gene genealogies. Mol Ecol 9: 1657–1660.
52. Wägelle JW, Rödding F (1998) A priori estimation of phylogenetic information conserved in aligned sequences. Molec Phylogen Evol 9: 358–365
53. Rambaut A, Drummond AJ (2007) Tracer v1.4. Available: http://beast.bio.ed.ac.uk/Tracer. Accessed 2014 Jan 21.
54. Silvestro D, Michalak I (2011) raxmlGUI: a graphical front-end for RAxML. Org Divers Evol 12: 335–337.
55. Shimodaira H (2002) An approximately unbiased test of phylogenetic tree selection. Syst Biol 51: 492-508.
56. Shimodaira H (2008) Testing regions with non-smooth boundaries via multiscale bootstrap. J Stat Plan Inference 138: 1227–1241.
57. Shimodaira H, Hasegawa M (2001) Consel: for assessing the confidence of phylogenetic tree selection. Bioinformatics 17: 1246–1247.
58. Baldwin BG, Sanderson MJ, Porter JM, Wojciechowski MF, Cambell CS, et al. (1995) The ITS region of nuclear ribosomal DNA: A valuable source of evidence on angiosperm phylogeny. Ann Missouri Bot Gard 82: 247–277.
59. Hudson RR (1990) Gene genealogies and the coalescent process. Oxford Surv Evol Biol 7: 1–44.
60. Jang CG, Müllner AN, Greimler J (2005) Conflicting patterns of genetic and morphological variation in European *Gentianella* section *Gentianella*. Bot J - Linn Soc 148: 175–187.
61. Flagel LE, Rapp RA, Grover CE, Widrlechner MP, Hawkins J, et al. (2008) Phylogenetic, morphological, and chemotaxonomic incongruence in the North American endemic genus *Echinacea*. Amer J Bot 95: 756–765.
62. Ramsey J., Robertson A., Husband B. (2008) Rapid adaptive divergence in new world *Achillea*, an autopolyploid complex of ecological races. Evolution 62: 639–653.
63. Fuertes Aguilar J, Rosselló JA, Nieto Feliner G (1999) nrDNA concerted evolution in natural and artificial hybrids of *Armeria* (Plumbaginaceae). Mol Ecol 8: 1341–1346.

64. Fuertes Aguilar J, Rosselló JA, Nieto Feliner G (1999b) Molecular evidence for the compilospecies model of reticulate evolution in *Armeria* (Plumbaginaceae). Syst Biol 48: 735–754.
65. Baack EJ, Rieseberg LH (2007) A genomic view of introgression and hybrid speciation. Curr Opin Genet Dev 17: 513–518.
66. Moody ML, Rieseberg LH (2012) Sorting through the chaff, nDNA gene trees for phylogenetic inference and hybrid identification of annual sunflowers (*Helianthus* sect. *Helianthus*). Molec Phylogen Evol 64: 145–155.
67. Álvarez I, Wendel JF (2003) Ribosomal ITS sequences and plant phylogenetic inference. Molec Phylogen Evol 29: 417–434.
68. Kučera J, Marhold K, Lihová J (2009) *Cardamine maritima* group (Brassicaceae) in the amphi-Adriatic area: A hotspot of species diversity revealed by DNA sequences and morphological variation. Taxon 59: 148–164.
69. Koopman WJM (2005) Phylogenetic signal in AFLP data sets. Syst Biol 54: 197–217.
70. Ennos RA (1994) Estimating the relative rates of pollen and seed migration among plant populations. Heredity 72: 250–259.
71. Malm JU, Prentice HC (2002) Immigration history and gene dispersal: allozyme variation in Nordic populations of the red campion, *Silene dioica* (Caryophyllaceae). Biol J Linn Soc 77: 23–34.
72. Tyler T (2002) Geographical distribution of allozyme variation in relation to post-glacial history in *Carex digitata*, a widespread European woodland sedge. J Biogeogr 29: 919–930.
73. Tyler T, Prentice HC, Widén B (2002) Geographic variation and dispersal history in Fennoscandian populations of two forest herbs. Pl Syst Evol 233: 47–64.
74. Rauscher JT, Doyle JJ, Brown AHD (2002) Internal transcribed spacer repeat-specific primers and the analysis of hybridisation in the *Glycine tomentella* (Leguminosae) polyploid complex. Mol Ecol 11: 2691–2702.
75. Rebernig CA, Schneeweiss GM, Bardy KE, Schönswetter P, Villaseñor JL, et al. (2010) Multiple Pleistocene refugia and Holocene range expansion of an abundant southwestern American desert plant species (*Melampodium leucanthum*, Asteraceae). Mol Ecol 19: 3421–3443.
76. Hewitt GM (2004) Genetic consequences of climatic oscillations in the Quaternary. Phil Trans R Soc B 359: 183–195.
77. Médail F, Diadema K (2009) Glacial refugia influence plant diversity patterns in the Mediterranean Basin. J Biogeogr 36: 1333–1345.
78. Kramp K, Huck S, Niketić M, Tomović G, Schmitt T (2009) Multiple glacial refugia and complex postglacial range shifts of the obligatory woodland plant species *Polygonatum verticillatum* (Convallariaceae). Pl Biol 11: 392–404.
79. Huck S, Büdel B, Kadereit JW, Printzen C (2009) Rangewide phylogeography of the European temperate-montane herbaceous plant *Meum athamanticum* Jacq.: Evidence for periglacial persistence. J Biogeogr 36: 1588–1599.
80. Rešetnik I, Frajman B, Bogdanović S, Ehrendorfer F, Schönswetter P (2014) Disentangling relationships among the diploid members of the intricate genus *Knautia* (Caprifoliaceae, Dipsacoideae). Molec Phylogen Evol 74: 97ec P
81. Schönswetter P, Solstad H, Garcia PE, Elven R (2009) A combined molecular and morphological approach to the taxonomically intricate European mountain plant *Papaver alpinum* s.l. (Papaveraceae) – Taxa or informal phylogeographical groups? Taxon 17: 1326–1343.
82. Wójcicky JJ (1991) Variability of *Prunus fruticosa* Pall. and the problem of an anthropohybridisaton. Veröff. Geobot. Inst. ETH, Stiftung Rübel, Zürich 106: 266–272.

14

A Nuclear DNA Perspective on Delineating Evolutionarily Significant Lineages in Polyploids: The Case of the Endangered Shortnose Sturgeon (*Acipenser brevirostrum*)

Tim L. King[1]*, Anne P. Henderson[2], Boyd E. Kynard[3], Micah C. Kieffer[4], Douglas L. Peterson[5], Aaron W. Aunins[6], Bonnie L. Brown[7]

1 U.S. Geological Survey (USGS), Leetown Science Center, Aquatic Ecology Branch, Kearneysville, West Virginia, United States of America, **2** USGS, Leetown Science Center, Fish Health Branch, Kearneysville, West Virginia, United States of America, **3** University of Massachusetts, Department of Environmental Conservation, Amherst, Massachusetts, United States of America, **4** USGS, Leetown Science Center, S. O. Conte Anadromous Fish Research Laboratory, Turners Falls, Massachusetts, United States of America, **5** University of Georgia, Warnell School of Forest Resources, Athens, Georgia, United States of America, **6** Cherokee Nation Technology Solutions, Leetown Science Center, Kearneysville, West Virginia, United States of America, **7** Virginia Commonwealth University, Department of Biology, Richmond, Virginia, United States of America

Abstract

The shortnose sturgeon, *Acipenser brevirostrum*, oft considered a phylogenetic relic, is listed as an "endangered species threatened with extinction" in the US and "Vulnerable" on the IUCN Red List. Effective conservation of *A. brevirostrum* depends on understanding its diversity and evolutionary processes, yet challenges associated with the polyploid nature of its nuclear genome have heretofore limited population genetic analysis to maternally inherited haploid characters. We developed a suite of polysomic microsatellite DNA markers and characterized a sample of 561 shortnose sturgeon collected from major extant populations along the North American Atlantic coast. The 181 alleles observed at 11 loci were scored as binary loci and the data were subjected to multivariate ordination, Bayesian clustering, hierarchical partitioning of variance, and among-population distance metric tests. The methods uncovered moderately high levels of gene diversity suggesting population structuring across and within three metapopulations (Northeast, Mid-Atlantic, and Southeast) that encompass seven demographically discrete and evolutionarily distinct lineages. The predicted groups are consistent with previously described behavioral patterns, especially dispersal and migration, supporting the interpretation that *A. brevirostrum* exhibit adaptive differences based on watershed. Combined with results of prior genetic (mitochondrial DNA) and behavioral studies, the current work suggests that dispersal is an important factor in maintaining genetic diversity in *A. brevirostrum* and that the basic unit for conservation management is arguably the local population.

Editor: Valerio Ketmaier, Institute of Biochemistry and Biology, Germany

Funding: The National Capital Region (NCR) of the U.S. National Park Service (NPS), and the U.S. Geological Survey's (USGS) Leetown Science Center provided principal funding for this project through the National Resources Preservation Program. The Northeast Regional Office, Protected Resources Division, NOAA Fisheries, Department of Commerce also provided partial funding. The funders had no role in study design, data collection and analysis, decision to publish, or preparation of the manuscript.

Competing Interests: The authors have declared that no competing interests exist.

* Email: tlking@usgs.gov

Introduction

Sturgeons (Acipenseridae) are one of two living groups of chondrostean fishes; the other group being the paddlefishes (Polydontidae). The fossil record suggests these were dominant fishes of the Permian period (~200 Myr [1]) and owing to morphological similarities to their extinct relatives, modern sturgeons often are described as being evolutionarily static [2], [3]. Acipenserids also are notable for their anadromous and amphidromous life histories, unique benthic and life-history specializations, and the propensity for inter-genus and inter-species hybridization, the latter resulting in various levels of polyploidy which is slightly at odds with the fact that estimated mutation rates within the mitochondrial and nuclear genomes of acipenserids are reduced compared to other fishes [4], [5], [6]. The continued existence of these relic fishes is in jeopardy throughout North America, Europe, and Asia where nearly all sturgeon species have experienced overfishing, habitat degradation or loss, and obstruction of spawning areas. Much effort has been directed at understanding ecological factors associated with sturgeon biology [7] and behavior [8], [9] to address the prevailing conservation biology tenet that management planning must be framed in terms of providing conditions that will facilitate potential adaptation. For resource managers to plan for an evolutionary future for such "trust species," they must have the means to identify evolutionarily distinct and significant lineages (e.g., species, metapopulations, populations, distinct population segments). However, some of the same characters that herald scientific interest (e.g., polyploidization, lengthy lifespans, long times to maturation, and intermittent semi-annual spawning) exacerbate efforts to identify and protect evolutionarily distinct lineages within each species.

Documenting heritable genetic information is a hallmark of contemporary conservation strategies for delineating management units and has previously been applied to sturgeons where recent molecular systematics studies have called into question the taxonomic foundations of the Acipenseriform classification, which has been historically based on morphological characters [10], [11], [12], [13]. Different gene regions, including some physically linked within the mitochondrial (mt) DNA molecule, have yielded differing phylogenetic interpretations of the Acipenseridae (see [13] for review). Efforts to resolve the molecular systematics of the Acipenseridae using nuclear (n) DNA sequences have either focused on a single gene region (18S rRNA, [5]) or on interspecific comparisons of repetitive DNAs observed as a result of genomic DNA digestion (*Hin*dIII [14], *Pst*I [6]). The former was complicated by the polyploid genome and provided no additional phylogenetic resolution, whereas the latter studies concur to some degree with molecular phylogeny observed with mtDNA sequences [6]. Unfortunately, not all sturgeon species have detectable levels of some satellite DNAs [14] and furthermore, the level of intraspecific variability has been shown to be greater than the interspecific divergence among species belonging to the same phylogenetic clade [6]. Given such equivocal results, some revision of Acipenseriform classification is needed to guide conservation efforts. Ultimately, this lack of genomic resources for sturgeons hinders mechanistic study (but see [15]).

Gene duplication and subsequent functional divergence is a fundamental process of adaptive evolution [16] and is particularly relevant in the Acipenseridae where the presence of evolutionary polyploidy ranging in a series from 4n-8n-12n times the ancestral haploid number [17] presents significant challenges for investigating the evolutionary processes shaping the nuclear genomes [4], [18], [19]. It is unclear whether these polyploid events resulted from complete genome duplications (autopolyploid), hybridization between species of different ploidy levels combined with genome doubling (allopolyploidy), or a combination of these processes [20]. Following the polyploid events that gave rise to extant sturgeon species, the random and gradual diploidization process [21] is assumed to have resulted in functional diploidy [22]; however, the degree to which their various polyploid nuclear genomes exhibit disomic inheritance is unknown.

The shortnose sturgeon *Acipenser brevirostrum* is an amphidromous species endemic to the large coastal rivers of eastern North America. This species is distinguished among all the living acipenserids by exhibiting the largest number of chromosomes, 372 [18]. *A. brevirostrum* was listed as an "endangered species" under the US Endangered Species Preservation Act of 1967 and remains so despite re-assessment in response to a 1994 petition to de-list populations in tributaries to the Gulf of Maine. Of significant note is that of 19 putative population units identified based on the species' perceived strong fidelity to natal rivers [23], [24] some river populations continue to exist, although much reduced, but in other rivers, the species has been extirpated. In most instances, spawning status is either unknown or indicated to be of limited extent [25], [23] further complicating the prediction of biological units that could respond to conservation measures.

To date, all published information on phylogeographic- and population-level structuring in *A. brevirostrum* has been assessed through nucleotide sequence variation detected in the maternally-inherited mtDNA. This is presumably due to the difficult nature of interpreting allelic data from the functionally polyploid (putatively hexaploid) nuclear genome [26]. The mtDNA research has primarily been focused on a moderately polymorphic 440 base pair segment of the control region (CR) adjacent to the tRNA proline gene. These findings are well documented in the peer-reviewed literature [27], [28], [29], [26], [30], [31] and are consistent both among studies and between researchers. Although results reflect a shallow gene genealogy (gene tree) for the *A. brevirostrum* mtDNA CR, analyses of haplotype frequencies at the level of putative individual populations showed significant differences among nearly all river/estuarine systems in which reproduction is known to occur. One prior study [31] concluded that although higher level genetic relationships exist (e.g., Northeast vs. Mid-Atlantic; Northeast vs. Southeast; Mid-Atlantic vs. Southeast; and other Mid-Atlantic regional subdivisions), *A. brevirostrum* appear to function in discrete populations, and that relatively low female-mediated gene flow exists between the majority of populations. This implies that effective dispersal among drainages within regions has been sufficient to prevent deep divergence within this species over evolutionary time scales.

Acipenser brevirostrum has been shown to possess the highest number of chromosomes (N = 362–372) among all the Acipenseriformes karyotyped to date [18]. These authors, however, could not determine the species' exact level of polysomy (hexaploid or dodecaploid). Contemporary cytogenetic techniques (including signals from fluorescent *in situ* hybridization) suggest *A. brevirostrum* is a hexaploid species [19]. While immensely complex, nuclear DNA-based approaches to *A. brevirostrum* conservation could identify significant levels of informative genetic variation because certain duplicated loci and repetitive DNA may lack functional constraints, thus allowing rapid accumulation of differentiation in DNA sequences [32]. Moreover, if the observed patterns of nuclear DNA diversity and variation differed from those empirically determined for the maternally-inherited mtDNA, this would inform biologists of the degrees of site philopatry or sex-biased dispersal for *A. brevirostrum*. However, no phylogeographic or population informative nuclear markers have been identified for *A. brevirostrum* [33].

To address this important research need and for the first time, allow an extensive assessment of the phylogeographic structure of *A. brevirostrum* from a multilocus nuclear DNA perspective, we characterized the inheritance of polysomic microsatellite DNA loci in shortnose sturgeon collected throughout the species' range using loci derived specifically from this species [34]. Because of the complex modes of inheritance underlying the putatively hexaploid genome, we scored each allele (fragment) as a dominant marker with two states, presence or absence, resulting in the production of a binary character matrix. We report on the findings of an extensive statistical comparison of the patterns in allelic variation to identify and assess the reproductive status of populations and to delineate functional units of management to aid in recovery planning.

Methods

Sample collection

While no tissue sampling was collected as part of this study, *Acipenser brevirostrum* ($n = 561$) were sampled from 17 river and estuarine systems representing the species' range (Figure 1) by researchers approved and permitted by the Institutional Animal Care and Use Committee of the U.S. Department of Commerce, National Oceanic and Atmospheric Administration, National Marine Fisheries Service. Sampling followed the guidelines mandated under NOAA Technical Memorandum NMFS-OPR-18 or NOAA Technical Memorandum NMFS-OPR-45. The mandated non-invasive procedures were that tissue (1.0 cm^2 fin-

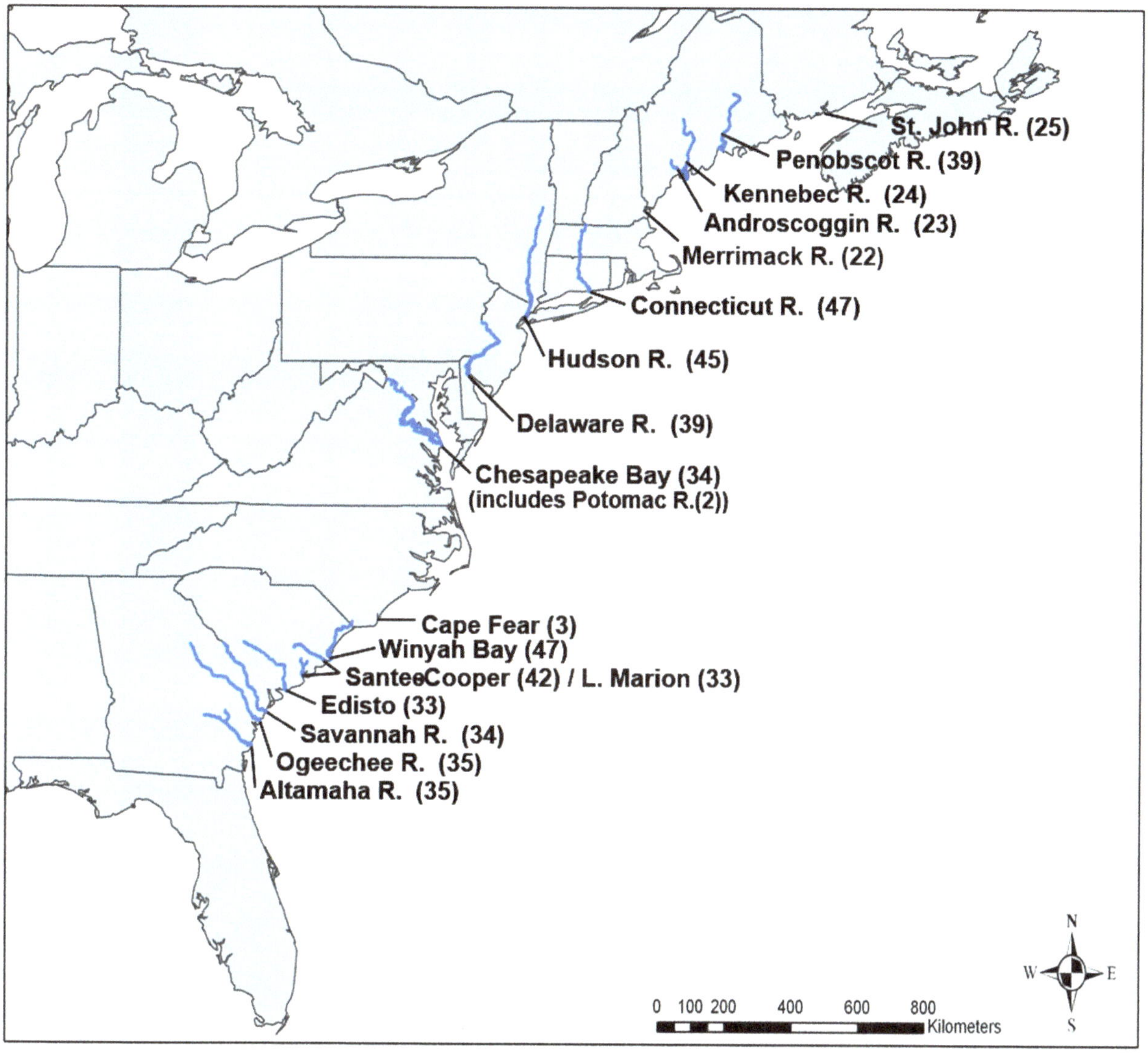

Figure 1. Map of a portion of the North American Atlantic Coast depicting the general location and sample size of 17 river and estuary collections of shortnose sturgeon (*Acipenser brevirostrum*) surveyed at 11 polysomic microsatellite DNA loci. Sample sizes are in parentheses.

clip) taken from soft pelvic fin was stored in 95% absolute ethanol or SDS/urea.

The ambiguous reproductive status of *A. brevirostrum* in the Potomac and Merrimack Rivers affected categorization of specimens. We chose to treat the Potomac River collection ($n = 2$) both as of unknown origin and as part of the large Chesapeake Bay-proper collection. The Merrimack River sample consisted of males collected at the same location and time; however, because eggs ($n = 4$) and embryos ($n = 2$) were collected in Spring 2009, we considered the Merrimack River collection as a reproducing population. Genomic DNA from ethanol preserved samples was extracted with the Gentra Puregene DNA kit (Qiagen, Valencia, CA) following the manufacturers guidelines for whole non-mammalian blood and resuspended in TE (10 mM Tris-HCl, pH 8.0, 1 mM EDTA). Previously extracted genomic DNA from other researchers was extracted using the methods described in [28]. DNA concentrations were determined by fluorescence assay [35], and the integrity of the DNA was visually inspected on 1% agarose gels [36]. SDS-urea preserved samples were also processed using these procedures with the exception that cell lysis was not necessary and samples were subjected directly to protein precipitation and alcohol purification. All DNA samples were quantified as described above and diluted to 100 ng/μl for use in PCR amplification.

A suite of 11 microsatellite loci previously identified from *A. brevirostrum* [34] was surveyed: *Abr*B438, *Abr*D10, *Abr*D114, *Abr*D135, *Abr*D141, *Abr*D193, *Abr*D236, *Abr*D332, *Abr*D345, *Abr*D379, and *Abr*D557. Optimized PCR mixes consisted of the following: 100–200 ng of genomic DNA, 1 × PCR buffer (10 mM Tris-HCl, pH 8.3, 50 mM KCl), 2.0 mM $MgCl_2$, 0.2 mM dNTPs, 0.4 mg/ml Bovine Serum Albumin, 2.0 U *Taq* DNA polymerase (Promega Corporation, Madison, WI, USA), 0.25 μM forward and 0.5 μM reverse primer, and 0.3 μM fluorescent labeled M13 primer in a total volume of 25 μl. Amplifications were carried out on either PTC-200 or PTC-225 Thermal Cyclers (Bio-Rad Laboratories, Hercules, CA) using the following cycling conditions: initial denaturing at 94°C for 15 min; 29 cycles of 94°C for 1 min; annealing temperature (56–66°C) for 45 sec,

72°C for 45 sec; 5 cycles of 94°C for 1 min; 53°C for 45 sec; 72°C for 45 sec; and final extension at 72°C for 10 min. Fluorescently labeled fragment analysis was performed using an ABI 3130XL Genetic Analyzer and binning of alleles was performed using GENESCAN 2.1 Analysis software and GENOTYPER 3.6 fragment analysis software (Applied Biosystems, Grand Island, NY). Each locus was independently scored, and each amplicon meeting the signal strength conditions specified (at least 10% relative to the strongest allele) and fitting into the appropriate size category (based on repeat motif and an assumed step-wise mutational pattern) was classified as an allele.

Statistical analyses of polyploid microsatellite markers

For polyploid individuals, gene duplication, multiple alleles, and the mode of inheritance can lead to practical and statistical complications in allelic identification and interpreting summary and population-level statistics [37]. Genetic stock identification studies on other Acipenserids using polysomic (tetrasomic) markers [38], [39] estimated gene dosages by relative peak intensity from electropherograms. Because of the large number of alleles (fragments) observed in the putatively hexaploid *A. brevirostrum*, allele dosage could not be reliably estimated from GENESCAN runs preventing the application of standard population genetic diversity statistics that require genotype or allele frequencies for their calculation. Two prior groups [39], [40] provided a validated approach to this dilemma and as in those studies, we scored each allele (fragment) as its own psuedodominant locus with one of two states, presence or absence, resulting in the production of a binary character (or allele) matrix. When codominant markers are screened in higher order polyploid species, and scored as psuedodominant loci (i.e., as binary character state), it is not possible to estimate either allele frequencies or heterozygosities directly. Allele (loci) frequencies at 11 polysomic microsatellite DNA markers were estimated on the *A. brevirostrum* allelotype matrix using the method of [41] as implemented by GenAlEx 6.3 [42]. Only alleles with an overall frequency of >1% were used for these and other statistical analyses.

Allelotypes were analyzed for patterns of population 'genetic' structure and for regions of genetic discontinuity at both the individual and collection levels. Pair-wise genetic distances of each individual fish to all other individuals (simple match coefficient; [43]) were calculated using the binary distance routine in GenAlEx. When calculated across multiple loci for a given pair of samples, this is equivalent to the tally of state differences between the two DNA allelotype profiles. Principal coordinate analyses (PCO) were used to graphically compare the individual distances without imposing the appearance of a bifurcating evolutionary history (ordinated with PAlaeontological STatistics ver. 1.76, PAST, [44]).

A second individual-based analysis designed to infer population structure among collections and identify genetic discontinuity in the allelotype matrix was performed using the model-based clustering method of the program STRUCTURE 2.3.2 [45]. Due to complex migration patterns assumed to exist among disjunct populations, a sequential method of inferring clusters (k) was used by first identifying the "uppermost" hierarchical level of population structure followed by subsequent analysis of each cluster to identify within-cluster structure [46]. In the initial phase, $k = 1$ to $k = 20$ clusters were considered for the 17 collections using a burn-in of 20,000 followed by 50,000 iterations, and 20 independent runs for each k. The optimum number of clusters in the initial phase was identified using Δk as described by Evanno et al. [46]. Subsequent analysis of each cluster tested $k = 1$ to $k = C+3$ (the number of collections (C) included in the subset plus three), with a burn-in of 20,000 followed by 50,000 iterations, and 20 runs for each k. In the within-cluster analyses, k also was determined using the Evanno et al. method [46].

Direct comparisons were made of two pair-wise population-scale distance measures, Jaccard's and Φ_{PT}, for assessing the underlying structure contained in the allelotype matrix. Jaccard's similarity coefficient [47] was calculated using PAST as it is one of the most commonly used and recommended [48] ecologically-based measures of similarity between populations. A non-parametric Analysis of Similarity (ANOSIM) [49] was performed on Jaccard's distance metric (1-Jaccard similarity) measured for 17 collections of *A. brevirostrum* using PAST. The test statistic, R (resulting from the distance values being converted to rank values) and its significance were computed by permutation of group membership, with 10,000 replicates. R values were presumed to be proportional to genetic distance. AMOVA Φ_{PT} calculated in GenAlEx using the binary data was considered as a coefficient of dissimilarity [43], [42]. Pair-wise Φ_{PT} values and significance levels of the variance components (H_0 = no genetic difference among populations; $\Phi_{PT} = 0$), based on 10,000 permutations, were measured for all collections. The relationship between the ecological-based (Jaccard's) distance and the population genetic distance (Φ_{PT}) matrices for all 17 collections was statistically assessed with a matrix regression analysis [50] performed by the MXCOMP routine in NTSYS-PC 1.8 [51].

Because a small number of effective migrants can have a profound effect on a small population, as the effects of drift can be large in the absence of balancing gene exchange, estimation of the number of migrants rather than simply the rate [52] is an important conservation consideration. We estimated the effective migrants per generation using Φ_{PT} for the amount of genetic differentiation at these selectively neutral microsatellite loci according to S. Wright's classical relationship [53]. Although this approximation of gene flow has been shown to be quite robust under a range of demographic conditions [54], [55] the empirical validity of substituting the binary character analog of F_{ST}, Φ_{PT}, into this equation is, to our knowledge, untested.

The evolutionary history among the 17 collections (geographic populations) of *A. brevirostrum* was inferred by analysis of the pair-wise population Φ_{PT} matrix using two methods. The non-metric multidimensional scaling (NMDS, [55]) option of PAST was used for visualization of the non-parametric monotonic relationship among the dissimilarity matrix, the Euclidean distance between collections, and the location of each collection in low-dimensional space. In addition, the Neighbor-Joining method [57], a clustering procedure based on the minimum-evolution criterion for phylogeographic trees (the topology resulting in the minimum total branch length at each step of the algorithm) was performed with MEGA4, [58]. The associated pair-wise Φ_{PT} distance matrix was subjected to clustering with 5000 bootstrap replicates using the program PAST.

To compare the genetic variation observed for nDNA (this study) with mtDNA [31] genomes of *A. brevirostrum*, two comparisons were made to visualize patterns among the 14 collections common to both studies. Ordination of the nDNA Φ_{PT} and mtDNA Φ_{ST} matrices was independently performed using the MDS option of PAST. A separate analysis of the relationship between the Φ_{ST} and Φ_{PT} pair-wise distance matrices was statistically assessed with the Mantel matrix regression analysis [50] performed by the MXCOMP routine in NTSYS-PC 1.8 [51].

Maximum likelihood assignment tests were used to determine the likelihood of each individual's multilocus allelotype being found in the collection from which it was sampled using the program AFLPOP 1.1 [59]. AFLPOP makes no assumption of

Hardy-Weinberg equilibrium and has been shown to be robust when applied to binary-coded polyploid allelotypes [40]. Allelotype patterns observed from other analyses were used to group collections into various management units and to test the appropriateness of these groupings using changes in assignment success. Constituencies of evolutionarily significant groupings of populations (e.g., regions, management units, distinct population segments) were ultimately investigated using hierarchical structuring of genetic variation measured (AMOVA) for numerous combinations of collections [60].

Results

Population structure of shortnose sturgeon

A total of 181 alleles were observed at the 11 loci resulting in a binary character matrix of 181 loci (columns) by 561 individuals (rows) (summarized in Table 1). The numbers of alleles with frequencies ≥1% observed at the 11 loci ranged from 55 (Cape Fear, $n = 3$) to 152 (Hudson, $n = 45$). Estimated heterozygosity (assuming Hardy-Weinberg equilibrium) ranged from 10.4% (Cape Fear) to 19.3% (Hudson). A lower estimate of heterozygosity was observed among southeastern populations compared to mid-Atlantic and northeastern populations.

At the individual level, the PCO scatter plot (Figure 2) indicated presence of three regional groups among the 17 surveyed river/estuarine systems. These regional groupings were: 1) *Northeast* - including five rivers from the Gulf of Maine (GOM), namely Saint John River, Canada; Penobscot, Kennebec, Androscoggin and Merrimack rivers; 2) *Mid-Atlantic* - including the Connecticut, Hudson, and Delaware rivers, and the Chesapeake Bay proper; and 3) *Southeast* – including the Cape Fear River, Winyah Bay, the Santee-Cooper, Edisto, Savannah, Ogeechee, Altamaha rivers, and Lake Marion. The number of inferred clusters (*k*) determined by STRUCTURE for the initial (uppermost hierarchical level) analysis was three, corresponding to the Northeast, Mid-Atlantic, and Southeast regional groupings identified by PCO (Figure 2). The PCO analysis clearly illustrated that the Southeastern cluster of individuals was the most divergent group of fish, separated from the other two regional groupings by a stronger zone of genetic discontinuity than the degree of separation between populations in the Northeast and Mid-Atlantic regions.

Due to the apparent complex migration patterns (zones of genetic discontinuity) existing among the three regions, a sequential method of inferring the number of clusters was employed to identify within-cluster structure (Figure 3; Evanno et al. 2005). Subsequent PCO analysis of the Northeast regional grouping of collections indicated a high degree of relatedness among the *A. brevirostrum* sampled from the Penobscot, Androscoggin, and Kennebec Rivers (Figure 3A), whereas fish sampled from the Saint John and Merrimack Rivers appeared well-differentiated from each other and from the other GOM rivers. The sequential STRUCTURE analysis of the Northeast cluster suggested panmixia among the *A. brevirostrum* sampled from the Penobscot, Androscoggin, and Kennebec Rivers and a moderate degree of differentiation between this group of collections and Saint John River to the north and Merrimack River to the south (Figure 3A). The level of differentiation of the Saint John and Merrimack Rivers from the other GOM collections did not appear as great as that seen among the mid-Atlantic collections or between the southeastern rivers and other collections (Figure 2).

PCO analysis of the Mid-Atlantic regional grouping indicated that *A. brevirostrum* sampled from Delaware River and Chesapeake Bay were genetically indistinguishable (mean and SE values

Table 1. Microsatellite allele (a.k.a. pseudodominant locus) counts, percentage of loci (alleles) polymorphic, number of private alleles, and number of common alleles across all loci for shortnose sturgeon (*Acipenser brevirostrum*) populations across the North American range.

Population	Saint John	Penobscot	Androscoggin	Kennebec	Merrimack	Connecticut	Hudson	Delaware	Chesapeake Bay	Cape Fear	Winyah Bay	Santee-Cooper	Lake Marion	Edisto	Savannah	Ogeechee	Altamaha
Sample size (n)	25	39	23	24	22	47	45	39	34	3	47	42	33	33	34	35	36
[1]No. Alleles (loci)	118	131	126	130	105	121	152	134	127	55	119	111	95	112	113	112	107
Polymorphism (%)	65.2	72.4	69.1	70.7	58.0	66.9	84.0	74.0	70.2	22.7	65.2	60.8	51.9	61.9	61.9	60.8	59.1
[2]No. Private Alleles	1	0	1	0	0	1	1	0	0	0	0	0	0	0	0	0	0
[3]No. Common Alleles (<=25%)	11	16	13	16	6	12	24	12	11	0	7	3	4	4	5	6	2

The analyses were conducted on the binary character matrix.
[1]Number of different fragments.
[2]Number of bands unique to a single population.
[3]Number of common alleles with frequency ≤25%.

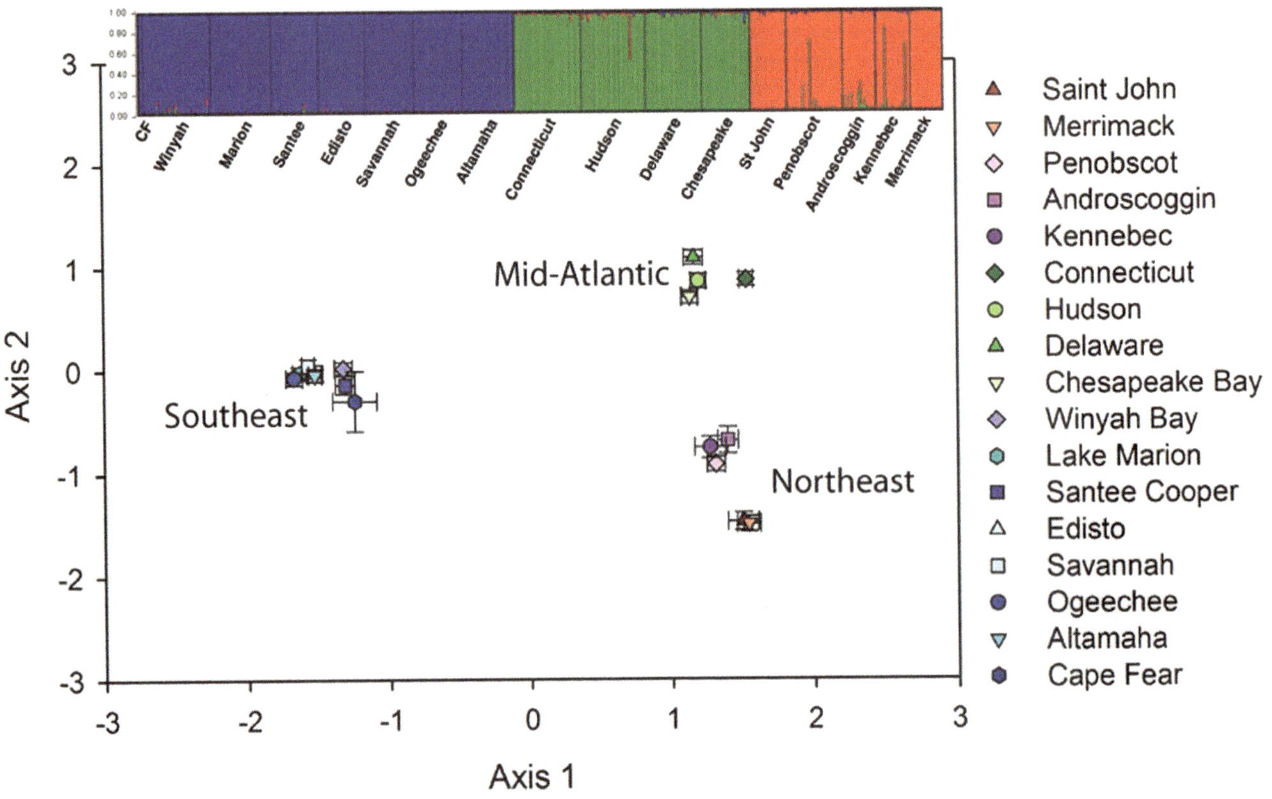

Figure 2. Combined graphical representation of principal coordinates (scatter plot) and STRUCTURE (histogram) analyses of 561 shortnose sturgeon (*Acipenser brevirostrum*) from 17 locations along the North American coast, surveyed at 11 polysomic microsatellite DNA loci. For the STRUCTURE histograms, each individual is represented by a single vertical bar, broken into *k* colored segments, the length of which is proportional to the membership fraction in each of the *k* clusters. Black lines partition the river samples.

were tightly clustered) and differentiated from Connecticut and Hudson River samples (Figure 3B). Likewise, the sequential STRUCTURE analysis also subdivided the mid-Atlantic cluster into three subclusters (Figure 3B). The two fish recently collected in Potomac River (during efforts to determine the existence of a reproducing population in this system) were genetically indistinguishable from other fish from Chesapeake Bay. This subsequent examination also revealed minimal overlap among the Connecticut River, Hudson River, and Delaware River/Chesapeake Bay collections; a pattern not readily discernible in the full analysis (Figure 2). Both PCO and sequential STRUCTURE analysis of *A. brevirostrum* from the Southeastern cluster of rivers suggested a low level of genetic differentiation among this group of populations, minimal structuring, and a moderate level of gene flow among collections throughout that region (Figure 3C).

At the population level, pair-wise Φ_{PT} values (Table 2, above diagonal) revealed that most (118/136; 87%) pair-wise comparisons among collections were statistically greater than zero ($P<0.0004$) indicating the presence of multiple populations displaying statistically significant differences in allele frequencies throughout the range of *A. brevirostrum*. Given the range and magnitude (0.040–0.249) of the Φ_{PT} values in question, the non-significant findings associated with this collection can likely be attributed to the inadequate sample size. The six remaining non-significant Φ_{PT} values were observed among geographically proximal collections within the Northeast and Southeast groupings (clusters). Pair-wise Φ_{PT} values were greatest among collections compared among the three major clusters and lowest among collections within these groupings. Low Φ_{PT} estimates were observed among collections in the Northeast (average 0.06) and Southeast (0.047) clusters. Moderately high Φ_{PT} estimates were observed among the four collections in the mid-Atlantic cluster (averaging 0.077) although the value between the Delaware River and Chesapeake Bay collections was 0.018. Considering the hierarchical AMOVA results (Table 3) in light of the individual-based PCO and STRCTURE analyses, seven distinct populations or groups of populations of *A. brevirostrum* were consistently supported: 1) St. John River in Canada, 2) three Maine rivers, 3) Merrimack River, 4) Connecticut River, 5) Hudson River, 6) Delaware River/ Chesapeake Bay; and 7) southeast (all collections between the Cape Fear and Altamaha Rivers).

Non-parametric analysis of similarity (ANOSIM) of Jaccard's (ecological) distance resulted in a similar pattern of genetic differentiation in the pairwise R values (Table S1; below diagonal) and probability values (above diagonal) as observed with Φ_{PT} values. This correspondence was further reflected in the Mantel matrix regression analysis of the relationship between Φ_{PT} and Jaccard's distance measures, which indicated a strong, nearly absolute, correlation between the two distance metrics ($r=0.98$, $P<0.0001$; Figure S1). As a result of the strength of this relationship, all subsequent distance-based analyses performed utilized the Φ_{PT} statistic because of its relationship to the standard diversity index used with codominant markers (F_{ST}).

Multivariate (NMDS) analysis of the underlying structure contained in the population-level pair-wise Φ_{PT} matrix (data not shown) revealed a congruent pattern among the 17 collections with that observed with the individual-based PCO and in the STRUCTURE analysis (Figure 2) resulting in three major

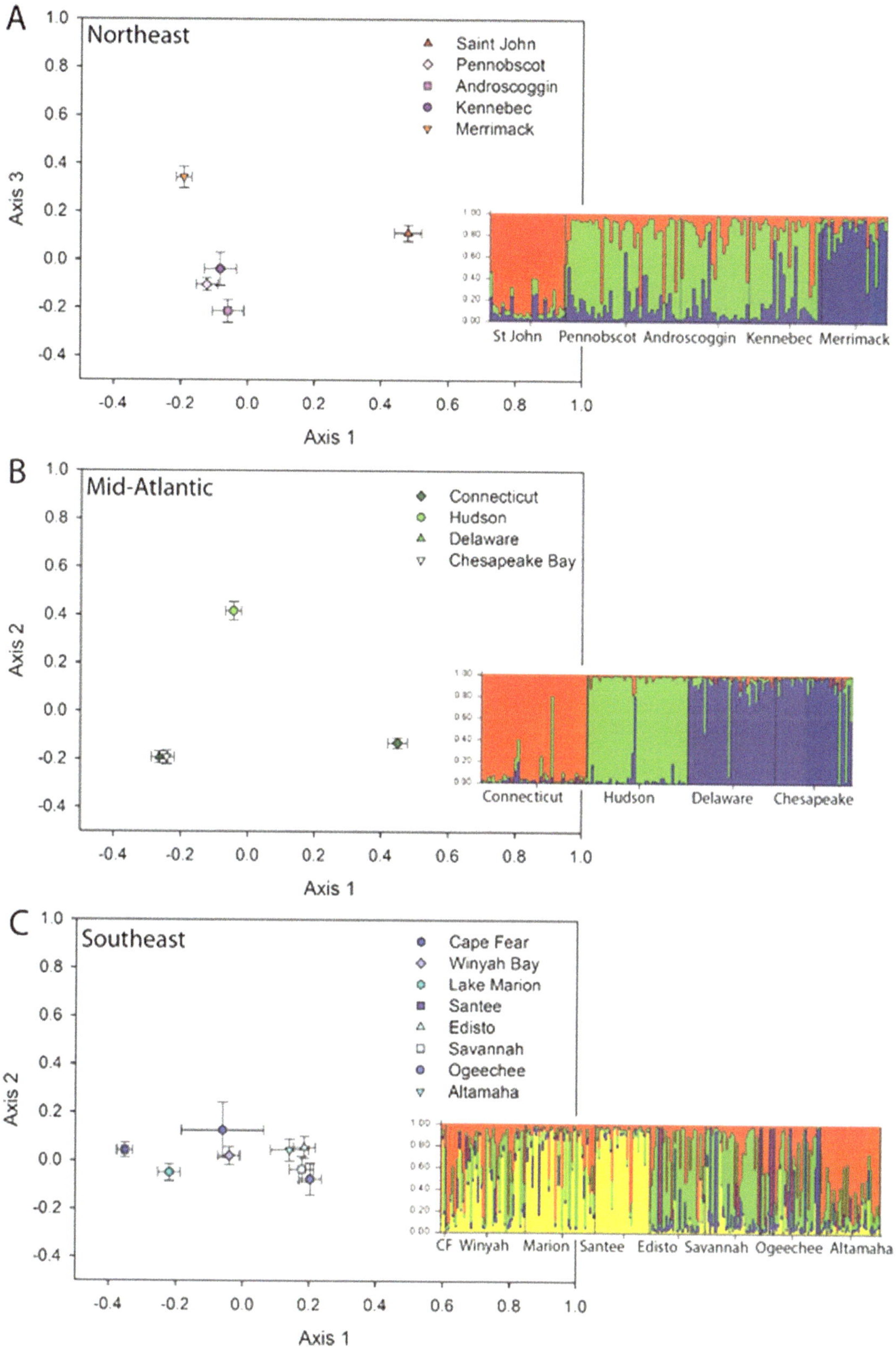

Figure 3. Combined graphical representation of sequential principal coordinates (scatter plots) and STRUCTURE (histograms) analyses of 561 shortnose sturgeon (*Acipenser brevirostrum*) surveyed at 11 polysomic microsatellite DNA loci. For the STRUCTURE histograms, each individual is represented by a single vertical bar, broken into *k* colored segments, the length of which is proportional to the membership fraction in each of the *k* clusters. Black lines partition the river samples. A) Northeast region collections; B) Mid-Atlantic region collections; C) Southeast region collections.

Table 2. Pair-wise Φ_{PT} among putative shortnose sturgeon populations (above diagonal) and estimates of the effective number of migrants per generation, $N_e m$ (below diagonal), for 17 collections of shortnose sturgeon (*Acipenser brevirostrum*) surveyed at 11 polysomic microsatellite loci.

Collection	Saint John	Penobscot	Androscoggin	Kennebec	Merrimack	Connecticut	Hudson	Delaware	Chesapeake Bay	Winyah Bay	Santee-Cooper	Lake Marion	Edisto	Savannah	Ogeechee	Altamaha
Saint John	-	0.068	0.077	0.068	0.100	0.191	0.162	0.175	0.155	0.253	0.289	0.269	0.269	0.278	0.280	0.277
Penobscot	3.43	-	***0.015***	***0.003***	0.065	0.116	0.094	0.107	0.095	0.189	0.219	0.207	0.189	0.201	0.205	0.203
Androscoggin	3.00	16.42	-	***0.013***	0.087	0.113	0.091	0.099	0.093	0.200	0.248	0.226	0.209	0.210	0.227	0.226
Kennebec	3.43	83.08	18.98	-	0.058	0.114	0.073	0.096	0.088	0.186	0.222	0.205	0.188	0.201	0.207	0.204
Merrimack	2.25	3.60	2.62	4.06	-	0.201	0.153	0.184	0.167	0.268	0.307	0.297	0.279	0.295	0.293	0.296
Connecticut	1.06	1.91	1.96	1.94	0.99	-	0.086	0.100	0.118	0.239	0.272	0.263	0.256	0.261	0.273	0.273
Hudson	1.29	2.41	2.50	3.17	1.38	2.66	-	0.067	0.075	0.179	0.217	0.208	0.188	0.196	0.210	0.201
Delaware	1.18	2.09	2.28	2.35	1.11	2.25	3.48	-	0.018	0.188	0.228	0.217	0.200	0.206	0.216	0.212
Chesapeake Bay	1.36	2.38	2.44	2.59	1.25	1.87	3.08	13.64	-	0.183	0.234	0.210	0.200	0.213	0.216	0.206
Winyah Bay	0.74	1.07	1.00	1.09	0.68	0.80	1.15	1.08	1.12	-	0.049	0.034	0.037	0.046	0.031	0.032
Santee-Cooper	0.62	0.89	0.76	0.88	0.56	0.67	0.90	0.85	0.82	4.85	-	0.044	0.043	0.043	0.046	0.069
Lake Marion	0.68	0.96	0.86	0.97	0.59	0.70	0.95	0.90	0.94	7.10	5.43	-	0.089	0.095	0.091	0.085
Edisto	0.68	1.07	0.95	1.08	0.65	0.73	1.08	1.00	1.00	6.51	5.56	2.56	-	***0.005***	***0.008***	0.020
Savannah	0.65	0.99	0.94	0.99	0.60	0.71	1.03	0.96	0.92	5.18	5.56	2.38	49.75	-	***0.007***	0.052
Ogeechee	0.64	0.97	0.85	0.96	0.60	0.67	0.94	0.91	0.91	7.81	5.18	2.50	31.00	35.46	-	0.020
Altamaha	0.65	0.98	0.86	0.98	0.59	0.67	0.99	0.93	0.96	7.56	3.37	2.69	12.25	4.56	12.25	-

Non-significant pair-wise Φ_{PT} probability values (H_0 = No genetic difference among populations; $\Phi_{PT} = 0$) based on 10,000 permutations values are in bold italics. Cape Fear River sample is not included due to inadequate sample size.

Table 3. Hierarchical AMOVA results for biogeographically relevant groups among 17 collections of shortnose sturgeon (*Acipenser brevirostrum*) surveyed at 11 polysomic microsatellite DNA markers.

	3-regions		5-regions		7-regions	
	Φ-statistic	% variance	Φ-statistic	% variance	Φ-statistic	% variance
Among regions (Φ_{RT})	0.158	16	0.164	16	0.170	17
Among populations within regions (Φ_{PR})	0.057	5	0.042	4	0.031	3
Within populations (Φ_{PT})	0.206	79	0.199	80	0.196	80

groupings of populations (Northeast, Mid-Atlantic, and Southeast) with varying degrees of clustering. The collections representing the southeastern populations and those from the GOM appeared tightly clustered (genetically similar). The Mid-Atlantic collections appeared to form three discrete groupings with the Delaware River and Chesapeake Bay collections the most closely related and more so than either was to the Hudson or Connecticut River collections. The Merrimack and Saint John River collections appeared moderately differentiated from the Maine collections.

The underlying genetic structure of the Φ_{PT} matrix also was depicted with an unrooted neighbor-joining (N-J) tree (Figure 4), which illustrated high levels of differentiation among *A. brevirostrum* collections that mirrored those identified by the sequential PCO and STRUCTURE analyses. The deep level of differentiation (genetic discontinuity) among the three major groupings of collections was strongly supported in the backbone of the phenogram and the evolutionary distinctiveness of the populations or groups of populations was confirmed by high bootstrap support, in particular, the absolute (100%) bootstrap support distinguishing the southeast clade from all other collections, and the high bootstrap support for clades containing the Northeast (98%) and Mid-Atlantic (89%) collections. The high degree of genetic similarity observed with other analyses among the Penobscot, Androscoggin and Kennebec Rivers collections and between the Delaware River and Chesapeake Bay collections (99% bootstrap support) was confirmed with high bootstrap support for these pairings.

The effective number of *A. brevirostrum* migrants per generation, N_em (based on Φ_{PT}), estimated between all pairs of collections (Table 2; below diagonal) was consistent with the patterns of differentiation observed with the individual- and population-based analyses (e.g., PCO, STRUCTURE, NMDS, and N-J). Clear zones of genetic discontinuity were evident as N_em among the three major regions was generally low, ranging from an average of 1.89 migrants between the Northeast and Mid-Atlantic collections to 0.89 between the Northeast and Southeast. The average N_em between the Mid-Atlantic and Southeast populations was 0.95. Estimates of N_em among collections within the three major regions were considerably higher ranging from 2.25–83.08 among the Northeast collections and 1.87–13.64 among the Mid-Atlantic collections. The range of N_em values among collections within the Southeast was 2.38–>100.

Quantitative estimates of hierarchical gene diversity among all collections also identified statistically significant genetic structuring; 16% ($P<0.001$) of the genetic variation occurred among collections (river/estuary populations) and 84% ($P<0.001$) was attributed to differentiation within collections. Of 11 hierarchical AMOVA analyses (Table S2), four models provided optimal delineation of genetic differentiation although the increase in variance among groupings over other models was minimal (1%). These four models all resulted in 17% ($P<0.001$) of the genetic variation occurring among groupings, 3% ($P<0.001$) occurring among populations within groupings, and 80% of the genetic variation was due to variation within collections. All four models, based on the subclusters identified by the PCO and STRUCTURE analyses, were variations either including or omitting the Saint John and Merrimack Rivers as distinct populations. All attempts to manipulate the southeast grouping of populations resulted in a decrease in variation among grouping components and an increase in the amount of within population variation.

The average correct assignment to collection of origin was 58.6% and ranged from 0% (Cape Fear River) to 97.8% (Connecticut River) (Table S3). With the exception of the Cape Fear River collection ($n = 3$), assignment to each collection was statistically greater than would be expected by chance ($P<0.05$). When the 17 rivers were pooled by the geographic regions identified in Figure 3, correct assignment to major grouping averaged 99.8% (Table 4). When pooled into two Northeast population groupings (i.e., Saint John separate from all other GOM collections), assignment to the GOM grouping was 100% (data not shown). A five-group model, (where the Northeast region included all GOM collections except the Saint John River) resulted in correct assignment to regional grouping in 99.1% of comparisons (Table S4).

Comparison of the patterns nDNA and mtDNA

Mantel analysis comparing the pair-wise Φ_{PT} (this study) and Φ_{ST} (from Table 5 of Wirgin et al. [31]) distance matrices for 14 shared collections of Atlantic coast collections of *A. brevirostrum*, identified a strong statistical relationship (correlation coefficient $r = 0.84$, $P<0.0001$; Figure S2) between the variation detected in the two genomes. Furthermore, nDNA and mtDNA data yielded concurrent depictions of the presence of three major groupings representing the northeastern, mid-Atlantic, and southeastern populations (Figure 5). Moreover, similar levels and patterns of genome differentiation were observed among the mid-Atlantic groups: Connecticut River, Hudson River, and Delaware River/Chesapeake Bay. The respective scatter plots also suggest the presence of at least three regional metapopulations: Northeast (i.e., Penobscot, Kennebec, and Androscoggin Rivers), Mid-Atlantic (Delaware River and Chesapeake Bay), and Southeast (Altamaha, Winyah Bay, Santee-Cooper, Edisto, Savannah, Ogeechee, and Cape Fear Rivers, and Lake Marion).

Discussion

Phylogeography

Allelotypes. This study represents the first report of nuclear DNA variation in the higher-order polyploid *Acipenser brevirostrum*. The markers uncovered sufficient diversity that each of the 561 individuals surveyed possessed a unique multilocus allelotype. Alleles detected at these loci, scored as binary allelotypes, allowed

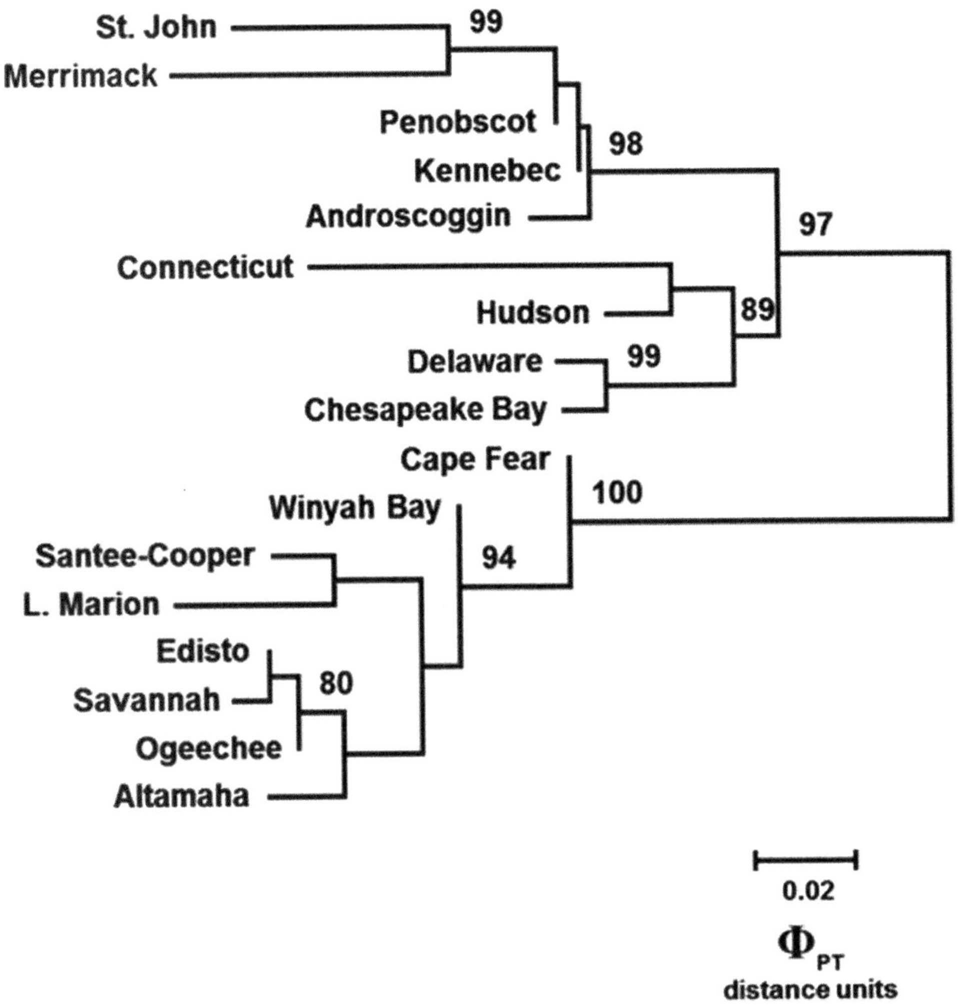

Figure 4. The evolutionary history among 17 collections of shortnose sturgeon (*Acipenser brevirostrum*) surveyed at 11 polysomic microsatellite DNA loci inferred from the pair-wise Φ_{PT} distance matrix using the Neighbor-Joining method [56]. Phylogenetic analyses were conducted in MEGA4 [57].

robust qualitative and quantitative assessment of population structuring, similar to that conducted on disomically-inherited codominant markers. This study applied multivariate ordination and Bayesian clustering, assessed statistical significance (i.e., value>zero) of pair-wise distance metrics, performed matrix regression analyses, estimated the evolutionary history among populations via phylogenetic algorithm (phenogram), conducted hierarchical partitioning of variation (AMOVA), and assessed the success of likelihood assignment testing. Notably, the results of all statistical approaches used to assess phylogeographic structure resulted in consistent findings.

Identifying units of management. Owing to their morphological similarity to lower Jurassic (~200 MYBP) Acipenseriformes [61], *Acipenser* species such as *A. brevirostrum* are considered to be evolutionarily static and are often referred to as phylogenetic relics. Intraspecific examination of the nuclear

Table 4. Assignment to three groupings of origin model consisting of 17 collections of shortnose sturgeon (*Acipenser brevirostrum*) surveyed at 11 polysomic microsatellite DNA markers.

allocated to	Northeast	Mid-Atlantic	Southeast
Northeast	132	0	0
Mid-Atlantic	1	165	0
Southeast	0	0	254
Assignment %	99.1	100	100

Mis-assigned individuals are distributed vertically.

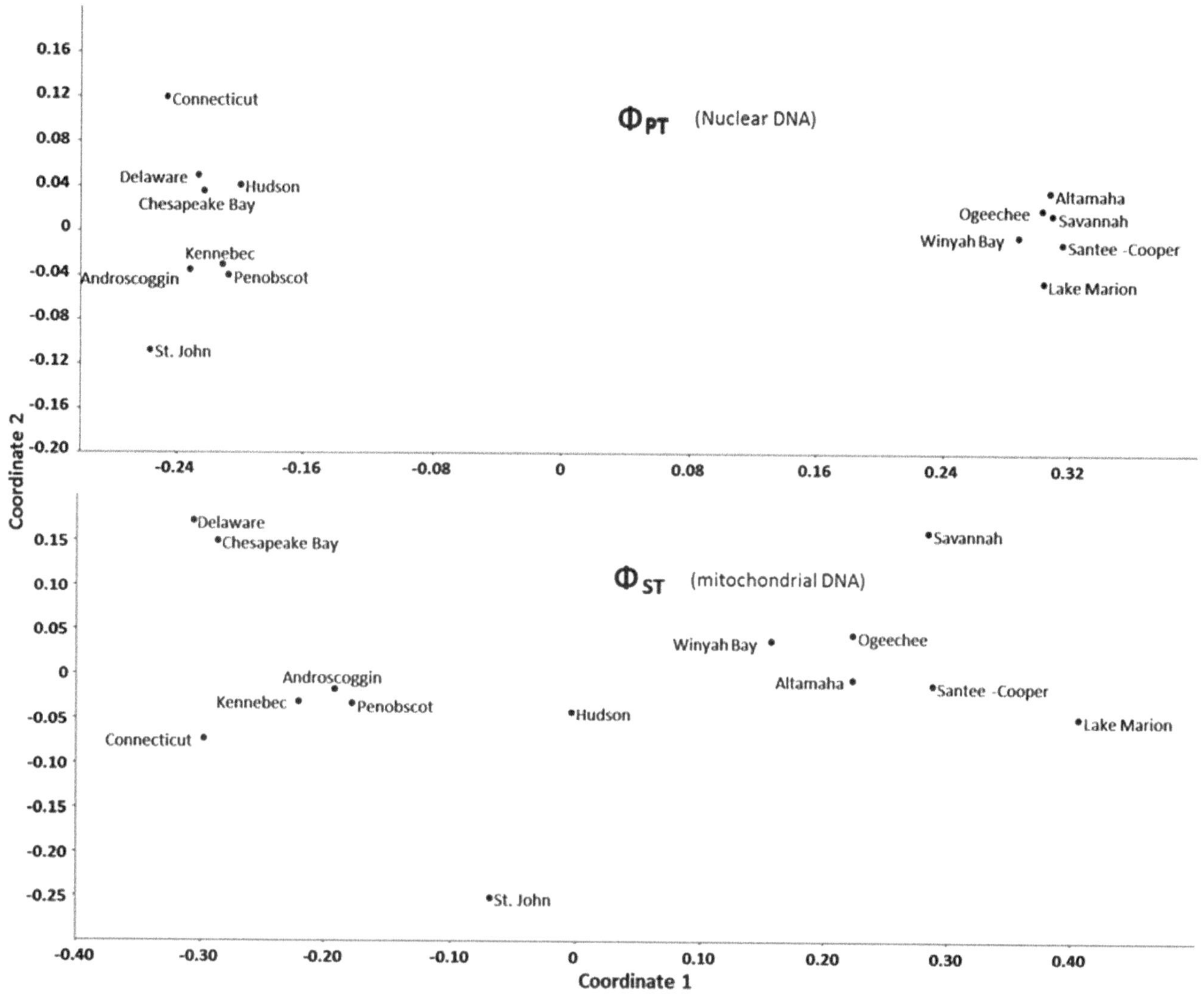

Figure 5. Results of independent multidimensional scaling analyses of pair-wise a) Φ_{PT} (nuclear DNA; Table 4, this study) and b) Φ_{ST} (mitochondrial DNA; Table 5 of Wirgin et al. [30]) matrices for 14 *Acipenser brevirostrum* collections that are in common between the two studies.

genome has revealed the presence of considerable allelic diversity and differentiation that appears to be reflective of the actions of various evolutionary processes. Phylogeographically, these findings suggest the presence of similar levels of genetic diversity and variation among the collections punctuated with a series of genetic discontinuities of varying "depth" across the species' range that could indicate demographic independence, regional adaptation, and reflect vicariant geographic events. Populations sampled within these regional groupings exhibited shallow but statistically significant differentiation, which were congruent with theoretical estimates of gene flow. Moreover, patterns of population relatedness were consistent with the observations of [23] that populations at both ends of the species' range are more dispersive than those in the middle. A possible explanation for elevated geneflow within the northern and southern collections could be the greater geographic proximity of rivers in these areas relative to those in the mid-Atlantic region leading to higher levels of straying.

Upon inspection of the patterns of allelotypic variation at the individual and population scales in the nDNA two major ("deep") zones of genetic discontinuity are inferred from all analyses conducted 1) separating the Northeastern and Mid-Atlantic collections and 2) delineating the Mid-Atlantic and Southeastern populations. Moreover, narrower ("shallow") zones of genetic discontinuity were evident among Saint John, Merrimack and the three Maine rivers in the Northeast region, and between the Connecticut and Hudson River collections and between Hudson River and an apparent Delaware River/Chesapeake Bay meta-population (unconfirmed) within the mid-Atlantic region. This implies there are seven demographically and evolutionary distinct lineages across the range and that within the United States portion of the *A. brevirostrum* range, six lineages are relevant in conservation considerations. In addition to support for recognition of these zones of discontinuity in clustering, the phylogeographic analysis, AMOVAs, and assignment testing, all suggested there are low levels of theoretical gene exchange between collections on either side of these genetic discontinuities.

The presence of demographically distinct and evolutionary significant lineages delineated by zones of genetic discontinuity is consistent with the findings of researchers assessing behavioral patterns in *A. brevirostrum*. Parker [9] and Parker and Kynard (unpublished data) found that under common garden conditions, *A. brevirostrum* were locally adapted to particular rivers. These researchers demonstrated differences in the innate dispersal patterns in early life stages of *A. brevirostrum* from Connecticut River versus sturgeon of Savannah River origin, suggesting that *A. brevirostrum* are likely behaviorally adapted to unique features of their watershed. Parker and Kynard [62] also found that *A. brevirostrum* from different rivers can have different migration strategies. Similar adaptive differences have been inferred for other sturgeon species including *Acipenser fulvescens* (Wolf and Menominee rivers [Kynard and Parker unpublished data]), *A. transmontanus* (Sacramento and Kootenai rivers [63]), and *A. oxyrinchus oxyrinchus*/*A. o. desotoi* (Hudson and Suwannee rivers [24]).

Inter-genomic comparison. Comparison of nuclear genetic diversity observed in this study with that identified previously in mtDNA sequence variation ([31] and references therein) illustrates that the patterns of variation in the two *A. brevirostrum* genomes were qualitatively consistent. Similarly, the observation from the NMDS scatter plots in the two studies (Figure 5) that pair-wise Φ_{ST} values on average were approximately twice the pair-wise Φ_{PT} values for the same suite of collections is consistent with hyper-polymorphism associated with microsatellite evolution ([64], [65]) and the expectation of allele size homoplasy associated with polyploidy and an artifact of scoring alleles as binary types. Ultimately, this is the quantitative 'penalty' realized because allelotypic diversity is likely to be an underestimation of the actual differentiation that exists among populations; particularly for those that have experienced extended reproductive isolation. Although quantitative variation and molecular variation are correlated, adaptive population structuring often far exceeds neutral population structuring even for populations diverging over contemporary time ([66], [67]). Therefore, the estimates of allelic differentiation detected at neutral loci in this study are likely an underestimation of the actual divergence. Lastly, a component of the greater mtDNA haplotype differentiation relative to nuclear DNA differentiation could reflect gender-mediated gene flow between adjacent populations and resulting in reduced philopatry (i.e., sex-biased dispersal) of males throughout the range.

The phylogeographic and genomically-congruent patterns predicting genetic structuring of *A. brevirostrum* have profound management implications, foremost being the strong indication of regional structure. Specific cases within each region provide evidence that the fundamental units of management generally are the populations located in rivers and estuaries. However, because interpretations of the delineation of those distinct population segments differ depending upon the genome under investigation (7–9 based on mtDNA versus 5 if based on nDNA), we are faced with the need to specify evolutionarily relevant thresholds for deciding statistical significance of haplotype or allelotype frequencies. Ultimately, the approach that may offer maximum flexibility in reaching recovery goals may be to allocate (or nest) biological units that are most likely to respond to management within regions. The following two sections examine specific cases and the suggestion that a metapopulation approach may serve best to illustrate the structure of gene flow in this species.

Cases in point

Saint John's River. Differentiation among the Northeast populations is on the whole less than observed for the Mid-Atlantic populations but greater than among the Southeast populations. Within the Northeast group, the Saint John River, Canada collection constitutes a population that is appreciably differentiated from the GOM collections at a number of levels. Foremost, the degree of gene differentiation between Saint John River and the GOM is "shallow". This "shallowness" of may be attributed to the fact that these populations are relatively young (due to recent deglaciation of the region) as compared to more southerly distributed populations based on the observed levels of divergence. In addition, designation as a distinct management unit may be warranted because 1) the Saint John River *A. brevirostrum* population's age-to-reproduction is different than other GOM sturgeon, 2) it is the northernmost reproducing population, and 3) it experiences differences in control of exploitation, management of habitat, and conservation status.

Gulf of Maine. The GOM collections analyzed in this study included a recently obtained sampling from the Merrimack River. These *A. brevirostrum* exhibited patterns of nDNA variation that suggest the collection is genetically differentiated from the other GOM collections. Such an interpretation should be made with caution because the sample consisted of 22 males collected at the same location and time, and because the level of differentiation is not as great as that observed among other collections proposed as distinct evolutionary lineages. Choosing to consider this collection as a distinct evolutionary lineage would serve the precautionary principle and conserve biodiversity.

Potomac River. Recent captures of adult *A. brevirostrum* in the Potomac ([68], [69]) and Merrimack Rivers (this study) have raised questions concerning the status of this endangered species and available habitat in these river systems. Fishery managers require empirically supported prioritization schemes to protect existing diversity. It is important, therefore to determine if these observations are indicative of a discrete natal population or simply the result of migratory foraging behavior of fish from nearby populations. Analysis of mtDNA sequences from four Potomac River *A. brevirostrum* identified haplotypes also found in fish from the Chesapeake Bay and Delaware River [28]. This suggested the fish (or their mothers) were genetically related to other collections made in the Chesapeake Bay proper and Delaware River. Grunwald et al. 2002 suggested the fish might have been migrants from the Chesapeake and Delaware Canal [28]. Individual-based analyses of the two Potomac River *A. brevirostrum* sampled for the present study strongly suggested the fish are genetically part of a Delaware River/Chesapeake Bay metapopulation (Figure 3B). It will require significant additional resources to determine whether a reproducing population of *A. brevirostrum* exists within the Chesapeake Bay proper or whether the collections made there simply represent fish on foraging forays from the Delaware River population. Without recolonization and reestablishment of a reproducing population somewhere among the Chesapeake Bay rivers, a large gulf will continue to exist among northern and southern *A. brevirostrum* populations.

Savannah River supplemental stocking program. Early resource management efforts for *A. brevirostrum* included the release of hatchery-reared juveniles to supplement Savannah River populations [70]. Although conducted as an experimental program from 1984–1992, approximately 97,000 juveniles were introduced. This management effort has potentially altered the population genetic structure of the targeted river system as well as the southeast metapopulation. Smith et al. [70] estimated that 39% of the Savannah River's 'breeding' adults were of hatchery origin. It has been suggested that because *A. brevirostrum* were released as larger juveniles during this supplemental stocking program to allow marking, the fish were not imprinted to the natal stream and may have effectively strayed to adjacent river systems, i.e., traveled to non-natal rivers for reproduction, thereby having a homogenizing effect on impacted river populations [29]. However, it is unknown if effective introgression of hatchery-origin fish into the wild populations of any southeast river has occurred. No strong signal supporting or refuting a discernible impact on the southeast metapopulation by the Savannah River supplemental stocking is apparent from this study. Due to the geographic proximity of the Edisto and Ogeechee rivers to the Savannah River, the degree of evolutionary similarity among these rivers (e.g., in assignment testing) could be the result of natural straying or a homogenizing effect of the supplemental stocking program.

Metapopulations

The current paradigm is that *A. brevirostrum* stray less often than the congeneric *A. oxyrinchus* [71]. However, at least two additional groupings involving adjacent river and estuarine systems have been identified in this study that exhibit strong signatures of functioning metapopulations (i.e., Maine rivers and Delaware River/Chesapeake Bay). Moreover, biologists tagging and tracking fish in Connecticut River recaptured two *A. brevirostrum* that originally were tagged and released in Hudson River [72]. Although this latter finding does not confirm effective movement (i.e., gene flow) between these river systems, evidence suggests some degree of straying and recolonization from adjacent southeastern rivers is possible. Indeed, if the strong signal of metapopulation structure observed among the southeastern rivers in this study resulted from the straying of stocked Savannah River *A. brevirostrum*, the effective population sizes of the affected rivers are indeed minuscule and the supplementation program must have been one of the most successful such efforts on record.

In addition to the demographically discrete and evolutionarily significant lineages identified for *A. brevirostrum* within the U.S., three metapopulations and other individual river populations delineated within each discrete lineage may be considered distinct management/recovery units for future recovery planning purposes. The three possible metapopulations are the: 1) Maine rivers (i.e., Penobscot, Kennebec, and Androscoggin rivers), 2) Delaware River and Chesapeake Bay proper, and 3) the Southeast assemblage (Cape Fear River, Winyah Bay, Santee-Cooper, Edisto, Savannah, Ogeechee, and Altamaha rivers, and Lake Marion). Population biology theory predicts that lower dispersal and associated gene flow leads to decreased genetic diversity in small isolated populations, which generates adverse consequences for fitness, and subsequently for demographic stability. Given recent tagging data suggesting *A. brevirostrum* migrate to adjacent rivers to a greater extent than previously believed [70], [71], [73], [74], [75], concomitant with the identification of at least three metapopulations within the species' range, this interpretation appears to bode well for the demographic fitness of some southern *A. brevirostrum* populations.

Conservation Implications

This study presents evidence that sufficient levels of genetic diversity are present in the *A. brevirostrum* nuclear genome to discriminate evolutionarily significant lineages of management relevance. The variation detected was highly phylogeographically congruent with the inferences based on the mtDNA control region ([31] and references therein). Moreover, these nDNA analyses detected statistically significant differences in allelotype frequencies among most collections. Four regional zones of genetic discontinuity were inferred from the patterns of genetic variation across the range of *A. brevirostrum* (two shallow zones in the Northeast and two deep zones dividing three regions) that likely delineate seven demographically discrete and evolutionarily significant lineages, each with differing adaptive potential for this species. These zones of inferred genetic discontinuity represented deeper levels of differentiation and a higher degree of reproductive isolation than that typically attributed to population-level differentiation; groups of populations of evolutionary significance which may warrant distinct conservation considerations. Perhaps more notable than the delineation of evolutionarily significant lineages, was the identification of at least one putative metapopulation within each of the three major regional groupings of populations, a finding that is encouraging as migration may help stave localized extinctions. It should be noted that while an increased level of gene flow is present within these putative metapopulations, a demographic connection has yet to be established and documentation of such a link should be considered a high priority research need. Moreover, many of the populations within an evolutionarily significant lineage were genetically differentiated to some degree and each geographic population is subjected to differing threats [75]. Based on patterns observed from these multilocus allelotypes, the basic unit for management and conservation (for recovery planning) of *A. brevirostrum* is arguably the individual (local) population.

Assuming conservation of local populations within a metapopulation is the acceptable focus, restoration of effective connectivity among currently fragmented *A. brevirostrum* populations should be a prominent recovery goal; the allelotypic patterns observed in this study provide guidance by facilitating understanding of where and how such efforts could be attempted. For example, the best available information suggests that individual rivers and estuaries along the North Carolina coast do not currently support reproducing populations of *A. brevirostrum*. If the distance to North Carolina (or elsewhere) rivers that could support a reproducing population exceeds the vagility of sturgeon inhabiting the southeast or Delaware River/Chesapeake Bay metapopulations, targeted translocations or restorative supplementation may represent plausible restoration strategies. In contrast, rivers geographically south of the Altamaha River historically occupied by shortnose sturgeon (i.e., Satilla, St. Marys, and St. John Rivers), no longer support reproducing *A. brevirostrum* populations. The shoreline distances of these rivers to the Altamaha River are similar to that observed among the major rivers comprising the Southeast metapopulation. Given the recent findings of moderate physical migration [74] and our implications for effective migration (gene flow) among *A. brevirostrum* inhabiting many of the rivers in the Southeast metapopulation, it seems logical that if one or more of these southernmost rivers provided suitable spawning and rearing habitat, *A. brevirostrum* would have effectively colonized this region. Therefore, habitat characterization and/or restoration in these southernmost rivers could facilitate range expansion.

Gene diversity estimates for *A. brevirostrum* have been shown to be moderately high in both nuclear (this study) and mitochondrial ([29], [30], [31]) genomes. Although rates of genetic diversity loss in polyploids versus diploids (functional sensu) has not been characterized for sturgeon, the nDNA and mtDNA studies performed to date suggest that dispersal is a very important factor maintaining genetic diversity in shortnose sturgeon. The gene diversity estimates may be indicative of larger effective population sizes than previously assumed. However, even at a very local spatial scale in a metapopulation consisting of moderate-density populations interconnected by considerable dispersal rates, genetic diversity can erode and directly affect the fitness of individuals. Gene flow estimates do not capture the intra-specific variation in individual behavior related to vagility, which is strongly affected by habitat fragmentation and population/metapopulation history. From a biodiversity conservation perspective, future success in *A.*

brevirostrum management could benefit from both in-depth demographic and genetic analyses. It should be considered a high-priority research need to better delineate population structure within the evolutionarily significant lineage framework suggested by these data.

Supporting Information

Figure S1 Scatter plot illustrating the significant correlation ($r = 0.98$; $P < 0.0001$; Mantel analysis) between Jaccard and Φ_{PT} pair-wise distances for 17 collections of shortnose sturgeon (*Acipenser brevirostrum*) surveyed at 11 polysomic microsatellite DNA loci.

Figure S2 Scatter plot depicting the Mantel matrix regression analysis comparing the mtDNA Φ_{ST} matrix for 14 Atlantic coast collections of shortnose sturgeon (*Acipenser brevirostrum*) (Wirgin et al. [30]) and the nuclear DNA Φ_{PT} pair-wise distance matrix (this study) for the same collections surveyed at 11 polysomic microsatellite DNA loci. The correlation coefficient (r) for this analysis was 0.84 ($P < 0.0001$).

Table S1 Pair-wise R values (below diagonal) and Bonferroni-corrected probability values (above diagonal) from the non-parametric Analysis of Similarity (ANOSIM) (Clark 1993) on Jaccard's distance metric (1-Jaccard similarity) measured among 17 collections of shortnose sturgeon (*Acipenser brevirostrum*) surveyed at 11 polysomic microsatellite loci.

Table S2 Hierarchical structuring of genetic variation was measured for numerous combinations of shortnose sturgeon (*Acipenser brevirostrum*) collections using analysis of molecular variance (AMOVA). Significance levels of the variance components were based on 1000 permutations. Abbreviations are as follows: NE = Northeast regional grouping includes Saint John River (SJ), Canada, Penobscot, Kennebec, Androscoggin and Merrimack rivers; Mid-Atlantic regional grouping includes the Connecticut (CT), Hudson (H), and Delaware (DE) rivers, and the Chesapeake Bay proper (CB); and the 3) SE = Southeast regional grouping includes the Cape Fear River (CF), Winyah Bay (WB), Santee-Cooper (S-C), Edisto (E), Savannah (S), Ogeechee (O), and Altamaha (ALT) rivers, and Lake Marion (LM).

Table S3 Assignment to collection of origin for 17 shortnose sturgeon (*Acipenser brevirostrum*) collections surveyed at 11 polysomic microsatellite DNA markers. Mis-assigned individuals are distributed horizontally.

Table S4 Assignment to proposed grouping (five groupings model) in shortnose sturgeon (*Acipenser brevirostrum*) surveyed at 11 polysomic microsatellite DNA markers. The overall correct assignment rate to proposed grouping was 99.1% (522/527). Mis-assigned individuals are distributed vertically. Northeast regional grouping includes Saint John River (SJ), Canada, Penobscot, Kennebec, Androscoggin and Merrimack rivers; and the Southeast regional grouping includes the Cape Fear River (CF), Winyah Bay (WB), Santee-Cooper (S-C), Edisto (E), Savannah (S), Ogeechee (O), and Altamaha (ALT) rivers, and Lake Marion (LM).

Acknowledgments

The authors wish to acknowledge the efforts of Diane Pavek (NPS-NCR) for being the driving force behind this research. We also acknowledge the NOAA-Fisheries funded Shortnose Sturgeon Status Review Team for their contributions to interpretations presented in this report, namely: Jeanette Bowers-Altman, New Jersey Division of Fish and Wildlife; Mark Collins, South Carolina Dept. of Natural Resources; Joel Fleming, Georgia Department of Natural Resources; Kathryn Hattala, New York Department of Environmental Conservation; Wilson Laney, USFWS – South Atlantic Fisheries Coordination Office, Raleigh, NC; Malcolm Mohead, NOAA Fisheries, Office of Protected Resources; Tom Squiers, Maine Department of Marine Resources; Dana Hartley, NOAA – Fisheries, NERO, Gloucester, MA; and Stephania Bolden, NOAA – Fisheries, SERO, St. Petersburg, FL. All tissue sampling followed the guidelines mandated by the U.S. Department of Commerce, National Oceanic and Atmospheric Administration, National Marine Fisheries Service under NOAA Technical Memorandum NMFS-OPR-18 or NOAA Technical Memorandum NMFS-OPR-45. Ike Wirgin, New York University College of Medicine, provided Connecticut, Androscoggin, Delaware, Hudson, and Kennebec rivers tissue and DNA samples. Douglas Peterson and Robert De Vries, University of Georgia, provided the Altamaha River tissue samples. NOS-NOAA archives provided Cooper, Edisto, Savannah, Santee-Cooper, Ogeechee, and Waccamaw rivers tissue samples. Morris Lee and Steve Fernandez, University of Maine contributed Penobscot River tissue samples. The U.S. Fish and Wildlife Service's Bear's Bluff National Fish Hatchery provided tissue samples of the parents and 30 progeny from each of seven captive-bred *A. brevirostrum* families, samples that were instrumental in determining the inheritance of the polysomic markers. The authors are grateful to Katharine Coykendall, USGS-Leetown Science Center for an IPDS review and Miranda McManus and Allan Strand, College of Charleston, for constructive suggestions on statistical analyses. The senior author is deeply indebted to reviewers, Mark Bain (deceased), Matthew Litvak, David Secor, and John Waldman for their helpful comments on an early version of this report in the form of the genetics section of the NOAA-Fisheries Shortnose Sturgeon Status Review Report. Mike Eackles (USGS-Leetown Science Center) kindly provided laboratory assistance for this project. Use of trade, product, or firm names does not imply endorsement by the U.S. Government.

Author Contributions

Conceived and designed the experiments: TLK. Performed the experiments: TLK APH. Analyzed the data: TLK AWA BLB. Contributed reagents/materials/analysis tools: BEK MCK DLP. Wrote the paper: TLK MCK AWA BLB.

References

1. Schultz RJ (1980) Role of polypoidy in the evolution of fishes. In: Lewis WH, editor. Polyploidy: Biological Relevance. New York, Plenum Press. pp. 313–340.
2. Gardiner BG (1984) Sturgeons as Living Fossils. In: Eldredge N, Stanley SM, editors. Living Fossils. New York: Springer-Verlag. pp. 148–152.
3. Bemis WE, Kynard B (1997) Sturgeon rivers: an introduction to acipenseriform biogeography and life history. Environ Biol Fishes 48: 167–183.
4. Birstein VJ, Hanner R, DeSalle R (1997) Phylogeny of the Acipenseriformes: cytogenetic and molecular approaches. Environ Biol Fishes 48: 127–155.
5. Krieger J, Fuerst PA (2002) Evidence for a slowed rate of molecular evolution in the order Acipenseriformes. Mol Biol Evol 19: 891–897.
6. Robles F, de la Herrán R, Ludwig A, Ruiz R, Ruiz R, et al. (2004) Evolution of ancient satellite DNAs in sturgeon genomes. Gene 338: 133–142.
7. Ludwig A (2006) A sturgeon view on conservation genetics. Eur J Wildl Res 52: 3–8.
8. Kynard B, Parker E (2005) Ontogenetic behavior and dispersal of Sacramento River white sturgeon, *Acipenser transmontatus*, with a note on body color. Environ Biol Fishes 71: 145–146.
9. Parker EL (2007) Ontogeny and life history of shortnose sturgeon (*Acipenser brevirostrum* Lesueur 1818): Effects of latitudinal variation and water temperature. PhD dissertation. Univ. of Massachusetts Amherst.

10. Birstein VJ, DeSalle R (1998) Molecular phylogeny of Acipenserinae. Mol Phylogenet Evol 9: 141–155.
11. Zhang S, Zhang Y, Zheng X, Chen Y, Deng H, et al. (2000) Molecular phylogenetic systematics of twelve species of Acipenseriformes based on mtDNA ND4L-ND4 gene sequence analysis. Sci China C Life Sci 43:129–137.
12. Birstein VJ, Doukakis P, DeSalle R (2002) Molecular phylogeny of Acipenseridae: nonmonophyly of Scaphirhynchidae. Copeia 2: 287–301.
13. Krieger J, Hett AK, Fuerst PA, Artyukhin E, Ludwig A (2008) The molecular phylogeny of the order Acipenseriformes revisited. J Appl Ichthyol 24: 36–45.
14. Lanfredi M, Congiu L, Garrido-Ramos MA, de la Herrán R, Leis M, et al. (2001) Chromosomal location and evolution of satellite DNA family in seven sturgeon species. Chromosome Res 9: 47–52.
15. Lazzari B, Mariani V, Malinverni R, Caprera A, Giuffra E (2008) A comparative gene index for the white sturgeon *Acipenser transmontanus*. Mar Genomics 1:15–21.
16. Ohno S, Wolf U, Atkin NB (1968) Evolution from fish to mammals by gene duplication. Hereditas 59:169–187.
17. Dingerkus G, Howell WM (1976) Karyotypic analysis and evidence of tetraploidy in the North American paddlefish, *Polyodon spathula*. Science 194(4267):842–844.
18. Kim DS, Nam YK, Noh JK, Park CH, Chapman FA (2005) Karyotype of North American shortnose sturgeon *Acipenser brevirostrum* with the highest chromosome number in the Acipenseriformes. Ichthyol Res 52: 94–97.
19. Fontana F, Congiu L, Mudrak VA, Quattro JM, Smith TIJ, et al. (2008) Evidence of hexaploid karyotype in shortnose sturgeon. Genome 51:113–119.
20. Ollermann LK, Skelton PH (1990) Hexaploidy in yellowfish species (Barbus, Pisces, Cyprinidae) from Southern Africa. J Fish Biol 37:105–115.
21. Furlong RF, Holland PWH (2002) Were vertebrates octoploid? Philos Trans R Soc Lond B Biol Sci 357:531–544.
22. Blacklidge KH, Bidwell CA (1993) Three ploidy levels indicated by genome quantification in acipenserformes of North America. J Hered 84: 427–430.
23. Kynard B (1997) Life history, latitudinal patterns, and status of the shortnose sturgeon, *Acipenser brevirostrum*. Environ Biol Fishes 48: 319–334.
24. Kynard B, Parker E (2004) Ontogenetic behavior and migration of Gulf of Mexico sturgeon, *Acipenser oxyrinchus desotoi*, with notes on body color and development. Environ Biol Fishes 70: 43–55.
25. Kieffer MC, Kynard B (1996) Spawning of the shortnose sturgeon in the Merrimack River, Massachusetts. Trans Am Fish Soc 125:179–186.
26. Waldman JR, Grunwald C, Stabile J, Wirgin I (2002) Impacts of life history and biogeography on the genetic stock structure of Atlantic sturgeon *Acipenser oxyrinchus oxyrinchus*, Gulf sturgeon *A. oxyrinchus desotoi*, and shortnose sturgeon *A. brevirostrum*. J Appl Ichthyol 18: 509–518.
27. Walsh MG, Bain MB, Squiers T Jr., Waldman JR, Wirgin I (2001) Morphological and genetic variation among shortnose sturgeon *Acipenser brevirostrum* from adjacent and distant rivers. Estuaries 24:41–48.
28. Grunwald C, Stabile J, Waldman JR, Gross R, Wirgin I (2002) Population genetics of shortnose sturgeon *Acipenser brevirostrum* based on mitochondrial DNA control region sequences. Mol Ecol 11: 1885–1898.
29. Quattro JM, Greig TW, Coykendall DK, Bowen BW, Baldwin JD (2002) Genetic issues in aquatic species management: the shortnose sturgeon (*Acipenser brevirostrum*) in the southeastern United States. Conserv Genet 3: 155–166.
30. Wirgin I, Grunwald C, Carlson E, Stabile J, Peterson DL, Waldman J (2005) Range-wide population structure of shortnose sturgeon *Acipenser brevirostrum* based on sequence analysis of the mitochondrial DNA control region. Estuaries 28:406–421.
31. Wirgin I, Grunwald C, Stabile J, Waldman JR (2010) Delineation of discrete population segments of shortnose sturgeon *Acipenser brevirostrum* based on mitochondrial DNA control region sequence analysis. Conserv Genet 11:689–708.
32. Wirgin II, Stabile JE, Waldman JR (1997) Molecular analysis in the conservation of sturgeons and paddlefish. Environ Biol Fishes 48:385–398.
33. Hett AK, Ludwig A (2005) SRY-related (Sox) genes in the genome of European Atlantic sturgeon (*Acipenser sturio*). Genome 48: 181–186.
34. Henderson AP, King TL (2012) Development of polysomic microsatellite markers for characterization of population structuring and phylogeography in the shortnose sturgeon (*Acipenser brevirostrum*). Conserv Genet Resour 4:853–859.
35. Labarca C, Paigen K (1980) A simple, rapid, and sensitive DNA assay procedure. Anal Biochem 102: 344–352.
36. Sambrook J, Fritsch EF, Maniatis T (1989) Molecular Cloning: a laboratory manual, second edition. Plainview: Cold Spring Harbor Laboratory Press.
37. Obbard DJ, Harris SA, Pannell JR (2006) Simple allelic-phenotype diversity and differentiation statistics for allopolyploids. Heredity 97:296–303.
38. Welsh A, Blumberg M, May B (2003) Identification of microsatellite loci in lake sturgeon, *Acipenser fulvescens*, and their variability in green sturgeon, *A. medirostris*. Mol Ecol Notes 3:47–55.
39. Rodzen JA, May B (2002) Inheritance of microsatellite loci in the white sturgeon (*Acipenser transmonatus*). Genome 45:1064–1076.
40. Israel JA, Bando KJ, Anderson EC, May B (2009) Polyploid microsatellite data reveal stock complexity among estuarine North American green sturgeon (*Acipenser medirostris*). Can J Fish Aquat Sci 66:1491–1504.
41. Lynch M, Milligan BG (1994) Analysis of population genetic structure with RAPD markers. Mol Ecol 3:91–99.
42. Peakall R, Smouse PE (2006) GENALEX 6: genetic analysis in Excel. Population genetic software for teaching and research. Mol Ecol Notes 6: 288–295.
43. Huff DR, Peakall R, Smouse PE (1993) RAPD variation within and among natural populations of outcrossing buffalograss *Buchloe dactyloides* (Nutt) Engelm. Theor Appl Genet 86: 927–934.
44. Hammer Ø, Harper DAT, Ryan PD (2001) PAST: Paleontological Statistics Software Package for Education and Data Analysis. Palaeontologia Electronica 4: 9.
45. Pritchard JK, Stephens M, Donnelly P (2000) Inference of population structure using multilocus genotype data. Genetics 155: 945–959.
46. Evanno G, Regnault S, Goudet J (2005) Detecting the number of clusters of individuals using the software structure: a simulation study. Mol Ecol 14:2611–2620.
47. Jaccard P (1901) Étude comparative de la distribution florale dans une portion des Alpes et des Jura. Bull Soc Vaudoise Sci Nat 37:547–579.
48. Kosman E, Leonard KJ (2005) Similarity coefficients for molecular markers in studies of genetic relationships between individuals for haploid, diploid, and polyploid species. Mol Ecol 14:415–424.
49. Clarke KR (1993) Non-parametric multivariate analysis of changes in community structure. Aust J of Ecol 18: 117–143.
50. Mantel N (1967) The detection of disease clustering and a generalized regression approach. Cancer Res 27:209–220.
51. Rohlf FJ (1993) NTSYS-pc: numerical taxonomy and multivariate analysis system. Version 1.8. New York: Exeter Publications.
52. Ryman N, Utter R, Laikre L (1995) Protection of intraspecific biodiversity of exploited fishes. Rev Fish Biol Fish 5:417–446.
53. Wright S (1969) Evolution and the Genetics of Populations, Vol. 2. The Theory of Gene Frequencies. Chicago: University of Chicago Press.
54. Nei M (1987) Molecular Evolutionary Genetics. New York: Columbia University Press.
55. Chakraborty R, Leimar O (1987) Genetic variation within a subdivided population. In Ryman N, Utter F, editors. Population Genetics and Fishery Management. Seattle: University of Washington Press. pp 89–120
56. Kruskal JB, Wish M (1978) Multidimensional Scaling, Sage University Paper series on Quantitative Application in the Social Sciences, 07-011. Beverly Hills and London: Sage Publications.
57. Saitou N, Nei M (1987) The neighbor-joining method: a new method for reconstructing phylogenetic trees. Mol Biol Evol 4:406–425.
58. Tamura K, Dudley J, Nei M, Kumar S (2007) MEGA4: Molecular Evolutionary Genetics Analysis (MEGA) software version 4.0. Mol Biol Evol 24:1596–1599.
59. Duchesne P, Bernatchez L (2002) AFLPOP: A computer program for simulated and real population allocation based on AFLP data. Mol Ecol Notes 3: 380–383.
60. Antunes A, Troyer JL, Roelke ME, Pecon-Slattery J, Packer C, et al. (2008) The evolutionary dynamics of the lion *Panthera leo* revealed by host and viral population genomics. PLoS Genet 4(11): e1000251. doi:10.1371/journal.pgen.1000251
61. Bemis WE, Findeis EK, Grande L (1997) An overview of Acipenseriformes. Environ Biol Fishes 48: 25–71.
62. Parker E, Kynard B (2005) Latitudinal differences in ontogenetic behavior between two populations of shortnose sturgeon (Acipenser brevirostrum): A laboratory study. Turners Falls, Massachusetts, USA: United States Geological Survey.
63. Kynard B, Parker E, Kynard BE (2010) Ontogenetic behavior of Kootenai River white sturgeon, *Acipenser transmontanus*, with a note on body color: a laboratory study. Environ Biol Fishes 88: 65–77.
64. Hedrick PW (1999) Perspective: Highly variable loci and their interpretation in evolution and conservation. Evolution 53:313–318.
65. Balloux F, Brüunner H, Lugon-Moulin N, Hausser J, Goudet J (2000) Microsatellites can be misleading: an empirical and simulation study. Genetics 54: 1414–1422.
66. Koskinen MT, Sundell P, Piironen J, Primmer CR (2002) Genetic assessment of spatiotemporal evolutionary relationships and stocking effects in grayling (*Thymallus thymallus*, Salmonidae). Ecol Lett 5: 193–205.
67. Stockwell CA, Hendry AP, Kinnison MT (2003) Contemporary evolution meets conservation biology. Trends Ecol Evol 18: 94–101.
68. Welsh SA, Mangold MF, Skjeveland JE, Spells AE (2002) Distribution and Movement of Shortnose Sturgeon (*Acipenser brevirostrum*) in the Chesapeake Bay. Estuaries 25 (1):101–104.
69. Kynard B, Breece M, Atcheson M, Kieffer M, Mangold M (2009) Life history and status of shortnose sturgeon (*Acipenser brevirostrum*) in the Potomac River. J Appl Ichthyol 25 (Suppl. 2):34–38.
70. Smith TIJ, McCord JW, Collins MR, Post WC (2002) Occurrence of stocked shortnose sturgeon *Acipenser brevirostrum* in non-target rivers. J Appl Ichthyol 18: 470–474.
71. Fernandes SJ, Kinnison MT, Zydlewski GB (2008) Investigation into the distribution and abundance of Atlantic sturgeon and other diadromous species in the Penobscot River, Maine: with special notes on the distribution and abundance of federally endangered shortnose sturgeon (*Acipenser brevirostrum*). Gloucester, Massachusetts, USA: National Oceanic and Atmospheric Administration.
72. Savoy T (2004) Population estimate and utilization of the lower Connecticut River by shortnose sturgeon. In Jacobson PM, Dixon DA, Leggett WC, Marcy BC Jr., Massengill RR, editors. The Connecticut River Ecological Study (1965–

1973) revisited: ecology of the lower Connecticut River 1973–2003. Am Fish Soc Mon 9. pp. 345–352.
73. Dadswell MJ (1979) Biology and population characteristics of the shortnose sturgeon, *Acipenser brevirostrum* LeSueur (1818) (Osteichthyes: Acipenseridae), in the Saint John River estuary, New Brunswick, Canada. Can J Zool 57:2186–2210
74. Fernandes SJ (2008) Population demography, distribution, and movement patterns of Atlantic and shortnose sturgeons in the Penobscot River estuary, Maine. Master's thesis. Univ. of Maine. 88 pp.
75. Shortnose Sturgeon Status Review Team. 2010. A Biological Assessment of shortnose sturgeon (*Acipenser brevirostrum*). Report to National Marine Fisheries Service, Northeast Regional Office. November 1, 2010. 417 pp.

Continental-Scale Footprint of Balancing and Positive Selection in a Small Rodent (*Microtus arvalis*)

Martin C. Fischer[1,2]*, Matthieu Foll[1,3,4], Gerald Heckel[1,4], Laurent Excoffier[1,4]

1 Computational and Molecular Population Genetics (CMPG), Institute of Ecology and Evolution, University of Bern, Bern, Switzerland, 2 Institute of Integrative Biology, ETH Zürich, Zürich, Switzerland, 3 School of Life Sciences, École Polytechnique Fédérale de Lausanne (EPFL), Lausanne, Switzerland, 4 Swiss Institute of Bioinformatics, Lausanne, Switzerland

Abstract

Genetic adaptation to different environmental conditions is expected to lead to large differences between populations at selected loci, thus providing a signature of positive selection. Whereas balancing selection can maintain polymorphisms over long evolutionary periods and even geographic scale, thus leads to low levels of divergence between populations at selected loci. However, little is known about the relative importance of these two selective forces in shaping genomic diversity, partly due to difficulties in recognizing balancing selection in species showing low levels of differentiation. Here we address this problem by studying genomic diversity in the European common vole (*Microtus arvalis*) presenting high levels of differentiation between populations (average $F_{ST}=0.31$). We studied 3,839 Amplified Fragment Length Polymorphism (AFLP) markers genotyped in 444 individuals from 21 populations distributed across the European continent and hence over different environmental conditions. Our statistical approach to detect markers under selection is based on a Bayesian method specifically developed for AFLP markers, which treats AFLPs as a nearly codominant marker system, and therefore has increased power to detect selection. The high number of screened populations allowed us to detect the signature of balancing selection across a large geographic area. We detected 33 markers potentially under balancing selection, hence strong evidence of stabilizing selection in 21 populations across Europe. However, our analyses identified four-times more markers (138) being under positive selection, and geographical patterns suggest that some of these markers are probably associated with alpine regions, which seem to have environmental conditions that favour adaptation. We conclude that despite favourable conditions in this study for the detection of balancing selection, this evolutionary force seems to play a relatively minor role in shaping the genomic diversity of the common vole, which is more influenced by positive selection and neutral processes like drift and demographic history.

Editor: Nadia Singh, North Carolina State University, United States of America

Funding: This study was supported by grants from the University of Bern Research Foundation and the Swiss National Science Foundation (grants 3100A0-126074 and 31003A-143393 to LE). The funders had no role in study design, data collection and analysis, decision to publish, or preparation of the manuscript.

Competing Interests: The authors have declared that no competing interests exist.

* Email: martin.fischer@env.ethz.ch

Introduction

Despite nearly six decades of genetic investigations, it remains unclear for most organisms to which extent the demographic history of populations, genetic drift or selection influences the pattern of genetic diversity of a species. Historically, the observation that many genes are genetically polymorphic within population was first explained by a selective advantage of heterozygotes [1]. This explanation was challenged by Kimura's neutral [2,3] and nearly neutral [4] theory of molecular evolution, which provided a competing explanation for the high frequency of genetic polymorphism. Nowadays it is generally accepted that a majority of the genetic variations evolved nearly neutrally, but that natural selection plays a decisive role in evolution and leaves footprints in the genome. Natural selection acts in at least three forms, which are positive, purifying and balancing selection. Positive selection can lower genetic diversity locally but increase it globally, to a level depending on the spatial and environmental heterogeneity [5–7]. Balancing selection maintains genetic variation within populations [8] and leads to generally low levels of differentiation between populations, even though it can contribute to increase population differentiation if selective pressures are spatially heterogeneous [5]. Finally, purifying selection generally decrease levels of genetic diversity, even though strong background selection can promote increased difference between populations by lowering their effective size [9]. In the past, balancing selection played an important role in evolutionary genetics in explaining the high level of genomic polymorphism observed among species or populations [8,10]. However, the effect of selection can be multifarious and the impact of each is still under debate [11], especially for balancing selection.

At least in humans a number of common genetic diseases have been proposed to be maintained in populations as a result of balancing selection, e.g. sickle-cell anaemia [12,13], glucose-6-phosphate dehydrogenase deficiency [14], thalassemia [15] and cystic fibrosis [16]. Other examples are the ABO blood group [17], polymorphisms of beta-globin [18], the major histocompatibility complex (MHC; [19]) including the human HLA-G promoter [20], CCR5 in humans [21], the complementary sex determination locus in bees [22], response to pathogens [23], high diversity genes in *Arabidopsis* [24] or self-incompatibility and nuclear-cytoplasmic gynodioecy in plants (see e.g. [25]). However, all these examples were identified by a candidate gene approach and not genome-wide scans. Hence they do not give any information about the importance of balancing selection in shaping genomic diversity. In this context, there are only few genome-wide studies of balancing selection in humans [26,27] or sticklebacks [28] and these studies remain inconclusive about the importance of balancing selection in shaping and maintaining genetic diversity, potentially due to methodological limitations (see below). Compared to balancing selection, the occurrence and influence of positive selection on an organism's genetic variation is much less questioned, as positive selection should allow the spread of advantageous traits and play a central role in the evolution of species (see e.g. [29,30]).

The prevalence of balancing selection is still highly debated, mainly due to missing evidence in organisms other than humans, but also because methods developed specifically to detect balancing selection are still few (see e.g. [27,31–34]). Moreover, the classical detection of balancing selection based on levels of differentiation between populations is difficult in organisms with low levels of differentiation (see [35]) like humans or *Drosophila* [36,37] and a decent number of populations need to be investigated to have the statistical power to detect balancing selection [35].

In order to better detect signals of balancing selection, we focused in this study on an organism showing particularly high levels of differentiation, which is the common vole (*Microtus arvalis*). This species has a very wide distribution in Europe, and it is found in most open grassland and farmland habitats up to 2,000 m altitude [38–40]. It ranges from the Atlantic coast of France to Central Russia, as well as from the Orkney archipelago in the North to the Mediterranean coast in Spain (Figure 1). In previous studies it has been shown that the vole populations have an overall high levels of differentiation for both mtDNA ($F_{ST} = 0.7$) and nuclear markers (STR, $F_{ST} = 0.17$) [41–43]. The widespread distribution of this species in different habitats and environments, and its peculiar pattern of genetic diversity makes it particularly suitable for the detection of markers with high or low levels of differentiation, and by extension for the determination of the respective roles of positive or balancing selection over a large geographic scale.

The aim of this study is to detect selective patterns in populations across the European mainland to disentangle the importance of balancing and positive selection in shaping the genetic diversity observed in the distribution range of the common vole. However, a major challenge in identifying genomic regions under selection is to separate the footprint of selection from that of population history and demography (e.g. [10,44,45]). Hence examining a large number of loci scattered throughout the genome is an effective way to tell apart the effect of selection from the confounding effects of population history and demography [10,46,47]. Cavalli-Sforza [48] and Lewontin and Krakauer [49] proposed that genetic drift and gene flow should affect all loci similarly, leading to some overall degree of differentiation between populations, but that selected loci would deviate significantly from this distribution. Indeed, positive selection acting on a given locus should increase population differentiation (and lead to high F_{ST}) whereas balancing selection should reduce it and lead to low F_{ST} (see e.g. [40,47,50,51]).

For non-model organisms Amplified Fragment Length Polymorphisms (AFLPs) allow the screening of thousands of randomly distributed loci in a genome [52,53]. To detect AFLP outliers, we used a recently developed extension of the Bayesian F_{ST}-based approach [35,54] based on the *F*-model [55]. BayeScan 2.1 [40] provides estimates of allele frequencies and *F*-statistics from AFLPs by incorporating for each individual the band intensity of a marker instead of simply using presence-absence patterns [56,57]. This procedure implicitly allows one to distinguish between homo- and heterozygotes, and significantly improves the detection of selection with AFLP markers, which nearly reach the power obtained with single nucleotide polymorphism (SNP) data for which individual genotypes are known [40].

Materials and Methods

Sample and DNA extraction

We analysed 21 vole populations across most of the distribution range of *M. arvalis* in Europe, with a total of 444 individuals (see Figure 1 and Table 1). The populations were spread over 2,500 km from Spain (EAv) to Poland (PSr), and over a 750 km latitudinal gradient from Belgium (BSt) to Italy (INa).

The samples for this study were obtained by strictly following the legislation on animal protection and experimentation of Switzerland and the other European countries involved. *Microtus arvalis* is not specifically protected by Swiss laws on animal protection (Tierschutzgesetz from December 16 2005) and hunting legislation (Verordnung zum Jagdschutzgesetz, February 29 1988) because of its role as an agricultural pest and general abundance. The use of snap traps for sampling *M. arvalis* is not a stress-inducing animal experiment (Schweregrad 0; Art. 137ff Swiss federal regulations on animal experimentation). However, Swiss samples analyzed in this study (some of them also covered in earlier studies; [40–43,56,58–62]) were obtained also under animal experimentation permits No. 55/02; 107/05; BE08/10; BE90/10 issued by the cantonal veterinary office of Bern according to federal law after ethical approval by the Bernese cantonal commission on animal experimentation. Additional samples were obtained from the researcher network on rodent-borne pathogens based at the German Federal Research Institute for Animal Health (FLI; http://www.fli.bund.de; GH is one of the coordinators) [63–65] and its international partners in the European projects EDEN and EDENext on biology and control of vector-borne infections in Europe (http://www.edenext.eu). Sample acquisition followed strictly the legislation of the relevant countries after approval by the according animal protection and ethics committees as required by the European Commission Seventh Framework Programme (FP7; http://cordis.europa.eu/fp7/home_en.html) [66,67].

Total genomic DNA was extracted from foot, tail or liver tissue stored in absolute ethanol and later deep-frozen using a standard phenol-chloroform protocol [68]. The quality and quantity of the DNA was determined on 0.8% agarose gels and with a spectrophotometer (NanoDrop ND-1000 Spectrophotometer, NanoDrop Technologies, Inc., Wilmington, USA). The DNA concentration was standardized to 100 ng DNA/µL for all individuals to ensure similar PCR yield across samples [40].

AFLP analyses

AFLP analyses were performed according to standard protocols as established by Vos et al. [52] and modified by Fink et al. [69]. Selective amplifications were performed using 21 primer combinations (Table S1). These primer combinations were then named according to the last two selective bases of each primer, e.g. the combination E01-AAC/M02-cag is referred to as *ACag*. Special care was taken to guarantee the reproducibility of AFLP marker

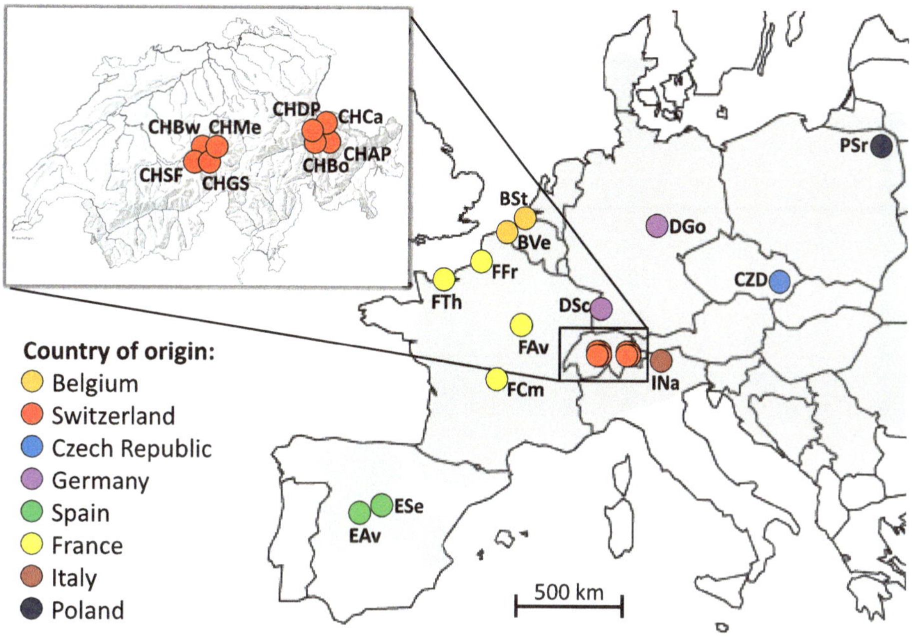

Figure 1. Geographic location of the 21 *Microtus arvalis* populations analyzed. The grey area corresponds to the European distribution of the species (after [98]). See Table 1 for sample abbreviations.

analyses: a liquid-handling robot (Microlab STAR, Hamilton Bonaduz AG, Bonaduz, Switzerland) was used for selective amplification, multiplexing of PCR products and loading of the 96-well sequencer plate, and 38 individuals (9%) were independently replicated for all 21 primer combinations (see [40] for more details).

AFLP fragment scoring and diversity

AFLP fragment scoring was performed with GeneMapper software version 3.7 (Applied Biosystems). Bin sets were created automatically and manually revised [40]. Two AFLP data matrices were produced, one with band intensity information and one with a standard binary presence-absence matrix. The AFLPs binary data matrix was used to estimate reproducibility, AFLP diversity estimates, and to run the first PCA analyses. A particular AFLP band intensity was scored as 'present' (1), if its value was larger than 10% of the 95% band intensity distribution quantile, or 'absent' (0), if its intensity was smaller than 10% of the 95% quantile value. AFLP marker frequencies, the number of variable markers per population and AFLP diversity were then calculated with the program AFLPDAT [70]. AFLP diversity was calculated as the average proportion of pairwise differences between individuals for each population, which is an index similar to Nei's gene diversity calculated from marker frequencies [71,72].

Outlier detection

A Bayesian genome scan approach (BayeScan) was used to detect markers under selection. This procedure is more efficient than classical outlier detection methods (like DETSELD, modified version of [73] or DFDIST, modified version of [74]) in the discovery of true selected loci, as it results in a lower number of false positives [75]. BayeScan 2.1 was specifically developed for AFLP markers. The inclusion of band intensity information makes the BayeScan analysis of dominant AFLPs almost as powerful as an analysis of the same number of codominant markers (e.g. reaching 92% of the power of a SNP data set) to detect selection (for more details see [40]). Moreover, this additional information makes it possible to infer population-specific inbreeding coefficient (F_{IS}) from AFLP data [56]. Band intensity information required by BayeScan 2.1 was obtained from the AFLP data matrix of marker band intensity provided by GeneMapper. Since markers with a low minor allele frequency systematically bias the F_{ST} estimates downwards [76], only markers with band frequencies between 5% and 95% were used for subsequent analyses. This procedure prevents an artificial increase in the number of inferred outlier markers under positive selection [76]. Note that markers having band frequencies higher than 95% were still considered as polymorphic if the distribution of band intensity across all individuals was bimodal [40] and if they did not exceed three-times the 95% quantile of the band intensity distribution for that marker to avoid artefacts of the sequencing machine. These markers are probably informative to infer F_{IS}, as they contain a high proportion of fixed and/or heterozygous individuals.

BayeScan assumes that allele frequencies within populations follow a multinomial-Dirichlet distribution [55,77,78] with F_{ST} parameters being a function of population-specific components shared among all loci and of locus-specific components shared among all populations. For a given locus, departure from neutrality is assumed when the locus-specific component is required to explain the observed pattern of diversity. BayeScan directly infers the posterior probability of each locus to be under the effect of selection by defining and comparing two alternative models: one model includes the locus-specific component, while the other excludes it [35]. The ratio of the model posterior probabilities is used to calculate then the posterior odds (PO), which measures how much more likely the model with selection is compared to the model without selection (see [40]). We used a threshold of PO>10 for a marker to be considered under selection, which refers to "strong evidence" for the alternative model (in this case the model with selection) as defined by Jeffreys [79]. For the Markov chain Monte Carlo (MCMC) algorithm we used 20 pilot runs of 5,000 iterations to adjust the proposal

Table 1. Location and properties of 444 *M. arvalis* samples from 21 populations across Europe.

ID	Country	Location	N	Latitude	Longitude	F_{ST} (CI)	F_{IS} (CI)	Variable AFLPs %	AFLP diversity
BSt	Belgium	Stalhille	21	51°13′N	3°05′E	0.304 (0.292–0.318)	0.0010 (0.0000–0.0029)	30.2	0.091
BVe	Belgium	Veurne	21	51°03′N	2°40′E	0.345 (0.331–0.359)	0.0008 (0.0000–0.0024)	25.3	0.078
CZD	Czech Republic	Drnholec	21	48°53′N	16°27′E	0.162 (0.151–0.172)	0.0431 (0.0288–0.0576)	46.3	0.132
FAv	France	Avallon	20	47°29′N	3°54′E	0.248 (0.236–0.261)	0.0008 (0.0000–0.0023)	37.3	0.115
FCm	France	Clérmont-Ferrand	21	45°47′N	3°05′E	0.262 (0.250–0.275)	0.0006 (0.0000–0.0018)	38.1	0.116
FFr	France	Fressenneville	22	50°05′N	1°34′E	0.302 (0.289–0.315)	0.0020 (0.0001–0.0058)	30.8	0.090
FTh	France	Thaon	21	49°16′N	0°26′W	0.294 (0.281–0.306)	0.0058 (0.0004–0.0137)	35.4	0.108
DGo	Germany	Gotha	20	50°57′N	10°42′E	0.224 (0.212–0.235)	0.0238 (0.0118–0.0382)	37.6	0.113
DSc	Germany	Schiltach	20	48°17′N	8°21′E	0.237 (0.225–0.250)	0.0007 (0.0000–0.0022)	33.8	0.107
INa	Italy	Naturno	21	46°39′N	11°01′E	0.349 (0.335–0.363)	0.0030 (0.0002–0.0083)	31.6	0.096
PSr	Poland	Srodkowy	21	53°35′N	22°47′E	0.297 (0.283–0.309)	0.0055 (0.0000–0.0136)	34.2	0.108
EAv	Spain	Avila	21	40°42′N	4°48′W	0.426 (0.411–0.440)	0.0009 (0.0000–0.0026)	27.8	0.084
ESe	Spain	Segovia	21	40°54′N	4°06′W	0.442 (0.429–0.457)	0.0006 (0.0000–0.0019)	26.1	0.078
CHAP[1]	Switzerland	Alp di Plaun	20	46°47′N	9°29′E	0.224 (0.212–0.236)	0.0129 (0.0001–0.0240)	39.3	0.123
CHBo[1]	Switzerland	Bonaduz	22	46°48′N	9°24′E	0.297 (0.285–0.310)	0.0050 (0.0000–0.0125)	32.3	0.098
CHBw[2]	Switzerland	Brienzwiler	21	46°45′N	8°07′E	0.328 (0.315–0.343)	0.0031 (0.0000–0.0085)	27.7	0.082
CHCa[1]	Switzerland	Calandahütte	22	46°53′N	9°29′E	0.257 (0.245–0.269)	0.0124 (0.0003–0.0232)	35.5	0.105
CHDP[1]	Switzerland	Domat/Ems	22	46°50′N	9°26′E	0.278 (0.265–0.290)	0.0061 (0.0000–0.0145)	32.6	0.097
CHGS[2]	Switzerland	Grosse Scheidegg	22	46°40′N	8°06′E	0.386 (0.372–0.400)	0.0011 (0.0000–0.0034)	22.6	0.068
CHMe[2]	Switzerland	Meiringen	22	46°43′′N	8°11′E	0.411 (0.396–0.424)	0.0007 (0.0000–0.0021)	19.9	0.061
CHSF[2]	Switzerland	Schreck-Feld	22	46°40′N	8°04′E	0.354 (0.341–0.368)	0.0019 (0.0001–0.0055)	25.1	0.076

We report the population ID, sampling site (country and location), number of individuals per population (N), coordinates, estimates of population specific F_{ST} and F_{IS} with 95% credibility intervals (CI), proportion of variable AFLP markers and AFLP marker diversity for each population.
[1]AFLPs of six primer combinations have been published in Foll *et al.* [56].
[2]AFLPs of 21 primer combinations have been published in Fischer et al. [40].

distribution to acceptance rates between 0.25 and 0.45 for the runs. A burn-in of 50,000 iterations was used and visually checked for convergence of the MCMC chains, followed by 50,000 iterations for estimation using a thinning interval of 10. False Discovery Rate (FDR) was used to control for multiple testing [40,80].

Inference of neutral genetic structure across Europe

We performed two principal component analyses (PCA) in R [81] to infer the patterns of neutral genetic structure in common voles across Europe. PCA analyses were performed on the neutral (excluding outlier loci) and evolutionary informative AFLP markers. Evolutionary informative AFLPs have band frequencies ranging between 5% and 95%, which excludes uninformative and rare markers [76]. One PCA analysis was done at the individual level using AFLP marker presence/absence data for all 444 individuals and the second analysis was done at the population level, on the basis of marker allele frequencies estimated by BayeScan [40,56] using band intensity information.

Inference of balancing selection

Markers detected under balancing selection were investigated in more detail, as heterozygosity information can be gained from the population-specific band intensity distribution for a specific AFLP marker. A marker under balancing selection should indeed have evenly distributed allele frequencies across most populations and heterozygous individuals should be observed within populations, leading to a bimodal band intensity distribution for this AFLP marker [56]. The markers inferred as under balancing selection were thus carefully examined for bimodality of band intensities. However, sex-chromosome linked markers may also show bimodal distributions and low differentiation between populations in samples with equal sex ratios, as males only have one X-chromosome. A t-test implemented in R was thus used to check for association between band intensity and gender, using a threshold of $p>0.05$ without correction for multiple testing, to be conservative in the identification of marker under balancing selection. We have used the same approach to test for any amplification difference among different 96-well PCR plates of the same primer pair (batch effect).

Inference of positive selection patterns across Europe

To infer the patterns of positive selection in common voles across Europe we performed scaled PCA in R of the population allele frequencies of loci inferred under positive selection by BayeScan.

To identify the strongest geographic patterns of selection across Europe, we used a locus-by-locus SAMOVA approach [82] to separate for each marker populations into groups ($k = 2$) leading to the highest level of genetic differentiation (F_{CT}). The three outlier loci showing the highest F_{CT} were identified and plotted onto the European map using the R package plotrix to visualize the population-specific allele frequencies of these patterns of selection. To find loci showing similar geographic patterns of selection across Europe, which could be the cause of multi-genic adaptation due to similar selective pressures on different loci or genetic linkage of markers, we computed a pairwise Pearson's correlation between the population-specific allele frequencies of the outlier loci using the R package psych and Holm's correction for multiple testing [83].

Results

AFLP variation and neutral genetic structure across Europe

The AFLP analyses of the 21 European vole populations provided 3,839 markers. The majority of these AFLP markers were polymorphic (3,318; 86%) and 2,054 (54%) showed informative band frequencies between 5% and 95% overall. For each individual, we obtained on average 2,342 AFLPs (range: 2,169–2,418) across all primer combinations, and the mean length of the fragments was 239 bp. An average of 183 AFLP markers was scored per primer combination across all individuals (range: 86–256; Table 2). The average proportion of variable AFLP bands per population was 31%, with an average AFLP diversity of 9.6%. F_{IS} estimates were low for all populations, ranging between 0.001 and 0.043 (Table 1). Average genetic differentiation among populations was globally high with an average population-specific F_{ST} of 0.31. The population from the Czech Republic (CZD) had the highest number of variable AFLP bands per population (46%), and consequently the lowest population-specific F_{ST} (0.16), whereas the lowest diversity was observed in a population of the Swiss Alps (CHMe), with only 20% of variable markers and hence a high population-specific F_{ST} (0.41).

The two "evolutionary neutral" PCAs were based on 1,843 neutral AFLP markers - these were the 2,054 evolutionary informative AFLP markers minus the 211 inferred outlier loci (see more details below). These neutral markers led to a clustering of individuals that approximately matches the geographic origin of the samples (Figure 1) except that the Swiss vole populations were somewhat farther apart than geography would suggest. The entire individual-based AFLP data set (Figure 2A) as well as the PCA from estimated population-specific allele frequencies (Figure 2B) show very similar patterns and allow a clear separation of the populations, which indicates the high information content of these AFLP markers.

Genome scan

The BayeScan analysis of the 2,054 informative AFLP markers in 21 populations across Europe revealed 211 markers with a PO for selection larger than 10 with an associated FDR of less than 1.4%. Among these markers under selection, 138 (6.7%) had high F_{ST} (mean F_{ST}: 0.52) indicative of positive selection, and 73 were associated with very low F_{ST} (mean F_{ST}: 0.08) indicative of balancing selection (Figure 3; Table 2).

Inference of balancing selection

Bimodal band intensity information of AFLPs (for more details see Figure 4A and B, or [40,56,57]) was used to identify prime candidates for balancing selection and to exclude false positives among the 73 low F_{ST} outliers. Among these, 40 markers were considered as unlikely to be under balancing selection, either because outliers showed significant band intensity differences between males and females (t-test, $p<0.05$) and were thus likely sex-chromosome linked (33 markers, Table 3, Figure 4C and D) or because of PCR amplification strength differences between 96-well plates (7 markers of the primer combination *GGtc*).

Among the remaining 33 markers with low F_{ST} values, 27 showed distributions that could be compatible with other factors than just balancing selection. Two markers (*CAta44* and *GCtc76*) had an overall bimodal distribution, but a clear bimodality was missing in individual populations. Thirteen markers had either a unimodal or multimodal band intensity distribution. Twelve markers had low allele frequencies (0.04–0.21) that could be a consequence of negative selection or frequency dependent

Table 2. AFLP markers detected under positive selection with BayeScan.

Primers	Markers	# BayeScan	Marker ID of specific primer combination
ACag	201	9	*33; 77; 84; 85; 119; 136; 186; 187; 200*
ACtc	175	2	*21; 94*
ACtt	157	5	*40; 76; 79; 115; 142*
AGaa	147	4	*5; 56; 60; 101*
AGac	200	7	*3; 13; 84; 144; 161; 188; 190*
AGtg	141	4	*97; 103; 106; 126*
CAat	210	5	*102; 103; 145; 166; 181*
CAta	185	8	*73; 84; 116; 155; 157; 158; 165; 169*
CCac	195	4	*17; 129; 153; 167*
CCta	219	3	*88; 153; 213;*
CCtt	256	12	*2; 34; 41; 42; 50; 67; 97; 158; 168; 169; 193; 208*
CGag	101	4	*67; 69; 70; 97*
CGtt	86	6	*24; 40; 50; 51; 79; 87*
CTaa	239	11	*3; 7; 78; 86; 166; 173; 184; 185; 187; 208; 232*
CTag	208	9	*64; 90; 102; 125; 132; 133; 147; 154; 176*
CTtg	211	10	*46; 93; 97; 104; 117; 133; 170; 174; 175; 187*
GCat	224	12	*15; 31; 60; 106; 169; 175; 180; 181; 192; 208; 212; 213*
GCta	199	6	*7; 20; 33; 82; 98; 187*
GCtc	181	7	*16; 22; 40; 43; 160; 166; 172*
GGac	168	4	*30; 31; 36; 164*
GGtc	136	6	*15; 18; 64; 85; 125; 129*
Total	3839	138	

Given are the 21 primer combinations investigated, the total number of scored AFLP markers per primer combination across all individuals, the number of markers with a posterior odds (PO)>10 and the primer combination specific IDs of the markers under selection.

selection, which is also form of balancing selection. Finally, six markers were identified as prime candidates for balancing selection (Table 3), as homozygous individuals had approximately twice the band intensity of heterozygous individuals (Figure 4A) and all populations showed intermediate allele frequencies across the European continent (see e.g. Figure 5A).

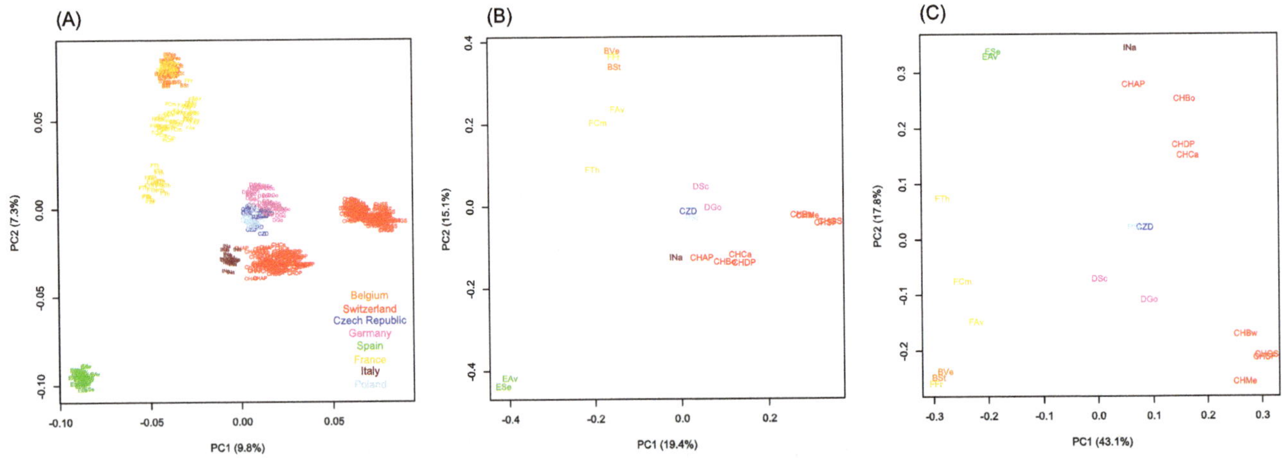

Figure 2. Principal component analysis (PCA) of (A) the neutral binary AFLP data matrix of 444 individuals from 21 populations across Europe using 1,843 neutral and evolutionary informative AFLP markers (see Material and Methods). (B) PCA of the 21 population-specific allele frequency estimates of neutral AFLP markers by BayeScan. The distribution of the populations on the plot roughly follows the geographic origin of the samples. (C) PCA of the estimated population-based allele frequencies of the 138 outliers probably under positive selection. For population IDs see Table 1. Colours correspond to country affiliation (see Figure 1).

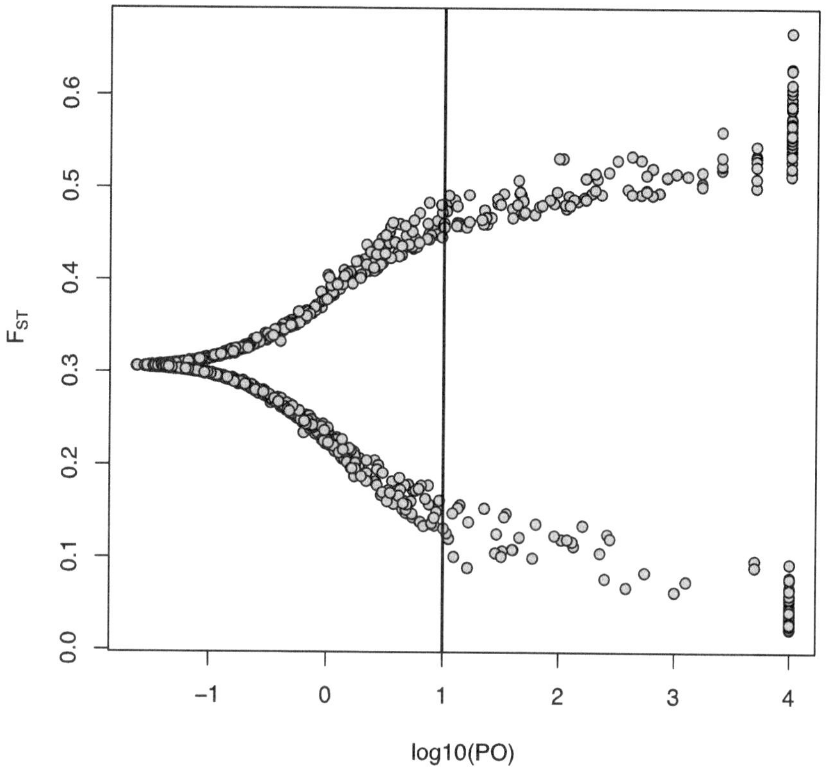

Figure 3. Results of BayeScan analysis for 2,054 informative AFLPs genotyped in 444 *M. arvalis* voles sampled in 21 populations. The marker-specific F_{ST} is plotted against the posterior odds (PO) of being under selection. The vertical line shows the critical PO of 10 used to identify outlier markers. Markers on the right side of the vertical line are outliers: 138 markers with high F_{ST} indicative of positive selection and 73 markers with low F_{ST} indicative of balancing selection were identified. Markers having a log10(PO)>4 were summarized in the category 4.

Inference of positive selection across Europe

We detected a total of 138 markers potentially under positive selection across Europe, with an average of 6.6 outliers per tested primer combination (range: 2–12; Table 2). For these outliers, strong allele frequency differences were always identified in three or more populations compared to the rest, showing that selection was inferred independently in multiple populations (see e.g. Figure 5 B–D).

The PCA based on allele frequencies estimated for the 138 loci potentially under positive selection revealed a different pattern than the neutral markers (Figure 2C). Especially the populations within the Swiss Alps (CHAP, CHBo, CHCa, CHDP, CHBw, CHMe, CHGS and CHSF) and Italian Alps (INa) are much more separated from the other populations and show larger extent of differentiation among themselves compared to the PCA on neutral loci (Figure 2A and B), which is potentially indicative of strong selection pressures in the alpine area.

SAMOVA allowed us to identify the outlier loci that produced the strongest splits between two groups of populations across Europe leading to the highest level of genetic differentiation (F_{CT}), which might be an indication of the strength of selection. The three loci that showed the strongest splits are *ACag119* with a F_{CT} of 0.93 (Figure 5B), *CTaa3* with a F_{CT} of 0.89 (Figure 5C) and *GGac31* with a F_{CT} of 0.87 (Figure 5D). The pairwise comparison of allele frequencies of outlier loci identified that *ACag119* showed a significant correlation with only three other loci, *CTaa3* with six and *GGac31* with 16 loci. Among the 138 loci under positive selection the average number of associations was 6.1 with a range of 0 to 24 associations. Additional information for all 138 outlier loci can be found in Table S2.

Discussion

The current study illustrates the capacity of Bayesian F_{ST} outlier approaches to identify the signature of positive and balancing selection in non-model organisms. The nearly 4,000 AFLP markers, of which 2,054 were evolutionarily informative, clearly allowed us to screen a representative part of the common vole genome for loci linked to recent adaptation on a continental scale in Europe.

Genetic structure across Europe inferred by AFLPs

The neutral AFLP markers allowed us to accurately resolve population genetic structure of the 21 vole populations across the European continent and the PCA led to a clustering of individuals and populations that corresponds approximately to the geographic origin of the samples (Figure 1 vs. 4A and B). Similar patterns were found in humans were genetic data also mirror geography in Europe [84]. This high resolution indicates the large information content present in this AFLP data set and is further supported by a very similar PCA-based clustering of populations inferred by 6,807 polymorphic SNPs (see Figure S2 in [60]), which were used to resolved the four evolutionary lineages present in Europe [43].

Pattern of selection across European continent

We scanned 21 vole populations across the European continent for evidence of selection. Overall slightly more than 8% of all markers were under positive or balancing selection. Despite the detection of some candidate loci for balancing selection (1.6%), more loci for local positive selection (6.7%) were identified. These results suggest that drift and the demographic history of vole

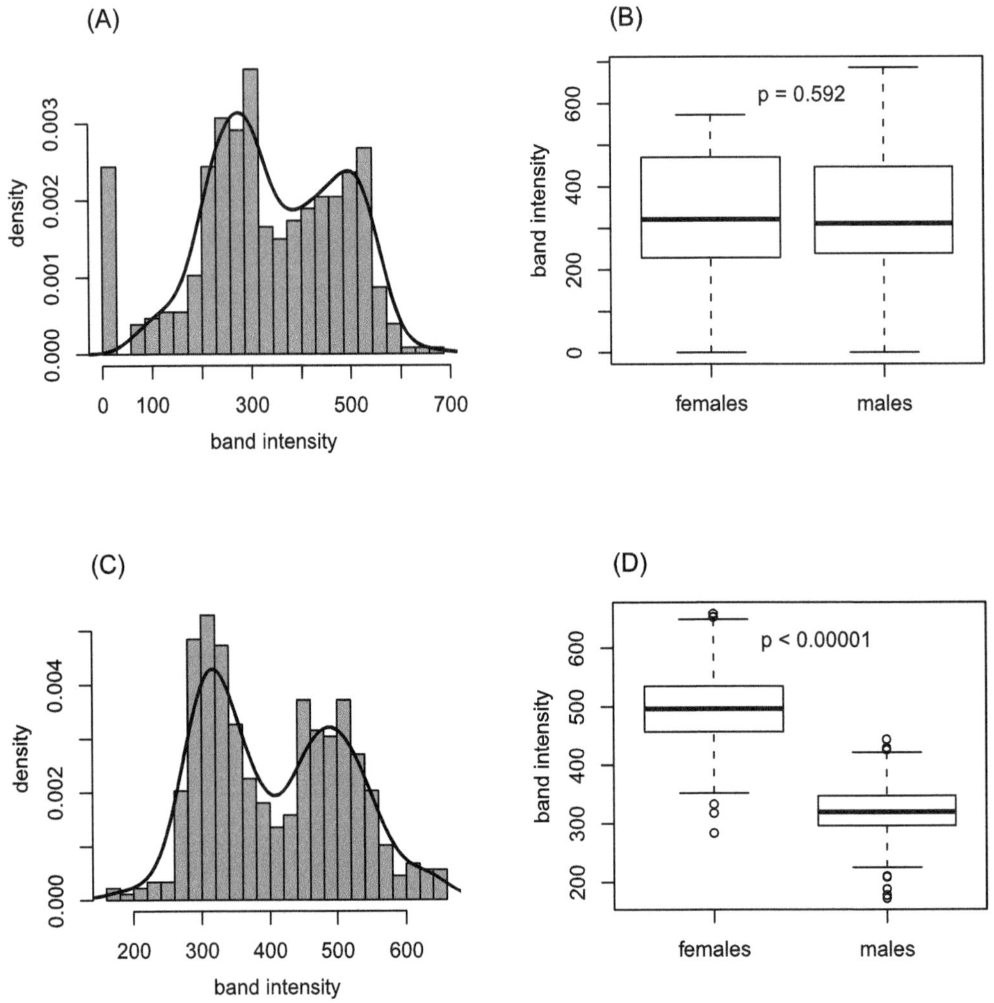

Figure 4. Bimodal band intensity distribution of two low F_{ST} outlier markers. (A) Bimodal distribution of a marker likely to be under balancing selection (*CTaa125*). The zero class of the distribution represents individuals not showing any band, the following first peak corresponds to heterozygous individuals and the second peak represents homozygous individuals. The black line represents a fitted density curve. (B) Comparison of band intensities in males and females for marker *CTaa125* shown in (A) revealing no statistical difference and suggesting it is an autosomal marker. (C) Bimodal distribution for marker *ACtt16*. (D) Corresponding box plot of sex-specific band intensities for marker *ACtt16* where females have band intensities about twice larger than males, hence suggesting it is an X-linked marker.

populations have strongly influenced the observed genetic diversity, but that also positive selection plays an important role in shaping the genetic diversity of vole populations, while balancing selection is less common. Nevertheless, the detection of several markers with multiple evidence of balancing selection is remarkable, especially the signature of a stabilizing evolutionary process on such a large geographic scale.

Contrasting to our results, balancing selection played in the past an important role in evolutionary genetics in explaining the high level of genomic polymorphism observed among species or populations [8,10]. Six decades ago Dobzhansky [1] suggested that genetic polymorphisms were maintained in populations by selection favouring heterozygotes, thus by balancing selection. Later Kimura [2,85] has shown that most polymorphisms in the genome should be selectively neutral after the action of purifying selection. It follows that clear examples of balancing selection in any organism should be quite limited and mainly inferred by a candidate gene approach (see Introduction and e.g. [12–25]), but little is known about the prevalence of balancing selection on a genome wide scale [26–28]. In humans balancing selection is thought to have a limited role in preserving genome-wide polymorphisms [26,86], as a specific survey of balancing selection in humans identified only 60 out of 13,400 genes [27]. In this study we identified 33 loci with significantly low levels of differentiation among populations, which represent slightly more than 1.5% of all informative markers and hence slightly more that the 0.5% inferred in humans [27]. Our findings, together with the human studies [27], indicate that large geographic scale balancing selection is probably not as frequent as previously suspected, and hence only plays a minor role in maintaining polymorphism in a population or in shaping the genetic diversity of a species.

The observation of evenly distributed allele frequencies across the whole European continent (e.g. see Figure 5A) despite extremely strong levels of differentiation among populations (average $F_{ST} = 0.31$) is quite remarkable, especially for a species with limited dispersal ability [43,87]. Such even allele frequencies across a large geographic range are difficult to explain in absence of strong stabilizing selection and hence good support for the presence of balancing selection.

Table 3. List of the 73 markers identified by BayeScan as being potentially under balancing selection.

ID	PO	F_{ST}	Frequency		Marker categorization
			Marker	Allele	
CCta185	226.3	0.105	0.991	0.843	bimodal (balancing selection)
CGag21	∞	0.057	0.918	0.764	bimodal (balancing selection)
CTaa125	63.1	0.136	0.927	0.699	bimodal (balancing selection)
CTaa235	14.0	0.157	0.813	0.618	bimodal (balancing selection)
GCat153	∞	0.094	0.723	0.476	bimodal (balancing selection)
GCtc42	118.0	0.120	0.904	0.770	bimodal (balancing selection)
CAta44	∞	0.032	1	0.738	bimodal [a]
GCtc76	∞	0.029	0.984	0.847	bimodal [a]
ACtc64	999.0	0.063	0.995	0.972	multimodal [b]
AGaa48	554.6	0.084	1	0.844	multimodal [b]
AGaa96	22.5	0.153	0.785	0.546	unimodal [b]
AGtg79	130.6	0.117	0.941	0.728	unimodal [b]
CAat185	249.0	0.078	0.969	0.812	unimodal [b]
CAta9	∞	0.035	1	0.940	unimodal [b]
CAta91	32.1	0.107	0.932	0.783	unimodal [b]
CCta61	28.6	0.126	0.934	0.761	unimodal [b]
GCat12	∞	0.033	1	0.982	unimodal [b]
GCat95	∞	0.054	0.913	0.739	unimodal [b]
GCat152	1249	0.075	0.93	0.740	unimodal [b]
GCta26	91.6	0.124	0.888	0.696	unimodal [b]
GCtc48	12.1	0.148	0.857	0.596	unimodal [b]
ACag100	134.1	0.113	0.226	0.131	low frequency [c]
CCac84	11.2	0.120	0.052	0.049	low frequency [c]
CCta81	27.9	0.105	0.074	0.044	low frequency [c]
CCta90	59.2	0.100	0.064	0.052	low frequency [c]
CCta180	31.5	0.102	0.06	0.037	low frequency [c]
CCta184	10.7	0.127	0.065	0.043	low frequency [c]
CCtt32	10.1	0.132	0.07	0.054	low frequency [c]
CCtt65	∞	0.078	0.33	0.206	low frequency [c]
CCtt131	39.3	0.109	0.119	0.082	low frequency [c]
CCtt145	383.6	0.068	0.075	0.074	low frequency [c]
CGtt4	16.5	0.139	0.204	0.123	low frequency [c]
CGtt6	40.0	0.109	0.098	0.069	low frequency [c]
ACtc47	∞	0.066	0.991	0.812	sex-linked [d]
ACtt16	∞	0.026	1	0.787	sex-linked [d]
ACtt18	105.4	0.120	0.881	0.734	sex-linked [d]
ACtt31	∞	0.024	0.998	0.762	sex-linked [d]
ACtt46	276.8	0.121	0.938	0.713	sex-linked [d]
AGac127	∞	0.079	0.998	0.829	sex-linked [d]
AGtg131	160.3	0.135	0.78	0.596	sex-linked [d]
CAat112	33.2	0.144	0.847	0.618	sex-linked [d]
CAta114	∞	0.024	0.986	0.744	sex-linked [d]
CAta142	∞	0.068	0.961	0.772	sex-linked [d]
CCac46	∞	0.038	0.995	0.747	sex-linked [d]
CCac57	∞	0.029	1	0.763	sex-linked [d]
CCta11	∞	0.024	1	0.752	sex-linked [d]
CCta29	34.7	0.148	0.66	0.477	sex-linked [d]
CCta73	12.4	0.101	0.998	0.983	sex-linked [d]

Table 3. Cont.

ID	PO	F_{ST}	Frequency		Marker categorization
			Marker	Allele	
CCtt100	∞	0.031	0.991	0.710	sex-linked [d]
CCtt150	∞	0.042	0.998	0.757	sex-linked [d]
CCtt191	∞	0.029	0.995	0.712	sex-linked [d]
CTaa27	∞	0.040	0.984	0.775	sex-linked [d]
CTaa102	∞	0.040	0.934	0.662	sex-linked [d]
CTaa113	∞	0.049	1	0.780	sex-linked [d]
CTag19	∞	0.045	1	0.772	sex-linked [d]
CTag94	∞	0.027	0.998	0.797	sex-linked [d]
CTtg10	∞	0.076	0.896	0.738	sex-linked [d]
CTtg105	∞	0.038	0.971	0.730	sex-linked [d]
CTtg155	∞	0.059	0.943	0.721	sex-linked [d]
GCat44	∞	0.068	0.946	0.778	sex-linked [d]
GCat69	∞	0.023	0.995	0.742	sex-linked [d]
GCat98	∞	0.044	0.995	0.792	sex-linked [d]
GCta14	∞	0.060	1	0.817	sex-linked [d]
GCta96	4999	0.090	0.895	0.632	sex-linked [d]
ACag54	262.2	0.125	0.576	0.831	sex-linked [d]
AGtg13	13.4	0.153	0.444	0.301	sex-linked [d]
GGtc1	∞	0.033	0.998	0.866	lab amplification difference [e]
GGtc9	45.3	0.122	0.988	0.748	lab amplification difference [e]
GGtc10	∞	0.047	0.995	0.981	lab amplification difference [e]
GGtc21	4999	0.097	0.846	0.714	lab amplification difference [e]
GGtc34	16.2	0.089	0.995	0.968	lab amplification difference [e]
GGtc37	∞	0.023	0.998	0.863	lab amplification difference [e]
GGtc55	∞	0.057	0.998	0.817	lab amplification difference [e]

The top six markers are prime candidates for being under balancing selection (underlined markers) due to their clear bimodal intensity distribution in all populations. Thirty-three additional markers are sex-lined. Given are marker ID, posterior odds (PO) for the marker, marker frequency, estimated allele frequency and marker categorization.
[a]clear bimodality was not found in single population.
[b]unimodal or multimodal band intensity distribution.
[c]low allele frequency.
[d]sex-chromosome linkage.
[e]bimodal distribution probably due to PCR amplification strength difference between 96-well plates of one specific primer pair (*GGtc*).

This study used a conservative *post-hoc* evaluation of AFLP marker band intensity distributions to provide further support for the authenticity of the signature of balancing selection, which allowed us to prioritize prime candidate loci for balancing selection. Six markers were characterized by low F_{ST} values, evenly distributed allele frequencies among populations (Figure 5A) and especially by the bimodal band intensity distribution, which clearly indicates the presence of heterozygous individuals in several populations (Figure 4A). Apart from these six loci, 27 markers showed peculiarities also compatible with other factors than only balancing selection. Twelve markers had low allele frequencies across Europe, maybe as a result of frequency-dependent selection, a selective mechanism that favours alleles when they are rare and might result in balanced genetic polymorphisms in populations [11]. But the observed low allele frequencies might also be explained by slightly negative selection [27]. For 15 markers no obvious bimodal band intensity distribution was observed, hence no clear signal of heterozygous individuals within populations could be identified, which might be explained by the stochasticity of slight technical variation in the sequencing machine that might have eroded the signal. However, especially the detection of 33 sex-chromosome linked markers (Figure 4B) clearly supports the use of AFLPs as a partially codominant marker system and indicates that heterozygous individuals or individuals carrying only one gene copy can reliably be estimated from the band intensity distribution in AFLP markers [56,57].

Compared to balancing selection the inference of directional selection is less questioned, even though some confounding demographic factors (e.g. surfing during range expansions; [88,89]) might produce some false positives. However, as we have used a quite stringent threshold for accepting a locus to be under selection (PO>10), our results suggest that we have here a very low false discovery rate of less than 1.4%. We detected that 6.7% of the informative markers probably evolved as a consequence of directional selection, which might be linked to adaptation to spatial heterogeneity of the environments of European vole populations. Given the wide distribution range and highly

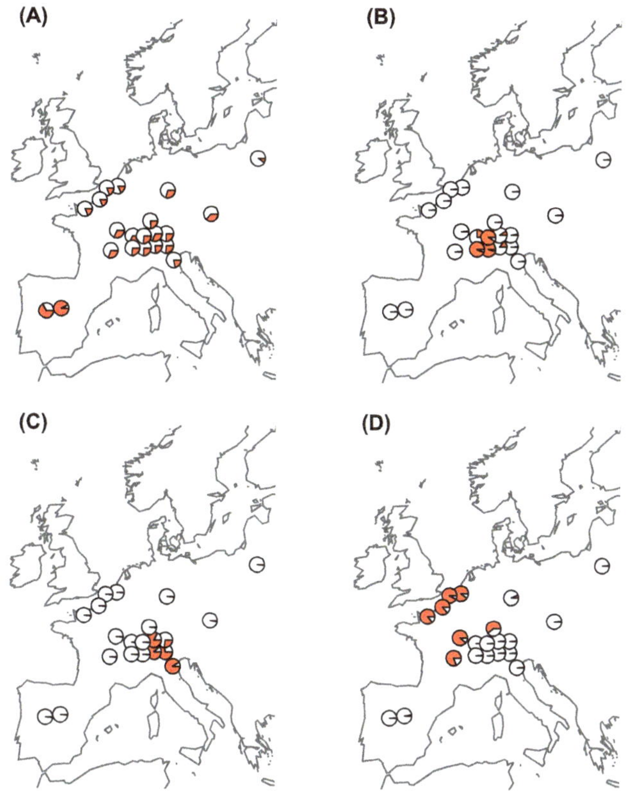

Figure 5. Allele frequency distributions of a locus potentially under balancing and three loci potentially under positive selection across Europe. Pie charts indicate the minor allele frequency. (A) Potential patterns of continental balancing selection of the locus *CTaa125*, which shows very homogenous allele frequencies across Europe. (B–D) Three loci under potential positive selection, which produced the strongest splits (F_{CT}) between two groups of populations across Europe identified by the locus-by-locus SAMOVA approach [82]. Shown are the loci (B) *ACag119* with a F_{CT} of 0.93, (C) CTaa3 with a F_{CT} of 0.89, and (D) GGac31 with a F_{CT} of 0.87.

heterogeneous environments where these voles are found, it is indeed expected that different polymorphisms might be selected in different populations and habitats [5,26]. The markers detected under positive selection in this study display a wide variety of allele frequency patterns across Europe. The PCA based on 138 markers under positive selection revealed a quite different structure (Figure 2C) than the PCA computed on 1,843 informative and neutral AFLPs (Figure 2A and B), indicating that selection acts differently on these loci than the interplay of drift and geographic separation. It is difficult to draw conclusions on the selection pressure from the allele frequency distribution of these markers; nevertheless there are some interesting patterns, which might be explained by environmental differences among populations. The two outlier loci that showed the strongest splits between two groups of populations across Europe (Figure 5B and C), were driven by populations from Alpine areas (some of the vole populations lived above 2,000 m asl). Hence they might be related to an adaptation to high elevation [40,90] or just to the highly heterogeneous environment observed at a small geographic scale, which is specific to Alpine regions [91]. These Swiss and Italian Alpine populations are much more separated in the PCA on loci under selection (Figure 2A) than in the neutral PCAs, indicating that probably many loci are under selection in this region. However, there are also patterns that are more difficult to interpret in environmental or geographic context, e.g. Figure 5D, but biotic interactions can be also very important for local adaption and are much more difficult to infer.

Outlook

AFLP genome scans enable us to detect markers under recent selection in the common vole genome, but it is unfortunately impossible to determine their function and location in the absence of a sequenced genome for this species. New high-throughput technologies make full genomes more accessible than before (for review see [92–94]), but target-capture sequencing of hundreds of individuals is still prohibitive for most non-model organisms [57] and full genome re-sequencing studies of pooled population data (Pool-Seq) is only possible for rather small genomes (see e.g. [91,95,96]). An alternative strategy would be the investigation of candidate loci for selection by direct high-throughput sequencing of AFLP fragments [60,97], which could be useful to further characterize candidate regions and genes linked with AFLP markers in this non-model organism.

Supporting Information

Table S1 The 21 selective primer combinations and their fluorescent labels used in the AFLP assay.

Table S2 SAMOVA results of the 138 loci probably under positive selection ranked by F_{CT} and the number of significant association with other loci having similar allele frequencies across Europe.

Acknowledgments

We are grateful to L. L. Bischoff, T. Hofer, E. Kindler and Y. Liu for discussions and comments during the preparation of the manuscript. We thank S. Tellenbach for technical assistance and I. Dupanloup for support with SAMOVA. We are very grateful to the following people for providing access to samples: J.B. Searle, M. Thoma, J. Thiel, J. Bryja, M. Ratkiewicz, B. Thorenz, N. Martinkova, M. Pascal, E. Jones and T. Cucchi.

Author Contributions

Conceived and designed the experiments: MCF MF GH LE. Performed the experiments: MCF. Analyzed the data: MCF MF. Wrote the paper: MCF GH LE.

References

1. Dobzhansky T (1955) A review of some fundamental concepts and problems of population genetics. Cold Spring Harbor Symposia on Quantitative Biology 20: 1–15.
2. Kimura M (1983) The neutral theory of molecular evolution. In: M.Nei, R.K.Koehn, editors.Evolution of Genes and Proteins.Sunderland, MA, USA: Sinauer. pp. 208–233.
3. Kimura M (1968) Evolutionary rate at the molecular level. Nature 217: 624–626.
4. Ohta T (1973) Slightly deleterious mutant substitutions in evolution. Nature 246: 96–98.
5. Huang Y, Wright SI, Agrawal AF (2014) Genome-wide patterns of genetic variation within and among alternative selective regimes. PLoS Genet 10: e1004527.
6. Levene H (1953) Genetic equilibrium when more than one ecological niche is available. The American Naturalist 87: 331–333.
7. Felsenstein J (1976) The theoretical population genetics of variable selection and migration. Annual Review of Genetics 10: 253–280.

8. Charlesworth D (2006) Balancing selection and its effects on sequences in nearby genome regions. PLoS Genetics 2: 379–384.
9. Maruki T, Kumar S, Kim Y (2012) Purifying selection modulates the estimates of population differentiation and confounds genome-wide comparisons across single-nucleotide polymorphisms. Molecular Biology and Evolution 29: 3617–3623.
10. Nielsen R (2005) Molecular signatures of natural selection. Annual Review of Genetics 39: 197–218.
11. Mitchell-Olds T, Willis JH, Goldstein DB (2007) Which evolutionary processes influence natural genetic variation for phenotypic traits? Nature Reviews Genetics 8: 845–856.
12. Aidoo M, Terlouw DJ, Kolczak M, McElroy PD, ter Kuile FO, et al. (2002) Protective effects of the sickle cell gene against malaria morbidity and mortality. Lancet 359: 1311–1312.
13. Piel FB, Patil AP, Howes RE, Nyangiri OA, Gething PW, et al. (2010) Global distribution of the sickle cell gene and geographical confirmation of the malaria hypothesis. Nature Communications 1: 104.
14. Verrelli BC, McDonald JH, Argyropoulos G, Destrol-Bisol G, Froment A, et al. (2002) Evidence for balancing selection from nucleotide sequence analyses of human *G6PD*. American Journal of Human Genetics 71: 1112–1128.
15. Allen SJ, O'Donnell A, Alexander NDE, Alpers MP, Peto TEA, et al. (1997) Alpha(+)-thalassemia protects children against disease caused by other infections as well as malaria. Proceedings of the National Academy of Sciences of the United States of America 94: 14736–14741.
16. Schroeder SA, Gaughan DM, Swift M (1995) Protection against bronchial asthma by *CFTR* delta F508 mutation: a heterozygote advantage in cystic fibrosis. Nature Medicine 1: 703–705.
17. Calafell F, Roubinet F, Ramirez-Soriano A, Saitou N, Bertranpetit J, et al. (2008) Evolutionary dynamics of the human ABO gene. Human Genetics 124: 123–135.
18. Baum J, Ward RH, Conway DJ (2002) Natural selection on the erythrocyte surface. Molecular Biology and Evolution 19: 223–229.
19. Garrigan D, Hedrick PW (2003) Perspective: detecting adaptive molecular polymorphism: lessons from the MHC. Evolution 57: 1707–1722.
20. Tan Z, Shon AM, Ober C (2005) Evidence of balancing selection at the HLA-G promoter region. Human Molecular Genetics 14: 3619–3628.
21. Wooding S, Stone AC, Dunn DM, Mummidi S, Jorde LB, et al. (2005) Contrasting effects of natural selection on human and chimpanzee CC chemokine receptor 5. American Journal of Human Genetics 76: 291–301.
22. Cho SC, Huang ZY, Green DR, Smith DR, Zhang JZ (2006) Evolution of the complementary sex-determination gene of honey bees: balancing selection and trans-species polymorphisms. Genome Research 16: 1366–1375.
23. Hohenlohe PA, Bassham S, Etter PD, Stiffler N, Johnson EA, et al. (2010) Population genomics of parallel adaptation in threespine stickleback using sequenced RAD tags. PLoS Genetics 6: e1000862.
24. Cork JM, Purugganan MD (2005) High-diversity genes in the *Arabidopsis* genome. Genetics 170: 1897–1911.
25. Delph LF, Kelly JK (2014) On the importance of balancing selection in plants. New Phytologist 201: 45–56.
26. Asthana S, Schmidt S, Sunyaev S (2005) A limited role for balancing selection. Trends in Genetics 21: 30–32.
27. Andres AM, Hubisz MJ, Indap A, Torgerson DG, Degenhardt JD, et al. (2009) Targets of balancing selection in the human genome. Molecular Biology and Evolution 26: 2755–2764.
28. Makinen HS, Cano M, Merila J (2008) Identifying footprints of directional and balancing selection in marine and freshwater three-spined stickleback (*Gasterosteus aculeatus*) populations. Molecular Ecology 17: 3565–3582.
29. Cutter AD, Payseur BA (2013) Genomic signatures of selection at linked sites: unifying the disparity among species. Nature Reviews Genetics 14: 262–274.
30. Barrett RDH, Schluter D (2008) Adaptation from standing genetic variation. Trends in Ecology and Evolution 23: 38–44.
31. Weedall GD, Conway DJ (2010) Detecting signatures of balancing selection to identify targets of anti-parasite immunity. Trends in Parasitology 26: 363–369.
32. Thomas JC, Godfrey PA, Feldgarden M, Robinson DA (2012) Candidate targets of balancing selection in the genome of *Staphylococcus aureus*. Molecular Biology and Evolution 29: 1175–1186.
33. Raj S, Pagani L, Gallego Romero I, Kivisild T, Amos W (2013) A general linear model-based approach for inferring selection to climate. BMC Genetics 14: 87.
34. Amambua-Ngwa A, Tetteh KKA, Manske M, Gomez-Escobar N, Stewart LB, et al. (2012) Population genomic scan for candidate signatures of balancing selection to guide antigen characterization in malaria parasites. PLoS Genet 8: e1002992.
35. Foll M, Gaggiotti O (2008) A genome-scan method to identify selected loci appropriate for both dominant and codominant markers: a Bayesian perspective. Genetics 180: 977–993.
36. Rosenberg NA, Pritchard JK, Weber JL, Cann HM, Kidd KK, et al. (2002) Genetic structure of human populations. Science 298: 2381–2385.
37. Caracristi G, Schlotterer C (2003) Genetic differentiation between American and European *Drosophila melanogaster* populations could be attributed to admixture of African alleles. Molecular Biology and Evolution 20: 792–799.
38. Niethammer J, Krapp F (1982) Handbuch der Säugetiere Europas. Wiesbaden: Akademische Verlagsgesellschaft. 649 p.
39. Hausser J (1995) Säugetiere der Schweiz: Verbreitung, Biologie, Oekologie. Basel: Birkhäuser Verlag.
40. Fischer MC, Foll M, Excoffier L, Heckel G (2011) Enhanced AFLP genome scans detect local adaptation in high-altitude populations of a small rodent (*Microtus arvalis*). Molecular Ecology 20: 1450–1462.
41. Braaker S, Heckel G (2009) Transalpine colonisation and partial phylogeographic erosion by dispersal in the common vole (*Microtus arvalis*). Molecular Ecology 18: 2518–2531.
42. Fink S, Excoffier L, Heckel G (2004) Mitochondrial gene diversity in the common vole *Microtus arvalis* shaped by historical divergence and local adaptations. Molecular Ecology 13: 3501–3514.
43. Heckel G, Burri R, Fink S, Desmet J-F, Excoffier L (2005) Genetic structure and colonization processes in European populations of the common vole *Microtus arvalis*. Evolution 59: 2231–2242.
44. Vitalis R, Dawson K, Boursot P (2001) Interpretation of variation across marker loci as evidence of selection. Genetics 158: 1811–1823.
45. Teshima KM, Coop G, Przeworski M (2006) How reliable are empirical genomic scans for selective sweeps? Genome Research 16: 702–712.
46. Schlotterer C (2003) Hitchhiking mapping - functional genomics from the population genetics perspective. Trends in Genetics 19: 32–38.
47. Storz JF (2005) Using genome scans of DNA polymorphism to infer adaptive population divergence. Molecular Ecology 14: 671–688.
48. Cavalli-Sforza LL (1966) Population structure and human evolution. Proceedings of the National Academy of Sciences of the United States of America 164: 362–379.
49. Lewontin RC, Krakauer J (1973) Distribution of gene frequency as a test of the theory of the selective neutrality of polymorphisms. Genetics 74: 175–195.
50. Beaumont MA (2005) Adaptation and speciation: what can Fst tell us? Trends in Ecology and Evolution 20: 435–440.
51. Luikart G, England PR, Tallmon D, Jordan S, Taberlet P (2003) The power and promise of population genomics: from genotyping to genome typing. Nature Reviews Genetics 4: 981–994.
52. Vos P, Hogers R, Bleeker M, Reijans M, Vandelee T, et al. (1995) AFLP - a new technique for DNA-fingerprinting. Nucleic Acids Research 23: 4407–4414.
53. Albertson RC, Markert JA, Danley PD, Kocher TD (1999) Phylogeny of a rapidly evolving clade: the cichlid fishes of Lake Malawi, East Africa. Proceedings of the National Academy of Sciences of the United States of America 96: 5107–5110.
54. Beaumont MA, Balding DJ (2004) Identifying adaptive genetic divergence among populations from genome scans. Molecular Ecology 13: 969–980.
55. Balding DJ, Nichols RA (1995) A method for quantifying differentiation between populations at multi-allelic loci and its implications for investigating identity and paternity. Genetica 96: 3–12.
56. Foll M, Fischer MC, Heckel G, Excoffier L (2010) Estimating population structure from AFLP amplification intensity. Molecular Ecology 19: 4638–4647.
57. Gaggiotti OE (2010) Bayesian statistical treatment of the fluorescence of AFLP bands leads to accurate genetic structure inference. Molecular Ecology 19: 4586–4588.
58. Beysard M, Heckel G (2014) Structure and dynamics of hybrid zones at different stages of speciation in the common vole (Microtus arvalis). Molecular Ecology 23: 673–687.
59. Hamilton G, Ray N, Heckel G, Beaumont M, Excoffier L (2005) Bayesian estimation of recent migration rates after a spatial expansion. Genetics 170: 409–417.
60. Lischer HEL, Excoffier L, Heckel G (2014) Ignoring heterozygous sites biases phylogenomic estimates of divergence times: implications for the evolutionary history of *Microtus* voles. Molecular Biology and Evolution 31: 817–831.
61. Martínková N, Barnett R, Cucchi T, Struchen R, Pascal M, et al. (2013) Divergent evolutionary processes associated with colonization of offshore islands. Molecular Ecology 22: 5205–5220.
62. Sutter A, Beysard M, Heckel G (2013) Sex-specific clines support incipient speciation in a common European mammal. Heredity 110: 398–404.
63. Schmidt-Chanasit J, Essbauer S, Petraityte R, Yoshimatsu K, Tackmann K, et al. (2010) Extensive host sharing of central European Tula virus. Journal of Virology 84: 459–474.
64. Ulrich RG, Heckel G, Pelz HJ, Wieler LH, Nordhoff M, et al. (2009) Rodents and rodent associated disease pathogen. Bundesgesundheitsblatt-Gesundheitsforschung-Gesundheitsschutz 52: 352–369.
65. Ulrich RG, Schmidt-Chanasit J, Schlegel M, Jacob J, Pelz HJ, et al. (2008) Network "Rodent-borne pathogens" in Germany: longitudinal studies on the geographical distribution and prevalence of hantavirus infections. Parasitology Research 103: S121–S129.
66. Ali H, Drewes S, Sadowska E, Mikowska M, Groschup M, et al. (2014) First molecular evidence for *Puumala* hantavirus in Poland. Viruses 6: 340–353.
67. Schmidt S, Essbauer S, Mayer-Scholl A, Poppert S, Schmidt-Chanasit J, et al. (2014) Multiple infections of rodents with zoonotic pathogens in Austria. Vector-Borne and Zoonotic Diseases, in press.
68. Sambrook J, Fritsch EF, Maniatis T (1989) Molecular cloning: a laboratory manual. Cold Spring Harbor, NY: Cold Spring Harbor Laboratory Press.
69. Fink S, Fischer MC, Excoffier L, Heckel G (2010) Genomic scans support repetitive continental colonization events during the rapid radiation of voles (Rodentia: *Microtus*): the utility of AFLPs versus mitochondrial and nuclear sequence markers. Systematic Biology 59: 548–572.
70. Ehrich D (2006) AFLPDAT: a collection of R functions for convenient handling of AFLP data. Molecular Ecology Notes 6: 603–604.

71. Nei M (1987) Molecular evolutionary genetics. New York: Columbia University Press.
72. Kosman E (2003) Nei's gene diversity and the index of average differences are identical measures of diversity within populations. Plant Pathology 52: 533–535.
73. Vitalis R, Dawson K, Boursot P, Belkhir K (2003) DetSel 1.0: a computer program to detect markers responding to selection. Journal of Heredity 94: 429–431.
74. Beaumont MA, Nichols RA (1996) Evaluating loci for use in the genetic analysis of population structure. Proceedings of the Royal Society B-Biological Sciences 363: 1619–1626.
75. Perez-Figueroa A, Garcia-Pereira MJ, Saura M, Rolan-Alvarez E, Caballero A (2010) Comparing three different methods to detect selective loci using dominant markers. Journal of Evolutionary Biology 23: 2267–2276.
76. Roesti M, Salzburger W, Berner D (2012) Uninformative polymorphisms bias genome scans for signatures of selection. BMC Evolutionary Biology 12: 94.
77. Rannala B, Hartigan JA (1996) Estimating gene flow in island populations. Genetical Research 67: 147–158.
78. Balding DJ (2003) Likelihood-based inference for genetic correlation coefficients. Theoretical Population Biology 63: 221–230.
79. Jeffreys H (1961) The theory of probability: Oxford University Press.
80. Benjamini Y, Hochberg Y (1995) Controlling the false discovery rate: a practical and powerful approach to multiple testing. Journal of the Royal Statistical Society Series B-Methodological 57: 289–300.
81. R Development Core Team (2011) R: a language and environment for statistical computing. Vienna, Austria: R Foundation for Statistical Computing.
82. Dupanloup I, Schneider S, Excoffier L (2002) A simulated annealing approach to define the genetic structure of populations. Molecular Ecology 11: 2571–2581.
83. Holm S (1979) A simple sequentially rejective multiple test procedure. Scandinavian Journal of Statistics 6: 65–70.
84. Novembre J, Johnson T, Bryc K, Kutalik Z, Boyko AR, et al. (2008) Genes mirror geography within Europe. Nature 456: 98–U95.
85. Kimura M (1968) Genetic variability maintained in a finite population due to mutational production of neutral and nearly neutral isoalleles. Genetical Research 11: 247–269.
86. Bubb KL, Bovee D, Buckley D, Haugen E, Kibukawa M, et al. (2006) Scan of human genome reveals no new loci under ancient balancing selection. Genetics 173: 2165–2177.
87. Schweizer M, Excoffier L, Heckel G (2007) Fine-scale genetic structure and dispersal patterns in the common vole *Microtus arvalis*. Molecular Ecology 16: 2463–2473.
88. Excoffier L, Foll M, Petit RJ (2009) Genetic consequences of range expansions. Annual Review of Ecology Evolution and Systematics 40: 481–501.
89. Klopfstein S, Currat M, Excoffier L (2006) The fate of mutations surfing on the wave of a range expansion. Molecular Biology and Evolution 23: 482–490.
90. Storz JF (2010) Genes for high altitudes. Science 329: 40–41.
91. Fischer MC, Rellstab C, Tedder A, Zoller S, Gugerli F, et al. (2013) Population genomic footprints of selection and associations with climate in natural populations of *Arabidopsis halleri* from the Alps. Molecular Ecology 22: 5594–5607.
92. Ellegren H (2008) Sequencing goes 454 and takes large-scale genomics into the wild. Molecular Ecology 17: 1629–1631.
93. Tautz D, Ellegren H, Weigel D (2010) Next generation molecular ecology. Molecular Ecology 19: 1–3.
94. Gibbons JG, Janson EM, Hittinger CT, Johnston M, Abbot P, et al. (2009) Benchmarking next-generation transcriptome sequencing for functional and evolutionary genomics. Molecular Biology and Evolution 26: 2731–2744.
95. Turner TL, Bourne EC, Von Wettberg EJ, Hu TT, Nuzhdin SV (2010) Population resequencing reveals local adaptation of *Arabidopsis lyrata* to serpentine soils. Nature Genetics 42: 260–263.
96. Rellstab C, Zoller S, Tedder A, Gugerli F, Fischer MC (2013) Validation of SNP allele frequencies determined by pooled next-generation sequencing in natural populations of a non-model plant species. PLoS One 8: e80422.
97. Paris M, Despres L (2012) Identifying insecticide resistance genes in mosquito by combining AFLP genome scans and 454 pyrosequencing. Molecular Ecology 21: 1672–1686.
98. Mitchell-Jones AJ, Amori G, Bogdanowicz W, Krystufek B, Reijnders PJH, et al. (1999) The atlas of European mammals. London: T. Poyser, A. D. Poyser.

16

Genome Duplication and Gene Loss Affect the Evolution of Heat Shock Transcription Factor Genes in Legumes

Yongxiang Lin[1,2]☯, Ying Cheng[1]☯, Jing Jin[1], Xiaolei Jin[1], Haiyang Jiang[1], Hanwei Yan[1], Beijiu Cheng[1]*

1 Key Lab of Crop Biology of Anhui Province, School of Life Sciences, Anhui Agricultural University, Hefei, Anhui, China, 2 Crop Research Institute, Anhui Academy of Agricultural Sciences, Hefei, Anhui, China

Abstract

Whole-genome duplication events (polyploidy events) and gene loss events have played important roles in the evolution of legumes. Here we show that the vast majority of Hsf gene duplications resulted from whole genome duplication events rather than tandem duplication, and significant differences in gene retention exist between species. By searching for intraspecies gene colinearity (microsynteny) and dating the age distributions of duplicated genes, we found that genome duplications accounted for 42 of 46 Hsf-containing segments in *Glycine max*, while paired segments were rarely identified in *Lotus japonicas, Medicago truncatula* and *Cajanus cajan*. However, by comparing interspecies microsynteny, we determined that the great majority of Hsf-containing segments in *Lotus japonicas*, *Medicago truncatula* and *Cajanus cajan* show extensive conservation with the duplicated regions of *Glycine max*. These segments formed 17 groups of orthologous segments. These results suggest that these regions shared ancient genome duplication with Hsf genes in *Glycine max*, but more than half of the copies of these genes were lost. On the other hand, the *Glycine max* Hsf gene family retained approximately 75% and 84% of duplicated genes produced from the ancient genome duplication and recent *Glycine*-specific genome duplication, respectively. Continuous purifying selection has played a key role in the maintenance of Hsf genes in *Glycine max*. Expression analysis of the Hsf genes in *Lotus japonicus* revealed their putative involvement in multiple tissue-/developmental stages and responses to various abiotic stimuli. This study traces the evolution of Hsf genes in legume species and demonstrates that the rates of gene gain and loss are far from equilibrium in different species.

Editor: Marc Robinson-Rechavi, University of Lausanne, Switzerland

Funding: This work was supported by grants from the Special Fund for Agro-scientific Research in the Public Interest (Grant No. 201303001) and the Genetically Modified Organisms Breeding Major Projects (Grant No. 2011ZX08003-002). The funders had no role in study design, data collection and analysis, decision to publish, or preparation of the manuscript.

Competing Interests: The authors have declared that no competing interests exist.

* Email: bjchengahau@163.com

☯ These authors contributed equally to this work.

Introduction

Whole-genome duplication, or polyploidy, is a common phenomenon in the evolution of plants and is particularly widespread in angiosperms [1,2]. Many modern diploid plants have experienced one or more episodes of polyploidy and possess vestiges of multiple rounds of polyploidy [3,4]. Recently, comparisons between legume genomes have revealed that legumes of the large Papilionoideae subfamily (papilionoids) have undergone whole-genome duplication [5]. This older shared polyploidy event is estimated to have occurred 56 to 65 million years ago (Mya) [6,7]. A second, more recent genome duplication event occurred only in the lineage leading to *Glycine* up to 13 Mya [8]. Over time, the genomes of these plants diploidized, accompanied by rearrangements and loss of genes and chromosomal segments; this eliminated much of the evidence of the original duplication. Genome duplication and subsequent fractionation have played key roles in shaping present-day legume genomes [9].

A gene family is a group of similar genes resulting from wholesale or partial gene duplication; the size of a gene family reflects the number of duplicated genes (paralogs) in each species [10]. Whole-genome duplication causes each gene in the genome to be present in two copies. However, after duplication, not all categories of genes respond the same way to polyploidy; gene loss often occurs independently in different gene families. Some gene families, such as NBS-LRR resistance genes, show rapid rates of turnover among family members, with the loss of major gene lineages in some plant families [11,12]. Alternatively, most genes are retained in other families with highly conserved amino acid sequences, such as transcription factors [13]. In the *Arabidopsis* genome, three whole-genome duplications that have occurred in the past 350 million years have brought about a greater than 90% increase in the number of transcription factor, signal transduction and developmental genes [14]. Genes retained after polyploidy may buffer critical functions, but over time, a gradual erosion of this capacity of duplicated genes may contribute to the cyclicality of genome duplication [15].

Here, we examine the evolution of the heat shock transcription factor (Hsf) family in legume species. Hsfs serve as the terminal components of signal transduction and are the central regulators of the expression of heat shock proteins and other heat shock-induced genes that confer thermotolerance to all eukaryotes [16–21]. Like many other transcription factors, Hsfs have a modular structure [22]. Hsf proteins share a well-conserved DNA binding domain (DBD) at their N termini, an adjacent bipartite oligomerization domain (HR-A/B region) composed of hydrophobic heptad repeats, a nuclear location signal (NLS), a nuclear export signal (NES) and a C-terminal activation domain (AHA motifs) [16,23–26]. In contrast to the small number of Hsf genes found in *Drosophila*, *Caenorhabditis elegans*, yeasts and animals [27–33], the Hsf system is more complex in plants than in any other organisms investigated thus far. To date, on the basis of genome-wide analysis, 21 [34], approximately 25 [35,36] and 25 [37] Hsf genes have been identified in the model plants *Arabidopsis*, rice and maize, respectively. On the basis of sequence divergence, three Hsf classes (A, B and C) and several subgroups are currently recognized [22]. Previous studies have shown that there are no apparent tandem duplications, and no clustered organization, in the Hsf families of several monocot and eudicot species [35,37]. How did the members of this gene family arise, and how are the copy numbers of genes in this family maintained? Multiple rounds of genome duplication and extensive gene loss in different plant lineages may have led to the generation of independent growth and evolution models of Hsf family genes.

In this study, we analyzed the Hsf gene families from six papilionoid legume species for which substantial information about genomes or transcriptomes was available, namely *Lotus japonicus* (birdsfoot trefoil), *Medicago truncatula* (barrel medic), *Cicer arietinum* (chickpea), *Glycine max* (soybean), *Cajanus cajan* (pigeonpea) and *Phaseolus vulgaris* (common bean). The aim of this investigation was to determine which genes were derived from genome duplication, subsequently giving rise to paralogs, which genes descended from speciation events, giving rise to orthologs and which genes have undergone gene loss. In addition to examining the phylogeny of the Hsf family, we performed a comprehensive examination of the legume genome structure anchored by Hsf genes. Furthermore, we searched for gene microsynteny within and between the genomes of the legumes to investigate the evolutionary history of the Hsf regions. Our data show that extensive synteny remains in the homeologous regions within/between legume species. In addition, most Hsf regions can be traced to ancient papilionoid-specific whole-genome duplication or recent *Glycine*-specific whole-genome duplication. However, different rates of gene loss in the Hsf family have occurred along separate lineages of legume. Our results may help facilitate the extrapolation of Hsf gene function from one lineage to another.

Materials and Methods

Data retrieval and sequence analysis

The most recent versions of genome, protein and cDNA sequences of each species were downloaded from the respective genome sequence sites as follows: *L. japonicus* (version 2.5) from the *L. japonicus* Genome Sequencing Project (http://www.kazusa.or.jp/lotus/), *M. truncatula* (version 3.5) from the *M. truncatula* Genome Sequencing Project (http://www.medicagohapmap.org/?genome), *G. max* (version 1.01) from the Soybean Genome Sequencing Project (http://www.phytozome.net/soybean.php), *C. cajan* (version 1.0) from the International Crops Research Institute for the Semi-Arid Tropics (http://www.icrisat.org/gt-bt/iipg/Genome_Manuscript.html), *P. vulgaris* (version 0.9) from the US Department of Energy Joint Genome Institute (http://www.phytozome.net/commonbean.php), *P. patens* (version 1.6) from the Joint Genome Institute (http://www.phytozome.net/physcomitrella), *S. moellendorffii* (version 1.0 filtered model 3) from the Joint Genome Institute (http://genome.jgi-psf.org/Selmo1/Selmo1.download.ftp.html) and *C. reinhardtii* (version 4.3) from the Joint Genome Institute (http://www.phytozome.net/chlamy.php). Although information about the whole genome of *C. arietinum* is not currently available, the *C. arietinum* transcriptome has been sequenced using next-generation sequence technology [38] and was obtained from the Chickpea Transcriptome Database (http://www.nipgr.res.in/ctdb.html). The downloaded nucleotide and protein sequences of each species were in turn used to build local databases using DNATOOLS software. Published Hsf protein sequences [34,35,39] were used to search the Pfam database [40], and an integrated and exactly conserved Hsf-type DBD domain sequence based on the Hidden Markov Model (HMM) was obtained. The Hsf domain (PF00447) in the Pfam HMM library was then used in BLASTP searches to identify Hsfs from the local databases. For the *C. arietinum* transcriptome, a TBLASTN search was performed, and the identified full-length cDNAs were translated in the correct frame. Only hits returning E-values of less than 0.001 were considered for further analysis. This step was crucial for finding as many similar sequences as possible. Moreover, on the basis of BLASTN search results in the genome databases using the predicted cDNA sequences of Hsf genes, information was obtained about the chromosome locations of these genes. Redundant sequences with different identification numbers and the same chromosome locus were eliminated from the data set. To confirm the presence of both Hsf-type DBD domain and HR-A/B regions in the sequences obtained, the predicted protein sequences of Hsf genes were analyzed in the Pfam HMM database and the SMART tool [41] to find the DBD domain, and proteins without these regions were excluded from the data set. Following this step, the remaining sequences were examined for the HR-A/B regions using the MARCOIL program [42] and the SMART tool, both of which can recognize the coiled-coil structure representing the core of the HR-A/B region; proteins without HR-A/B regions were removed from the data set.

In addition to the DBD and HR-A/B domains, many Hsfs also contain an NLS and an NES domain, and most plant class A Hsfs contain one or several AHA motifs. To identify the NLS domain in the Hsfs, the program PredictNLS [43] (from the website) was used. In addition, the NetNES 1.1 server [44] was used to detect the NES domain in all of the Hsfs. Moreover, since the highly conserved amino acid sequence of AHA motifs has been elaborated previously, and detailed investigations of these motifs have been reported [25,26,34], the AHA motifs could be predicted based on sequence comparisons and their characteristics. Information about the AHA motifs was further verified by alignments with published Hsf sequences.

Multiple sequence alignments and phylogenetic analysis

Multiple sequence alignments using ClustalX (version 1.83) [45] were performed on the N-terminal domains of the Hsfs obtained, including the DBD domains, the HR-A/B regions and parts of the linker between these regions. The alignment was then adjusted manually by Jalview. A phylogenetic tree was constructed with the aligned protein sequences using MEGA (version 4.0) [46] using the NJ method with the following parameters: Poisson correction, pairwise deletion and bootstrap (1,000 replicates; random seed). The Hsf of *Saccharomyces cerevisiae* (ScHsf1), and the Hsfs of *C. reinhardtii*, *S. moellendorffii* and *P. patens*, were used as the outgroup. In order to analyze the classes and subgroups of the legume Hsf families, 25 maize Hsfs (ZmHsfs) [37], 25 rice Hsfs (OsHsfs) [35] and 21 *Arabidopsis* Hsfs (AtHsfs) [34] were included in the phylogenetic analysis by generating a NJ tree (Poisson correction, pairwise deletion and bootstrap = 1,000 replicates). To confirm the robustness of the NJ tree, we built the ML tree using maximum likelihood method (MEGA 6.0; bootstrap = 1,000 replicates, amino acid substitution model, Jones-Taylor-Thornton matrix).

Intraspecies microsynteny analysis

To categorize the expansion of the Hsf gene families, the physical locations of all members of this family were examined in *L. japonicus*, *M. truncatula*, *G. max* and *C. cajan*. Tandem duplication is characterized by multiple gene family members occurring within either the same or neighboring genomic regions. Tandem duplicated genes were defined as genes in any gene pair, T1 and T2, that (1) belong to the same gene family, (2) are located within 100 kb each other, and (3) are separated by zero, one or fewer, five or fewer, or 10 or fewer nonhomologous (not in the same gene family as T1 and T2) spacer genes [47]. A method similar to that of Maher *et al.* [48] and Zhang *et al.* [49] was implemented to identify large-scale duplication events. To classify two Hsf genes as residing within a duplicated block, their neighboring protein-coding genes must be highly similar at the amino acid level. First, all Hsf genes in each family were used as the original anchor points. Next, 15 protein-coding sequences upstream and downstream of each anchor point were compared by pairwise BLASTP analysis to identify duplicated genes between two independent regions. The software then counted the total number of protein-coding genes flanking any anchor point that had the best nonself match (*E-value* $<10^{-10}$) with a protein-coding gene neighboring another anchor point. When four or more such gene pairs with syntenic relationships were detected, the two regions were considered to have originated from a large-scale duplication event.

Interspecies microsynteny analysis

The analysis of microsynteny across species was based on comparisons of the specific regions containing Hsf genes. Similarly, the Hsf genes of *L. japonicus*, *M. truncatula*, *G. max* and *C. cajan* were set as the anchor points according to their physical locations. The protein-coding sequences assigned to the flanking regions of each Hsf gene in one species were aligned with those in the other species by pairwise comparisons. A syntenic block is defined as the region in which three or more conserved homologs (BLASTP *E-value* $<10^{-20}$) were located within a 100 kb region between genomes [50].

Duplication event dating and adaptive evolution analysis

The duplicated gene pairs within each duplicated block were used to calculate Ks and to analyze Ka/Ks ratios. Protein sequences of the gene pairs were aligned using MUSCLE [51], and the results were used to guide the codon alignments by PAL2NAL [52]. The generated codon alignments were subjected to computation of Ks and divergence levels (Ka/Ks ratios) using DnaSP software (version 5.10). A sliding window analysis of Ka/Ks ratios was performed with the following parameters: window size, 150 bp; step size, 9 bp.

When dating large-scale duplication events, Ks can be used as the proxy for time. For each pair of duplicated regions, the mean Ks of the flanking conserved genes were calculated, and these values were then translated into divergence time in millions of years assuming a rate of 6.1×10^{-9} substitutions per site per year. The divergence time (T) was calculated as $T = Ks/(2\times6.1\times10^{-9})$ 10^{-6} Mya [53].

Codeml program is available under PAML (phylogenetic analysis maximum likelihood) V. 4.7 software [54]. To further assess whether positive selection acts upon specific sites, six site models that allow ω ratios (Ka/Ks ratios) to vary among sites, as implemented in the program Codeml, were used based on the coding sequences of Hsf genes [55]. These models are the one-ratio model (M0), the nearly neutral model (M1a), the positive-selection model (M2a), the discrete model (M3), the β model (M7) and the β & ω model (M8). The likelihood ratio tests (LRT) were performed to compare the corresponding models with and without selection (ie, M0 vs M3, M1a vs M2a, and M7 vs M8) [55]. M0–M3 comparison can be used to test whether ω values vary among sites. Both M1a–M2a and M7–M8 comparisons can be used to test positive selection acting on sites [55]. The Bayes empirical Bayes (BEB) were used in the M2a and M8 models to calculate the Bayesian posterior probability (BPP) of the codon sites under a positive selection [56].

Microarray data

The data for evaluating Hsf gene expression in various tissues of *L. japonicus* acquired with the Lotus 52 K Affy chip were obtained from the Lotus Transcript Profiling Resource [57]. The locus names of Hsf genes in the *L. japonicus* Genome Sequencing Project were used to query the corresponding probe set IDs in the GeneChip. The log transformed expression values for the retrieved probe sets were then used to perform cluster analysis by Cluster [58].

Plant material and stress treatments

L. japonicus plants (Miyakojima MG-20) were grown in a greenhouse at 25±2°C with a 14/10 h (light/dark) photoperiod. Four-week-old seedlings were prepared for abiotic stress treatments. For temperature treatments, the uniform-sized seedlings were transferred to the temperature-controlled growth chambers, which were maintained at 42±1°C for heat stress and at 4±1°C for cold stress. For oxidative stress, seedling leaves were sprayed with 10 mM H_2O_2 solution. After each treatment, the leaves of the seedlings were harvested at 0, 1 h and were immediately frozen in liquid nitrogen and kept at −80°C pending the extraction of RNA.

Quantitative real-time PCR

Total RNA was extracted from the collected samples using Trizol reagent (Invitrogen, USA), followed by Dnase I digestion to remove residual genomic DNA contamination. The quality and quantity of the total RNA was measured by electrophoresis on 1% (w/v) agarose gels and examined with a NanoDrop ND-1000 UV-Vis spectrophotometer (NanoDrop Technologies, Inc.). For each sample, the first strand cDNA was reverse transcribed from 1 μg total RNA using the QuantiTect Rev. Transcription Kit (Qiagen, Germany) according to the manufacturer's instructions. Quantitative real-time PCR was conducted on an ABI PRISM 7300 real-time PCR system (Applied Biosystems, USA). All the gene-specific primer sequences for quantitative real-time PCR were designed by Primer Express Version 3.0 software (Applied Biosystems, USA) and are listed in Table S1. Each PCR reaction mixture contained 2.0 μL transcription product, 400 nM primers, and 12.5 μL 2×SYBR Green Master Mix Reagent (Applied Biosystems, USA) in a total volume of 25 μL. The thermal cycle used was as follows: 50°C for 2 min, 95°C for 10 min, 40 cycles of 95°C for 15 s, and 60°C for 1 min. Melting curve analysis was then used to verify the identity of the amplicons and the specificity of the reaction. To normalize the variance among samples, *β-tubulin* was used as an endogenous control. The relative expression of each gene was calculated as the $\Delta\Delta C_T$ value in comparison to unstressed samples (Applied Biosystems, USA). These experiments were independently replicated at least three times for each sample.

Results

Hsf genes form a complex family in legume genomes

Papilionoids represent all major legume crops and model legume species. Most papilionoid species fall into one of two large clades, i.e., the temperate galegoid clade (cool season legumes) and the Millettioid clade (tropical season legumes). To determine the number of full-length Hsf proteins in the six legumes, BLAST and HMM searches were performed against the annotated genomes of *L. japonicus*, *M. truncatula*, *G. max*, *C. cajan* and *P. vulgaris* as well as transcriptome data for *C. arietinum*. A total of 11, 19 and 13 Hsfs were identified in the cool season legumes *L. japonicus*, *M. truncatula* and *C. arietinum*, respectively, while 46, 22 and 29 Hsfs were identified in the tropical season legumes *G. max*, *C. cajan* and *P. vulgaris*, respectively (Table 1, Table S2). To obtain a broader perspective on the evolutionary history of the legume Hsf family, we also searched for Hsf genes in the single-celled green alga *Chlamydomonas reinhardtii*, the lycophyte *Selaginella moellendorffii* and the bryophyte *Physcomitrella patens*, which contain only two, one and seven Hsfs, respectively (Table 1, Table S2). Sequence alignment and domain analysis of the deduced Hsfs showed that the highly structured N-terminal DBD domain of each Hsf is the most conserved region, and the adjacent HR-A/B region, with a heptad pattern of hydrophobic amino acid residues, leads to the formation of a helical coiled-coil structure (Figure S1 and Figure S2; Table S3). These data indicate that the Hsf gene family has expanded in legumes relative to the basal plant taxa analyzed here, and to a greater extent in tropical season legumes than in cool season legumes.

Combining the six legume protein sequences, we constructed a phylogenetic tree using neighbor-joining analysis (Figure 1). The legume Hsfs were grouped into three major classes, A, B and C. Class A and B were further divided into nine (A1–9) and five (B1–5) subgroups with well-supported bootstrap values (Figure S3). Accordingly, previously defined classes and subgroups [22,34] were identified from legume Hsfs (Figure S3).

Furthermore, phylogenetic analysis showed that the Hsf genes from the six legume species could be delineated into 18 well-supported ancient gene lineages (clades 1–18 in Figure 1), and strong amino acid sequence conservation was proven from the short branch lengths at the tips of the clades, indicating close evolutionary relationships among members. In each clade, branches with more than one Hsf gene from the same species are likely to have undergone gene duplication events, whereas the absence of representatives in some species is probably attributable to gene losses. In most cases, Hsf genes from tropical season legumes were more abundant than Hsf genes from cool season legumes. It is worth noting that in almost every clade, at least one extra copy of the Hsfs from *G. max* was present compared with that from the other species. On the other hand, in each clade, the members of different species may have evolved from a common ancestral gene by divergence of the lineage. Therefore, the 18 defined clades provide a framework for inferring parologs and orthologs of Hsfs. The phylogenetic relationships based on the ML tree were largely consistent with these results (Figure S4).

Genome duplication played an important role in the expansion of the Hsf family

To examine the relationship between the genetic divergences within each legume Hsf family and the corresponding expansion patterns, we further surveyed gene duplication events in the legume Hsf families (Figure 2; Table 2). *P. vulgaris* and *C. arietinum* were excluded from this analysis due to the lack of information about the locations of their Hsf genes. We character-

Table 1. The number of Hsfs identified in the legume species and lower plants.

Class	LjHsf	MtHsf	CaHsf	GmHsf	CcHsf	PvHsf	SmHsf	PpHsf	CrHsf
A	9	12	8	26	14	17	0	3	1
B	2	7	4	19	7	11	1	4	1
C	0	0	1	1	1	1	0	0	0
Total	11	19	13	46	22	29	1	7	2

Lj = L. japonicus, Mt = M. truncatula, Ca = C. arietinum, Gm = G. max, Cc = C. cajan, Pv = P. vulgaris, Sm = S. moellendorffii, Pp = P. patens, Cr = C. reinhardtii.

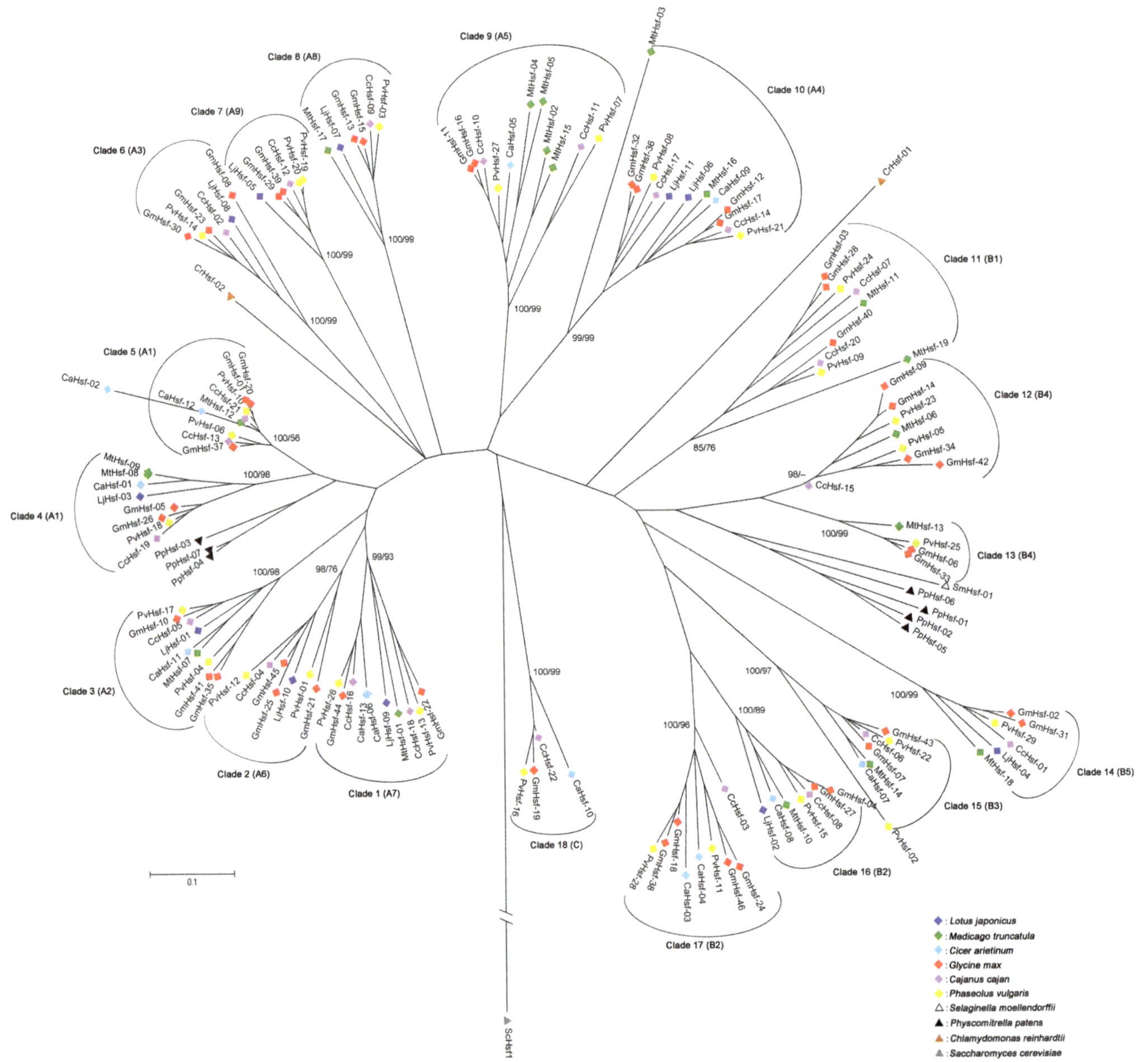

Figure 1. Phylogenetic tree of 140 Hsf proteins from *L. japonicus*, *M. truncatula*, *C. arietinum*, *G. max*, *C. cajan* and *P. vulgaris*. This tree was constructed based on amino acid sequence comparison of the conserved N-terminal regions of Hsfs including the DNA-binding domain, the HR-A/B region and parts of the linker between them, using the neighbor-joining method with 1,000 bootstrap replicates. The Hsf of *Saccharomyces cerevisiae* (ScHsf1) and the Hsfs of *C. reinhardtii*, *S. moellendorffii* and *P. patens* were used as the outgroup. The colors indicate the species background of the Hsfs. The tree was divided into 18 shared clades (Clades 1–18) according to evolutionary distances. The bootstrap values of both neighbor-joining (NJ) tree (first number; 1000 replicates) and maximum likelihood (ML) tree (second number; 1000 replicates) were shown on the branches leading to each of the clades. The clades were supported by high bootstrap values in neighbor-joining and maximum likelihood analyses. Different subclasses of Hsfs are indicated in brackets. Gene names are presented in Table S2. The scale bar represents 0.1 amino acid changes per site.

ized Hsf paralogs as being cluster or scattered. Chromosomal location analyses showed that the majority of Hsf genes are randomly scattered in the genomes, with tandemly clustered genes occurring in several places (Table S2). Legumes have experienced one or more polyploidy events. Thus, large-scale duplication events may have played an important role in the evolution of the legume Hsf families.

To investigate this possibility, we searched for gene similarity in the Hsfs flanking regions. If four or more of the 15 up- and downstream genes flanking two Hsf genes achieved a best non-self match using BLASTP (*E-value* $<10^{-10}$), we considered these gene pairs to be conserved and defined these two regions as derived from a large-scale duplication event. We also defined a flexible set as a set of genes in which the flanking regions of an Hsf pair contained two or three conserved genes to avoid the possibility that pairs of Hsf genes resided within more divergent blocks.

We identified three conserved genes flanking the pair *LjHsf-06/LjHsf-11* in *L. japonicus*. Therefore, this pair is considered to have evolved from large-scale duplication, based on our flexible set. In *C. cajan*, one gene pair (*CcHsf-10/CcHsf-11*) was found to have involved large-scale duplication. However, it should be noted that approximately 45% of the Hsf family could not be assigned to

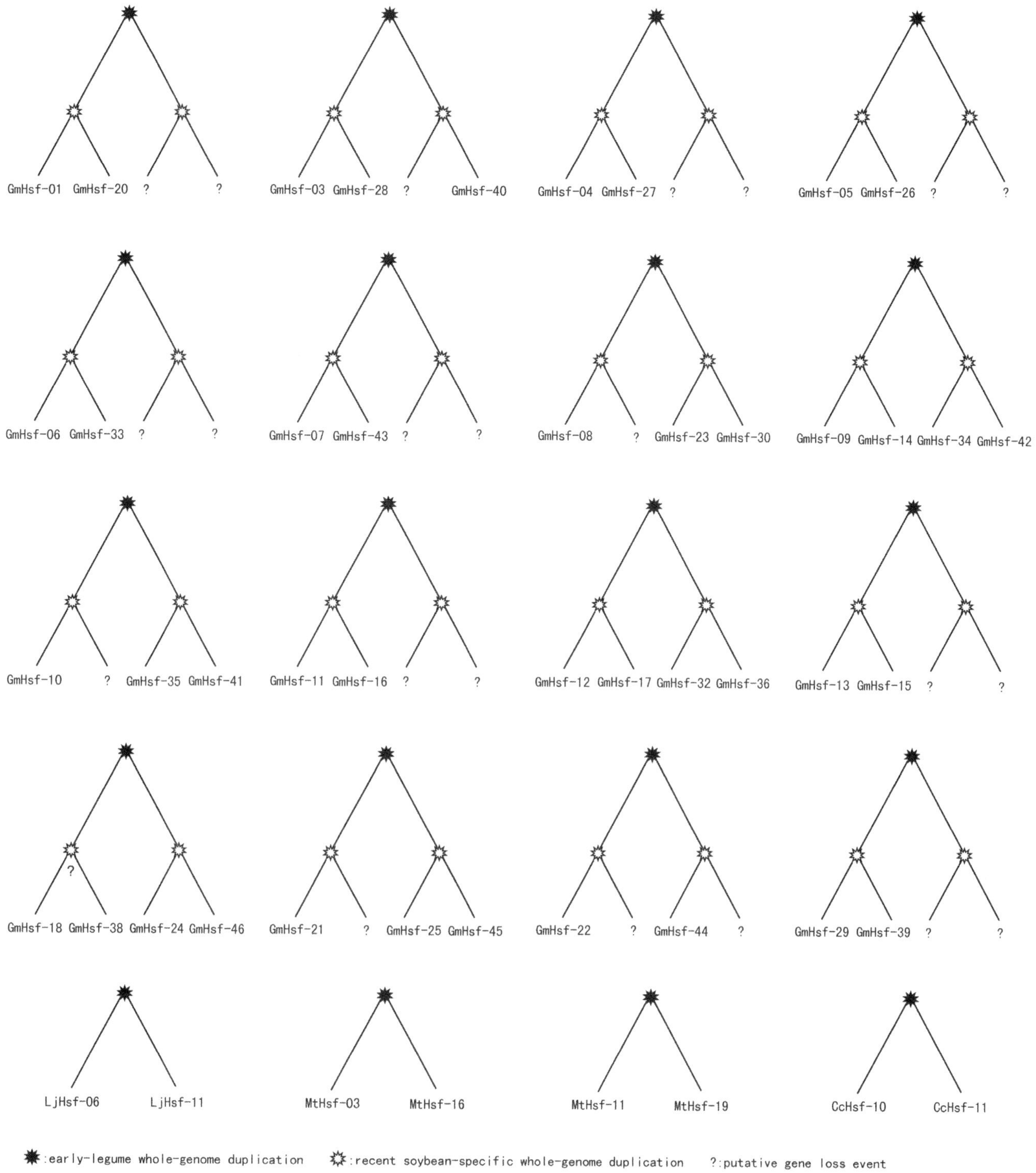

Figure 2. Idealized gene trees of the duplication groups of Hsf genes in *G. max, L. japonicus, M. truncatula* and *C. cajan*. Each tree represents a duplication group from large-scale gene duplication. As shown in the trees, every Hsf gene of *G. max* was expected to be present in four copies after two rounds of whole-genome duplication (early and recent). Similarly, the number of Hsf genes in *L. japonicus, M. truncatula* and *C. cajan* will have doubled after the early-legume whole-genome duplication. The five duplicated gene pairs (*LjHsf-06/LjHsf-11, GmHsf-09/GmHsf-34, GmHsf-18/GmHsf-24, GmHsf-18/GmHsf-46* and *GmHsf-21/GmHsf-45*) were classified in the flexible set. The question marks indicate possible gene loss events. The *GmHsf-18/GmHsf-38* pair could be formed by a segmental duplication that predated the recent whole-genome duplication, and both the *GmHsf-18* and *GmHsf-38* lost homoeologs from the recent whole-genome duplication.

any chromosome. In *M. truncatula*, genes flanking both pairs, *MtHsf-03/MtHsf-16* and *MtHsf-11/MtHsf-19*, were found to be conserved. In addition, two gene pairs (*MtHsf-04/MtHsf-05* and *MtHsf-08/MtHsf-09*) were located near each other on chromosomes 2 and 4, and thus most likely resulted from tandem duplication. Moreover, we found that the DNA sequences for

Table 2. Estimates of the dates for the large scale duplication events in legume species.

Duplicated Hsf gene pairs	Number of conserved flanking protein-coding genes	Ks (mean ± s.d.)	Date (mya)
GmHsf-05 & *GmHsf-26*	14	0.12±0.02	9.84
GmHsf-25 & *GmHsf-45*	15	0.12±0.01	9.84
GmHsf-06 & *GmHsf-33*	16	0.13±0.01	10.66
GmHsf-13 & *GmHsf-15*	16	0.13±0.02	10.66
GmHsf-04 & *GmHsf-27*	14	0.14±0.02	11.48
GmHsf-01 & *GmHsf-20*	16	0.15±0.01	12.30
GmHsf-11 & *GmHsf-16*	15	0.15±0.02	12.30
GmHsf-24 & *GmHsf-46*	16	0.15±0.02	12.30
GmHsf-34 & *GmHsf-42*	12	0.16±0.02	13.11
GmHsf-32 & *GmHsf-36*	11	0.16±0.01	13.11
GmHsf-12 & *GmHsf-17*	14	0.16±0.01	13.11
GmHsf-23 & *GmHsf-30*	7	0.17±0.02	13.93
GmHsf-29 & *GmHsf-39*	16	0.17±0.03	13.93
GmHsf-03 & *GmHsf-28*	16	0.18±0.02	14.75
GmHsf-07 & *GmHsf-43*	6	0.18±0.03	14.75
GmHsf-09 & *GmHsf-14*	16	0.18±0.03	14.75
GmHsf-35 & *GmHsf-41*	7	0.18±0.06	14.75
GmHsf-18 & *GmHsf-38*	5	0.38±0.10	31.15
GmHsf-08 & *GmHsf-23*	4	0.57±0.02	46.72
GmHsf-10 & *GmHsf-41*	7	0.58±0.11	47.54
GmHsf-10 & *GmHsf-35*	6	0.59±0.10	48.36
GmHsf-09 & *GmHsf-42*	4	0.62±0.04	50.82
GmHsf-21 & *GmHsf-25*	5	0.62±0.08	50.82
GmHsf-08 & *GmHsf-30*	4	0.65±0.07	53.28
GmHsf-14 & *GmHsf-34*	4	0.65±0.06	53.28
GmHsf-22 & *GmHsf-44*	8	0.65±0.07	53.28
GmHsf-14 & *GmHsf-42*	5	0.66±0.04	54.10
GmHsf-24 & *GmHsf-38*	5	0.71±0.10	58.20
GmHsf-38 & *GmHsf-46*	6	0.72±0.10	59.02
GmHsf-12 & *GmHsf-32*	6	0.77±0.04	63.11
GmHsf-17 & *GmHsf-32*	6	0.77±0.06	63.11
GmHsf-12 & *GmHsf-36*	6	0.79±0.04	64.75
GmHsf-17 & *GmHsf-36*	6	0.80±0.06	65.57
GmHsf-28 & *GmHsf-40*	4	0.85±0.17	69.67
GmHsf-03 & *GmHsf-40*	4	0.91±0.12	74.59
MtHsf-11 & *MtHsf-19*	5	0.77±0.10	63.05
MtHsf-03 & *MtHsf-16*	5	0.80±0.03	65.66
CcHsf-10 & *CcHsf-11*	5	0.74±0.14	60.66

The Hsf gene pairs from flexible sets were not used for calculation.

MtHsf-08 and *MtHsf-09*, as well as their four flanking genes (within an approximately 40-kb region) were completely identical to each other. In *G. max*, no tandem duplication was identified, but 42 out of 46 Hsf genes (approximately 91.3%) were arranged into duplicated chromosomal regions. These 42 genes were classified into 16 duplication groups; each group had two to four members that had conservation between their flanking genes (Figure 2). In the duplication groups of *G. max* Hsfs, the relationships between four putative duplicated gene pairs (*GmHsf-09/GmHsf-34*, *GmHsf-18/GmHsf-24*, *GmHsf-18/GmHsf-46* and *GmHsf-21/GmHsf-45*) were judged according to the flexible set.

Assuming that synonymous silent substitutions per site (Ks) occur at a constant rate over time, the conserved flanking protein-coding genes were used to estimate the dates of the large-scale duplication events [48]. In this analysis, the duplicated blocks (excluding the flexible set) were used to date duplication events. The mean Ks values for each duplication event, and the estimated date, are shown in Table 2. The duplicated regions of *G. max* were divided into two groups (except for *GmHsf-18* and *GmHsf-38*) based on the Ks values of paralogs flanking the Hsf pair (Table 2). The Ks values distribution of each pair of genes in duplicated blocks is shown in Figure 3A. In one group, the paralogs flanking 17 Hsf pairs yielded a mean Ks value of 0.155 (the first peak in Figure 3A), corresponding to an event approximately 13 Mya. This estimate is consistent with the timing of a recent *Glycine*-lineage-specific tetraploidization event [8]. In the other group, the paralogs flanking 17 pairs had a mean Ks value of 0.701 (the second peak in Figure 3A), corresponding to an event roughly 57 Mya, concordant with the early-legume duplication that occurred near the origin of the papilionoid lineage [59]. For *GmHsf-18* and *GmHsf-38*, the duplication event was estimated to have occurred approximately 31 Mya, which is between the two rounds of genome duplication. In addition, *GmHsf-38* and *GmHsf-24* or *GmHsf-46* are all related via the ancient genome duplication, but the relationships between *GmHsf-18* and *GmHsf-24* or *GmHsf-46* are uncertain. Therefore, *GmHsf-18* may be the product of a segmental duplication of *GmHsf-38*. From these results, we conclude that two whole-genome duplications played a key role in the expansion of the *G. max* Hsf family.

In the *G. max* Hsf duplicated network, when two duplicated genes from recent duplication could not be found simultaneously, we reasoned that a possible ancient gene loss event occurred. As shown in Figure 2, in the ancestor of *G. max* lineage, ancient genome duplication should have produced at least 32 Hsf genes. However, eight of these lines lacked both copies, which would have been obtained from recent genome duplication, suggesting that approximately 25% of the ancient duplicates were lost over millions of years. Moreover, taking ancient gene losses into account, the number of *G. max* Hsf genes derived from recent genome duplication should be 50. Among these, eight pairs lost one copy of the gene, indicating that only about 16% of recent duplicates were lost. On the contrary, in *L. japonicas*, *M. truncatula* and *C. cajan*, only a few Hsf-containing segments could be matched in duplicated pairs. What is the origin of the remaining Hsf genes in these species?

Massive losses of duplicated Hsf genes in *L. japonicas* and *M. truncatula*

To identify the evolutionary origins and orthologous relationships within the Hsf genes of legumes, Hsf family members were used as anchor genes to study the molecular history of the chromosomal regions in which they reside. Using a stepwise gene-by-gene reciprocal comparison of the regions hosting the Hsf genes, we observed strongly conserved microsynteny among these regions across *L. japonicus*, *M. truncatula*, *G. max* and *C. cajan* (Figure 4 and Figure S5). After this interspecies microsynteny analysis, we were able to assemble 78 out of 86 Hsf-containing genomic segments from these four species into 17 groups (Figure S5 A–Q). We propose that all of the segments within a group descended from a single Hsf-containing segment in the genome of the last common ancestor of the legumes, and thus, we refer to these groups as orthologous groups. All of the groups contain at least one cool season legume and one tropical season legume segment with an Hsf gene. A total of 79 Hsfs (11 from *L. japonicus*, 16 from *M. truncatula*, 40 from *G. max* and 12 from *C. cajan*) were present in the 17 orthologous groups of segments. A representative synteny diagram for three of these groups is shown in Figure 5. *L. japonicas*-*M. truncatula*-*G. max*-*C. cajan* microsynteny also allowed us to verify the 18 ancient gene lineages inferred from the phylogenetic analysis. These results demonstrate that there is a one-to-one correspondence between syntenic orthologous groups and ancient gene lineages, except for the ancient gene of clade 18 (class C Hsfs).

Each orthologous group of segments includes between four and 24 orthologous groups of genes (average of 11; including Hsfs) with representation in at least two species. These groups are shown connected by black lines in Figure 5 and Figure S5; most of these groups include genes that obtain 'best hits' in BLASTP searches of entire genomes across species. To estimate the extent of conserved gene content and order, synteny quality was counted for those genes falling into syntenic intervals in *L. japonicus*, *M. truncatula*, *G. max* and *C. cajan*. Synteny quality was calculated as twice the number of matches divided by the total number of genes in both segments; this process discounts gene amplification but counts conservation of genes between species [5]. The average synteny quality of regions orthologous across these four species was 61.32% (Table 3). The lowest synteny quality, 48.97%, was between *M. truncatula* and *G. max* syntenic regions. The *G. max* and *C. cajan* comparison exhibited the highest conservation, 70.91%. These results support the orthology of the segment groups used in this study.

In each orthologous group, high levels of microsynteny were maintained between the members of three legume species (*L. japonicus*, *M. truncatula* and *C. cajan*) and networks of duplicated regions in *G. max*, each anchored by the Hsf gene. Within an orthologous group, segments of different legume species are thought to have shared the ancient legume whole-genome duplication that occurred outside of the papilionoid lineage. In many groups, only one region of *L. japonicus*, *M. truncatula* and *C. cajan* was comparable to homoeologous regions of *G. max*, suggesting that one member of the Hsf gene pair produced from ancient genome duplication was lost in their ancestral lineages. For example, the *MtHsf-17/LjHsf-07/CcHsf-09* anchored regions showed microsynteny with two *G. max* duplicated regions containing *GmHsf-13/GmHsf-15* (Figure 5A). In only a few groups, two duplicated regions of *L. japonicus*, *M. truncatula* or *C. cajan* were syntenic to the *G. max* duplicate regions. In one instance (Figure 5B), where the *LjHsf-06/LjHsf-11* anchored regions were putative duplicated regions in *L. japonicus*, and the *MtHsf-03/MtHsf-16* anchored regions were duplicated regions in *M. truncatula*, these four regions could be aligned with four duplicated regions in *G. max* that contained syntenic counterparts of Hsf genes (*GmHsf-12/GmHsf-17/GmHsf-32/GmHsf-36*). The four *G. max* segments arose from two rounds of whole-genome duplication. Moreover, the Hsf orthologs were usually found in the syntenic regions of the three legume species, and there were no counterparts in one species of *L. japonicus*, *M. truncatula* or *C. cajan*, indicating that two Hsf copies produced from ancient genome duplication were lost in its ancestral lineage. We also uncovered four cases in which the regions containing Hsf orthologs was syntenic between only two legume species. *MtHsf-12* were conserved with those of *G. max* (*GmHsf-01/GmHsf-20*), while the orthologs of *MtHsf-12/GmHsf-01/GmHsf-20* were missing in *L. japonicus* and *C. cajan* (Figure 5C).

Because the 17 orthologous groups of Hsf-containing segments indicate that there are at least 17 Hsf genes in this ancestor, after ancient whole-genome duplication, 34 Hsf genes should have been produced in their progenitor. In nine orthologous groups, only one Hsf-containing region of *L. japonicus* showed microsynteny with the regions of other legumes, and in seven groups, the syntenic intervals anchored by Hsf genes were missing in *L. japonicus*. This indicates that 23 out of 34 ancient duplicated genes (approximately 68%) were lost in the *L. japonicas* Hsf family. Moreover, there were 11 orthologous groups with the single orthologous region in *M. truncatula* and four groups without sharing microsynteny with Hsf anchored regions in *M. truncatula*. This suggests that 19 out of 34 ancient duplicated genes (approximately 56%) were lost in

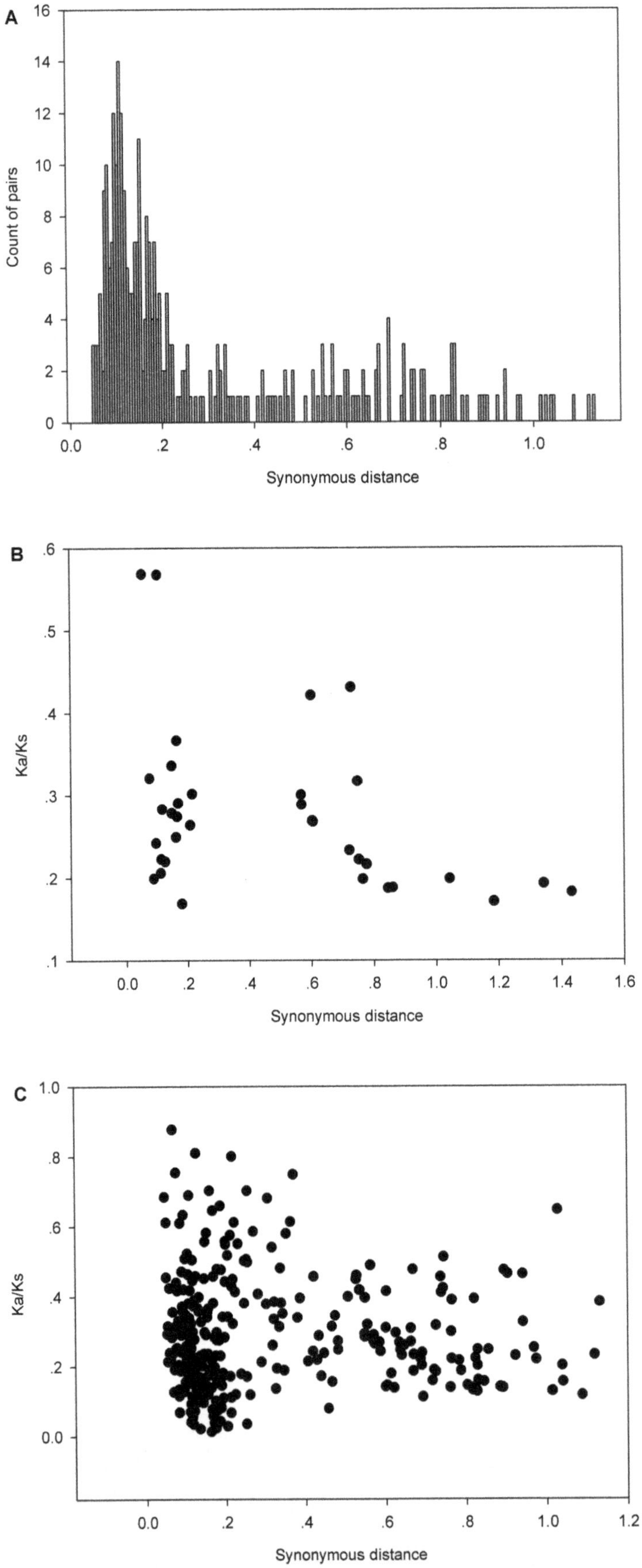
A
Count of pairs
0
2
4
6
8
10
12
14
16
0.0
.2
.4
.6
.8
1.0
Synonymous distance
B
Ka/Ks
.1
.2
.3
.4
.5
.6
0.0
.2
.4
.6
.8
1.0
1.2
1.4
1.6
Synonymous distance
C
Ka/Ks
0.0
.2
.4
.6
.8
1.0
0.0
.2
.4
.6
.8
1.0
1.2
Synonymous distance

Figure 3. Estimates of Ks and Ka/Ks ratios in pairwise comparisons. (A) Distribution of synonymous distances (Ks) between paralogous genes flanking duplicated Hsf genes in *G. max*. The histogram shows the number of duplicate gene pairs (y-axis) versus synonymous distance between pairs (x-axis). The Ka/Ks ratios of the duplicated Hsf genes (B) and their flanking paralogs (C) in *G. max* are shown in the scatter plots; the y and x axes denote the Ka/Ks ratio and synonymous distance for each pair, respectively.

the *M. truncatula* Hsf family. In *C. cajan*, all 12 mapped Hsf genes were found to possess conserved microsynteny among the species investigated, and Hsf orthologs were found in nine groups located in the single syntenic region compared with other legume species, but 10 other Hsfs in this species could not be localized on the genome, and their regions were not used for comparison. Therefore, the number of duplicated Hsfs that remain in *C. cajan* is uncertain.

Strong purifying selection for Hsf genes in *G. max*

Almost the entire Hsf family of *G. max* has been expanded by two genome duplications. To better understand the evolutionary constraints acting on this gene family, we measured the Ka/Ks ratios for 35 unambiguous pairs of Hsf paralogs (not including paralogs from the flexible set) in the network of duplicated regions of *G. max*. The resulting pairwise comparison data showed that all the paralog pairs have Ka/Ks ratios <1 (Figure 3B), suggesting that the Hsf family has mainly undergone strong purifying selection, and the Hsf genes are slowly evolving at the protein level. Given the important role of the two rounds of whole-genome duplication in the evolution of the *G. max* Hsf family, the significance of changes in the strength of selection over evolutionary time was also stressed, and the Ka/Ks ratios were sorted into two sets on the basis of the Hsf paralogs that arose from either the recent or earlier whole-genome duplication. The average Ka/Ks ratio for the recent -duplicated Hsfs (0.30) was higher than that of the early-duplicated Hsfs (0.25), but there was no significant difference between these ratios (t-test, P>0.05). Moreover, the variance of the Ka/Ks ratios for the recent-duplicated Hsfs (0.013) was not significantly different from that of the early-duplicated Hsfs (0.006; F-test, P>0.05). This indicates that the younger and older proteins in the Hsf family are under similarly stable evolutionary constraints, which supports the notion that this family is essential for the regulation of cellular processes in *G. max*.

To assess the potential for selection on the regions surrounding Hsfs, pairwise Ka/Ks ratios were also calculated for the duplicated

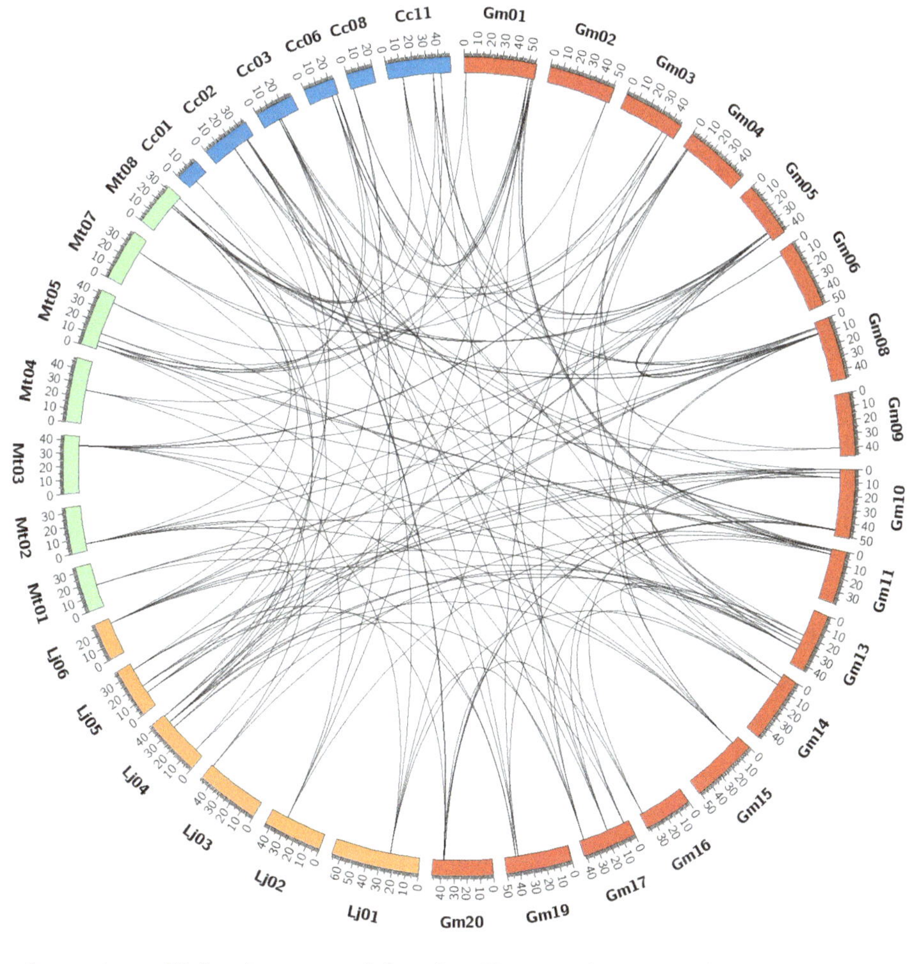

Figure 4. Extensive microsynteny of Hsf regions across *L. japonicus, M. truncatula, G. max* and *C. cajan*. *G. max* chromosomes, labeled Gm, are indicated by red boxes. The *L. japonicus, M. truncatula* and *C. cajan* chromosomes, shown in different colors, are labeled Lj, Mt and Cc, respectively. Numbers along each chromosome box indicate sequence lengths in megabases. The whole chromosomes of these four legumes, harboring Hsf regions, are shown in a circle. Black lines represent the syntenic relationships between Hsf regions.

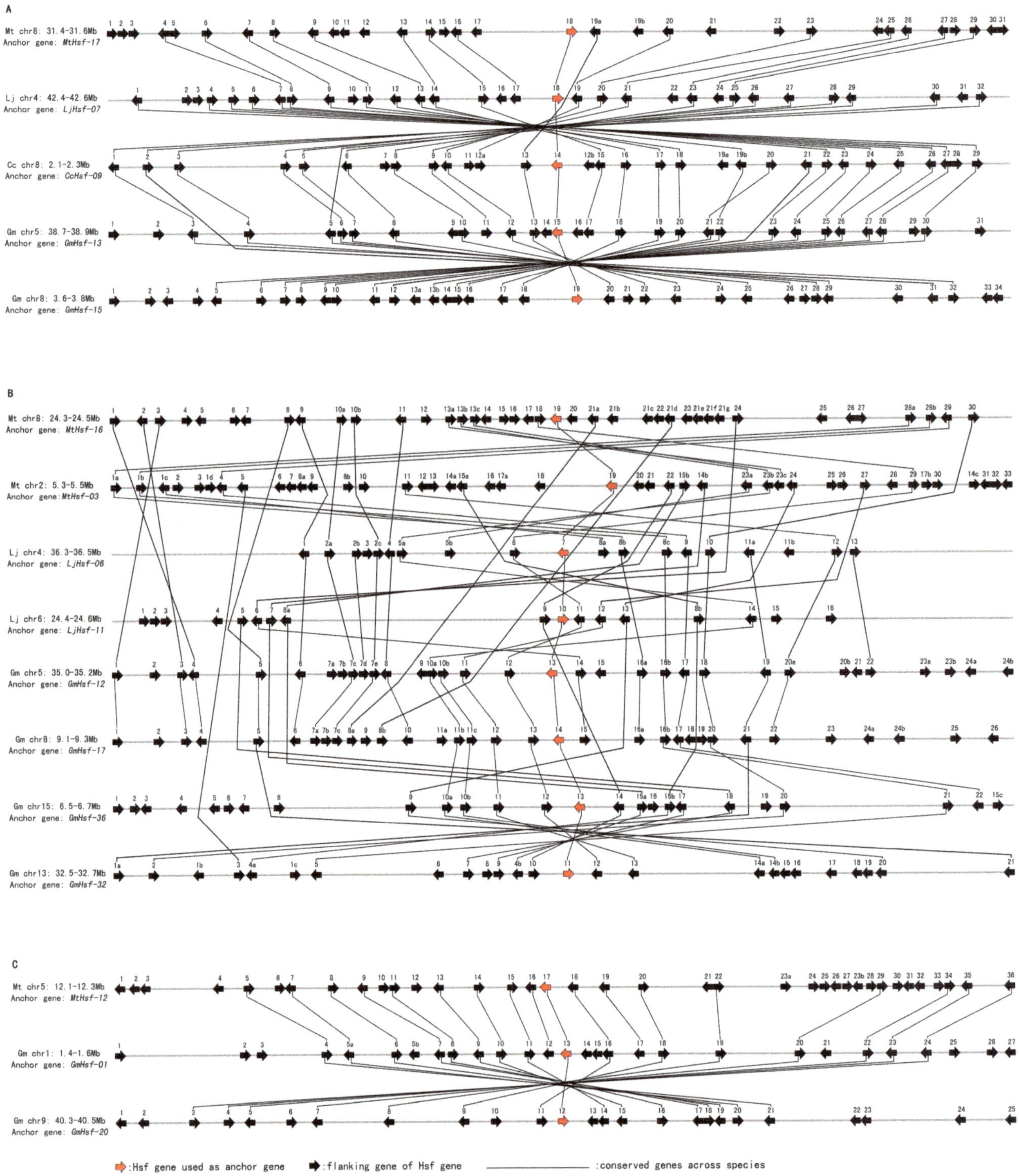

Figure 5. Comparative maps of representative Hsf genes and their flanking genes within syntenic chromosomal intervals across selected legume species. The relative positions of all flanking protein-coding genes were defined by the anchored Hsf genes, highlighted in red. The chromosome segments are shown as gray horizontal lines, with arrows corresponding to individual genes and their transcriptional orientations. All genes are numbered from left to right, in order, for each segment. Where several duplicated genes were present within a region, these genes were given the same number, with the letters a, b, c... appended in order. Conserved gene pairs among the segments are connected with lines. (A) The syntenic chromosomal intervals containing *MtHsf-17*, *LjHsf-07*, *CcHsf-09*, *GmHsf-13* and *GmHsf-15* across *M. truncatula*, *L. japonicus*, *C. cajan* and *G. max*. (B) The syntenic chromosomal intervals containing *MtHsf-03*, *MtHsf-16*, *LjHsf-06*, *LjHsf-11*, *GmHsf-12*, *GmHsf-17*, *GmHsf-32* and *GmHsf-36* across *M. truncatula*, *L. japonicus*, and *G. max*. (C) The syntenic chromosomal intervals containing *MtHsf-12*, *GmHsf-01* and *GmHsf-20* across *M. truncatula* and *G. max*. The full microsynteny maps of the regions containing Hsf genes within *M. truncatula*, *L. japonicus*, *C. cajan* and *G. max* are shown in Figure S5.

Table 3. The synteny quality of regions orthologous across *L. japonicus*, *M. truncatula*, *G. max* and *C. cajan*.

	L. japonicus	*M. truncatula*	*G. max*	*C. cajan*
L. japonicus				
M. truncatula	58.06%			
G. max	68.00%	48.97%		
C. cajan	70.79%	51.16%	70.91%	

non-Hsf genes (flanking genes) between the duplicated regions containing Hsfs in *G. max*. Interestingly, all Ka/Ks values for 322 pairs of duplicated non-Hsf genes were lower than 1 (Figure 3C), clearly indicating that these genes are evolving under purifying selection. There was no significant difference in average Ka/Ks ratio between the recent-duplicated non-Hsf genes (0.29) and the early-duplicated non-Hsf genes (0.27; t-test, P>0.05). However, the variance of these ratios for the recent-duplicated non-Hsf genes (0.032) was significant greater than that for the early-duplicated ones (0.013; F-test, P<0.01). Duplicated non-Hsf genes have likely evolved in a more "dynamic" regime than that of the Hsf genes.

Since positive selection at a few individual codon sites can be masked by overall strong purifying selection, we performed a sliding-window analysis of Ka/Ks between each pair of Hsf paralogs, which were derived from gene duplication events in *G. max*. As expected from the basic Ka/Ks analysis, sliding window analysis clearly showed that numerous sites/regions are under moderate to strong negative selection (Figure S6). As shown in Figure 6, the conserved domains of Hsfs, such as the DBD domains, HR-A/B regions and NLS motifs, are mainly subjected to strong purifying selection, with Ka/Ks ratios <<1. Moreover, the domains of Hsfs generally had lower Ka/Ka ratios (valleys) than the regions outside of them (peaks), which is consistent with functional constraint being dominant in these domains. There were a few exceptions to the generally low Ka/Ka ratios in domains. For instance, the comparison between *GmHsf-25* and *GmHsf-45* revealed sites with Ka/Ka ratios >>1 in the DBD domain, indicating positive selection in this region.

To further identify possible positive selection acting at specific sites, six site models that allow ω ratios to vary among sites were used based on the coding sequences of *G. max* Hsf family. To detect whether some sites along particular Hsf classes were under positive selection, the hypothesis testing on the class A and B *G. max* Hsfs was also performed by site models. In *G. max* Hsf family, the discrete model M3 fit better than the one-ratio model M0, suggesting that ω ratios vary among sites (Table S4; LRT, P<0.01). Both M1a–M2a and M7–M8 comparisons suggested that the most codon sites were under a strong constraint and no reliable positive selection sites were detected in *G. max* Hsf family (Table S4; LRT, P<0.01). In class A and B *G. max* Hsfs, M3 model also appears to be a better fit to the data than the M0 model, and the models M2a and M8 were not significantly better than the null hypothesis models M1a and M7 (Table S4; LRT, P<0.01). Only one positively selected site, listed in Table S4, was detected based on posterior probability in class A Hsfs of *G. max*. The results showed that *G. max* Hsf genes were highly conserved and the majority of sites were dominated by purifying selection.

The expression patterns of Hsf genes in *L. japonicus*

In order to gain insight into the possible functions of Hsf genes, we comprehensively examined information about the expression of all *L. japonicus* Hsf genes using microarray data and quantitative real-time PCR analysis. We first analyzed the expression of *L. japonicus* Hsf genes in nodule, root, stem, leaf and flower from the microarray data (Figure 7). Out of 11 of these genes, the expression data for *LjHsf-05* were not included in the database. The ten remaining genes were expressed in all the tissues investigated, but they exhibited differential patterns in terms of both specificity and expression level. According to their expression profiles, *L. japonicus* Hsf genes can be classified into four types. The transcripts of the first type (*LjHsf-03* and *LjHsf-11*) were highly accumulated in both underground (nodule and root) and aerial (stem, leaf and flower) parts, but the expression level was higher in the underground parts (Figure 7A). In the second type, *LjHsf-04* showed maximum expression in the root and *LjHsf-06* had a similar pattern. However, *LjHsf-06* was much more highly expressed than *LjHsf-04* (Figure 7B). The genes of the third type (*LjHsf-02* and *LjHsf-10*) were expressed preferentially in stem and flower, and *LjHsf-02* showed higher expression than *LjHsf-10* (Figure 7C). The fourth type has four members (*LjHsf-01*,

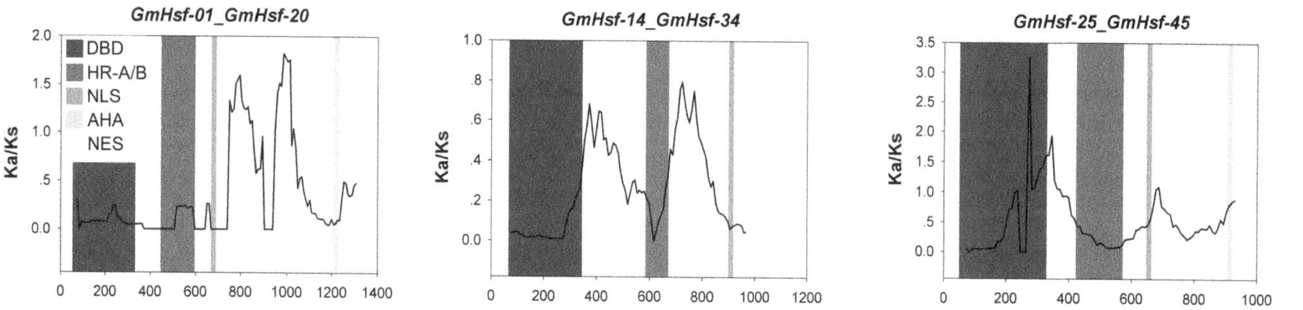

Figure 6. Sliding window plots of representative duplicated Hsf genes in *G. max*. As shown in the key, the gray blocks, from dark to light, indicate the positions of the DBD domain, HR-A/B region, NLS, NES and AHA motifs, respectively. The window size is 150 bp, and the step size is 9 bp. The data for all pairs of duplicated Hsf genes of soybean are shown in Figure S6.

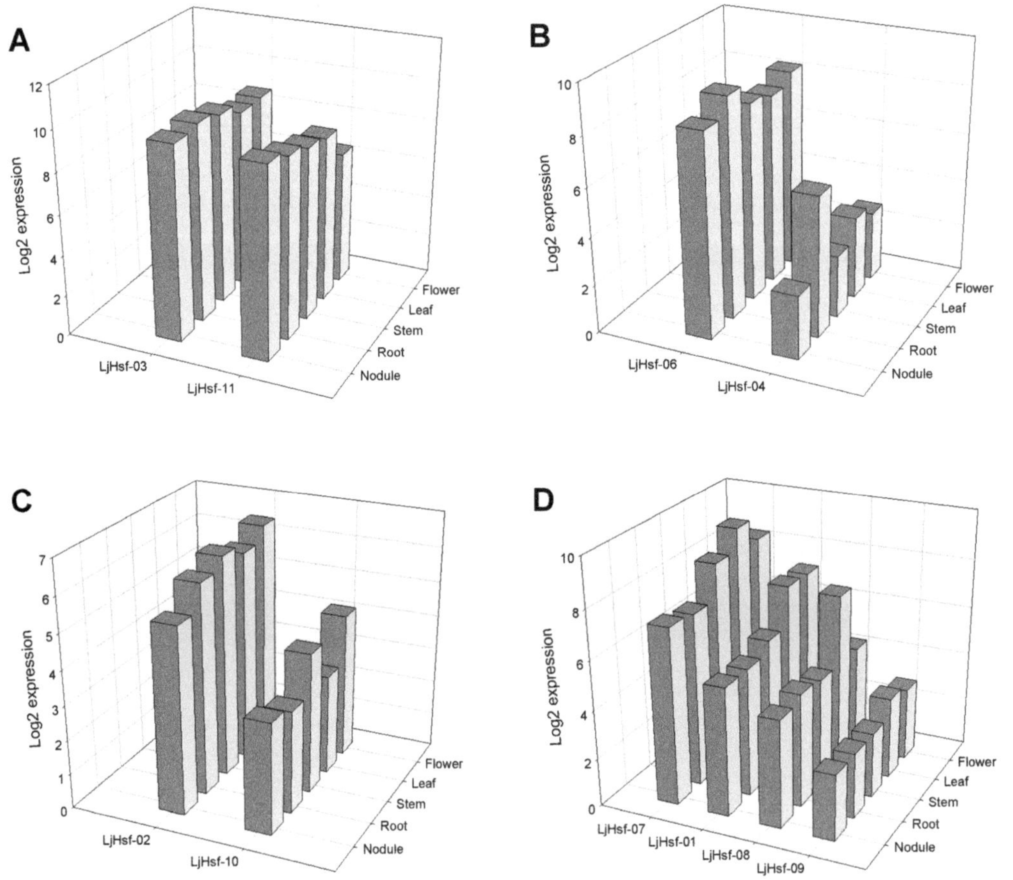

Figure 7. *L. japonicus* Hsf genes expression in various plant tissues. The type of tissue (nodule, root, stem, leaf and flower) and the gene name are shown on the y-axis and x-axis, respectively. Hierarchical clustering based on average log signal values in various tissues grouped 10 of the *L. japonicus* Hsf genes into four types (A–D).

LjHsf-07, *LjHsf-08* and *LjHsf-09*), the genes predominantly expressed in leaves (Figure 7D). *LjHsf-01* and *LjHsf-09* were also expressed at higher levels in flowers than nodule, root and stem. Although the Hsf genes of the fourth type had similar expression pattern across a range of tissues, their transcript levels were quite diverse. *LjHsf-07* was the most highly expressed gene and *LjHsf-09* the lowest.

Quantitative real-time PCR analysis was then performed to evaluate the response of the *L. japonicus* Hsf gene family to abiotic stress. RNA was isolated from the leaves of 4-week-old *L. japonicus* seedlings subjected to heat, cold and H_2O_2 stress treatment and was used for the experiments. The results showed that a total of ten genes were significantly up- or down-regulated compared to controls (>2 or <0.5) in at least one of the stress conditions examined (Figure 8). Among these genes, most were responsive to more than one stress treatment. Two genes (*LjHsf-04* and *LjHsf-11*) were significantly up-regulated by all three stresses. Three (*LjHsf-01*, *LjHsf-02* and *LjHsf-09*) were expressed at remarkably high levels in response to both heat and H_2O_2 stress. *LjHsf-08* was induced by heat stress but was suppressed by H_2O_2 stress. *LjHsf-07* was significantly down-regulated upon exposure to heat and H_2O_2 stresses. A few genes were primarily responsive to one stress treatment. *LjHsf-05* and *LjHsf-10* responded specifically to heat stress, while *LjHsf-06* was distinctively up-regulated under H_2O_2 stress. In contrast, *LjHsf-03* showed minor fluctuations during all three stresses. It is worth noting that five genes (*LjHsf-01*, *LjHsf-02*, *LjHsf-04*, *LjHsf-09* and *LjHsf-10*) were strongly heat-inducible in our experiments, suggesting that they could have important roles in the heat shock regulatory network.

Discussion

In this study, we identified 11, 19 and 13 Hsfs in the cool season legumes *L. japonicus*, *M. truncatula* and *C. arietinum*, respectively, and 46, 22 and 29 Hsfs in the tropical season legumes *G. max*, *C. cajan* and *P. vulgaris*, respectively. Before reconstructing the gene gain/loss history of legume Hsf families, it is necessary to trace the Hsf genes in different legume genomes back to a common ancestor. Phylogenetic trees are quite informative for inferring the number of Hsfs in the most recent common ancestor of the six legume species analyzed in this study [60]. There are 18 well-supported clades representing legume Hsfs (Figure 1), although the representatives of one or two species are missing from some clades. These clades are perceived as shared, and the genes in a shared clade are assumed to be descendants of an ancestral Hsf gene. Therefore, there are at least 18 Hsf genes in the most recent common ancestor of legumes. A large fraction of variability in the clades suggests lineage-specific gene gain and loss. For instance, in almost every clade, Hsfs from *G. max* are present in at least an extra copy compared with the other species, and the extra copy is very close to its potential paralog. These results are consistent with the well-documented fact that *G. max* has undergone an additional WGD not shared with by the other

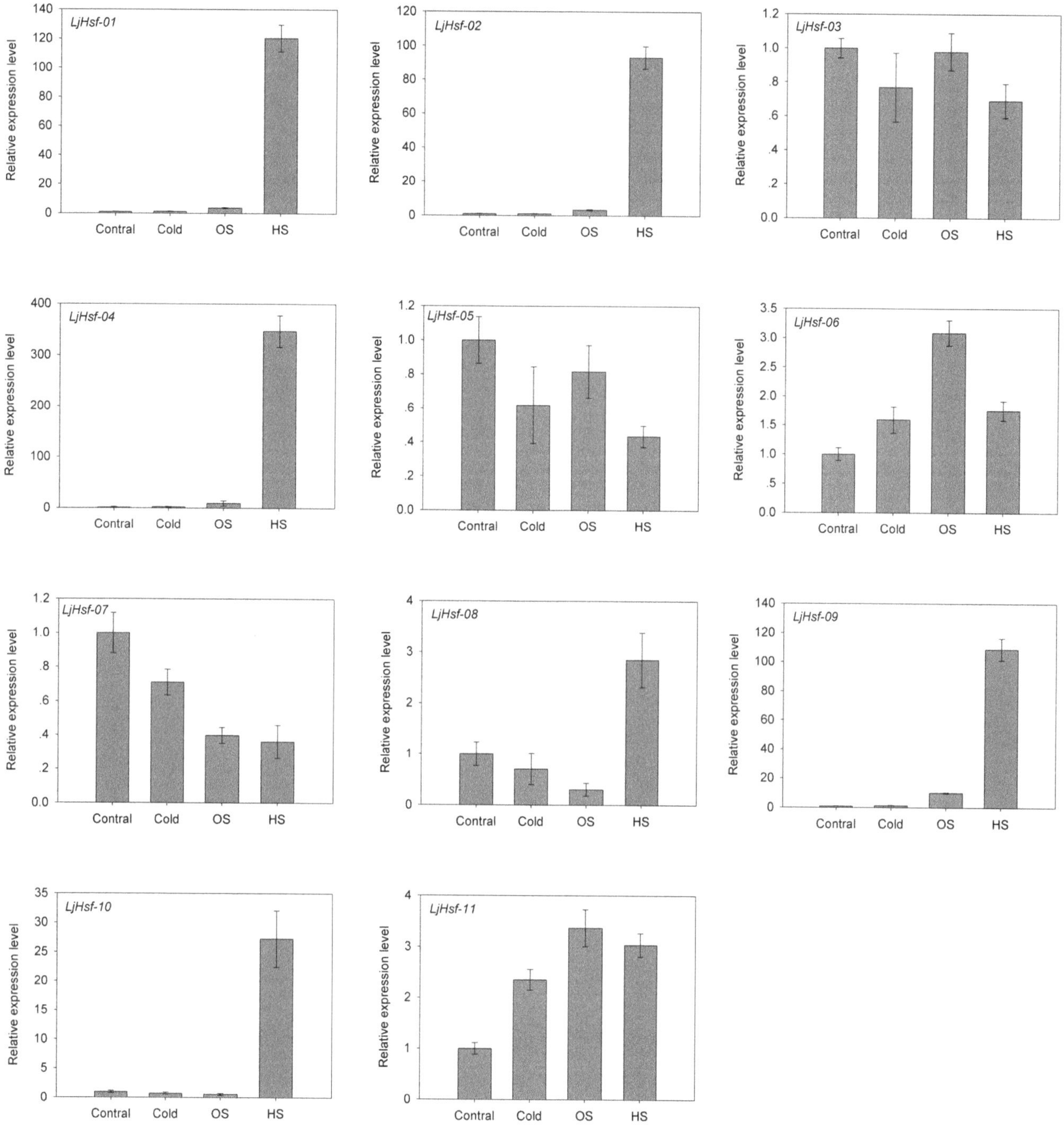

Figure 8. Expression of *L. japonicus* Hsf genes in response to abiotic stress measured by quantitative real-time PCR. The mRNA level of each gene in *L. japonicus* seedlings given heat (HS: 42°C), cold (4°C) and oxidative (OS: 10 mM H_2O_2) stress for 1 h was plotted relative to the value obtained for the unstressed contral. Error bars represent standard errors.

three species [8]. However, there are not simply twice as many Hsf genes in *G. max* vs. the other legumes, which indicates that differential gene loss events may have occurred in different species. Furthermore, all of the clades were confirmed by the following interspecies microsynteny analysis (except for the shared clade representing class C Hsfs), indicating most duplicates of class C Hsfs have been lost in legumes. When we compared the number of ancestral genes with those in the extant legumes species, it appeared as though the Hsf family had contracted in most of the cool season legumes and expanded in the tropical season legumes. For example, compared with the number of ancestral genes, the number of Hsf genes was nearly halved in *L. japonicus* but increased approximately 2.5-fold in *G. max*. These results most likely reflect the complex evolutionary history of the Hsf family in legume species.

In *G. max*, most Hsf genes were assigned to the duplicated segments, and these segments could be divided into two groups based on Ks values. The members of one group were supported with an average Ks value of 0.701, which corresponds to the second peak in Figure 3A. In *L. japonicus*, *M. truncatula* and *C.*

cajan, a few Hsf-containing segments were matched in pairs. The Ks values for these pairs were included in Table 2, such as *MtHsf-03/MtHsf-16* (0.80) and *MtHsf-11/MtHsf-19* (0.77). All of these Ks values from *G. max* and other legumes probably represent the same polyploidy that occurred in all members of Papilionoideae subfamily approximately 59 Mya. Thus, the rate of synonymous substitution in *M. truncatula* appears to be greater than that in *G. max*. In a previous study of *Rxp* regions, higher rates of synonymous substitution were also detected in *M. truncatula* compared with *G. max* [61].

After the shared polyploidy event and following divergence from the other legumes, the lineage leading to present-day *G. max* is known to have undergone a second whole-genome duplication approximately 13 Mya; this duplication did not occur in *L. japonicus*, *M. truncatula* or other legume lineages. Two whole-genome duplications in the ancestor of *G. max* lead to the expectation of a maximum of four homoeologs in *G. max* genome [62,63]. As illustrated in Figure 2, Hsf genes doubled twice in *G. max* and formed two pairs of paralogs, accordingly. In some cases, these four homoeologous Hsf regions were retained. For instance, *GmHsf-12*, *GmHsf-17*, *GmHsf-32* and *GmHsf-36* are located in four homoeologous regions, respectively. More conserved flanking protein-coding genes were identified between the two paralogous Hsf-containing segments derived from the recent duplication event than those from the ancient duplication (Table 2). This suggests that high levels of sequence conservation were maintained between recently duplicated regions in *G. max*, which is similar to previous observations [62,64]. Furthermore, 17 pairs of Hsf genes of *G. max* generated by recent genome duplication were retained, while in eight pairs, one copy was lost (Figure 2). A higher, sharp peak was produced (the first peak in Figure 3A) when the number of gene pairs within these 17 pairs of recent duplicated regions was plotted against Ks values. These results indicate that many younger Hsf genes from recent genome duplications tended to be retained in *G. max* genome. By contrast, it has been demonstrated that massive gene losses occurred after tetraploidization in the maize ancestral lineage, and approximately 50% of the duplicated copies of genes have been removed or severely damaged over the past 12 million years [65,66]. In *Glycine*, the large number of duplicate genes in homeologous regions indicates that the process of diploidization is a slow and ongoing process [67].

In light of the two above-mentioned rounds of genome duplication and the fact that genome duplication should double the number of genes, the *G. max* and other papilionoid legume genomes should show a 2:1 relationship regarding the Hsf-containing regions. Each pair of *G. max* regions should have a corresponding orthologous region in *L. japonicas*, *M. truncatula* and *C. cajan*. In this study, by aligning Hsf-containing regions into paralogous pairs produced by ancient polyploidy, only one pair was detected in *L. japonicas* and *C. cajan*, with two pairs in *M. truncatula*. However, more than 90% of the chromosome regions hosting Hsf genes in *G. max* fell into pairs, triples or quadruples. Further analyses including estimates of the dates of duplications indicated that these duplicated regions arose from two polyploidy events in the *Glycine* lineage. A high degree of microsynteny between the genomes of the two model legumes has previously been found in a comparison of the genomic regions around the apyrase genes [68]. The analysis of microsynteny can help unravel the actual evolutionary relationships between Hsf regions among the legume species by taking advantage of the surrounding genomic sequences. When the Hsf-containing regions in *L. japonicas*, *M. truncatula* and *C. cajan* were compared to those in *G. max*, we found that significant synteny was maintained, although small insertions/deletions and inversions were observed between regions (Figure 4 and Figure S5). Almost all Hsf-containing regions in *L. japonicas*, *M. truncatula* and *C. cajan* showed close relationships with the orthologous duplicated regions in *G. max*. Syntenic regions are thought to share a common origin derived from ancient legume duplication. In most cases, a single region of *L. japonicus*, *M. truncatula* and *C. cajan* was syntenic to two or three duplicated regions in *G. max*, and in many cases, two paralogons appeared to be missing in *L. japonicus* or *M. truncatula*. This indicates that the duplicated copies of Hsf genes in these genomes have been removed or severely damaged after the genome duplications occurred. Approximately half of the Hsf genes of *C. cajan* were not included in the map-based analyses because these genes could not be localized to the genome owing to their unknown chromosome positions. These results may therefore not hold true for *C. cajan*.

One possible mechanism behind this phenomenon, that many Hsf genes produced by the ancient genome duplication have been lost in *L. japonicus* and *M. truncatula*, is diploidization following the early legume genome duplication. Previous studies have demonstrated that there was substantially less conservation within internal duplications in either *M. truncatula* or *L. japonicus* than in synteny blocks between the two genomes [5]. The fate of these duplicate genes is more likely to be under the control of natural selection (nonrandom loss) than for genes that are not dosage dependant [69].

Duplicated genes undergo a short period of shared "relaxed" selection during their early evolutionary lives, evolving in a neutral way. Following this selection, most paralogs are lost within a few million years, and only a few paralogs are preserved and undergo purifying selection [10,53,70]. Although many gene loss events were found to have occurred in legume Hsf families, two Hsf regions of *L. japonicus* (*LjHsf-06/LjHsf-11*) and the two Hsf regions of *M. truncatula* (*MtHsf-03/MtHsf-16*) were found to have conserved microsynteny to the four Hsf regions of *G. max* (*GmHsf-12/GmHsf-17/GmHsf-32/GmHsf-36*; Figure 5B). This suggests that these sets of Hsf genes may perform a basic, important role in legumes and have remained intact after genome duplications.

The observation that multiple copies of Hsf were retained in *G. max* is reasonable from an evolutionary perspective because Hsfs confer various abiotic and biotic resistance traits to plants [71–77]. The dosage of protein may have increased due to the presence of numerous Hsf genes, thereby leading to a corresponding increase in resistance. Previous studies have demonstrated the advantages of increased dosage of genes involved in plant resistance, for instance, resistance to glyphosate in plants [78]. Furthermore, strong purifying selection was detected between the paralogs produced from the recent genome duplication and the ancient duplication in *G. max*. Purifying selection probably plays a key role in maintaining the long-term stability of biological structures of *G. max* Hsfs by removing deleterious mutations, thus ensuring that gene functions are maintained as long as they are needed.

Moreover, *G. max* is a member of tropical season legumes; natural selection may have played a role in determining the number of duplicates within the Hsf gene family in the tropical season legumes, which are better adapted to more tropical climates. On the contrary, *L. japonicus* and *M. truncatula* are cool season legumes, and they may therefore require fewer Hsf genes. Thus, many copies of these genes may have been lost during the long-term evolutionary process.

Although the legume species differ in genome size, basic chromosome number and ploidy level, comparative genomics can be used for a bridging model with other legume species in view of their close phylogenetic relationships and extensive genome conversation [79]. *L. japonicus* was selected as a model system for gene function characterization because of its small genome size and improving knockout and over-expression techniques, and efforts were made to translate information gained from it into exploration to other legume species. *L. japonicus* maintained only 11 Hsf genes and this number is probably the lowest among the reported Hsf gene families of higher plants. In our study, the orthologs of the 11 Hsf genes were unambiguously defined in the other legumes through microsyntenic analysis. These directed our interest towards further understanding of the expression characteristics of these Hsf genes in different tissues and their responses to abiotic stresses. In several plants, Hsf gene expression has been found to be tissue- and stage-specific [80–82]. Our study revealed that none of the *L. japonicus* Hsf genes examined was expressed only in a particular tissue type, suggesting they play regulatory roles at multiple developmental stages; nevertheless, selected *L. japonicus* Hsf genes exhibited higher levels of expression in particular tissue types, indicating that members of this family might take part in different biological processes in this species. In particular, the member of class B and the member of class A showed similar expression patterns in tissues, e.g. *LjHsf-02* and *LjHsf-10* were expressed at a higher level in stem and flower than other tissues, supporting the assumption that they could co-operate with each other. In *Arabidopsis* and rice, the expression of Hsf genes was strongly induced by heat, cold, salt and osmotic stress [81,83]. In our study, we also found that the Hsf genes of *L. japonicus* were responsive to diverse abiotic stresses, and heat induced their expression more strongly than oxidation and cold. *LjHsf-01* of subclass A2, *LjHsf-02* of subclass B2 and *LjHsf-04* of subclass B1 were strongly and transiently up-regulated by heat shock. Among all Hsf genes in *Arabidopsis*, *HsfA2* was most highly expressed under high temperature conditions and was identified as a key positive regulator of the heat stress response [84,85]. *Arabidopsis HsfB1* and *HsfB2b* repressed the general heat shock response in the absence of excessive heat but were necessary for the development of acquired thermotolerance under heat stress conditions [86]. It is noteworthy that many Hsf genes of *L. japonicus* induced by heat stress were also induced by oxidative stress, and *LjHsf-04* and *LjHsf-11* were induced by all three stresses tested. These findings could support the notion that Hsfs serve as important sensors for H_2O_2 in plants and could be pivotal in linking the heat shock response with other stress-responsive signaling networks [87]. The expression pattern described can provide a basis for identifying the roles of the retained members of the *L. japonicus* Hsf family, and would give clues for studying their syntenic counterparts in other legume species.

Supporting Information

Figure S1 DBD domain alignment of Hsf proteins from *L. japonicus*, *M. truncatula*, *C. arietinum*, *G. max*, *C. cajan*, *P. vulgaris* and three lower plants.

Figure S2 HR-A/B region alignment of Hsf proteins from *L. japonicus*, *M. truncatula*, *C. arietinum*, *G. max*, *C. cajan*, *P. vulgaris* and three lower plants.

Figure S3 NJ phylogeny of *A. thaliana*, *O. sativa*, *Z. mays*, *L. japonicus*, *M. truncatula*, *C. arietinum*, *G. max*, *C. cajan* and *P. vulgaris* Hsf proteins.

Figure S4 ML phylogeny of *L. japonicus*, *M. truncatula*, *C. arietinum*, *G. max*, *C. cajan* and *P. vulgaris* Hsf proteins.

Figure S5 The full syntenic maps of chromosome regions containing Hsf genes across *L. japonicus*, *M. truncatula*, *G. max* and *C. cajan* genes.

Figure S6 Sliding window analysis employed to estimate selective pressures on 35 pairs of duplicated Hsf genes in *G. max*.

Table S1 Primers used in quantitative real-time PCR.

Table S2 Information about Hsfs in *L. japonicus*, *M. truncatula*, *C. arietinum*, *G. max*, *C. cajan*, *P. vulgaris*, *S. moellendorffii*, *P. patens* and *C. reinhardtii*.

Table S3 Domain and motif survey of legume and three lower plant Hsfs.

Table S4 Likelihood ratio tests and parameter estimations for the six site models based on the coding sequences of *G. max* Hsf genes.

Acknowledgments

We thank the members of the Key Lab of Crop Biology of Anhui Province and Dr. Xin Zhang for discussing and editing the manuscript. We are also grateful to Dr. Jia-Xing Yue (Department of Ecology and Evolutionary Biology, Rice University, Houston, USA) for helpful comments on the manuscript.

Author Contributions

Conceived and designed the experiments: YL YC BC. Performed the experiments: YL JJ HY. Analyzed the data: YL XJ HJ. Wrote the paper: YL YC.

References

1. Cui L, Wall PK, Leebens-Mack JH, Lindsay BG, Soltis DE, et al. (2006) Widespread genome duplications throughout the history of flowering plants. Genome Res 16: 738–749.
2. Jaillon O, Aury JM, Noel B, Policriti A, Clepet C, et al. (2007) The grapevine genome sequence suggests ancestral hexaploidization in major angiosperm phyla. Nature 449: 463–467.
3. Blanc G, Wolfe KH (2004) Widespread paleopolyploidy in model plant species inferred from age distributions of duplicate genes. Plant Cell 16: 1667–1678.
4. Adams KL, Wendel JF (2005) Polyploidy and genome evolution in plants. Curr Opin Plant Biol 8: 135–141.
5. Cannon SB, Sterck L, Rombauts S, Sato S, Cheung F, et al. (2006) Legume genome evolution viewed through the Medicago truncatula and Lotus japonicus genomes. Proceedings of the National Academy of Sciences 103: 14959–14964.
6. Fawcett JA, Maere S, Van de Peer Y (2009) Plants with double genomes might have had a better chance to survive the Cretaceous-Tertiary extinction event. Proc Natl Acad Sci U S A 106: 5737–5742.
7. Cannon SB, Ilut D, Farmer AD, Maki SL, May GD, et al. (2010) Polyploidy did not predate the evolution of nodulation in all legumes. PLoS One 5: e11630.
8. Schmutz J, Cannon SB, Schlueter J, Ma J, Mitros T, et al. (2010) Genome sequence of the palaeopolyploid soybean. NATURE 463: 178–183.

9. Young ND, Bharti AK (2012) Genome-enabled insights into legume biology. Annual review of plant biology 63: 283–305.
10. De Grassi A, Lanave C, Saccone C (2008) Genome duplication and gene-family evolution: the case of three OXPHOS gene families. Gene 421: 1–6.
11. Li J, Ding J, Zhang W, Zhang Y, Tang P, et al. (2010) Unique evolutionary pattern of numbers of gramineous NBS–LRR genes. Molecular Genetics and Genomics 283: 427–438.
12. Cheng Y, Li X, Jiang H, Ma W, Miao W, et al. (2012) Systematic analysis and comparison of nucleotide-binding site disease resistance genes in maize. FEBS J 279: 2431–2443.
13. Cannon SB, Mitra A, Baumgarten A, Young ND, May G (2004) The roles of segmental and tandem gene duplication in the evolution of large gene families in Arabidopsis thaliana. BMC Plant Biol 4: 10.
14. Maere S, De Bodt S, Raes J, Casneuf T, Van Montagu M, et al. (2005) Modeling gene and genome duplications in eukaryotes. Proc Natl Acad Sci U S A 102: 5454–5459.
15. Chapman BA, Bowers JE, Feltus FA, Paterson AH (2006) Buffering of crucial functions by paleologous duplicated genes may contribute cyclicality to angiosperm genome duplication. Proc Natl Acad Sci U S A 103: 2730–2735.
16. Baniwal SK, Bharti K, Chan KY, Fauth M, Ganguli A, et al. (2004) Heat stress response in plants: a complex game with chaperones and more than twenty heat stress transcription factors. J Biosci 29: 471–487.
17. Kotak S, Larkindale J, Lee U, von Koskull-Doring P, Vierling E, et al. (2007) Complexity of the heat stress response in plants. Curr Opin Plant Biol 10: 310–316.
18. Morimoto RI (1998) Regulation of the heat shock transcriptional response: cross talk between a family of heat shock factors, molecular chaperones, and negative regulators. Genes Dev 12: 3788–3796.
19. Nover L, Scharf KD (1997) Heat stress proteins and transcription factors. Cell Mol Life Sci 53: 80–103.
20. Schoffl F, Prandl R, Reindl A (1998) Regulation of the heat-shock response. Plant Physiol 117: 1135–1141.
21. Wu C (1995) Heat shock transcription factors: structure and regulation. Annu Rev Cell Dev Biol 11: 441–469.
22. Scharf KD, Berberich T, Ebersberger I, Nover L (2012) The plant heat stress transcription factor (Hsf) family: structure, function and evolution. Biochim Biophys Acta 1819: 104–119.
23. Lutz N (1987) Expression of heat shock genes in homologous and heterologous systems. Enzyme and Microbial Technology 9: 130–144.
24. Peteranderl R, Rabenstein M, Shin YK, Liu CW, Wemmer DE, et al. (1999) Biochemical and biophysical characterization of the trimerization domain from the heat shock transcription factor. Biochemistry 38: 3559–3569.
25. Doring P, Treuter E, Kistner C, Lyck R, Chen A, et al. (2000) The role of AHA motifs in the activator function of tomato heat stress transcription factors HsfA1 and HsfA2. Plant Cell 12: 265–278.
26. Kotak S, Port M, Ganguli A, Bicker F, von Koskull-Doring P (2004) Characterization of C-terminal domains of Arabidopsis heat stress transcription factors (Hsfs) and identification of a new signature combination of plant class A Hsfs with AHA and NES motifs essential for activator function and intracellular localization. Plant J 39: 98–112.
27. Clos J, Westwood JT, Becker PB, Wilson S, Lambert K, et al. (1990) Molecular cloning and expression of a hexameric Drosophila heat shock factor subject to negative regulation. Cell 63: 1085–1097.
28. Hsu AL, Murphy CT, Kenyon C (2003) Regulation of aging and age-related disease by DAF-16 and heat-shock factor. Science 300: 1142–1145.
29. Sorger PK, Pelham HR (1988) Yeast heat shock factor is an essential DNA-binding protein that exhibits temperature-dependent phosphorylation. Cell 54: 855–864.
30. Wiederrecht G, Seto D, Parker CS (1988) Isolation of the gene encoding the S. cerevisiae heat shock transcription factor. Cell 54: 841–853.
31. Xiao X, Zuo X, Davis AA, McMillan DR, Curry BB, et al. (1999) HSF1 is required for extra-embryonic development, postnatal growth and protection during inflammatory responses in mice. EMBO J 18: 5943–5952.
32. Xing H, Wilkerson DC, Mayhew CN, Lubert EJ, Skaggs HS, et al. (2005) Mechanism of hsp70i gene bookmarking. Science 307: 421–423.
33. Fujimoto M, Izu H, Seki K, Fukuda K, Nishida T, et al. (2004) HSF4 is required for normal cell growth and differentiation during mouse lens development. EMBO J 23: 4297–4306.
34. Nover L, Bharti K, Doring P, Mishra SK, Ganguli A, et al. (2001) Arabidopsis and the heat stress transcription factor world: how many heat stress transcription factors do we need? Cell Stress Chaperones 6: 177–189.
35. Guo J, Wu J, Ji Q, Wang C, Luo L, et al. (2008) Genome-wide analysis of heat shock transcription factor families in rice and Arabidopsis. J Genet Genomics 35: 105–118.
36. Xiong Y, Liu T, Tian C, Sun S, Li J, et al. (2005) Transcription factors in rice: a genome-wide comparative analysis between monocots and eudicots. Plant Mol Biol 59: 191–203.
37. Lin YX, Jiang HY, Chu ZX, Tang XL, Zhu SW, et al. (2011) Genome-wide identification, classification and analysis of heat shock transcription factor family in maize. BMC Genomics 12: 76.
38. Garg R, Patel RK, Jhanwar S, Priya P, Bhattacharjee A, et al. (2011) Gene discovery and tissue-specific transcriptome analysis in chickpea with massively parallel pyrosequencing and web resource development. Plant Physiol 156: 1661–1678.
39. Czarnecka-Verner E, Yuan CX, Fox PC, Gurley WB (1995) Isolation and characterization of six heat shock transcription factor cDNA clones from soybean. Plant Mol Biol 29: 37–51.
40. Punta M, Coggill PC, Eberhardt RY, Mistry J, Tate J, et al. (2012) The Pfam protein families database. Nucleic Acids Res 40: D290–301.
41. Letunic I, Doerks T, Bork P (2012) SMART 7: recent updates to the protein domain annotation resource. Nucleic Acids Res 40: D302–305.
42. Delorenzi M, Speed T (2002) An HMM model for coiled-coil domains and a comparison with PSSM-based predictions. Bioinformatics 18: 617–625.
43. Cokol M, Nair R, Rost B (2000) Finding nuclear localization signals. EMBO Rep 1: 411–415.
44. la Cour T, Kiemer L, Molgaard A, Gupta R, Skriver K, et al. (2004) Analysis and prediction of leucine-rich nuclear export signals. Protein Eng Des Sel 17: 527–536.
45. Thompson JD, Gibson TJ, Plewniak F, Jeanmougin F, Higgins DG (1997) The CLUSTAL_X windows interface: flexible strategies for multiple sequence alignment aided by quality analysis tools. Nucleic acids research 25: 4876–4882.
46. Tamura K, Dudley J, Nei M, Kumar S (2007) MEGA4: Molecular Evolutionary Genetics Analysis (MEGA) software version 4.0. Mol Biol Evol 24: 1596–1599.
47. Hanada K, Zou C, Lehti-Shiu MD, Shinozaki K, Shiu SH (2008) Importance of lineage-specific expansion of plant tandem duplicates in the adaptive response to environmental stimuli. Plant Physiol 148: 993–1003.
48. Maher C, Stein L, Ware D (2006) Evolution of Arabidopsis microRNA families through duplication events. Genome research 16: 510–519.
49. Zhang X, Feng Y, Cheng H, Tian D, Yang S, et al. (2011) Relative evolutionary rates of NBS-encoding genes revealed by soybean segmental duplication. Mol Genet Genomics 285: 79–90.
50. Sato S, Nakamura Y, Kaneko T, Asamizu E, Kato T, et al. (2008) Genome structure of the legume, Lotus japonicus. DNA Res 15: 227–239.
51. Edgar RC (2004) MUSCLE: multiple sequence alignment with high accuracy and high throughput. Nucleic Acids Res 32: 1792–1797.
52. Suyama M, Torrents D, Bork P (2006) PAL2NAL: robust conversion of protein sequence alignments into the corresponding codon alignments. Nucleic acids research 34: W609–W612.
53. Lynch M, Conery JS (2000) The evolutionary fate and consequences of duplicate genes. Science 290: 1151–1155.
54. Yang Z (2007) PAML 4: phylogenetic analysis by maximum likelihood. Mol Biol Evol 24: 1586–1591.
55. Yang Z, Nielsen R, Goldman N, Pedersen AM (2000) Codon-substitution models for heterogeneous selection pressure at amino acid sites. Genetics 155: 431–449.
56. Yang Z, Wong WS, Nielsen R (2005) Bayes empirical bayes inference of amino acid sites under positive selection. Mol Biol Evol 22: 1107–1118.
57. Hogslund N, Radutoiu S, Krusell L, Voroshilova V, Hannah MA, et al. (2009) Dissection of symbiosis and organ development by integrated transcriptome analysis of lotus japonicus mutant and wild-type plants. PLoS One 4: e6556.
58. Eisen MB, Spellman PT, Brown PO, Botstein D (1998) Cluster analysis and display of genome-wide expression patterns. Proc Natl Acad Sci U S A 95: 14863–14868.
59. Bertioli DJ, Moretzsohn MC, Madsen LH, Sandal N, Leal-Bertioli SC, et al. (2009) An analysis of synteny of Arachis with Lotus and Medicago sheds new light on the structure, stability and evolution of legume genomes. BMC Genomics 10: 45.
60. Nam J, Kim J, Lee S, An G, Ma H, et al. (2004) Type I MADS-box genes have experienced faster birth-and-death evolution than type II MADS-box genes in angiosperms. Proceedings of the National Academy of Sciences of the United States of America 101: 1910.
61. Kim KD, Shin JH, Van K, Kim DH, Lee SH (2009) Dynamic rearrangements determine genome organization and useful traits in soybean. Plant Physiol 151: 1066–1076.
62. Innes RW, Ameline-Torregrosa C, Ashfield T, Cannon E, Cannon SB, et al. (2008) Differential accumulation of retroelements and diversification of NB-LRR disease resistance genes in duplicated regions following polyploidy in the ancestor of soybean. Plant Physiol 148: 1740–1759.
63. Shin JH, Van K, Kim DH, Kim KD, Jang YE, et al. (2008) The lipoxygenase gene family: a genomic fossil of shared polyploidy between Glycine max and Medicago truncatula. BMC Plant Biol 8: 133.
64. Van K, Kim DH, Cai CM, Kim MY, Shin JH, et al. (2008) Sequence level analysis of recently duplicated regions in soybean [Glycine max (L.) Merr.] genome. DNA Res 15: 93–102.
65. Lai J, Ma J, Swigonova Z, Ramakrishna W, Linton E, et al. (2004) Gene loss and movement in the maize genome. Genome Res 14: 1924–1931.
66. Messing J, Bharti AK, Karlowski WM, Gundlach H, Kim HR, et al. (2004) Sequence composition and genome organization of maize. Proc Natl Acad Sci U S A 101: 14349–14354.
67. Schlueter JA, Scheffler BE, Schlueter SD, Shoemaker RC (2006) Sequence conservation of homeologous bacterial artificial chromosomes and transcription of homeologous genes in soybean (Glycine max L. Merr.). Genetics 174: 1017–1028.
68. Cannon SB, McCombie WR, Sato S, Tabata S, Denny R, et al. (2003) Evolution and microsynteny of the apyrase gene family in three legume genomes. Mol Genet Genomics 270: 347–361.
69. Blanc G, Wolfe KH (2004) Functional divergence of duplicated genes formed by polyploidy during Arabidopsis evolution. Plant Cell 16: 1679–1691.

70. Lynch M, Conery JS (2003) The evolutionary demography of duplicate genes. J Struct Funct Genomics 3: 35–44.
71. Lohmann C, Eggers-Schumacher G, Wunderlich M, Schoffl F (2004) Two different heat shock transcription factors regulate immediate early expression of stress genes in Arabidopsis. Mol Genet Genomics 271: 11–21.
72. Charng Y, Liu H, Liu N, Chi W, Wang C, et al. (2007) A heat-inducible transcription factor, HsfA2, is required for extension of acquired thermotolerance in Arabidopsis. Plant Physiology 143: 251.
73. Davletova S, Rizhsky L, Liang H, Shengqiang Z, Oliver DJ, et al. (2005) Cytosolic ascorbate peroxidase 1 is a central component of the reactive oxygen gene network of Arabidopsis. The Plant Cell Online 17: 268–281.
74. Sakuma Y, Maruyama K, Qin F, Osakabe Y, Shinozaki K, et al. (2006) Dual function of an Arabidopsis transcription factor DREB2A in water-stress-responsive and heat-stress-responsive gene expression. Proceedings of the National Academy of Sciences 103: 18822.
75. Ogawa D, Yamaguchi K, Nishiuchi T (2007) High-level overexpression of the Arabidopsis HsfA2 gene confers not only increased themotolerance but also salt/osmotic stress tolerance and enhanced callus growth. Journal of experimental botany 58: 3373.
76. Kumar M, Busch W, Birke H, Kemmerling B, Nurnberger T, et al. (2009) Heat shock factors HsfB1 and HsfB2b are involved in the regulation of Pdf1.2 expression and pathogen resistance in Arabidopsis. Mol Plant 2: 152–165.
77. Czarnecka-Verner E, Pan S, Salem T, Gurley WB (2004) Plant class B HSFs inhibit transcription and exhibit affinity for TFIIB and TBP. Plant Mol Biol 56: 57–75.
78. Widholm JM, Chinnala AR, Ryu JH, Song HS, Eggett T, et al. (2001) Glyphosate selection of gene amplification in suspension cultures of 3 plant species. Physiol Plant 112: 540–545.
79. Zhu H, Choi HK, Cook DR, Shoemaker RC (2005) Bridging model and crop legumes through comparative genomics. Plant Physiol 137: 1189–1196.
80. Kotak S, Vierling E, Baumlein H, von Koskull-Doring P (2007) A novel transcriptional cascade regulating expression of heat stress proteins during seed development of Arabidopsis. Plant Cell 19: 182–195.
81. Swindell WR, Huebner M, Weber AP (2007) Transcriptional profiling of Arabidopsis heat shock proteins and transcription factors reveals extensive overlap between heat and non-heat stress response pathways. BMC Genomics 8: 125.
82. Giorno F, Wolters-Arts M, Grillo S, Scharf KD, Vriezen WH, et al. (2010) Developmental and heat stress-regulated expression of HsfA2 and small heat shock proteins in tomato anthers. J Exp Bot 61: 453–462.
83. Mittal D, Chakrabarti S, Sarkar A, Singh A, Grover A (2009) Heat shock factor gene family in rice: genomic organization and transcript expression profiling in response to high temperature, low temperature and oxidative stresses. Plant Physiol Biochem 47: 785–795.
84. Nishizawa A, Yabuta Y, Yoshida E, Maruta T, Yoshimura K, et al. (2006) Arabidopsis heat shock transcription factor A2 as a key regulator in response to several types of environmental stress. The Plant Journal 48: 535–547.
85. Schramm F, Ganguli A, Kiehlmann E, Englich G, Walch D, et al. (2006) The heat stress transcription factor HsfA2 serves as a regulatory amplifier of a subset of genes in the heat stress response in Arabidopsis. Plant molecular biology 60: 759–772.
86. Ikeda M, Mitsuda N, Ohme-Takagi M (2011) Arabidopsis HsfB1 and HsfB2b Act as Repressors of the Expression of Heat-Inducible Hsfs But Positively Regulate the Acquired Thermotolerance. Plant Physiology 157: 1243–1254.
87. Miller G, Shulaev V, Mittler R (2008) Reactive oxygen signaling and abiotic stress. Physiol Plant 133: 481–489.

PERMISSIONS

All chapters in this book were first published in PLOS ONE, by The Public Library of Science; hereby published with permission under the Creative Commons Attribution License or equivalent. Every chapter published in this book has been scrutinized by our experts. Their significance has been extensively debated. The topics covered herein carry significant findings which will fuel the growth of the discipline. They may even be implemented as practical applications or may be referred to as a beginning point for another development.

The contributors of this book come from diverse backgrounds, making this book a truly international effort. This book will bring forth new frontiers with its revolutionizing research information and detailed analysis of the nascent developments around the world.

We would like to thank all the contributing authors for lending their expertise to make the book truly unique. They have played a crucial role in the development of this book. Without their invaluable contributions this book wouldn't have been possible. They have made vital efforts to compile up to date information on the varied aspects of this subject to make this book a valuable addition to the collection of many professionals and students.

This book was conceptualized with the vision of imparting up-to-date information and advanced data in this field. To ensure the same, a matchless editorial board was set up. Every individual on the board went through rigorous rounds of assessment to prove their worth. After which they invested a large part of their time researching and compiling the most relevant data for our readers.

The editorial board has been involved in producing this book since its inception. They have spent rigorous hours researching and exploring the diverse topics which have resulted in the successful publishing of this book. They have passed on their knowledge of decades through this book. To expedite this challenging task, the publisher supported the team at every step. A small team of assistant editors was also appointed to further simplify the editing procedure and attain best results for the readers.

Apart from the editorial board, the designing team has also invested a significant amount of their time in understanding the subject and creating the most relevant covers. They scrutinized every image to scout for the most suitable representation of the subject and create an appropriate cover for the book.

The publishing team has been an ardent support to the editorial, designing and production team. Their endless efforts to recruit the best for this project, has resulted in the accomplishment of this book. They are a veteran in the field of academics and their pool of knowledge is as vast as their experience in printing. Their expertise and guidance has proved useful at every step. Their uncompromising quality standards have made this book an exceptional effort. Their encouragement from time to time has been an inspiration for everyone.

The publisher and the editorial board hope that this book will prove to be a valuable piece of knowledge for researchers, students, practitioners and scholars across the globe.

LIST OF CONTRIBUTORS

Fenglou Mao
Computational Systems Biology Laboratory, Department of Biochemistry and Molecular Biology and Institute of Bioinformatics, University of Georgia, Athens, Georgia, United States of America

Chan Zhou
Computational Systems Biology Laboratory, Department of Biochemistry and Molecular Biology and Institute of Bioinformatics, University of Georgia, Athens, Georgia, United States of America
Gastrointestinal Unit, Massachusetts General Hospital, Harvard Medical School, Boston, Massachusetts, United States of America

Yanbin Yin
Computational Systems Biology Laboratory, Department of Biochemistry and Molecular Biology and Institute of Bioinformatics, University of Georgia, Athens, Georgia, United States of America
Department of Biological Sciences, Northern Illinois University, DeKalb, Illinois, United States of America

Ying Xu
Computational Systems Biology Laboratory, Department of Biochemistry and Molecular Biology and Institute of Bioinformatics, University of Georgia, Athens, Georgia, United States of America
College of Computer Science and Technology, Jilin University, Changchun, China

Jinling Huang
Department of Biology, East Carolina University, Greenville, North Carolina, United States of America

Johann Peter Gogarten
Department of Molecular and Cell Biology, University of Connecticut, Storrs, Connecticut, United States of America

Stefanie Kaufmann
Department of Genome Oriented Bioinformatics, Technische Universität München, Freising, Bavaria, Germany

Dmitrij Frishman
Department of Genome Oriented Bioinformatics, Technische Universität München, Freising, Bavaria, Germany
Institute of Bioinformatics and Systems Biology, German Research Center for Environmental Health, Neuherberg, Bavaria, Germany
Department of Bioinformatics, St Petersburg State Polytechnical University, St Petersburg, Russia

Annie Rodolakis
INRA, UR1282 Infectiologie Animale et Santé Publique, Nouzilly, France

Khalil Yousef Mohamad
INRA, UR1282 Infectiologie Animale et Santé Publique, Nouzilly, France
Laboratory of Animal Production - Pathology, Faculty of Agriculture, University of Al-Furat, Deir Ezzor, Syria

Bernhard Kaltenboeck and Kh. Shamsur Rahman
Department of Pathobiology, College of Veterinary Medicine, Auburn University, Auburn, Alabama, United States of America

Simone Magnino
Istituto Zooprofilattico Sperimentale della Lombardia e dell'Emilia Romagna "Bruno Ubertini", National Reference Laboratory for Animal Chlamydioses, Sezione Diagnostica di Pavia, Pavia, Italy

Konrad Sachse
Friedrich-Loeffler-Institut Jena, OIE and National Reference Laboratory for Chlamydiosis, Jena, Germany

Daniel H. Chitwood
Donald Danforth Plant Science Center, St. Louis, Missouri, United States of America

Asav P. Dharia, Ajay Obla, Matthew D. Gajdosik, Amanda Simon and Craig E. Nelson
University of Connecticut Department of Molecular and Cell Biology, Storrs, Connecticut, United States of America

Dong-Qiang Cheng
State Key Laboratory of Genetic Resources and Evolution, Kunming Institute of Zoology, Chinese Academy of Sciences, Kunming, China
Kunming College of Life Science, University of Chinese Academy of Sciences, Beijing, China

Ying Li
State Key Laboratory of Genetic Resources and Evolution, Kunming Institute of Zoology, Chinese Academy of Sciences, Kunming, China
Institute of Animal Genetics and Breeding, College of Animal Science and Technology, Sichuan Agricultural University, Yaan, China

Jing-Fei Huang
State Key Laboratory of Genetic Resources and Evolution, Kunming Institute of Zoology, Chinese Academy of Sciences, Kunming, China
Kunming Institute of Zoology-Chinese University of Hongkong Joint Research Center for Bio-resources and Human Disease Mechanisms, Kunming, China

Roberta Moschetti and Ruggiero Caizzi
Dipartimento di Biologia, Università degli Studi di Bari, Bari, Italy

Emanuele Celauro, Fulvio Cruciani and Patrizio Dimitri
Dipartimento di Biologia e Biotecnologie "Charles Darwin" and Istituto Pasteur Fondazione Cenci-Bolognetti, Sapienza Università di Roma, Roma, Italy

Sylvie Juliant, Marie-Luce Violet, François Monjaret, Pierre Cérutti, Annick Ozil and Martine Duonor-Cérutti
CNRS UPS3044 Baculovirus et Thérapie, Saint Christol Lèz Alès, France

Anne Harduin-Lepers
CNRS UMR8576, Unité de Glycobiologie Structurale et Fonctionnelle, Université Lille Nord de France, Lille1, Villeneuve d'Ascq, France

Béatrice Catieau
CNRS UMR8576, Unité de Glycobiologie Structurale et Fonctionnelle, Université Lille Nord de France, Lille1, Villeneuve d'Ascq, France
Laboratoire Français du Fractionnement et des Biotechnologies de Lille, Lille, France

Charu Lata
National Research Centre on Plant Biotechnology, New Delhi, India
CSIR-National Botanical Research Institute, Lucknow, Uttar Pradesh, India

Awdhesh Kumar Mishra, Mehanathan Muthamilarasan, Venkata Suresh Bonthala, Yusuf Khan and Manoj Prasad
National Institute of Plant Genome Research, New Delhi, India

Rafe M. Brown, Karen V. Olson and Anthony J. Barley
Biodiversity Institute and Department of Ecology and Evolutionary Biology, University of Kansas, Lawrence, Kansas, United States of America

Jennifer A. Weghorst
Biodiversity Institute, University of Kansas, Lawrence, Kansas, United States of America

Mariano R. M. Duya
Institute of Biology, University of the Philippines Diliman, Quezon City, Philippines

Melizar V. Duya
Diliman Science Research Foundation Inc., University of the Philippines Diliman, Quezon City, Philippines

Myron Shekelle
Research Institute of EcoScience, Ewha Womans University, Seoul, Republic of Korea

Irene Neri-Arboleda
Thomas Glen Erin Animal Hospital, Mississauga, Ontario, Canada

Jacob A. Esselstyn
Museum of Natural Science & Department of Biological Sciences, Louisiana State University, Baton Rouge, Louisiana, United States of America

Nathaniel J. Dominy
Department of Anthropology, Dartmouth College, Hanover, New Hampshire, United States of America,

Adrian Luczon and Perry S. Ong
Institute of Biology, University of the Philippines, Diliman, Quezon City, Philippines

Gillian L. Moritz
Department of Biological Sciences, Dartmouth College, Hanover, New Hampshire, United States of America

Mae Lowe L. Diesmos
University of Santo Tomas, Espana, Manila, Philippines

Arvin C. Diesmos
Herpetology Section, Zoology Division, Philippine National Museum, Ermita, Manila, Philippines

Cameron D. Siler
Sam Noble Oklahoma Museum of Natural History and Department of Biology, University of Oklahoma, Norman, Oklahoma, United States of America

Srinivas Akula and Lars Hellman
Department of Cell and Molecular Biology, Uppsala University, The Biomedical Center, Uppsala, Sweden

Sayran Mohammadamin
Department of Cell and Molecular Biology, Uppsala University, The Biomedical Center, Uppsala, Sweden
Salahaddin University, College of Science, Biology Department, Erbil, Iraqi Kurdistan

Daniel Vitales and Joan Vallés
Laboratori de Botànica - Unitat associada CSIC, Facultat de Farmàcia, Universitat de Barcelona, Barcelona, Catalonia, Spain

Oriane Hidalgo
Laboratori de Botànica - Unitat associada CSIC, Facultat de Farmàcia, Universitat de Barcelona, Barcelona, Catalonia, Spain
Jodrell Laboratory, Royal Botanic Gardens, Kew, Richmond, Surrey, United Kingdom

Teresa Garnatje
Institut Botànic de Barcelona (IBB-CSICICUB), Barcelona, Catalonia, Spain

Alfredo García-Fernández
Institut Botànic de Barcelona (IBB-CSICICUB), Barcelona, Catalonia, Spain
Área de Biodiversidad y Conservación, Universidad Rey Juan Carlos, Móstoles, Madrid, Spain

Jaume Pellicer, Robyn S. Cowan and Michael F. Fay
Jodrell Laboratory, Royal Botanic Gardens, Kew, Richmond, Surrey, United Kingdom

Arnoldo Santos-Guerra
Unidad de Botánica-ICIA, Puerto de la Cruz, Canary Islands, Spain

Marek Slovák and Jaromír Kučera
Institute of Botany, Slovak Academy of Sciences, Bratislava, Slovakia

Eliška Záveská
Department of Botany, Charles University, Praha, Czech Republic

Peter Vd'ačný
Department of Zoology, Comenius University, Bratislava, Slovakia

Tim L. King
U.S. Geological Survey (USGS), Leetown Science Center, Aquatic Ecology Branch, Kearneysville, West Virginia, United States of America

Anne P. Henderson
USGS, Leetown Science Center, Fish Health Branch, Kearneysville, West Virginia, United States of America

Boyd E. Kynard
University of Massachusetts, Department of Environmental Conservation, Amherst, Massachusetts, United States of America

Martin C. Fischer
Computational and Molecular Population Genetics (CMPG), Institute of Ecology and Evolution, University of Bern, Bern, Switzerland
Institute of Integrative Biology, ETH Zürich, Zürich, Switzerland

Matthieu Foll
Computational and Molecular Population Genetics (CMPG), Institute of Ecology and Evolution, University of Bern, Bern, Switzerland
School of Life Sciences, E´cole Polytechnique Fédérale de Lausanne (EPFL), Lausanne, Switzerland
Swiss Institute of Bioinformatics, Lausanne, Switzerland

Gerald Heckel and Laurent Excoffier
Computational and Molecular Population Genetics (CMPG), Institute of Ecology and Evolution, University of Bern, Bern, Switzerland
Swiss Institute of Bioinformatics, Lausanne, Switzerland

Ying Cheng, Jing Jin, Xiaolei Jin, Haiyang Jiang, Hanwei Yan and Beijiu Cheng
Key Lab of Crop Biology of Anhui Province, School of Life Sciences, Anhui Agricultural University, Hefei, Anhui, China

Yongxiang Lin
Key Lab of Crop Biology of Anhui Province, School of Life Sciences, Anhui Agricultural University, Hefei, Anhui, China
Crop Research Institute, Anhui Academy of Agricultural Sciences, Hefei, Anhui, China

Micah C. Kieffer
USGS, Leetown Science Center, S. O. Conte Anadromous Fish Research Laboratory, Turners Falls, Massachusetts, United States of America

Douglas L. Peterson
University of Georgia, Warnell School of Forest Resources, Athens, Georgia, United States of America,

Aaron W. Aunins
Cherokee Nation Technology Solutions, Leetown Science Center, Kearneysville, West Virginia, United States of America

Bonnie L. Brown
Virginia Commonwealth University, Department of Biology, Richmond, Virginia, United States of America

Index

M

N

O

P

R

S